Footprints in the Soil

People and Ideas in Soil History

Cover illustration from an original watercolor by Jane Anne Lehr

International Union of Soil Sciences

Footprints in the Soil

People and Ideas in Soil History

edited by

Benno P. Warkentin
Crop and Soil Sciences
Oregon State University
Oregon, USA

Stories about the accumulation of soil knowledge over the millennia

ELSEVIER

AMSTERDAM • BOSTON • HEIDELBERG • LONDON • NEW YORK • OXFORD
PARIS • SAN DIEGO • SAN FRANCISCO • SINGAPORE • SYDNEY • TOKYO

Elsevier
Radarweg 29, PO Box 211, 1000 AE Amsterdam, The Netherlands
The Boulevard, Langford Lane, Kidlington, Oxford OX5 1GB, UK

First edition 2006

Library of Congress Cataloging-in-Publication Data
A catalog record for this book is available from the Library of Congress

British Library Cataloguing in Publication Data
A catalogue record for this book is available from the British Library

ISBN-13: 978-0-444-52177-4
ISBN-10: 0-444-52177-1

For information on all Elsevier publications
visit our website at books.elsevier.com

Transferred to digital print 2007
Printed and bound by CPI Antony Rowe, Eastbourne

Working together to grow
libraries in developing countries

www.elsevier.com | www.bookaid.org | www.sabre.org

ELSEVIER   BOOK AID   Sabre Foundation
                International

Contents

List of Contributors

Addiscott, Tom M. Rothamsted Research, Harpenden, Herts, AL5 2JQ, UK.

Amundson, Ronald Division of Ecosystem Sciences, University of California, Berkeley, CA 94720.

Babel, Ulrich c/o Institute of Soil Science, University Hohenheim D 70593 Stuttgart 70, Germany.

Barrera-Bassols, Narciso Instituto de Geografía, Universidad Nacional Autónoma de México.

Berthelin, Jacques Laboratoire des Interactions Microorganismes-Minéraux-Matiéres Organiques dans les Sols (LIMOS) UMR 7137 C.N.R.S., Université H. Poincaré-Nancy I. Faculté des Sciences B.P.239, 54506 Vandoeuvre-les-Nancy Cedex, France.

Binkley, Dan Department of Forest, Rangeland and Watershed Stewardship, Colorado State University, Ft. Collins, CO 80523 USA.

Blanchart, E. Institut de Recherche pour le Développement (IRD), Laboratory «Matière Organique des Sols Tropicaux (MOST)», 911 Avenue Agropolis, BP 64501, 34394 MONTPELLIER Cedex 5, France.

Blum, Winfried E.H. University of Natural Resources and Applied Life Sciences (BOKU), Vienna, Peter Jordan-Str. 82, A 1190 Vienna, AUSTRIA.

Bockheim, J.G. Department of Soil Science, University of Wisconsin, Madison, USA.

Evtuhov, Catherine Department of History, Georgetown University, Washington, DC, USA.

Feller, C. Institut de Recherche pour le Développement (IRD), Laboratory «Matière Organique des Sols Tropicaux (MOST)», 911 Avenue Agropolis, BP 64501, 34394 MONTPELLIER Cedex 5, France.

Finck, Arnold Institute of Soil Science and Plant Nutrition, University on Kiel, D-24098 Kiel (Germany).

Gennadiyev, A.N. Faculty of Geography, Moscow State University, Moscow, Russia.

Gregorich, E.G.   Agriculture Canada, Ottawa, Canada.

Gregorich, L.J.   Gregorich Research, Ottawa, Canada.

Hasegawa, Shuichi   Laboratory of Soil Amelioration, Graduate School of Agriculture, Hokkaido University, Kita 9 Nita 9, Sapporo, Hokkaido 060-8589 Japan.

Holliday, Vance T.   Departments of Anthropology & Geosciences, University of Arizona, Tucson, USA.

Sandor, Jonathan, A.   Agronomy Department, Iowa State University, Ames, IA, 50011-1010 USA.

Showers, Kate B.   Senior Research Associate, Centre for World Environmental History, University of Sussex.

Sparks, Donald L.   Department of Plant and Soil Sciences, University of Delaware, Newark, Delaware, 19717–1303, USA.

Sparling, G.P.   Landcare Research, Hamilton, New Zealand.

Toutain, François   Laboratoire des Interactions Microorganismes-Minéraux-Matiéres Organiques dans les Sols (LIMOS) UMR 7137 C.N.R.S., Université H. Poincaré-Nancy I, Faculté des Sciences B.P.239, 54506 Vandoeuvre-les-Nancy Cedex, France.

Warkentin, Benno   Department of Crop and Soil Sciences, Oregon State University, Corvallis, Oregan 97331.

Williams, Barbara J.   Dept. of Geog. and Geol., Univ. of Wisconsin Colleges-Rock, 1206 S. River Rd., Janesville, WI 53546, USA.

Winiwarter, Verena   APART – Fellow of the Austrian Academy of Sciences, University of Natural Resources and Applied Life Sciences (BOKU), Vienna, Peter Jordan-Strasse 82, A 1190 Vienna, Austria.

WinklerPrins, Antoinette M.G.A.   Geography Dept., Michigan State University, USA.

Yaalon, D.H.   Hebrew University, Institute of Earth Sciences, Jivat Ram campus, Jerusalem, 91904 Israel.

Zinck, J. Alfred   Dept. of Earth Systems Analysis, International Institute for Geo-Information Science and Earth Observation (ITC), The Netherlands.

Preface

Soil, the primary basis of life on the globe, has drawn increasing attention at the turn of the 21st century from an ever widening audience. The popular press, writing for a lay audience, discusses sustainability of soil resources; the Smithsonian Museum in the USA—following activities in other countries—is mounting a major exhibit on soil; scientists from ecology to geology are including soils knowledge in their studies; and historians of science are interested in the social and cultural background of soils knowledge. Now is the time for soil scientists to present the stories they know best, to tell the stories of their understanding of soils.

The focus of this book is the development of ideas and concepts in soil science. It aims to bring to a broad audience from many disciplines and backgrounds an understanding of how interesting and vital soils are. We have a long and fascinating story of gathering soils knowledge, forming this into the discipline of soil science, and applying it to improve the human condition.

The idea for this book arose from discussions within the Soil Science Society of America and the International Union of Soil Sciences about the need to recognize the people who left "footprints in the soil," to trace the development of our present understanding of how soils function in ecosystems, and to show how they have been and currently are being applied to solve problems. In short, how do we know what we know?

The chapters concentrate on soil knowledge in the western world and draw primarily on written accounts available in English and European languages. Insights from other disciplines have added to this process of gathering and synthesizing knowledge. The history of science discipline is contributing valuable knowledge of the culture of soil understanding, of the conditions in society that fostered the ideas, of why they developed in certain ways. Until 100 years ago, western soil scientists considered their discipline in a holistic perspective. The early books in Europe and North America contained ecological statements about soil in the physical, biological and cultural world around them. This was largely lost as the technological approach dominated in much of the 20th century; it was regained in the last quarter of that century. This concept that soil is more than an object to study, but also is central to our ideas about nature, is an important recent development in soil science. The modern holistic vision of the soil is nothing new—it is the legacy of the ages and shares a direct lineage with early Greek, Islamic, and Chinese thought and is primary in indigenous cultures.

Experimentation and practice, science and application, have proceeded together in gathering and synthesizing soils knowledge. The authors of <u>Footprints in the Soil</u> wield an authority that they earned by long practice in their fields of study. Their experience and expertise generates a confidence that allows them to speak clearly and forcefully about the flow of ideas and methods across generations of soil scientists, and national or cultural boundaries. Here they explore the changing and diverse approaches to investigating and managing the soil.

<u>Footprints in the Soil</u> examines the history of soil science in four sections: treating sequentially, the "Early Understanding of Soils," "Soil as a Natural Body," its "Properties and Processes," and finally its "Utilization and Conservation." In chapter 1, Verena Winiwarter draws on critical editions of early Roman agricultural texts (in their original Latin) to survey practical efforts to improve soil fertility. Through these early works she provides an understanding of a central feature of Roman soil knowledge. For early agriculturalists such as Cato, Varro, and Columella, the soils were dynamic, living entities that played an interactive role in nature equal to that of humans. This interaction required proper management.

In chapter 2, Barbara Williams examines agriculture in pre-Columbian Mesoamerica. Using a combination of 16th century pictorial manuscripts such as the Codex Vergara, the Florentine Codex—the most comprehensive textual encyclopedia of Aztec soil knowledge, early dictionaries and accounting writings, and recent ethnopedological works, she portrays a farming culture which viewed the soil as part of a holistic system that we today call ecological. Aztec farmers constructed a complex, magical relationship between humans, plants, and soils; the proper understanding and management of the latter was fundamental to their agricultural system. Williams situates this in the broader history of Aztec culture and its eventual collision with European invasion.

In chapter 3, Jon Sandor, Antoinette WinklerPrins, Narcis Barrera-Bassols, and Alfred Zinck explore how diverse, non-modern cultures developed original approaches to soil and land use. Drawing on an extensive and growing literature from ethnopeodological investigations of indigenous cultures, they discuss the potential applications of this legacy to current efforts to manage and develop the land. Central to this approach is the holistic character of indigenous agriculture—whether African, American, or Asian. In several case studies the authors present aspects of indigenous knowledge organized into three categories: concepts, religious beliefs, and attitudes about soils; soil properties and processes; and finally how various cultures applied this soil knowledge.

The appearance of a self-conscious discipline of soil science in the late 19th century occurred on the foundation of previous work. In chapter 4 Christian Feller, Eric Blanchart and Dan Yaalon survey the work of five scientists from the 16th, 18th, and 19th centuries. Through selected writings in the voices of the scientists themselves, these authors provide evidence of the early development of tools (Palissy's "auger" and Thaer's soil mapping), concepts (Buffon's ideas of natural cycles and "horizons," and Müller's "natural forms of humus"), and methods.

Verena Winiwarter and Winfried Blum examine, in chapter 5, the soil as an object of myth, religion, and belief. Surveying recent literature that treats ancient and current mythology, environmental ethics, ethnoecology, and ecotheology, the authors discuss the relationship between philosophical views of the soil and practices related to soil management. Central to this investigation is the specific role of soils in belief systems. In analyses of ancient Greek, Chinese, and Jewish writings, they place soils in a religious and cultural context, showing that later holistic concepts of the soil—including the Gaia concept of Lovelock and Margulis, and some of those within the Deep Ecology movement—are rooted in the distant past.

Section II on the soil as a natural body begins with Catherine Evtuhov's discussion of Dokuchaev's scientific contributions to the emergence of soil science as a discipline (chapter 6). She places Dokuchaev's holistic view of the soil—as an organic body in constant interaction with its environment—in its specific cultural and historical context in post-reform Russia. She explores how Dokuchaev's soil science related to his practical concerns (agricultural reforms), to his scientific community, and institutional setting. The intense scientific and practical exchange that characterized the Russian scientific world led to Dokuchaev's juxtaposing the study of the soil with other disciplines, and in investigating soils in the totality of their natural environment.

In chapter 7, Ronald Amundson discusses how institutional arrangements in the USA influenced the development of soils research. In excerpts from their correspondence, he shows how two scientists with different backgrounds contributed to soils knowledge. Eugene Hilgard from his experience in the western USA and his university position, championed an emphasis on soil chemistry to solve soil fertility problems. Milton Whitney, at the U.S. Department of Agriculture and experience in the eastern part of the continent, championed soil physics to understand how soil fertility could be increased.

Alexander Gennadiyev and James Bockheim in chapter 8 discuss two key concepts in soil classification and in understanding soils in their landscapes: soil cover pattern and the soil catena. The understanding of these regularities observed in the field developed in Russia and then in western countries.

The section concludes with Vance Holliday's comprehensive history of soil geomorphology in the United States. The author traces the institutional and scientific developments chronologically across five periods. Focusing at times on the contributions of individual scientists such as Marbut and Jenny, and at times on methods such as soil stratigraphy, studies and concepts such as soil catena, this story explains the importance of geomorphology as a background for soil classification.

Section III begins with a strong statement of the essential nature of soils in ecology (chapter 10). Dan Binkley addresses the interactions between soil science and ecology, a topic that has been relatively ignored in the history of ecology literature. Ecology and soil science both became self-conscious disciplines—with new monographs, journals, courses, and even institutes under those titles—at the beginning of the 20th century. From soil science, ecology found a new object of investigation that extended beyond the limits of individual organisms. From ecology, soil science learned a method and language for expressing the holistic, interactive nature of the soil. Throughout the 20th century, the efforts of these soil scientists and ecologists were united in Jenny's school in the United States, and Vernadsky's in Russia and continuing in the Soviet Union. Binkley's survey of the intellectual developments in ecological soil science tracks the role of soils from mid-19th century plant geography to late 20th century highly technological and collaborative investigations.

In chapter 11, Jacques Berthelin, Ulrich Babel, and François Toutain collaborate to tell the story of biological life in the soil from the mythological views of prehistoric civilizations and of the ancient Greek philosophers, to those of the early modern scientists who first began experimental investigations of soil's vegetative powers. The authors trace the beginning of modern soil biology to the late 19th century work of agricultural chemists who discovered the role of life—and then microorganisms—in soil processes, and those who further explored it, such as Winogradsky. As the 20th century progressed, soil microbiology became explicitly ecological with the added study of soil fauna, leading to expanded possibilities for understanding and investigating the soil as a biological, ecological phenomenon.

In chapter 12, Donald Sparks traces the development of soil chemistry in the 19th century when European agricultural chemists of several nationalities began chemical analyses to understand plant nutrition. In a selection of case studies on the history of ion exchange, clay mineralogy, soil acidity, sorption processes, and the kinetics of soil chemical processes, the author presents a survey of the central personalities, their investigations, and most significant contributions to the rise and development of soil chemistry.

Suichi Hasegawa and Benno Warkentin focus on two aspects of the soil physics story: the transmission of water through unsaturated soil, and the changing concepts in soil structure from emphasis on solids to the importance of voids. Darcy's 1856 physical quantification of saturated water flow was extended in the early 20th century to unsaturated soils, the usual condition in surface soils. The second section examines soil structure or architecture related to achieving tilth—the preparation of soils for planting. When soil structure became the domain of scientific investigation, the focus on plowing and other mechanical methods for managing the consistency and interaction of soil aggregates gave way to an emphasis on the pores as habitat for soil life.

In the final section, soil utilization and conservation, Kate Showers devotes chapter 14 to the history of soil conservation in its international context. To understand the interaction between human societies and the soil, she analyzes the technical elements of soil conservation, the relative merits of various methods, the fundamental issue of how social activities influence, positively or negatively, soil erosion, and how they might be managed. The author surveys the inherently soil conserving aspects of traditional land use systems, suggests a popular world history of soil erosion and conservation experiences, and traces the rise of professional soil conservation in the 20th century. Drawing on case studies of the implementation in British Colonial Africa of soil conservation methods developed in the United States, she draws some conclusions about ongoing and future soil preservation efforts.

In chapter 15, Edward Gregorich, Graham Sparling, and Joan Gregorich examine how the health of the soil in ancient civilizations, during the rise of scientific soil science in the 19th century, and remaining today, is tied directly to human prosperity. Common notions of stewardship evolved into scientific studies which recognize the interactive relationships between living beings and the environment, and developed it into a recent sophisticated ecosystem approach. There is, however, still a discontinuity between these theoretical perspectives and current practice. The authors show that whether it was a spiritual, religious, philosophical, or scientific conceptualization, the connection between the soil and human society has been intimate, abiding, and prophetic.

In chapter 16, Arnold Finck relates the history of soil nutrient management for plant growth in agricultural production. He examines the increased use of inputs of plant nutrients, from early gatherer societies to agricultural societies that concentrate on increasing soil productivity. With the rise of scientific agriculture and of experimental investigations of plant nutrition came an understanding and an increased use of chemical fertilizers. The author then discusses how practical experience in the 20th century led to a preference for convenient and low-cost chemical fertilizers over organic products such as composts.

Tom Addiscott in chapter 17 addresses the influence of environmental issues on soil research. He examines the impact of soil processes and human activity on the soil in its environmental context. While recognizing that many kinds of productive activity (both industrial and agricultural) and a range of organisms (from livestock to microbes) can lead to environmental pollution, he focuses his discussion on human soil use. Addiscott shows how the increasing use of chemical fertilizers led, in part, to the rise of environmental considerations as a scientific and social issue. He examines management strategies developed in agricultural practice to decrease undesirable effects of nitrogen and heavy metals.

In the final chapter, Jon Sandor looks to the history of agricultural terraces for insights into the changing soil uses in ancient and modern traditional cultures. Through an analysis of an extensive literature he examines the diverse geographies, forms, and impacts of terrace agriculture in relation to the soil. Ancient agricultural terraces, their landscapes, soils, and farmers are a rich resource for investigating the history of soil use in varied cultural contexts. Such histories provide examples of sustainable agriculture, which could lead to improvements in current soil management practices.

The diversity of topics and of authors' backgrounds meld into a mosaic of stories about accumulation of soils knowledge.

Lloyd Ackert
Benno P. Warkentin
August, 2005

Acknowledgments

This book depends, first and foremost, upon the enthusiastic effort by the chapter authors to present the knowledge and interpretations of their fields of expertise. They patiently dealt with suggested revisions, requests for more information, and occasional suggestions for deletions. Their responses were gratifying and made the editor's tasks a pleasurable one.

Reviewers, with their important role in assuring quality and scientific integrity, also gave graciously of their time. Special recognition should go to Lloyd Ackert, Kate Showers, Guenther Stotsky, Verena Winiwarter, and Dan Yaalon.

This book would not have happened without the help of Tracy Mitzel, Department of Crop and Soil Sciences at Oregon State University, who patiently showed me the simpler details of sending, receiving and storing manuscripts electronically, and who performed the more complex details herself. My thanks for her facility with the programs. She also made the many time-consuming corrections and revisions in the manuscripts.

Many colleagues provided suggestions and insights on the topics in this book. I learned from them and I now thank them. The book committee provided guidance in planning and evaluation: Lloyd Ackert, Stephen Nortcliff, Donald Sparks, John Tandarich, and Dan Yaalon.

Support from the SSSA, the IUSS, and from Don Sparks, President of the 2006 World Congress of Soil Science (WCSS 2006) and from Lee Sommers and Larry Wilding, co-chairs of the Organizing Committee for the WCSS 2006 is gratefully acknowledged.

My personal thanks to Judith Taylor at Elsevier for the enthusiasm she has shown for the book and for her suggestions. Betty Daniels at Elsevier waited patiently for the manuscripts and the information required for publication, and provided gentle encouragement.

Finally to the soil scientists who, over the millennia, left their footprints in the soil for us to learn from and to build the present soil science.

Benno P. Warkentin
Corvallis, September 2005

Special Acknowledgment and Appreciation to Dan Yaalon

Dan H. Yaalon, Professor Emeritus at the Institute of Earth Sciences, Hebrew University of Jerusalem, Israel, organized a Working Group, later the Standing Committee on the History, Philosophy and Sociology of Soil Science in the International Society of Soil Science in 1985. He chaired the committee for many years, organizing conferences and writing papers that made soil scientists aware of the history of their science. He deserves recognition for his role in furthering history of soil science studies.

He was involved in the early discussions within the Soil Science Society of America that led to the planning for this book. The decision was finally taken by the SSSA to proceed with an edited book of chapters written by authors who were experts in the chapter topics. This book was begun. Originally acting as co-editor, Dan Yaalon continued his interest, using his experience in helping to choose the topics for this book, and then the appropriate authors. Unfortunately illness prevented him from completing the task and participating in the final evaluation of peer reviews of the chapters, but the influence of his knowledge and experience can be seen in the chapters.

A special acknowledgment and thanks.

Benno P. Warkentin
Corvallis, September 2005

Section I
Early Understanding of Soils

The footprints in the soil were already being made in ancient times when practitioners and philosophers accumulated knowledge about soils that eventually became part of soil science. The story in this book begins when communication by writing in Latin languages became available.

Verena Winiwarter picks up the story from the time Romans were writing about their experiences in husbandry. They combined earlier Greek ideas with Roman practical knowledge to leave their legacy of soils knowledge. Some of their writing survived, and was added to, in Moorish writings of a millennium later. Other cultures contributed their understanding of the nature of soils and how to manage them for food production. Barbara Williams tells how the Aztecs in Central and South America set down their ideas in glyphs to indicate both practical information and the beauty of soils in nature. Jon Sandor and colleagues detail some local case studies on soils and agricultural practices in non industrialized societies of the Americas. The term "ethnopedology" is being applied to the knowledge and perspectives contributed by these cultures. Modern soil science is looking to these experiences for help in such issues as management of soils for sustainability of food production.

While soil studies as soil science are generally considered to have begun about 1850, many of the ideas, and indeed methods still used for examining soils, were developed by naturalists before that time. Christian Feller and colleagues describe some of the contributions of these "forerunners" in western Europe, often in the naturalists own words. Feller et al. hint that there must be more of these stories waiting to be found and interpreted, waiting for the footprints to be identified.

Religion has a major influence on human attitudes toward nature, how we see ourselves in the natural world, and our approaches to soils as ethical issues. Verena Winiwarter and Winfried Blum draw attention to the world views of different religions and the threads of this complex topic. Is environmental ethics a basis for thinking ethically about soils? How does this contribute to knowledge about soils?

1
Soil Scientists in Ancient Rome

Verena Winiwarter[1]

Introduction

Soil fertility is a common concern of all agricultural societies. Fertility management is a decisive factor in the long-term development of any such culture at a given locale. Soil fertility however, is not a given. It is a function of natural factors such as parent mineral, climate, surface water regime, natural vegetation and morphology on one hand and of the agricultural techniques employed on a given piece of land on the other hand. This functional coupling of natural factors and human intervention has long been known, as ancient agricultural texts from all over the world show. A multitude of interventions into soils is known to have been conducted in ancient times, among them physical, chemical and biological treatments. The first soil scientists were those that were concerned with the land's fertility, be they farmers or land-owners.

From around 500 BCE to about the 5th century CE the Mediterranean and later on western and central Europe were increasingly dominated both politically and culturally by the Romans. Latin became the *lingua franca* in a large part of this empire, and it certainly makes sense to discuss it as an entity in some respects, although over the long time span of roughly a millennium, and given the considerable diversity of bio-regions, very different techniques to work the soil were developed and crops and tillage methods were highly diversified.

An applied soil science, in the sense of a systematic experimental investigation into the properties and processes of the earth's living skin, and the development of measures to alter its properties, was developed by experts on agriculture, in close connection with the practical use of soil. Among all the ancient agricultural texts, those of Latin writers are by far the most detailed and practical ones, not only with respect to their knowledge of soils but in their general treatment of agricultural matters. The reason for that is not just the 'practical sense' of the Romans as

---

[1] APART – Fellow of the Austrian Academy of Sciences, University of Natural Resources and Applied Life Sciences (BOKU), Vienna, Peter Jordan-Strasse 82, A 1190 Vienna, Austria.

opposed to their more 'philosophical' Greek antecedents, but the acknowledged
political relevance of an agricultural surplus for the expanding empire.

The history of written Roman soil knowledge starts as early as the 3rd century
BCE, with Cato the Elder. The author was a well-known statesman, orator, and the
first Latin prose writer of importance. Cato was born of a plebeian family and had
a quite remarkable career. Among other offices he held was that of a military tribune
in the Second Punic War. He pleaded for simple, rustic ideals, and against the – in
his view – pompous and decadent Hellenistic culture which was fashionable in
Rome during his lifetime. Thus his treatise on agriculture has to be understood not
just as a practical manual, but also as a political statement about the relevance of
profitable farming and the re-connection of urban Romans to their rural basis.
Attachment to the soil is a political program as much as it is a necessity of
successful cultivation, and not just in Cato's book.

Cato is but one, and the earliest, of a wealth of authors. All works transmitted
through time in their entirety can be found in Table 1, which is sorted according to
the age of the texts. Of the fragments, Gargilius Martialis' Treatise on Arboriculture
shall be mentioned as one example, as he connects each of the trees he describes
with a description of the right soil for it to thrive. Many more agricultural manuals
have existed, according to the extant treatises who mention many names of experts
of which this reference is the single trace we have (cf. Winiwarter, 2002, for a
comprehensive list).

One of the most mysterious works on agriculture, and one who had a profound
effect on Roman agricultural writing, is a treatise written by a Phoenician named
Mago. He is sometimes called 'the father of agricultural textbooks', although we
have only very blurred traces of his work. Mago's treatise is reported to have comprised of 28 volumes. It was rescued by the Roman conquerors during the fall of
Carthage so it must be older than that. The Roman Senate even commissioned its
translation, as Pliny the Elder reports in the book on agriculture of his encyclopedia,

Table 1. Important Roman writers on agriculture

Author	Title	Author's lifetime
M. Porcius CATO	De Agri Cultura	234–149 BCE
M. Terentius VARRO	Res Rusticae	116–27 BCE
P. VERGILIUS Maro	Georgica	70–19 BCE
L. Iunius Moderatus COLUMELLA	De Re Rustica	first century CE
C. PLINIUS Secundus (the Elder)	Naturalis Historia	CE 23–79
Rutilius Taurus Aemilianus PALLADIUS	Opus Agriculturae	fourth/fifth century CE

where Mago is mentioned alongside with other non-Latin writers (Plin. n.h. XVIII, V, 22f). Two of the later Roman writers on agriculture, Varro and Columella, inform about the subsequent history of Mago's work: Varro gives an account that shows great interest in Mago's work and reports that the translator was not a mere translator, but had added Greek knowledge and left out part of the original work. '*All these are surpassed in reputation by Mago of Carthage, who gathered into 28 books, written in the Punic tongue, the subjects they had dealt with separately. These Cassius Dionysius of Utica translated into Greek and published in twenty books, dedicated to the Praetor Sexti(li)us. In these volumes he added not a little from Greek writers whom I have named, taking from Mago's writing an amount equivalent to eight books. Diophanes, in Bithynia, further abridged these in convenient form into six books, dedicated to king Deiotarus.*' (Varro, r.r. 1,1,10). Varro compresses a history of about 100 years into these few sentences. Carthage was destroyed in 146 BCE, and the translation was made soon thereafter, around 140 BCE. The abridged, six volume edition can be dated at around 60 BCE because of the mention of king Deiotarus. The Greek translation was made between these two dates. It is suggested to have been commissioned in connection with the colonization projects of C. Gracchus, because in several of the projected settlement areas Greek would have been the only common language of the people assembled from different parts of the Empire (Mahaffy, 1898–1890, 33f). Columella, who refers to the same incident, is far less elaborate but tells basically the same story (Col. r.r. 1,V,10).

Neither the translation Columella refers to, nor its abbreviated version have been passed on to us. So, unfortunately, we have no direct knowledge of Mago's books. Mago is of particular interest because through the references to his work the geographical extension of the world in which agricultural knowledge was exchanged can be anticipated – it spanned the entire Mediterranean basin. The active interest by the Roman Senate is an important proof of the high value agriculture had, and the complicated history of translation, expansion and compression makes visible that the authors were aware of the body of knowledge that had been accumulated and that they made use of it. There is, however, no way to tell what Mago said about soils, the few quotations we have are about other themes.

A century after Mago's contemporary, Cato, another distinguished Roman patrician, M. Terentius Varro, wrote the treatise *Res rusticae* (On Agriculture) in the style and form in which a learned man of his time would write, as a dispute among three estate owners discussing a variety of subjects. He devoted a considerable part of his book to etymological aspects of agricultural vocabulary – among them many learned jokes, too – a part of his interest we have come to belittle. But he has a fully developed soil theory.

For modern standards, the most original and scientific work ever written on agriculture in pre-industrial times is surely Columella's multi-volume treatise on agriculture (*De re rustica*). Nevertheless, we do not know much about him. He lived in the first century CE, was born in Spain and passed through a typical Roman career. He owned several estates, thus he had a grasp of the regional differentiation of agriculture from his own experience. We can infer from his invention of a new implement for the grafting of trees that he did take part in the practical running of his estates. His agricultural work consists of twelve books, amongst which the one on horticulture is written in verse. The other books are organized by topics; one entire book (No. 2) is on soils. Why would Columella, the most practical and organized writer, include a poem on gardening in his treatise? It is done, so he says, in commemoration of a work of verse which had been written about a generation earlier by a very prominent Roman poet, Virgil. The poem, called the *Georgica* (Georgics) deal with the various subjects of agriculture. In the first book, Virgil explains the main tasks of fieldwork. His work is first and foremost a poem, a literary work, but he goes practical so far as to explain e.g. soil test methods. In this 'educational' work, Virgil aims to show that agriculture was a subject worth writing poetry about. Virgil is important in particular for the tradition of knowledge. He was very widely known in later, Medieval and Renaissance, times.

Contemporary to Columella, Pliny the Elder, like Virgil not primarily an agricultural writer, compiled the very famous encyclopedia *Naturalis Historia*, summing up the knowledge of his age. In books 17 and 18 a great deal of agricultural knowledge and a calendar of agricultural tasks can be found. Pliny, his Latin full name Gaius Plinius Secundus, was born 23 CE in Novum Comum (today: Como, Italy), and died on Aug. 24, 79 CE in Stabiae, near Mt. Vesuvius, while studying the eruption of the volcano. Pliny came from a prosperous family, and could pursue studies in Rome, he was a well traveled man. Like so many of the agricultural authors he was aware of natural differences in the vast span of the Roman Empire.

In late antiquity, 4th–5th century CE, a man named Rutilius Taurus Aemilianus Palladius, writing somewhere in southern France, completed a treatise which is most important not just for the new knowledge it presents, but for the new structure by means of which knowledge is presented. Palladius was a *vir inlustris*, a person of highest social status, had estates in Italy and Sardinia and presumably came from southern Gaul. We do not know if he was Christian or pagan, his work does not allow conclusions about that. He wrote at a time when rural life had again become desirable among the literate. Palladius wrote the first agricultural work which is almost entirely structured by time, in calendar style. Only two introductory books precede the twelve books which deal with agricultural work by the month. It is notable that Palladius was able to organize all knowledge he possessed into a calendar with the sole exception of soils, to which he devoted the second book of this work, the first being an introduction. He must have understood the overarching importance of soils clearly to give them a full book length treatment.

These are glimpses of the context of soil knowledge, which has to be taken into account when discussing it as a precursor to soil science. Agriculture was of high esteem in Ancient Rome, as is documented from the 2nd century BCE until the 4th century CE. It figured as the theme of practical treatises written by authors of standing and rank for an audience of estate owners, was considered an integral part of encyclopedic attempts, and fit for poetry. The authors engaged in a discourse with each other across space and time, using the parts of older works that they considered useful, while adding to and refining this body of knowledge with their own bits, gained from experience and experiment (a distinction unknown to them, because it was first made for the English language in the 18th century).

We are in the lucky position that all the aforementioned texts are available in critical editions, so that the modern reader can work from a (relatively) good text, in which the differences between the extant manuscripts are made visible. A word of caution should be said at this point about translations: While the translations usually are good in grammatical terms, they were not made with a detailed interest in soils, nor can one expect philologists to understand in all detail all the concepts they were translating. If possible, the youngest and/or best edition should be used. Translations, especially if used in comparison, give a first overview, but any serious work has to be done with the texts in their original language.

Soil Terminology and Soil Concepts

Terminology is the prerequisite for taxonomy, and thus offers very important information about the conceptual framework in which the Roman authors operated. Important Latin vocabulary used to describe soils by the authors of Roman Antiquity is compiled in Table 2. The table is a compilation of the treatises of Cato, Varro, Columella, Pliny and Palladius. The reader is referred to K.D. White's seminal volume on Roman Agriculture for separate lists of soil vocabularies for Cato, Varro, Columella, and Pliny (White, 1970, Appendix II).

White sorted the vocabulary into the categories 'Mineral Content', 'Moisture or Dryness', 'Structure and Texture', 'Warm and Cold', 'Richness or Leanness', and 'Heavy or Light'. This paper takes a different approach. Categories here were built to be compatible with qualities of soil which modern soil science discerns. Therefore, all words that are about 'Grain Size', 'Density and Structure', 'Humidity' and 'Color', which correspond to modern categories, are pulled together, and 'Fertility', 'Taste', 'Temperature' and 'Special properties', which do not, or not directly, translate into modern categories, are given apart from the others.

Table 2. Soil classification in Roman Antiquity

Grain size		Density and structure	
(h)arenosus	sandy (fine sand)	crassus	dense
sabulosus	sandy (coarse sand)	densus	compacted
glareosus	gravelly	gluttus	mellow
saxus	rocky	putris	crumbly
		gravis	heavy
		levis	light
		solutus	loose
		spissus	dense
		subactus	worked through
		tenuis	delicate
Humidity		*Color*	
aquosus	watery	niger	black
aridus	draughty	pullus	dark
(h)umidus	wet	rubicundus	reddish
limosus	muddy	rubidus	dark-red (or red-brown)
siccus	dry		
sucosus	juicy		
Taste		*Fertility*	
amarus	bitter	laetus	lush
salsus	salty	macer	meagre
		pinguis	fat, fertile
		sterilis	infertile
		note on Fertility: infecundus (infertile)	
		is used only by Pliny.	
[*Temperature*]:		sitiens	thirsty*
aestivus	summerly (hot)	*probably not with reference to water	
calidus	hot		
frigidus	cold		
Special Properties			
cariosus	carious		
creteus	clayey (or chalky)		
cretosus	chalky (or clayey)		
lutosus	loamy		
rimosus	riven		

Temperature is not to be mistaken for degrees Fahrenheit or Centigrade. A hot soil for Columella should not be given too much manure, as it would be burned, whereas a cold soil needs a lot of it. The Roman scholars used words derived from Greek philosophy, where hot and cold were properties not simply denoting temperature, but more the temperament like that of a human being. A cold soil then has to be understood more like one of cold temper, one with to much 'phlegm', a phlegmatic soil. Of course manuring means to add chemical energy to a soil, so we can understand Columella today using the energy concept. Whether he himself had a concept of manure as added energy, one cannot say.

Of the special properties, cariosus shall be discussed in more detail. The idea of the notion is clear for Columella, a practitioner, and he is able to explain it to us, unlike his predecessors, above all Cato, who only warns to touch it with no explanation given (Cato, agr. 5, 6; 34, 1; 37, 1). Terra cariosa is not a soil, but a state of soil which can result from treatment at the wrong time. Soil that is ploughed when the upper layer is wet from rain and the lower layers have not been humidified becomes 'cariosa', it looses its fertility for years through plowing under adverse conditions.

Word and concept come from the 'rustici', the rural people, as Columella informs his readers, and is probably understood by those scholars that have practical experience (Col. r.r. II,4,5). The various authors are more or less successful in explaining it (see also Pall. op.agr. X, VIII,10, who uses the noun instead). Terra cariosa is not understood by Pliny, the encyclopedist, as can be clearly seen in his text (Plin. n.h. XVII,34–35). He misunderstands it as a type of soils, whereas for Columella it is a result of wrong treatment and a (temporary) condition (Winiwarter, 2000).

There is more to concepts than just taxonomy. Varro uses a theory about soils which holds that the different types are generated by mixing of eleven kinds of (mineral) substances. The passage in Varro (I,9, 2–3) reads:

'In illa enim cum sint dissimili vi ac potestate partes permultae, in quis lapis, marmor, rudus, arena, sabulo, argilla, rubrica, pulvis, creta, glarea, carbunculus, id est quae sole perferve ita fit, ut radices satorum comburat, ab iis quae proprio nomine dicitur terra, cum est admixta ex iis generibus aliqua re, cum dicitur aut cretosa sic ab illis generum discriminibus mixta.'

For there are many substances in the soil, varying in consistency and strength, such as rock, marble, rubble, sand, loam, clay, read ochre, dust, chalk, ash, carbuncle (that is, when the ground becomes so hot from the sun that it chars the roots of plants); and soil if it is mixed with any part of the said substances, is e.g. called chalky, as well as according to other differences as mixed (my translation compiled from Loeb and the German translation by Flach). Note that sabulo is translated as loam in Loeb, whereas Flach has 'potter's clay' – translation is interpretation. The

Table 3. Classifications used by L. Iunius Moderatus Columella in his second book

Classification driven by perception	Classification driven by interaction (aims)
Morphological framing of soil:	Threefold dichotomy of qualities of soil:
plains	fertile vs. *poor* – pinguis vel macer
hilly terrain	friable vs. *dense* – solutus vel spissus
mountainous terrain	humid vs. *dry* – umidus vel siccus

Roman authors were not trying to be consistent with all their predecessors, and each work has its distinct use of words. To give another example, we cannot judge for sure, if cretosus and creteus are synonyms, and overall, the difference between clayey and chalky seems to be blurred. Columella did not use the systematic treatment of different minerals as soil constituents, as can be found in Varro. It is a conscious decision by Columella not to follow Varro's suggestions. Instead, Columella set up a systematic soil classification system, which rests on dichotomies as detailed in Table 3.

This classification scheme is remarkable, because it is a morphological classification, designating flatland, hilly land, and mountainous land. In this first attempt to classify what he perceives, Columella includes landscape features in his discussion (see left column of Table 3). The other parts of the scheme are meagre versus fertile, friable vs. dense, and wet vs. dry soils (Col. II,2). Together these qualities are sufficient to describe almost any type of land one may encounter. Interestingly enough, however, particle size is not part of the classification, but after all, no agriculturist really wants to classify rocky areas, and for agricultural land morphology, density, water content, and fertility can in fact prove to be sufficient criteria.

The dichotomies given in the right column are related much more directly to the cultivators' interaction with the natural system, the classification arises from aims, and not from the point of perception as such. The words highlighted in italics in Table 3, poor, dense and dry, are the main characteristics of a bad, infertile soil for Columella.

He does, however, give this very systematic matrix only to dismiss it, as (he says) it is not important from a practical point of view to arrange the knowledge in this framework. He then goes on to discuss soils in much more detail, but from an applied, interactive viewpoint: questions of fertility management and suitability are the basis of his thinking.

Methods for Testing Soils

The authors wrote for an audience of estate owners who often sold or bought land. Therefore, books on agriculture usually included a section of advice on buying new land. In theory everything seemed very easy: an estate should have good earth (*solum bonum*) and a good climate (*caelum bonum*). Clearly, however, a potential buyer needed to know more specific details about the land than the *topos* of good soils and good climate would provide in order to make economically sound decisions. One of the measures suggested is to visit a piece of land one wants to buy more than once. At first glance one will be able to see neither the benefits nor the problems of a certain place. If one comes again, such relatively hidden features are more easily noticed (Col. I,4,1). Or, as Cato puts it, a good piece of land will please more at each visit (Cato I,2). Even after one has bought it and learned about the land, one should continue to look at the land and its soil several times a year. Columella advises to have a close look in spring and then again when the fruits are ripening in order to understand what will really grow best on any particular piece of ground. As estate owners seldom spent all the year on their farms, this advice from Columella can be seen as placing quite a strong demand on landlords. Columella firmly believed that absenteeism had a deteriorating effect on agriculture. He also believed that there was no bad land, only mistreated one – this allows us a glimpse of a central feature of all Roman soil knowledge: Soils were seen as part of an interaction, humans were as active a part of this relation as were soils, and the latter were respected as being dynamic, living entities, which could be ruined by human mismanagement.

In addition to the general criteria, several experimental methods are described that enable one to test for important soil properties: Salinity or bitterness of the soil is important for viticulture, since the wine will acquire the taste of the ground. To test for this, one takes clear, fresh water and mixes it thoroughly with a soil sample. Once the mixture is filtered, through an unglazed earthenware or a sieve as used in wine-making, the details vary depending on the author, one can cautiously taste the water, which will have taken on the taste of the soil (Col. II,2,20; Verg. Georg. II 238–247; Pall I, V,3).

Natural land-cover also serves as a good indicator, and several indicator species are named that allow one to distinguish between sweet and saline grounds, and especially between grounds fit for grain or not. Columella gives a long list of plants (Col. II,2,21); Pliny remarks generally on the possibility of plant cover as a soil quality indicator (Plin. n.h. XVIII,34) and refers to a list of plants that Cato had already given. But Columella, who is generally the most cautious author, also warns against restricting oneself to the use of plant cover as the sole indicator.

Ancient Roman authors even suggest a test for soil fertility, although – according to Columella – it should not be used for black soils. After digging a hole into the ground, one should try to refill the hole. If the earth increased in volume, leaving a small hill, it was considered fertile. If the volume had not changed and the hole could be filled evenly, the earth was considered middle quality. The soil quality was considered meagre if the soil volume decreased (Col. II,2,18; Verg. Georg. II

226–237). Columella compares the volume change with the effect of yeast on dough. It was known that soil consists of several layers, and that this was of practical importance especially in planting orchards, as trees root deeper than grains (Col. II,2,21). For example, Pliny gives a very detailed description of the special features of the soil in Campania, a region of Italy, elaborating on the properties of the various layers of soil (Plin. n.h. XVIII,110).

Soils were also tested for cohesion. One should take a small clod of soil, add a few drops of water to moisturize it and then mould it in the palm of the hand. If it stuck to the skin, the soil was considered to be of good quality (explained in terms of 'natural humidity' and 'richness'). (Verg.Georg. II 250; Col. II,2,2,18, also Pall I, V,3). Pliny remarks, that as potter's clay, which is infertile, also shows this stickyness, he considers the test inconclusive (Plin. n.h. XVII,27).

The last indicator for soil quality to be mentioned in this context is smell. Nowadays it is known that the typical smell of earth is due to the activity of fungi, in particular the Actinomyces. Fungi play a crucial role in soil biology, and a soil that smells of them is obviously a healthy soil even by modern standards. Pliny (Plin. n.h. XVII,39) already mentions soil's smell, which, according to him, is strongest when rain moistens a surface that has dried out, and also becomes stronger when the soil is worked.

Looking at these analytical methods, one can summarize that salty or bitter water-soluble minerals were a known threat to cultivation, and a test for them was available, as were tests for main structural soil features, such as cohesion, density, water absorbing capacity and soil biology, as measured by means of the activity of fungal life. Further, indicator plants had been identified and were used in assessments.

Generally, authors were interested in soil fertility. The splitting of the phenomenon 'fertility' into sub-qualities has its limitations, as fertility is not solely a quality as such. It is the result of an interaction between natural conditions and cultivation measures. This was, as has been pointed out, common knowledge of all authors.

Suitability Descriptions

A good descriptive system allows one to work with a taxonomy which concentrates on those features of the described system relevant to the user. The plant-relative description system shall be demonstrated using the sweet chestnut tree, for which we also have Gargilius Martialis as a reference. But the reader should be aware that most authors give soil information with each and every crop or legume or tree they discuss.

Ea pullam terram et resolutam desiderat; sabulonem humidum vel refractum tofum non respuit; opaco et septentrionali clivo laetatur; spissum solum et rubricosum reformidat. (Col. r.r. 4,33,2) It likes a black and loose soil: does not refuse a damp, gravelly soil or crumbling tufa; delights in a shady slope with a northern exposure; and fears a heavy soil that is full of red ochre. (transl. Loeb, Vol. I, 457)

Note: I cannot understand why the Loeb edition translates 'pullus' as black here; I would prefer 'dark'.

The following texts discussed below are the Latin originals

Quaerit solum facile nec tamen harenosum, maximeque sabulum umidum aut carbunculum vel tofi etiam farinam, quamlibet opaco septentrionalique et praefrigido situ, vel etiam declivi. recusat eadem glaream, rubricam, cretam omnemque terrae fecunditatem. (Plin. n.h. XVII, 147)

Solum castanea desiderat molle et subactum; non tamen arenosum patitur et sabulum. Si umor adfuerit, pulla terra vel maxime prodest: sed et carbunculus, itemque tofus optime servit diligenter infractus. Spissum rubricosum locum non vult: vix provenit in argilla: in glarea cretaque nec nascitur. (Garg. Mart, De Arboribus Pomiferis, IV, 6)

Amant solum molle et solutum, non tamen harenosum. In sabulone proueniunt, sed humecto. Nigra terra illis apta est et carbunculus et tofus diligenter infractus. In spisso agro et rubrica uix prouenit: in argilla et glarea non potest nasci. (Pall., Op.Agr., XII, 19)

Amant solum molle et solutum, non tamen arenosum; et in sabulone proveniunt, sed humecto. Terra nigra illis est apta et carbunculus et tofus diligenter infractus. In agro spisso et rubrica vix proveniunt: in argilla et glarea non possunt nasci. (Petrus de Crescentiis, Ruralia Commoda, V, 6, 1)

If one compares the five statements, which are in general quite similar, one can still see differences, as well as different degrees of detail. What becomes clear is that the soil a chestnut tree likes is definitely NOT sandy or gravelly, and that a good, fertile, soft soil is best suited for chestnuts. The second thing the authors agree on is that a dense soil (*spissus*) is not suitable for chestnuts. All authors but Columella agree that *argilla* (clay) is hardly (*vix*) suitable, if we consider that Plinius uses the word '*creta*' which also means clay in this context. The only two statements which are completely identical (only changes in word order) are those of Palladius and Petrus de Crescentiis. It is sure that the latter used Palladius as his source and found no reason to change any of the substance. Between these two and Gargilius Martialis, only a few words are different: While the later texts have *solutus*, Gargilius gives *subactus* and instead of *amare* uses *desiderare* for 'to like'. Palladius used Gargilius, but left out the dependency on humidity, which his source gives: The part '*Si umor adfuerit*' is left out in Palladius chapter, and hence, in Petrus de Crescentiis. Petrus de Crescentiis is no author of Antiquity, but the last one to write an agricultural textbook in Latin, done in 1300 in Italy. The stability with which knowledge about suitable soils for chestnuts was preserved over about 1200 years is astounding. The manuscripts (as far as the editions I use give information on them) are not corrupt for the cited parts of texts, so one can conclude that the similarity was not produced by the editors but is original.

Concluding Remarks

Soil treatment knowledge was vital for economic success. The Roman authors knew about manure, about the fertilizing effect of leguminous plants and about the fertility increase which comes with fallowing. They knew that some plants cannot be planted on the same field in consecutive years, and Pliny gives a detailed report on marl, a soil fertility treatment which he reports to have been brought to Italy by Britons and the Gauls (Plin. n.h. XVII, IV, 42ff). Drainage was also used, as was irrigation, at least for meadows, which were inundated once a year for a period of several weeks. While historians of agriculture have looked into these measures in detail, soils as such do not figure prominently in their accounts.

The history of systematic, applied soil knowledge does not come to an abrupt halt at the end of antiquity, as the single example of an author writing in 1300 hopefully indicates. The Middle Ages had many soil experts, most of their work is not available in translations, thus it has gone fairly unnoticed so far (cf. Winiwarter, forthcoming, 2006). The body of soil knowledge which was produced over millennia was in general both sound and practical, and all the authors deserve respect for their largely successful attempts to translate practice into textbooks. Their attitude combined respect (and even worship of its luminous aspects) for a system of known complexity and fragility with systematic practical efforts to use it for the sustained benefit of humans. They were neither ecologists nor conservationists. If they can be compared to any modern type of scholarship (which has to be done with caution) they most closely resemble the ecological economists of today.

Sources: Editions and Abbreviations

Cato, agr. = M. Porcius Cato, De Agri Cultura, lat.-engl./ed. and transl. William Davis Hooper, Harrison Boyd Ash, Cambridge MA, Loeb Classical Library 1. Ed. 1934, Reprint 1979.
Col. r.r. = L. Iunius Moderatus Columella, De Re Rustica, lat.-dt./ed. and transl. Will Richter (Vol. 1 München: Tusculum 1981, Vol. 2 München: Tusculum 1982, Vol. 3, München: Tusculum 1983).
Gargilius Martialis, De Arboribus Pomiferis, Gargilii Martialis Quae Supersunt, ed. Angelus Maius, Lunaeburgi apud Heroldum et Wahlstabium (1832).

Pall. Op.agr. = Rutilius Taurus Aemilianus Palladius, Opus Agriculturae/Ed. Rodgers Robert H. (Leipzig 1975).

Pall. Op.agr. = Rutilius Taurus Aemilianus Palladius, Opus Agriculturae I and II lat.-frz. / ed. and transl. Martin René (Paris 1976).

Petrus de Crescentiis; Ruralia Commoda: das Wissen des vollkommenen Landwirts um 1300 ed. Will Richter, zum Druck vorbereitet von Reinhilt Richter-Bergmeier, Heidelberg: C. Winter, 1995–2002. (Series: Editiones Heidelbergenses; 25–27, 30.)

Plin. n.h. XVII = C. Plinius Secundus, Naturalis Historiae Liber XVII, lat.-dt./Ed. and transl. Roderich König, Joachim Hopp (München: Artemis und Winkler/Tusculum 1994).

Plin. n.h. XVIII = C. Plinius Secundus, Naturalis Historiae Liber XVIII, lat.-dt./Ed. and transl. Roderich König, Joachim Hopp, Wolfgang Glöckner (München: Artemis und Winkler/Tusculum 1995).

Varro, r.r. = M. Terentius Varro, Res Rusticae lat.-dt./ed. and transl. Dieter Flach, Gespräche über die Landwirtschaft (Bd. 1: Darmstadt 1996, Bd. 2: Darmstadt 1997).

Verg. Georg. = P. Vergilius Maro, Georgica lat.-dt./Hg. und übers. J. und M. Götte (München: Tusculum 5. Aufl. 1987).

References

Mahaffy, J. P., 1889/1890, The Work of Mago on Agriculture. In: Hermathena 7, 29–35.

White, Kenneth Douglas, 1970, Roman Farming. London, Thames and Hudson.

Winiwarter, Verena, 1998, Der Boden bei den Antiken Agrarschriftstellern. Attitudes to soil by ancient Roman scholars. Attitude au sujet de sol des Auteurs Romain en Antiquité. CD-ROM-16. World Congress of Soil Science, 20.-26.8.1998, Montpellier, F.

Winiwarter, Verena, 1999, Böden in Agrargesellschaften: Wahrnehmung, Behandlung und Theorie von Cato bis Palladius. In: Rolf Peter Sieferle & Helga Breuninger (Hg.), Natur-Bilder. Wahrnehmungen von Natur und Umwelt in der Geschichte. Campus, Frankfurt/M., 181–221.

Winiwarter, Verena, 2000, Soils in Ancient Roman Agriculture: Analytical Approaches to Invisible Properties. In: Novotny, H. & M. Weiss (Hg.), Shifting boundaries of the real: making the invisible visible. vdf Hochschulverlag, Zürich 2000, 137–156.

Winiwarter, Verena, 2002, Landwirtschaftliches Wissen vom Boden. Zur Geschichte der Konzepte eines praktischen Umgangs mit der Erde. In: Erde,

Kunst- und Ausstellungshalle der Bundesrepublik Deutschland (Hg), Wissenschaftliche Redaktion: Bernd Busch, (Schriftenreihe Forum/Band 11, Elemente des Naturhaushalts III), Wienand, Köln, 2002, 221–232.

Winiwarter, Verena, 2006, Prolegomena to a History of Soil Knowledge in Europe. In: John R. McNeill, Verena Winiwarter, (eds.), A world environmental history of soils, The White Horse Press, Cambridge and Isle of Harris, forthcoming 2006.

2
Aztec Soil Knowledge: Classes, Management, and Ecology

Barbara J. Williams[1]

Pre-Columbian Mesoamerica was one of two culture hearths in the Western Hemisphere, extending from Central Mexico and the Yucatan Peninsula, south through Guatemala and western Central America. Beginning as a center of plant domestication and incipient agriculture, in the course of six millennia it became the locus of hydraulic agrarian societies of diverse cultures, among them the Mayas, Mixtec-Zapotecs, Tarascans, and Aztecs. Underpinning these high civilizations was the highly-productive Mesoamerican farmer.

What Mesoamerican farmers knew about soils, soil management, and soil–plant relationships is best documented for the Nahuatl-speaking people of the Basin of Mexico and its immediate environs, for convenience labeled here as "Aztecs." In this discussion, Aztec ethnopedology is framed by contemporary soil science concepts, but it is important to note that Aztec perceptions and activities in the soils domain were conditioned by their world view and ideology, which lie beyond the scope of this chapter.

Sources

Ethnohistorical materials from which to discern Aztec soil knowledge and management include both indigenous pictorial manuscripts (codices) and documents written in Nahuatl, Spanish, and Latin. None of the extant native pictorials was written before the Conquest. However, during the Contact Period, native scribes continued to produce hieroglyphic documents following pre-Hispanic formats. Two such painted manuscripts record Aztec household populations and landholdings around 1540. These, the Codex Vergara (CV) and the Códice de Santa María

[1] Dept. of Geog. and Geol., Univ. of Wisconsin Colleges-Rock, 1206 S. River Rd., Janesville, WI 53546, USA.

Asunción (CSMA; Williams and Harvey, 1997), are very important documents for Aztec soil knowledge because hundreds of land parcels are drawn with perimeter and area measurements, and hieroglyphs indicating field soil types. Of lesser importance are the Humboldt Fragment VIII (Seler, 1904) and the Cadastral Fragment of the Ramírez Collection, which are smaller sections of similar land-holding registers with soil glyphs.

These Aztec codices have no accompanying text. Other primary sources meld the native pictorial tradition with explanatory, alphabetic text. One example is the Florentine Codex: General History of the Things of New Spain (FC). To compile this encyclopedia of Nahua culture (and its earlier versions), Fray Bernardino de Sahagún worked with native informants and painters from the Basin of Mexico for over four decades (ca. 1540–1585). With texts in both Nahuatl and Spanish as well as illustrations, it is the most comprehensive extant source on Aztec soil knowledge. Two other illustrated texts, important mainly for their information on Aztec pharmacopoeia and medicine, include depictions of plants and soils and describe characteristic ecological settings. One is Martín de la Cruz' Libellus de medicinalibus indorum herbis (1964 [1552]) [LM], the other Historia Natural de la Nueva Espana by Francisco Hernández (1577). Unfortunately, illustrations in the published versions of Hernández are the work of European artists who expunged indigenous details and perspectives. The only faithful copies of selected drawings—the originals burned in the 1671 fire of the royal library at El Escorial, Spain—are to be found in Juan Eusebio Nieremberg's Historia Naturae Maxime Peregrinae [1635] (Somolinos d'Ardois, 1960, p. 303, 309).

Source materials also include dictionaries, e.g. Alonso de Molina's Vocabulario en lengua castellana y mexicana (M), and Rémi Siméon's Diccionario de la lengua náhuatl o mexicana (RS), and toponyms, such as those depicted in the Codex Mendoza (CM) and its counterpart, the Matrícula de Tributos. Aztec soil infor-mation is found also in wills and testaments, property descriptions, and other legal documents, indicating that indigenous soil knowledge had socioeconomic rele-vance as a property descriptor well into the 16th century.

Interestingly, little about Aztec soil knowledge is found in either the native or Spanish chronicles of the Colonial period. This oversight of the soils domain con-tinued unabated even into the 20th century. For example, with few exceptions, ethnographies written as recently as the 1950s, which describe indigenous or tra-ditional agriculture and plant knowledge in detail, fail to include local soil knowl-edge. Fortunately, in the last two decades ethnopedology has begun to receive the attention it merits, as attested in a comprehensive, annotated bibliography by Barrera-Bassols and Zinck (2000). Finally, while there are methodological and theoretical problems inherent in ascribing information from contemporary Nahuatl-speaking farmers to pre-Conquest origins, such continuities appear to exist in some—if not many—respects. Thus, field data also are helpful in historical reconstruction of Aztec soil knowledge and management (e.g. see Sanders, 1957; Williams and Ortiz Solorio, 1981; Williams, 1982; Wilken, 1987).

Aztec Soil Descriptions: Texts and Glyphs

The most detailed written descriptions of Aztec soil knowledge come from three sections of the Florentine Codex (Book 11: Chap. 12, Para. 3, 4, and 5). These contain word lists, first for agricultural soils, then for "useless" non-agricultural soils, and finally terms mostly for special pottery clays and dyes. For each lexeme, Nahua informants described its linguistic derivation, identifying characteristics, similarity to other lexemes, and other brief observations. The format anticipated Sahagún's goal to compile a Nahuatl dictionary, which explains the linguistic bias (López Austin, 1974a, pp. 113–114) and the stilted phrases of the entries. Unless otherwise noted, soil descriptions below are quoted from paragraphs 3 and 4 of the FC (1963, Book 11, pp. 251–255), augmented when possible by information from other sources.

Aztec Agricultural Soil Classes

Deducing from FC informant commentaries, some agricultural soil lexemes label generic soil classes, while other terms are best characterized as descriptive attributes. Soil texture, color, organic content, moisture, topographic location, genesis, typical plants, fertility, and farmer practices often are noted. An analysis of these entries helps elucidate the defining attributes of Aztec soil classes, Aztec understanding of soil processes, and perceived relationships between soil qualities, plant growth, and crop production. Some examples are shown on the illustrations.

Soils Classed by Physical Properties

The FC and other sources suggest that the primary attribute of four or five Aztec generic soil classes was soil texture or consistence: stony, sandy, clayey, silty (?), and tepetate.

Stony/gravelly soil (tetlalli): "It is land on the mountain-rocky, gravelly, loose-graveled. It is very rocky, gravelly, rough, dry, dry deep down; a productive place, the growing place of tecintli maize. It is dry deep down; it is dry. It hardens; it becomes wet; it produces." As indicated by the reference to "producing when it becomes wet" [meaning after the spring rains], such soil was dry-farmed (<u>temporal</u>

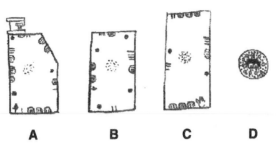

Figure 1. Aztec depiction of sandy soil, *xalalli*, in agricultural fields (A-C, CSMA, f. 63v) and in toponyms, *Xalac*, "Where There Is Much Sand" (D, CM f. 40r).

in Spanish). The glyph for stony soil is composed of the conventional symbol for rock [tetl]. Glyphic variants of the rock sign in the CV and CSMA suggest that stony soil was partitioned into sub-classes, perhaps according to degree of stoniness, or composition and size of gravels.

Sandy soil (xalalli): "It is [so] called from xalli [sand] and tlalli [earth], because it is sandy, of fine sand. It is not fertile; it is just a producer of maize stalks. It produces straight, slender maize stalks. It is not esteemed, not of much substance. It does not germinate much; it has no substance." Sandy soil is consistently glyphed in the codices as a circular array of small dots (Fig. 1).

Clayey soil (tezoquitl): "Its name comes from tetl [rock] and zoquitl [mud], because it is firm, gummy, hard; dark, blackish, bitumen-like . . . ." The stickiness of clay was recognized in a special term "tlalzacutli. It is just the same as clayey soil, because it is sticky, viscous, gluey . . . ." Today, many farmers refer to "black soil," which has attributes similar to the Aztec clay soil class but with the additional attribute of high fertility. In the CSMA and CV the glyph for clayey soil is composed of a rock [tetl] pierced by a spine, which conveys the sound of zo, thus reading phonetically te-zo(quitl). Glyphic variations in these codices suggest that generic clays were partitioned into species-level subordinate classes based on particle size, sandy clay and gravelly clay. In CM toponyms and FC illustrations, clays are indicated by a pictograph/ideograph of an amorphous, sticky substance (Fig. 2).

Loess (teuhtlalli): Wind-borne silt is identified as an Aztec soil type in dictionaries and the FC: "Teuhtli—From nowhere does its name come. This is earth which is very fine—that which swirls up, which sweeps up. It blackens things; it soils things . . . ." Teuhtlalli illustrated in the FC (#864) emphasizes the defining characteristics of wind-transported sediment by volutes composed of small dots rising from the earth's surface. The drawing resembles the conventional glyph for wind in name glyphs.

While the Aztecs recognized a particle size transported by wind ["earth which is very fine"], silt-sized particles are not mentioned nor apparently depicted as a soil

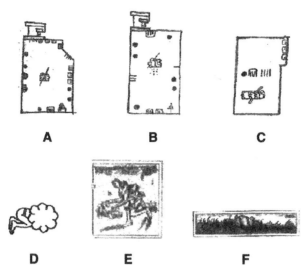

Figure 2. Aztec clay soil glyphs. (A) generic clay soil, *tezoquitl*, composed of rock-*tetl* and spine-*zo* signs (CSMA, f. 56v) (B) sandy-clay sub-class, read from the clay sign and dots, *xalli*-sand (CSMA, f. 63v) (C) gravelly-clay sub-class, read from the clay sign and reduplicated rock sign (CSMA, f. 69r) (D) "On The Little Clay," *Zoquitzinco*, a toponym in the CM, f. 33r. Clay stickiness is suggested by the massive lump of gray material, reading *zoquitl*. The bottom half of a man's body phonetically reads "small, little"- *tzinco*. (E) clay depicted in the FC, Book 11 Illus. 873 shows a man mixing a sticky substance. (F) Illus. 850 of clay in the FC shows a rock sign atop a lumpy, dense, earthy mass.

constituent. Further, the FC word list and Nahuatl dictionaries contain no specific term for loams, i.e. texture classes based on relative amounts of sand, silt, and clay. However, the glyphic variants of clay suggest that Aztec farmers took into account relative amounts of clay, sand, and gravel in a soil as the basis for partitioning clays into subordinate classes, conceptually similar to the idea of loams. And, although "loam" may not have been in the Nahuatl lexicon, Nahuatl syntax allows lexeme strings, or uses conjunctives, to easily express the notion of admixtures. For example, "tepetlatl xaltlalneliuqui," lit. tepetate-sand-mixed together, or "tepetlatl ihuan quauhtlalli tlalneliuqui," lit. tepetate and woodland soil mixed together, are linguistic constructions commonly used by Nahuatl-speaking farmers in the eastern piedmont of the Basin of Mexico (Williams, 1982, p. 210).

Tepetate. Tepetate is an indurated substrate of volcanic soil exposed at the surface through erosion of overlying horizons. The term derives from rock [tetl] and mat[petlatl], which best translates figuratively as "soft rock," that is, earth material of hard but friable consistence, material which can be crushed in the hand (Williams, 1972, pp. 619–620). In Aztec times tepetate was classed both as ". . . a rough stone. It is whitish, porous . . . I break up tepetlatl. I dig up tepetlatl" (FC, p. 265), and also

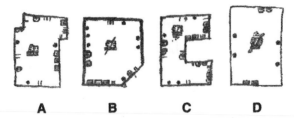

A   B   C   D

Figure 3. Tepetate soil classes. (**A**) generic tepetate, composed of rock-*tetl* and mat-*petlatl* glyphs (CSMA, f. 57v). (**B**) clayey-tepetate sub-class, composed of rock, mat, and spine signs (CSMA, f. 56v). (**C**) sandy-tepetate sub-class, composed of rock, mat, and dot signs (CSMA, f. 57v). (**D**) clayey-sandy tepetate sub-class, depicted by rock, mat, spine, and dot signs (CSMA, f. 67v).

as an agricultural soil, called tepetatlalli [tepetate soil] by farmers today (Williams, 1982, p. 210). Tepetate soil glyph variants suggest that Aztec farmers partitioned generic tepetate into sandy, clayey, and clayey-sandy sub-types (Fig. 3).

Soil Classes and Color

Many soils recognized by farmers today in Mexico and Guatemala are labeled by color terms. Although the class labels suggest discrimination on a single dimension, in fact they are multi-dimensional, generic classes. That is, "Black Soil" connotes properties other than merely black color (Williams and Ortiz Solorio, 1981, p. 341, 358). Interestingly, of the Aztec agricultural soils listed in dictionaries and the FC, only one is labeled by color. A generic class, yellow soil, tlalcoztli ". . . is named from tlalli [earth] and coztic [yellow], because yellow soil is good, fine, fertile, fruitful, esteemed." Glyphs which apparently refer to yellow soil in the CV and CSMA are composed of two elements, a rock [tetl] and dots along side or appearing to flow from the tetl sign. Its phonetic reading of "yellow soil" probably utilized near homonyms (Williams and Harvey, 1997, p. 33).

Soil Classes and Topography

Upland soil. The physiography of the central basins of the Mexican Plateau—flat-floored valleys interrupted by volcanic hills and mountains—creates a topographic contrast set of uplands and plains. Aztecs described the mountains and slopes with the term tepetlalli, meaning hill [tepetl] land [tlalli]: "This is the top soil of a mountain, the upland, the slope. It is also called ximmilli. It is dry, clayey, ashen, sandy soil, ordinary soil, rain-sown. It is the growing place of mature maize, of amaranth, of beans. It is the sprouting place of the tuna, the <u>nopal</u>, the maguey; of the American cherry; of trees . . . ." The FC illustration of tepetlalli (#855) depicts an expansive upland landscape with a farmer cultivating a plot with his hoe. This scene, as well as the written entry, implies that hill soil was not an Aztec generic class but rather a topographic location where various soils were under rain-fed cultivation. The list of cultigens includes the major dry-farmed subsistence crops of the semi-arid highlands. With the exception of amaranth, these crops continue to dominate traditional upland agriculture today.

Hillslope soil. In the CSMA a combination of hill [tepetl] and eye [ixtli] glyphs indicates a soil type for which no obvious referent in texts has been found. Most likely it reads "hillslope"-tlaixtli, a term for slope lands used in Atla, Puebla (Montoya Briones, 1964, p. 24, 47), or "soils on the surface of the hill" as in the place name Tepetlixpa, from tepetl-hill, and ixpa-on the surface (Olagüibel, 1975, p. 34). Its prevalence in the CSMA and CV suggests it was a generic class.

Soil Classes and Organic Matter

As attested in the following texts, Aztec farmers clearly observed the influence of organic matter on soil fertility, and they associated decomposition processes with soil genesis. Organic content defined several soil classes, which quite likely were generic-level. One is woodland soil: "quauhtlalli: Its name comes from quauitl [wood] and tlalli [earth]. It is rotten wood or oak leaves; humus; or topsoil with [rotten] wood. It is black or yellow. It is a fertile place (FC)," and excellent for cultivation of wheat and maize (RS).

Another soil of organic genesis is reedy soil: "tollalli: It is reedy land, or the rotting of the very reeds; the reed, the small reed converted into soil. It becomes reedy, full of small reeds, grassy. It produces; it becomes fertile; it has true substance."

Aztec farmers had a specific term for humus, which they linked to organic matter decomposition and soil genesis. The concept is attested in the definition of the term tlazollali (from compost pile, tlazolli): "This is humus which turns into soil. It is [a soil] which is fertile."

Soil Classes and Water

The Aztecs recognized water as an agent of soil genesis, attested in the generic class of alluvial soil—"atoctli: Its name comes from atl [water] and totoca [it runs]; that is, water-borne yellow soil, water-borne sand. It is soft, porous, very porous, good, good smelling. It is that which is fertile, esteemed, well considered; it is food-producing (FC)," deep and moist (RS). Atoctli still is recognized in the countryside with its attributes of exceptional fertility and workability, and it appears in place names, such as San Pedro Actopan. The glyph is composed of a green-maize stalk [toctli] and the water [atl] glyph. Modified atoctli glyphs in the CSMA suggest distinction of several species-level alluviums: clayey, gravelly, and sandy. Sandy alluvium is described in the FC: "Xalatoctli: This is sand borne by the water. It is very loose. Borne by the water, it is porous, very porous," and very fertile (RS). A sub-ordinate class of sandy alluvium, gravelly-sandy alluvium, appears to be indicated glyphically as well, which suggests that alluvium had three levels of taxonomic depth (Fig. 4).

Given the semi-arid environment of Aztec cultivators, one would expect particular attention was accorded to soil moisture conditions. We indeed see this focus, for example, in the term "techiauitl," which refers to the moisture retention characteristic of stone mulch, or of stony land: "when it gets water, it does not dry off quickly; it just lies moist, it lies completely workable, it lies wet . . . ." Consistently moist soils (tierra de humedad in Spanish) are labeled "chiauhtlalli: Its name is [so] called from chiaua [it becomes soggy] and tlalli[land]. This always lies wet, although unirrigated. It has substance; it is food-producing. When it rains heavily, [the maize] perishes. But when the rain is not heavy, it brings joy, it brings contentment . . . ."

Thus, Aztec soil knowledge included the concepts of permeability and moisture retention. Very impermeable soils were classed as useless. Nantlalli belongs to this category: "The water extends on top; it does not soak in; it does not reach below. As it is wet it just forms a thin film, goes on the surface . . . ."

Just as today, in Aztec times irrigated land was of premium value, as attested in the following: "Irrigated land – atlalli .is [so called] from atl [water] and tlalli [earth]. This is the irrigated field. it is a watered garden, one which can be irrigated; it is irrigable; [land] which becomes wet, becomes mud. It is good, fine, precious; a source of food; esteemed; a place of fertility. It is a fertile place; it has substance. It is [a place] to be planted to maize, to be planted to beans, to be harvested."

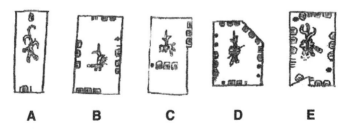

A   **B**   **C**   **D**   **E**

Figure 4. Alluvium classes. (A) generic alluvium, *atoctli*, shown by glyphs of cornstalk-*toctli* and water-*atl* (CSMA, f. 18r). (B) sandy-alluvium sub-class, read from cornstalk, water, and dots, *xalli*-sand, signs (CSMA, f. 43r). (C) gravelly-alluvium sub-class, depicted by cornstalk, water, and rock signs (CSMA, f. 45r). (D) clayey-alluvium sub-class, read from cornstalk, water, rock, and spine signs (CSMA, f. 56r). (E) gravelly, sandy-alluvium sub-class, shown by cornstalk, water, rock, and sand symbols (CSMA, f. 61r).

Soil Classes and Human Agency

Archaeologists have long associated elevated levels of soil fertility with residential sites and midden heaps. Aztec farmers made similar observations, as seen in the description of "house land:" "Callalli: Its name comes from calli [house] and tlalli [earth]. This is the land upon which a house has rested, and also the surrounding houses. It is fertile; it germinates." Human activity also was perceived capable of decreasing soil fertility, as in the description of axixtlalli, from atl [water] and xixtli [human feces]: "This is land which has been urinated upon, which is greasy. Its characteristics are just the same as those of nitrous land . . . ."

Nahuatl terminology for these classes and for other productive solis was not adopted by European settlers of New Spain. However, two terms for unproductive soils did become part of the Mexican-Spanish lexicon, and are still used today (Gibson, 1964, p. 300). Tepetate is one of these Nahuatl loan words. Another loan refers to the Aztec class of saltpeter/nitrous land, tequesquite in Spanish, tequixquit-lalli in Nahuatl: "It is salty, bitter, corrosive, that which is leached [of its salt]; unwanted, undesirable. It is waste, disregarded . . ." (FC, p. 254). The conventional glyph is a daisy-shape symbol in toponyms (e.g. CM, f. 4v, 29r).

Control of the Soil Environment

Aztec farmers purposefully modified the soil environment in numerous ways typical of intensive agriculturalists. The evidence from which to identify specific pre-Hispanic practices is meager, however. Some of it is direct, as in lexical items and archaeological remains. More often it is indirect, or inferred from accounts of Colonial or more recent traditional practices. The following summarizes some selected Aztec management practices for which we have the strongest direct evidence. It relies heavily on the work of ethnohistorian Teresa Rojas Rabiela, who has conducted a decades-long exhaustive search for, and meticulous analysis of, primary sources on indigenous agriculture (e.g. Rojas R., 1983, 1984, 1988, 1990; Rojas Rabiela and Sanders, 1985). Also, many of the practices mentioned below are reported in a comprehensive regional synthesis of Mesoamerican agriculture by geographers Whitmore and Turner (2001).

Soil Preparation and Management

Using hand tools, Aztec farmers worked the soil surface to improve structure and permeability. In an oft-quoted entry in the FC, the native informants describe the activities of a good farmer: ". . . He is bound to the soil; he works—works the soil, stirs the soil anew, prepares the soil; . . . breaks up the clods, hoes, levels the soil . . . breaks up the soil . . ." (Book 10, Chap. 12, pp. 41–42). Land which was worked was called tlahuitectli: "This is land which is worked, worked down, packed. I beat the land. I work the land down with blows" (Fig. 5A).

The practice of pulverization was essential for creating a tillable soil from an eroded tepetate hardpan. In tepetate reclamation, once the soil structure was improved, organic matter was added to raise nutrient levels and improve tilth. More recently, gun powder and modern earthmoving equipment have been used to break up tepetate.

References to preparation for planting suggest that sometimes holes were excavated in the soil surface and subsequently filled with fine, pulverized soil and perhaps fertilizers. The holes also functioned as sediment traps, and collected rain or manual irrigation water. More commonly, soil was heaped into ridges, cuemitl, creating an aerated planting surface. In swampy areas and in shallow lake beds, ridges were constructed on a large scale by excavating and piling up bottom muck and aquatic vegetation, forming raised fields of Anthrosols, called chinampas. The resultant adjacent ditches allowed canoe transportation into and out of the chinampa zones. In the southern part of the Basin, chinampa agriculture still functions, but within an area much smaller than that of its archaeological vestiges.

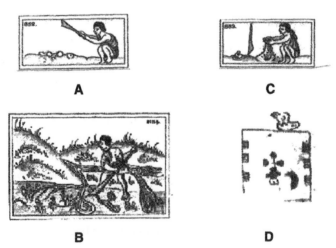

A   **C**

B   **D**

Figure 5. Aztec Management Practices. (A) pulverization of soil clods and hardpans—
tlahuitectli (FC, Illus. 852). (B) water control and diversion for irrigation (FC, Illus. 854).
(C) fertilization [*tlalahuiac*], here probably with night soil [*cuitlatl*] shown to the left
of the farmer's hoe (FC, Illus. 853). (D) fertilization of clay with woodland soil. From
top to bottom, Seler (1904: p. 206) reads the tree-jar-teeth glyphs in the field center as
cuauh-con-tla, to be resolved as *cuauhtlalli, contlalli*—"woodland soil mixed into clay."
A "fallow" sign, grass-*zacatl* appears at the lower right.

The artificial soils of the chinampas were the most intensively managed of the Basin, and their enormous productivity is thought largely to account for the support of the 150,000–200,000 people living in the Aztec island capital of Tenochititlan at the time of European contact (Calnek, 1976, p. 288; Parsons, 1991, pp. 39–40). Almost continuous cropping was typical. A key component for high agricultural output was the practice of using space-saving seed beds from which plants would be transplanted at the appropriate time. Even maize was grown in this fashion. To reduce frost risk, plants were covered with shed-like structures of organic materials.

Slope Management

Various types of planting platforms were constructed on very gentle as well as steep slopes. Soil embankments [metepantles], some held in place by economic plants such as maguey and nopal, and terraces faced with stone, tepetate blocks, or other material, functioned to control sheetwash and rilling and to trap sediments, which increased soil depth, permeability, water retention, and fertility. Also, new planting surfaces were created by siltation behind permanent or temporary check-dams constructed across intermittent stream courses.

Water Management

Many of the soil descriptions cited above reflect the importance Aztec farmers ascribed to conditions of soil moisture. Irrigated land was the most esteemed, and numerous Nahuatl words and phrases relate specifically to irrigation (Lameiras and Pereyra, 1974). Their temporary and permanent dams diverted water from perennial and intermittent streams to fields with or without using canals (Doolittle, 1990). Large hydraulic constructions, the causeway-dikes, controlled water and salinity levels in the lake system necessary for the functioning of the chinampas. Springs in the mountains and melting snows were distributed throughout the terraced slopes in small-scale irrigation systems, which in the aggregate covered thousands of hectares of cultivated land (Fig. 5B). Through irrigation, Aztec farmers advanced the time of summer planting, thereby reducing the risk of frost damage in the fall, and they were able to plant more productive varieties of cultigens.

Fertilization

Aztec knowledge about the role of organic matter in soil genesis led to practices aimed at maintaining or increasing soil fertility through application of soil amendments. This is attested by the terms "tlalauiyac: This is the land which is good, which produces, which is mellow. I fertilize it. I add humus to it. I make it mellow. I make it good," and "tlazotlalli: fertile land that contains fertilizer" (Fig. 5C).

Fertilizers mentioned in documentary sources include refuse [tlazolli] (green manure? compost? or organic wastes?), human excrement [cuitlatl], bat dung, organic ash, and alluvium [atoctli]. Sediments suspended in irrigation and flood waters also served as fertilizer, and when canals were cleaned, the water-borne mud [azoquitl] and other sediments [zoquitl] were returned to fields. As noted, the highly productive Anthrosols of the chinampas were composed almost completely of nutrient-rich organic matter. A possible pictorial reference to fertilizing with organic-rich soil is the compound soil glyph in the Humboldt Fragment VIII (Seler, 1904, p. 206), which suggests that woodland soil (quauhtlalli, depicted by a tree sign) was used to fertilize clay (depicted by a jar sign) (Fig. 5D).

Fallowing as a strategy for maintaining or restoring fertility was practiced pre-Hispanically, but in the Basin of Mexico and environs, the early accounts suggest that all arable land in late Aztec times was under cultivation. Fallow fields in the CSMA and the Humboldt VIII are indicated by a grass glyph [zacatl] and shading. Rather than indicating fallow as a deliberate management strategy, however, these documents quite likely reflect land abandonment due to death or illness of farmers engendered by the disastrous epidemics of the Contact Period.

Problem Soils

Worn out soil. Apparently some circumstances resulted in irreclaimable (?) worn out land: "It is called tlalzolli, bad soil, because nothing can be grown there. It is the growing place of nothing—useless, productive of nothing. It is worked in vain; it fails. It is worn out land . . . ." Since Aztec techniques to maintain or enhance soil fertility were numerous, perhaps the category of worn out land had more to do with historical/ social circumstances of farming communities than with shortcomings in soil knowledge and management practices.

Salinization

In the Basin of Mexico, the northern lakes of Zumpango, Xaltocan and Texcoco were saline. This fostered the industry of salt-making, but presented major problems for agriculture along the shorelines. Sahagún's informants categorized these saline/alkaline soils as "useless lands." However, archaeological work at the former island site of Xaltocan in the last decade has uncovered chinampa cultivation dated ~800 AD (Nichols and Frederick, 1993, p. 141), showing that farmers had developed technologies to counteract saline conditions. One method to reduce water salinity was to channelize freshwater streams from the drainage basin into

Lake Xaltocan agricultural zones. Another especially ingenious hydraulic technique is indicated by a recently discovered canal cut into the lake bottom from the eastern shore across the lakebed northwestward toward the island Xaltocan chinampa area. Apparently the canal was constructed during a period of seasonal or short-term lake desiccation, and may have used density differences to float or direct freshwater through the saline water to the chinampa fields (Nichols and Frederick, 1993, p. 142).

After the Conquest, progressive soil salinization occurred in the Basin, and alkali dust storms in the dry season became common (Gibson, 1964, pp. 305–306). In the chinampas bordering Lake Texcoco, farmers in the early 19th century were reported to remove the layers of salt-impregnated soil into canoes and dump the soil into canals. New sediments from deeper parts of the lake bottom at a different place would then be excavated and brought to a chinampa as replacement soil. Alternatively, after a period of time the farmer would use the same soil he had dumped previously: "They know that a soil impregnated with salts, when washed, as they say, it lacks salts, and with this the soil becomes good for successful cultivation" (Alzate y Ramírez [1831] reprinted in Rojas Rabiela, 1983, pp. 18–19). Tylor (1861) suggested that soil flushing was a pre-Conquest technology. He reported that ancient Mexicans gathered mud from along the shores of Lake Texcoco to place on their floating islands, or reed rafts [Note that he probably mistook people transporting plants as actual floating agricultural plots]. "The mud at first was too filled with salt and sodium to be useful for planting; however, by sprinkling water from the lake over the mud and letting it filter through, they were able to dissolve the greater part of the salts, which permitted cultivation of the island" (reprinted in Rojas Rabiela, 1983, p. 40).

Soils and Plants

To the Aztecs, as well as contemporary soil scientists, plants were known to be a factor in soil formation. Evidence of these perceived soil-plant connections are: a) the two generic-level, organic soil classes—woodland soil [quauhtlalli] and reedy soil (peat?) [tollalli]; b) compost [tlazolli] described as turning into soil; and c) the practice of amending soils with plant residues.

The enormous labor invested in irrigation, terracing, fertilization, water-control dikes and canals, land reclamation, and construction of chinampas shows that Aztec farmers had identified critical plant growth requirements and knew how to manipulate soil conditions to increase crop yields. The soil's ability to sustain crops often was a stated attribute of soil classes. To the Aztecs, productive soils were a) irrigated or constantly moist, b) high in organic matter – woodland humus, compost, house plots, peat or muck, c) fertilized soils, d) alluviums, and e) yellow soil. Infertile classes were sandy, ashen, nitrous, urinated, and worn out.

Soil-crop Correlates

Certain cultigens were associated with specific soil classes by FC informants: stony soil with upland or dry-farmed maize; woodland soil with maize and wheat (the latter a European introduction); mountain/slope land with maize, beans, amaranth, tuna, nopal, maguey; irrigated land with maize and beans. It is doubtful that references to these soil/plant associations should be taken literally, however. Because they are all ubiquitous subsistence crops, it is more likely that the soil-cultigen references were to be understood metaphorically to mean "food-producing," as in the description of alluvium—"it is food producing."

Soil Indicator Plants

Interestingly, the presence of some non-productive soils was thought to be marked by specific plants. For example, couch grass (tequixquizacatl) was an indicator plant for both nitrous soil (tequixquitlalli) and ashen soil (nextlalilli) (FC, p. 193; Illus. 634). The presence of ashen soil also was indicated by grama grass. Poor soil was marked by the zacateteztli plant: ". . . it lies scattered in small clumps. It is a little white. [Its growing place is in the poor lands.] Wherever it lies, it is a sign that the land is not good" (FC, p. 194; Illus. 637).

Plant Habitats

Aztec knowledge of soil–plant relationships is indirectly expressed in their classification of ecological habitats, within which specific plants were said to grow. For example, many plants were associated with an aquatic or riparian habitat. Also, location of plants recorded in the FC and by Hernández suggests an Aztec classification of ecological zones into vegetative, topographic, climatic, or edaphic contrast sets. Some plants were said to grow in forests, others in grasslands, some in mountains, others on plains, in hot versus temperate climates and good versus poor lands. Other plant environments mentioned include deserts/dry lands, crags, gardens and fields, and specific geographic areas, such as the "Papaloquitl: it grows in the east" (FC, p. 139), or specific sites: "Chichipiltic. It grows in Xaltenco" (FC, p. 144). Multiple habitats characterize some plant distributions, e.g. "its growing place is everywhere, in the mountains, on the plains," or "its growing place is in the mountains, in the forests."

Soil and Plant Nomenclature

Nahuatl plant nomenclature often reflects characteristic ecological or edaphic environments in which the plant is found. Garibay (1964, p. 359) noted that plant names beginning with cuauh-, tepe- and te- (the initial syllables of tree, hill, and rock, respectively) often translate as "growing in the hills or mountains," as in for example, tepechian, from tepe-[tepetl, hill] and chia-[chian, salvia], to read "mountain salvia." The initial syllable tlal-[from tlalli, soil/earth] signifies growing on the plains or interior of the land, and the syllable a-[from atl, water] often identifies an aquatic plant. In addition to specifying a mountainous habitat, the syllable "te-" may specify a stony soil environment, as in "stone moss:" teamoxtli [tetl-amoxtli]. In some instances a mineral or soil associated with a plant becomes the name of the plant itself, as in tetlahuitl [tetl-tlahuitl], "red stone." The "black earth" plant, tlayapaloni xiuhtontli, derives its name from the dark color of the humic soil in which it grows, which was thought to be the source of the dark color of the plant (Maldonado-Koerdell, 1964, p. 296).

Soil Depictions in the Martín de la Cruz Herbal

A very interesting and beautiful Aztec painted manuscript (1552) is the Libellus de medicinalibus indorum herbis (LM), The Book of Indian Medicinal Herbs. The codex consists of 185 oil color illustrations of medicinal plants which a native Aztec doctor, Martin de la Cruz, prescribed in his medicinal remedies (Viesca, 1995). The accompanying text is a contemporaneous translation into Latin by Juan Badiano, of what probably was an original Nahuatl text describing the therapeutic recipes. The Nahuatl text has never been found.

The LM has drawn special acclaim because of the artistry of the drawings and because, unlike many herbals of the period, the plant drawings not only depict above-ground foliage but also plant roots and surrounding environmental substrate. Notwithstanding, the LM is of limited value as a primary source to reconstruct Aztec perceptions of soil–plant relationships. The text makes few such references, and the use of Latin hinders linguistic analysis to reveal Nahua cognition. Although the still-unidentified artist was almost surely a native Nahua, the drawings are strongly influenced by European naturalism (Fernández, 1964, p. 240). As Ortiz de Montellano explains (1990, p. 20), because the manuscript was intended as a gift to Charles V of Spain to gain royal support for the Indian school in Tlatelolco, "the herbal was to be a tool lobbying for the view that Indians were 'human,' capable of being educated, and possessors of a worthwhile culture." Hence, the Latin text and Europeanized artistic expression were to show native sophistication, as well as Aztec culture.

The purpose of the drawings apparently was to aid native doctors in identification of pharmaceutical plants. The Latin text would not have been helpful in that regard (see Viesca, 1997, pp. 45–47). In the opinion of many scholars, the artist augmented the information with which to make such identifications by depicting not only plant foliage but also the ecological or edaphic setting (Emmart, in Cruz, 1940, pp. 37–38; Robertson, 1959, pp. 157–158; Fernández, 1964, p. 240; Maldonaldo-Koerdell, 1964, p. 293; Pozo, 1964, p. 334; Somolinos, 1964, pp. 321–322; Barrera-Bassols and Zink, 2000, pp. 377–378). To evaluate the edaphic/ecological representations, the artist's source of information should be considered. For plant foliage, details in the manuscript suggest that plant specimens from which to make drawings were delivered to the artist, although not always in the sequence in which they were to appear in the book (Pozo, 1964, p. 335). In contrast, nothing apparent in the manuscript suggests how ecological information was transmitted to the painter—whether orally or in written form, by soil samples, from his field observations, or deduced from plant nomenclature. The drawings, as Maldonado-Koerdell (1964, p. 293) suggests, "perhaps were made without direct contact with reality." That is, they may reflect only the artist's own ideas, or his aesthetic sense. The painter used classic Nahuatl hieroglyphs sparingly, understandable in light of the manuscript's intended readership. Instead, he may have reworked glyphic symbols into more naturalistic expression. A detailed study of the plant-environment drawings applying glyphic decipherment techniques coupled with Nahuatl linguistics is needed to test this hypothesis. Thouvenot's analysis (1982) of "precious stones" illustrations in Book 11 of the FC demonstrates the efficacy of this methodology.

Water glyphs, rock glyphs, and snake and ant pictographs are the most obvious indigenous elements in the ecological drawings. The LM confirms Aztec association between certain plants and various hygric environments. Flowing water is shown by the conventional sign for water-atl in blue, the color symbol for water in Aztec codices. Of the two occurrences of the sign, one (f. 9r) is semi-naturalistic, showing European influence, the other (f. 61r) is the classic Nahuatl water-atl hieroglyph (Fernández, 1964, p. 239) [Fig. 6]. A riparian environment (f. 44r) or stagnant water, such as a pond or swamp (ff.8v, 18r, 44r) are conveyed by roots immersed in a background of blue (Maldonaldo-Koerdell, 1964, p. 297; Emmart, in Cruz, 1940, p. 37). Roots emerging from a small patch of blue (e.g. f. 20r) suggest moist soil (Emmart, in Cruz, 1940, p. 37), perhaps chiauhtlalli. The shells and circles of the atl sign attached to the roots of the tolpatlactli (from tollin-rush or reed) (f. 18r) symbolize the marshy environment of rushes.

We know that Aztecs recognized a number of soil classes defined by soil texture, and that these classes were expressed by conventional glyphic signs. The artist of the LM seems to have minimally drawn upon this glyphic corpus. In some instances, a bouldery soil (tetlalli) seems to be indicated by realistically drawn

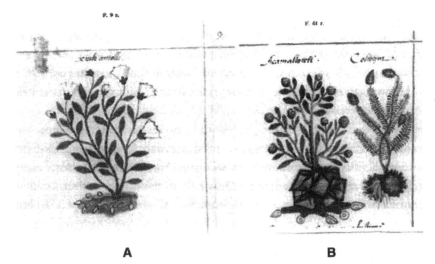

A   **B**

Figure 6. Illustrations of medicinal plants in the Libellus de medicinalibus indorum herbis (1552) show that soil and plant domains were closely related in Aztec ecology. Not only are key features of plant foliage depicted but also the roots and their surrounding environment. The drawings blend indigenous ideographic art with evolving European naturalism. For example, flowing water characterizing the habitat of certain plants is rendered by a semi-naturalistic representation of the Aztec water glyph in A (LM, f. 9r) in contrast to the classic form of the water symbol in B (LM, f. 61r). The vividly-colored, blocky material within which the Acamallotetl roots are wedged is abstract but in a distinctively Europeanized painting style. Acamallotetl labels both a plant and a stone in the herbal and is of uncertain etymology. Color version (see the color plate section at the end of the book).

yellowish brown lumps (ff. 36v, 37v, 43v, 44v, 45r, 48r). The conventional Nahuatl sign for rock-tetl occurs in twelve drawings (ff. 21v, 22v, 24r, 26r, 25v, 27v, 34r, 38v, and 49r). The tetl sign conveys various meanings, depending on the plant: a) as in stony soil-tetlalli, and te-xiyotl, hard or stone shell (f. 22v), referring to "stone crop," Sedum (Emmart, in Cruz, 1940, p. 41); b) as in tex-calli, a plant growing in the crags, or in the mountains; c) in an ideograph for drupaceous fruit plants, such as cherries or plums, suggesting their interior pits (f. 49r tlalcacapol, earth-cherry, Emmart, in Cruz, 1940, p. 41); d) a dry environment (e.g. f. 24r, tlalquiquitzal-Achillea Millefolium, usually found on dry hillsides or roadsides (Emmart, in Cruz, 1940, p. 41). The variety of forms and of colors of the tetl sign suggests perhaps many as-yet-undeciphered additional meanings.

Material surrounding plant roots in many drawings appears to be in angular or sub-angular blocks with black-dotted, diffuse frames (unlike the native painting style which employed a continuous, unwavering-width frame line). Interior sections of the blocks exhibit varying colors and shading, again unlike traditional manuscript painting (Robertson, 1959, pp. 65–66). The impression is one of peds or blocky structure, a soil attribute not recorded in other sources on Aztec soil properties. As Ortiz de Montellano (1990, p. 32) noted, information is more credible if expressed in several different sources. Thus, in the absence of corroborating data, here we may be seeing the aesthetic sense of the LM artist rather than expression of Aztec concepts of soil structure.

One of the issues in Aztec ethnopedology is the extent to which Aztecs conceptualized the existence of soil horizons and horizonation processes. Several references suggest awareness of changes in soil properties with depth, for example in stony soil "it is dry, dry deep down." Drawings in the LM have been cited also as showing some Aztec cognizance of soil horizons (Williams, 1975, p. 118; Barrera-Bassols and Zinck, 2000, pp. 377–378).

Six LM drawings are most important in this regard (ff7r, 18r/v, 19r, 30r, and 47r). Unlike any other drawings in the codex, plants are depicted atop an elongated column of material bordered by a frame line. Within the frame are various colors in mottles and shades giving the illusion of horizons in a soil profile. Closer inspection, however, reveals details which shed doubt on this interpretation. One drawing (7r) shows stone moss growing on top and at the bottom of the column. In another drawing of the same plant (f. 18r), the word eztetl-jasper seems to label the material in the column, which correlates with the reddish tinge in the illustration. In her commentary on the LC, Emmart concluded that these two columnar drawings are of rock, whereas the other four represented soils or earth (in Cruz, 1940, p. 210, 235, 297). The shape of the columns and boundedness probably led Fernández (1964, p. 239) and Maldonado-Koerdell (1964, p. 297) to conclude that the drawings are of plants in containers (seed-beds) planted in a soil good for their growth. In other words, these paintings are not of natural soil profiles.

The columnar drawings are strongly Europeanized in their coloring and in the treatment of the plants, which are depicted in three-dimensional perspective at the top of the column as nowhere else in the codex. On the other hand, Aztec glyphic elements also appear. Indigenous unconcern with perspective is seen in moss growing at the bottom as well as the top of the column (f. 7r); indigenous color coding uses reddish-purple to read "eztetl"—jasper (f. 18r), and "chichiltic tlalli"—red earth (f. 19r). Also, a pictograph of the "river gravel of diverse colors," mentioned in the medicinal text, is depicted at the bottom of the column (f. 18v) (Fig. 7A). Finally, whether of rock, actual soil, or artificial planting medium, the columnar illustrations seem to attempt to convey substrate structure: columnar (f. 7r, 18r, 18v 19r) and granular (f. 30r, f. 47r) (Fig. 7B). But, since they lack horizontal zonation from top to bottom, these drawings do little to elucidate Aztec conceptions of soil horizons.

The LM artist used indigenous pictographs in two salient instances. One illustrates the ecological association of certain plants with ant hills. In one

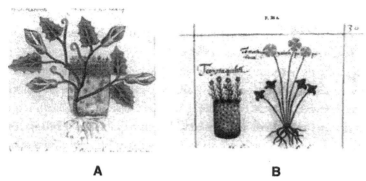

Figure 7. Illustrations of plants growing atop framed soil columns seem to depict soil horizons. Rather than modeled from actual soil profiles, these drawings quite likely represent plants growing in seed beds or planters in an artificially-produced plant medium. In A, at the base of the soil column is a depiction of small pebbles, pictographically alluding to the accompanying medicinal remedy ingredient, "river gravels of diverse colors" (LM, f. 18v). In B, the plant medium appears to be of granular structure or sandy/gravelly texture (LM, f. 30r). Other than the rock and possibly sand signs, the LM artist did not use the corpus of native graphemes for Aztec soil classes. Color version (see the color plate section at the end of the book).

illustration, yellow ants drawn in profile in plant roots climb the plant stalk (f. 13v), and in another case we see red ants from a top view in an ant hill from which a plant emerges (f. 28r). Phoneticism is the intent in the second example of pictograph use. In drawing the serpent fruit [couaxocotl: from couatl-snake and xocotl-fruit] the artist depicts two snakes climbing the plant and eating the fruit (f. 38v).

Sixteenth century Europeans marveled at the extent and detail of Aztec botanical knowledge. As the information in the FC, in Hernández, and in the LM suggest, Aztec botany embodied a strong ecological component, just as did Aztec pedology. Apparently, it would have seemed quite unnatural to Aztecs to view soils and plants as isolated systems. Even their lexicon required acknowledgment of the inter-relationships between soils and plants, a kind of ecological thinking that is often considered to be modern concept.

Non-agricultural Application of Aztec Soil Knowledge

Soil knowledge was integrated into Aztec culture beyond agricultural pursuits. Aztec geographic naming practices often reflected soil environments in toponyms for land tracts and settlements, as in Chiauhtlan (Place of Abundant Moist Soil), Tlacozauhtitlan (Place of Abundant Yellow Soil), or Actopan (Place on the Alluvium) (Peñafiel, 1885; CM, 1992). Also, settlement patterns were influenced by soil distributions. For example, a study of houseplot soil types recorded in the CSMA showed a clear preference to site houses on the highest quality soils of the household landholdings (Williams, 1994).

In Aztec medicine, certain soils were specified in therapeutic remedies. Among many examples, white salty earth-iztatlalli, was used for ailments of sore throat, swollen cheeks, excessive heat, whirlwind, and fear, and nitrous earth-tequixquitlalli [potassium nitrate] in potions and lotions (Emmart, in Cruz, 1940, p. 64). White, purple, and reddish earths, and dark substances such as black clay and soot, also had medicinal properties, whose recommendations were linked to their colors (Maldonaldo-Kordell, 1964, p. 296; López Austin, 1974b, p. 222).

In the Aztec political economy, the soil classification system was used in property descriptions, and quite likely soil quality was a basis to assess property taxes (tribute). Martín Cortés, son of the conquorer Hernán Cortés, noted that ". . . he who has a piece of land paid a tribute; and the one with two, two; and the one with three, three; and he who had a piece of irrigated land, paid double that of one who had dry land" (1865, vol. 4, p. 443). Soil quality assessment also played a role in the Aztec political economy in that Aztec elites often controlled the most productive land by right of office, inheritance, or service to the state.

Among a host of other, non-agricultural applications of soil knowledge in Aztec culture are identification of suitable construction materials, such as tepetate and clay for adobe bricks, of clay for pottery production, of dying substances for paints, whether for writing inks or body paint used by priestly deity impersonators, and probably even for human geophragy, as described to me by Nahua informants in the 1970s.

Retrospect

The extant sources indicate Aztec culture was unique in its development of an indigenous written soil classification, not only in Mesoamerica but also in the entire Western Hemisphere in pre-Columbian times. The properties ascribed to different soils in many respects portend those observed by contemporary soil scientists, among them texture, consistence, color, moisture retention, permeability, organic content, chemistry (alkalinity), and fertility. Many of the Aztec soil types were defined by multiple factors, and some classes exhibit hierarchical (taxonomic) structure. Their soil descriptions suggest that the Aztecs were aware of the factors of soil formation and soil genesis. While they were clearly cognizant of changes in soil attributes with depth, there is little evidence that Aztec agriculturalists systematically conceptualized soil horizons or horizonation processes. They focused instead on the qualities of the surficial soil as a two-dimensional body, and on its management for crop production.

Centuries before the arrival of Europeans, massive labor investment by Aztec farmers and their predecessors transformed the soil environment of the Basin of Mexico and its environs. Technology was applied to keep the soil in place (not always effectively during periods of depopulation), to provide water, to maintain and enhance soil nutrients, to treat problem soils, to reclaim eroded soils, and to create new soil for crop production. All of these efforts attest to a working knowledge of soil productivity and specific plant requirements for growth. Such was undoubtedly gained through millennia of management experience passed down through generations with probable experimentation along the way, the latter certainly by the mid-15th century in the royal gardens of Aztec kings (Emmart, in Cruz, 1940, pp. 71–81; Ortiz de Montellano, 1990, p. 181; Nuttal, 1992).

The empiricism of Aztec soil (and more broadly, ecological) knowledge was undoubtedly impacted by the over-arching belief system of the culture. Concepts such as hot and cold, lucky and unlucky days, the symbolism of color and the four directions, and of the cause and effect of the activities of deities quite likely were part of Aztec ethnopedology and ecology. The sources reviewed here are mute in this respect, perhaps because of Spanish or self-imposed indigenous censorship, as was the case in other knowledge domains (e.g. in medicine [López Austin, 1974b, p. 217]). Spanish treatises on sorcery, superstitions, and idolatry may prove fruitful sources to explore the mystical, magical, and religious aspects of Aztec soil knowledge.

As indigenous elite culture was discredited, and linguistic acculturation and rural depopulation occurred in the Colonial period, Spanish influences seem to have fostered a simplification of Aztec soil knowledge (Williams, 1992, p. 491). Simultaneously, adaptations to European agricultural introductions apparently brought about additions to local soil knowledge, including such concepts as soil "vitamins," and arguably the Galen "hot/cold" system applied to fertilizers and soils (e.g. Ortiz Solorio, 1990, p. 19). Notwithstanding adoption of European cultigens, livestock, and plows, many Aztec intensive soil management practices survived five centuries. In the last decades of the 20th century, traditional farmers of Nahua ancestry in the Basin of Mexico still applied some of the knowledge and practices of their Aztec forebearers. The demographic and economic dynamics of the Basin today mitigate against perpetuation of this ancient knowledge system as the traditional farming lifeway disappears, and the soils "of the plains, the mountains, the crags, the grasslands, the forests, the gardens and fields" are adapted to the needs of 21st century urban life.

References

Barrera-Bassols, N. and J.A. Zinck. 2000. Ethnopedology in a worldwide perspective: an annotated bibliography. Intern. Inst. for Aerospace Survey and Earth Sci., Enschede, Netherlands.

Cadastral Fragment of the Ramírez collection. Museo nacional de antropología, Archivo histórico. Col. antigua 213. Opúsculos históricos, vol. 25, pp. 499–500. Mexico D.F.

Calnek, E.E. 1976. The internal structure of Tenochtitlan. pp. 287–302. In E.R. Wolf (ed.) The Valley of Mexico: studies in pre-hispanic ecology and society. School of Amer. Res., Santa Fe, NM.

Codex Mendoza. 1992. F.F. Berdan and P.R. Anawalt (ed.) Univ. of California Press, Berkeley, CA.

Codex Vergara. Bibliothèque nationale de France, manuscrit mexicain no. 37–37. Paris, France.

Códice de Santa María Asunción. 1997. B.J. Williams and H.R. Harvey (ed.) Univ. of Utah Press, Salt Lake City, UT.

Cortés, M. 1865 [1563]. Carta de d. Martín Cortés . . . al rey d. Felipe II. In Colección de documentos inéditos . . . en América y Oceania. L. Torres de Mendoza (comp.). vol. 4, pp. 440–462. Imprenta de Frías y Cía., Madrid, Spain.

Cruz, M. de la. 1940. The Badianus manuscript (Codex Barberini, Latin 241) Vatican library: an aztec herbal of 1552. E.W. Emmart (ed., trans., annot.) The Johns Hopkins Press, Baltimore, MD.

Cruz, M. de la. 1964. Libellus de medicinalibus indorum herbis: manuscrito azteca de 1552. A.M. Garibay Kintana (trans.) Inst. Mex. del Seguro Social, Mexico D.F.

Doolittle, W.E. 1990. Canal irrigation in prehistoric mexico: the sequence of technological change. Univ. of Texas Press, Austin, TX.

Fernández, J. 1964. Las miniaturas que ilustran el códice. pp. 237–242. In M. de la Cruz, Libellus de medicinalibus, Inst. Mex. de Seguro Social, Mexico D.F.

Florentine Codex, general history of the things of New Spain. Fray Bernardino de Sahagún. 1961, Bk. 10, The people. 1963, Bk. 11, Earthly things. C.E. Dibble and A.J.O. Anderson (ed. and trans.) School of Amer. Res., and Univ. of Utah Press, Santa Fe, NM.

Garibay K., A.M. 1964. Nombres nahuas en el códice de la Cruz-Badiano, sentido etimológico. pp. 359–369. In M. de la Cruz, Libellus de medicinalibus. Inst. Mex. de Seguro Social, Mexico, D.F.

Gibson, C. 1964. The aztecs under spanish rule. Stanford Univ. Press, Stanford, CA.

Hernández, F. 1959 [1577]. Historia natural de nueva españa. J. Rojo Navarro (trans.) Univ. Nac. Autón. de Mexico, Mexico D.F.

Lameiras, B.B. de and A. Pereyra. 1974. Terminología agrohidráulica prehispánica nahua. Inst. Nac. de Antro. e Hist., Col. Sci. 13, Mexico, D.F.

López Austin, A. 1974a. The research method of fray Bernardino de Sahagún: the questionnaires. pp. 111–149. *In* M.S. Edmonson (ed.) Sixteenth-century mexico: the work of Sahagún. School of Amer. Res., and Univ. of New Mexico Press, Santa Fe and Albuquerque, NM.

López Austin, A. 1974b. Sahagún's work and the medicine of the ancient nahuas: possibilities for study. pp. 205–224. *In* M.S. Admonson (ed.) Sixteenth-century mexico: the work of Sahagún. School of Amer. Res., and Univ. of New Mexico Press, Santa Fe and Albuquerque, NM.

Matrícula de tributos. Museo nacional de antropología 35–52. Mexico, D.F.

Maldonado-Koerdell, M. 1964. Los minerales, rocas, suelos y fósiles del manuscrito. pp. 291–299. *In* M. de la Cruz, Libellus de medicinalibus. Inst. Mex. de Seguro Social, Mexico, D.F.

Molina, A. de. 1970 [1571]. Vocabulario en lengua castellana y mexicana. Editorial Porrúa, Mexico, D.F.

Montoya Briones, J. 1964. Atla: etnografía de un pueblo náhuatl. Univ. Nac. Autón. de México, Dept. de Invest. Antro. 14, Mexico, D.F.

Nichols, D. and C.D. Frederick. 1993. Irrigation canals and chinampas: recent research in the northern Basin of Mexico. pp. 123–150. *In* V.L. Scarborough and B.L. Isaac (ed.) Economic aspects of water management in the prehispanic new world. Res. in Econ. Anthro. 7, JAI Press, Greenwich, CT.

Nuttal, Z. 1992. Los jardines del antiguo México. pp. 41–61. *In* C.J. Gonzalez (ed.) Chinampas prehispánicas. Antologiás, Ser. Arquel., Inst. Nac. de Antro. e Hist., Mexico, D.F. Orig. publ. in Memorias y revista de la sociedad cientifica 'Antonio Alzate,' 1920, 37:193–213.

Olagüibel, M. de. 1975. Onomatología del Estado de México [1894]. M. Colin (ed.) Biblioteca enciclopédica del Estado de México, Toluca, Mexico.

Ortiz de Montellano, B.R. 1990. Aztec medicine, health, and nutrition. Rutgers Univ. Press, New Brunswick, NJ.

Ortiz Solorio, C.A. 1990. Desarrollo de la etnoedafoloía en México. Centro de edafología, Colegio de postgraduados, Montecillo, Edo. de Mexico.

Parsons, J.R. 1991. Political implications of prehispanic chinampa agriculture in the Valley of Mexico. pp. 17–41. *In* H.R. Harvey (ed.) Land and politics in the Valley of Mexico: A two-thousand-year perspective. Univ. of New Mexico Press, Albuquerque, NM.

Peñafiel, A. 1885. Nombres geográficos de México. Secretaría de Fomento, Mexico, D.F.

Pozo, E.C. del. 1964. Valor médico y documental del manuscrito. pp. 329–343. *In* M. de la Cruz, Libellus de medicinalibus, Inst. Mex. de Seguro Social, Mexico, D.F.

Robertson, D. 1959. Mexican manuscript painting of the early colonial period: The metropolitan schools. Univ. of Oklahoma Press, Norman, OK.

Rojas Rabiela, T. (ed.) 1983. La agricultura chinampera:compilación histórica. Univ. Autón. Chapingo, Mexico.

Rojas Rabiela, T. 1984. Agricultural implements in mesoamerica. pp. 175–204. *In* H.R. Harvey and H.J. Prem (ed.) Explorations in ethnohistory. Univ. of New Mexico Press, Albuquerque, NM.

Rojas, Rabiela, T. 1988. Las siembras de ayer: la agricultura indígena del siglo XVI. Centro de Invest. y Estudios Sup. en Antro. Social, Mexico, D.F.

Rojas Rabiela, T. (coord) 1990. Agricultura indígena: pasado y presente. Centro de Invest. y Estudios Sup. en Antro. Social, Ediciones de la Casa Chata 27, Mexico, D.F.

Rojas Rabiela, T. and W.T. Sanders (ed.) 1985. Historia de la agricultura época prehispánica: siglo XVI. Colección biblioteca, Inst. Nac. de Antro. e Hist., Mexico, D.F.

Sanders, W.T. 1957. Tierra y agua: a study of the ecological factors in the development of mesoamerican civilizations. Ph.D. diss. Harvard Univ., Cambridge, MA.

Seler, E. 1904. Mexican picture writing of Alexander von Humboldt. pp. 127–229. *In* C.P. Bowditch (ed. and trans.) Mexican and Central American antiquities. Bur. of Am. Ethno., Bull. 28. U.S. Gov. Print. Office, Washington, D.C.

Siméon, R. 1977. Diccionario de la lengua náhuatl o mexicana. J. Oliva de Coll (trans.) Siglo XXI, Mexico, D.F.

Somolinos d'Ardois, G. 1960. Vida y obra de Francisco Hernández. pp. 97–482. *In* Obras completas del Dr. Francisco Hernández. Tomo 1. Univ. Nac. Autón. de México, Mexico, D.F.

Somolinos d'Ardois, G. 1964. Estudio histórico. pp. 301–327. *In* M. de la Cruz, Libellus de medicinalibus. Inst. Mex. de Seguro Social, Mexico, D.F.

Thouvenot, M. 1982. Pierres précieuses glyphées: étude de quelques vignettes du codex florentino de fray Bernardino de Sahagún (mexique ancien). L'Ethnographie 78 (86):31–102.

Tylor, E.B. 1861. Anahuac, or Mexico, and the mexicans, ancient and modern. Longman, Green, Longman, and Roberts, London.

Viesca, C. 1995. Y Martín de la Cruz, autor del códice de la Cruz Badiano, era un médico tlatelolca de carne y hueso. Estudios de cultura nahuatl 25:479–498.

Viesca, C. 1997. Ticiotl: conceptos médicos de los antiguos mexicanos. Monografías de hist. y filos. de la medicina 2, Univ. Nac. Autón. de México, Mexico, D.F.

Whitmore, T.M. and B.L. Turner II. 2001. Cultivated landscapes of Middle America on the eve of conquest. Oxford Univ. Press, New York, NY.

Wilken, G.C. 1987. Good farmers: traditional agricultural resource management in Mexico and Central America. Univ. of California Press, Berkeley, CA.

Williams, B.J. 1972. Tepetate in the Valley of Mexico. Annals of the Assoc. of Am. Geog. 62:618–626.

Williams, B.J. 1975. Aztec soil science. Boletín del Inst. de Geog. 7:115–120. Univ. Nac. Autón. de Mexico.

Williams, B.J. 1982. Aztec soil glyphs and contemporary nahua soil classification. pp. 206–222. *In* M.E.R.G.N. Jansen and T.J.J. Leyenaar (ed.) The indians of Mexico in pre-columbian times. Rijksmuseum voor volkenkunde, Leiden, Netherlands.

Williams, B.J. 1992. Tepetate in 16th century and contemporary folk terminology Valley of Mexico. Terra, Suelos volcánicos endurecidos, 10:483–493.

Williams, B.J. 1994. Sixteenth century nahua soil classes and rural settlement in Tepetlaoztoc. pp. 354–366. *In* Trans. World Cong. of Soil Science, 15th, Acapulco, Mex. 10–16 July. ISSS and the Mex. Soc. of Soil Science, Mexico, D.F.

Williams, B.J. and H.R. Harvey. 1997. The códice de Santa María Asunción: households and lands in sixteenth-century Tepetlaoztoc. Univ. of Utah Press, Salt Lake City, UT.

Williams, B.J. and C.A. Ortiz Solorio. 1981. Middle american folk soil taxonomy. Annals of the Assoc. of Am. Geog. 71:335–358.

3

The Heritage of Soil Knowledge Among the World's Cultures

Jonathan A. Sandor[1], Antoinette M.G.A. WinklerPrins[2],
Narciso Barrera-Bassols[3], and J. Alfred Zinck[4]

Introduction

Soil science has ancient roots, yet is relatively young as a distinct and recognizable scientific discipline, emerging in the mid-1800s and rapidly developing in the 20th century. It encompasses a major and growing body of knowledge that is a facet of the combined knowledge of the larger scientific community. Soil and agricultural sciences are fundamentally linked, and soil science is strongly tied with basic and applied branches of physics, chemistry, biology, earth sciences, ecology, and engineering. As far as soil science has advanced within a short period, it primarily reflects the knowledge residing in modern industrialized nations, which comprise only a small fraction of the world's cultures in both time and space. In addition to the body of current scientific knowledge, many past and contemporary knowledge systems have evolved in cultures across the globe, and within individuals and groups within modern countries. Much of this knowledge is hidden, overshadowed by modern science and overwhelmed by global economic development. There is growing recognition that the knowledge and perspectives of other cultures are valuable, and can contribute to solving current major societal, environmental, and natural resource challenges.

Living on the vast array of earth's soils and ecosystems are a diversity of cultures that over generations of experience have developed substantial knowledge and a keen sense of their land; commonly referred to as *indigenous* or *local* knowledge (Fig. 1). There are estimated to be 3000 indigenous groups with a total population between

[1] Agronomy Dept., Iowa State University, USA.

[2] Geography Dept., Michigan State University, USA.

[3] Instituto de Geografía, Universidad Nacional Autónoma de México.

[4] Dept. of Earth Systems Analysis, International Institute for Geo-Information Science and Earth Observation (ITC), The Netherlands.

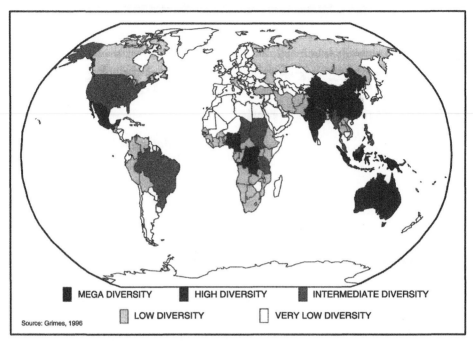

Figure 1. Worldwide linguistic diversity as an indicator of ethnic diversity.
Source: Narciso Barrera-Bassols, based on Grimes, 1996.

300 and 700 million people, and more if all Mestizo groups are considered (Toledo, 2001). Cultures have applied their knowledge in many ways as they use and manage land for their subsistence and construction needs, with a range of outcomes and degrees of success and failure. The main purpose of this chapter is to present a glimpse into the concepts, knowledge, and related use of soils by the world's diverse cultures. Another goal is to discuss the importance of indigenous knowledge of soils and land use in the context of its potential and realized applications in critical areas in which soil science can play a key role, such as agricultural development, cultural and environmental protection, and sustainable land use.

What does indigenous or local knowledge mean in the context of soils? Indigenous knowledge includes the knowledge of soil properties, processes, and management developed and held by people who have lived in a particular environment for a significant time. In many cultures this knowledge is transmitted orally from generation to generation; however, written or symbolic soil knowledge is recorded in ancient cultures such as in China (Needham and Gwein-Djen, 1981; Gong et al., 2003), Egypt (Chadefaud, 1998), India (Abrol and Nambiar, 1997), the Mediterranean region (Butzer, 1994), the Near East (Yaalon and Arnold, 2000), and Mesoamerica (see Williams' research on Aztec soil glyphs in this volume; and review in WinklerPrins and Barrera-Bassols, 2004). Indigenous knowledge is complex, multifaceted, and often subtle in its expression. A common misconception about traditional cultures is that their knowledge is static. While it is true that these cultures are conservative and have long-standing traditions, there is abundant evidence that on the contrary, indigenous land use systems are dynamic and innovative, constantly adapting to new circumstances and uncertainties (WinklerPrins, 1999). To be viable over thousands of years, cultures would have to be resilient and responsive to changes that are inherent to nearly all environments. Local soil knowledge is often of a practical nature; a cross or mix between beliefs, knowledge and practice. It relies mainly on sensory perception, but also on careful observation and interpretation of soil and geomorphic processes, and ecological relationships, such as using certain plants as indicators of kinds or condition of soil (e.g., Marten and Vityakon, 1986). While it is commonly considered to be mainly empirical, or based on trial and error, indigenous knowledge also includes scientific processes and experimentation (Prain et al., 1999). Although indigenous soil knowledge differs greatly from modern soil science in scope, scale, technology, analytical and quantitative capability, and transmission, there are shared traits.

Indigenous knowledge also contains important elements that are lacking or minor in modern science. Indigenous knowledge is time-tested; it has operated over much longer time periods than modern soil science. Its local nature means that it is also attuned to subtle variation in the physical environment. Indigenous knowledge systems are holistic and integrated within cultures, including religious and spiritual aspects that are often interwoven throughout people's lives (Posey, 1999). A special relationship of humans to land develops when multiple generations become rooted in a place. People develop strong ties to their homeland, and this is especially true of traditional cultures who have inhabited the same land for a long time and depend upon it for sustenance. Indigenous peoples view their land as sacred, though this does not imply that they have not significantly altered their land, nor run into environmental degradation problems. Many people in modern societies, through displacement or deliberate mobility, have lost the close connection with their native land. And with that change often comes a loss of knowledge, accumulated over many generations, about care for land resources.

The scientific discipline in which indigenous knowledge of environment is studied is ethnoecology (Toledo, 2001). Ethnopedology (Williams and Ortiz-Solorio, 1981) is a branch of ethnoecology that studies soil knowledge and land use systems of the world's cultures. Barrera-Bassols and Zinck (2000) have reviewed current concepts and trends in ethnopedology and compiled a bibliography of over 430 ethnopedology studies, covering research on 217 ethnic groups in 61 countries in Africa, America, and Asia. Ethnopedology is interdisciplinary, involving the fields of soil science, agricultural science, anthropology, archaeology, geography, sociology, economics, history, and others. It is a young but rapidly growing area of study, with important applications (WinklerPrins and Sandor, 2003; WinklerPrins and Barrera-Bassols, 2004).

Significance of Indigenous Knowledge

There are compelling reasons to become aware of the knowledge of soils among non-modern cultures. One is that aspects of modern soil science are derived from traditional knowledge; for example, a number of soil taxonomic classes in Russian, Dutch, and other modern soil classifications are derived from folk soil knowledge (e.g., Krasilnikov and Tabor, 2003; Siderius and de Bakker, 2003). Indigenous knowledge is at the same time the precursor of modern scientific understanding of soil, and the basis of parallel systems of knowledge about soils in traditional cultures. Important perspectives on modern-day soil science theory and practice can be gained by considering the history and nature of soil knowledge in the world's past and present cultures. Further, cultures' world views can fundamentally differ. This includes viewing soils and land use in different ways, though there are certainly common threads (Davis, 1993). Few of us realize that human perception and interpretation of reality is in many ways a function of our cultural upbringing and education; that reality is filtered through our cultures. Although it is quite difficult to see the world through different cultural "glasses," making the effort to do so can broaden perspectives, and thereby spark new ideas and improve cross-cultural communication. The potential exists for productive dialogue, cross-fertilization, and synergy among holders of different soil knowledge systems, and this potential has been partially realized in a few cases. Anthropologists and others who work in multiple cultures have come to realize that it is important to evaluate soil knowledge and land use within its cultural and social-economic context (Sillitoe, 1998). Regarding soil use, people not only adapt to existing soil conditions, but also transform them, through deliberate management for their purposes, or unintentionally. How this comes about and becomes manifested can vary significantly across cultures.

In the history of conquest and colonization, such as in the past few centuries, traditional cultures and their knowledge base have been deliberately suppressed and in many instances destroyed. While not as overt in intent, global economic development in recent years has also resulted in the continuing erosion of non-modern cultures. Indigenous and small-farmers' (e.g., Mestizo, Mulato, and others) knowledge of soil, environment, and land management has been largely undervalued up to the present day. Attitudes have begun to change with increasing recognition that indigenous cultures carry valuable, time-tested knowledge about sustainable management of land resources, accumulated through experience and transmitted across many generations. Along with this growing awareness arises concern about the loss of knowledge as so many traditional cultures, and their agricultural and other land use systems, have been disrupted and displaced. Time Magazine ran a feature story on indigenous knowledge in 1991, entitled "Lost Tribes, Lost Knowledge," with the subheading: "When native cultures disappear, so does a trove of scientific and medical wisdom." Steps to address this concern and to help protect traditional peoples have been undertaken by the international community through efforts such as the United Nations Environmental Programme's Convention on Biological Diversity. Efforts to acknowledge and conserve indigenous knowledge are crucial not only to the survival of the world's cultural and biological diversity and heritage, but also because this knowledge is highly relevant to major social, economic, food, environment, and natural resource issues we face today. Indigenous knowledge and experience can contribute significantly in the search for solutions to these problems (Prain et al., 1999).

An element unique to indigenous knowledge that can provide important scientific data and feedback to modern land use is that it is time-tested. The environmental knowledge embedded in traditional cultures offers a perspective on land use and management not otherwise available. The long-term nature of indigenous people's land use strategies, commonly on the order of many centuries to millennia, contrasts with the rapid changes, on the order of a century or less, of soils and soil use practices under intensive modern agriculture (e.g., Gupta, 1998; Sandor et al., 2005). The oldest scientific experiments monitoring soil quality and response to land use practices have been operating for about one century as well. Indigenous soil knowledge and associated ancient human-influenced landscapes constitute a resource to test land use effects on soils at time scales envisioned in the concept of sustainability. The long-term experience of traditional cultures with resource use and management, including successes and failures, can help to evaluate land use in relation to soil quality and sustainable agriculture (Sandor and Eash, 1991; Altieri, 2004).

Local soil knowledge can also provide long-term insights about human responses to environmental change and uncertainties, and so is relevant to global environmental change, such as climate warming and desertification. For example, fundamental strategies about farming under more arid conditions may be learned from cultures that developed agricultural systems adapted to arid environments, or those that have long-term experience in coping with droughts. There is also much to be learned from past soil and other environmental degradation, and how traditional cultures or those from the archaeological past responded (Hillel, 1991; Sandor and Eash, 1991; Redman, 1999; Diamond, 2005). In some cases, cultures responded in time to stem degradation and survive, while in others soil degradation in such forms as salt accumulation or severe erosion contributed to their downfall. Although soil use differs between modern and traditional cultures in terms of technology, fundamental soil processes and changes involved are similar; therefore information from past and traditional contemporary cultures is relevant to modern land use.

In addition to offering longer time perspectives on land use and management, traditional cultures also offer a different set of spatial scales, which also has important implications for sustainable agriculture. Indigenous knowledge is often also referred to as *local* knowledge, in that cultures through time have become attuned to the complexity of specific environments and microenvironmental variation (WinklerPrins, 1999). Their knowledge often relates to highly specific spatial scales. This knowledge is relevant to at least two major issues in agriculture today. A major theme in modern intensive agriculture is "precision agriculture," that is, tailoring management practices such as fertilizer inputs, irrigation, disease, insect, and weed control, and crop variety selection to specific parts of fields based on soil, topographic, and other environmental criteria. The rationale is that more specific management of crops will bring both economic and environmental benefits; for example, by only applying fertilizers where they are needed in optimal amounts. The long-term experience of farmers in traditional cultures in managing specific landscape areas in a range of climatic and geographic regimes could provide valuable information for developing precision agriculture in modern contexts. Local knowledge may also be used as a basis for appropriate agricultural development both within a culture, and for transferring indigenous local technology to other regions with similar environments (Altieri, 2001). For example, efforts have been made in the Andes to restore ancient agricultural terraces and canals within areas where they were abandoned centuries before (Denevan, 2001). Indigenous terracing techniques could also be used to help people who have recently arrived in mountainous areas to implement much needed soil conservation measures as they farm steep slopes. A particularly interesting example of "practical archaeology" is the redevelopment of raised field agriculture in the Lake Titicaca region. Raised field or waru-waru agriculture involves construction of elevated soil mounds with waterways in between, providing an integrated and effective strategy to overcome drainage, thermal, and fertility constraints to growing crops in a high-altitude environment. Knowledge of this indigenous agroecological system had been lost for centuries, until research by archaeologist Clark Erickson (1992) and others uncovered remains of the ancient fields. A cooperative project initiated among scientists, agricultural extension workers, and local farmers revived the use of raised fields for growing potatoes, as well as other indigenous tuber and grain crops.

A critical and immediate reason to study and record local soil knowledge concerns its link to cultural survival, and indigenous rights and protection. Modern transformations, assimilations, and globalization in general threaten the integrity of many cultures. Certainly a number of cultures and their heritage of knowledge have already been lost, and many others are at risk. It is important that the rights of indigenous peoples be recognized and protected (Davis, 1993; Posey, 1999; Razak, 2003). Paralleling the importance of language in cultural identity, the heritage of land and environmental knowledge is essential to cultural viability. A signal of the catastrophic decline in indigenous cultures and environments is the tragic loss of linguistic diversity as many of the world's 6,000 extant languages are becoming extinct (e.g., Razak, 2003). Indigenous cultures have strong ties to land, and the soil knowledge base and land use are integrally linked. Displacement and disconnection from land and land use translate to substantial loss of cultural heritage. Documenting the storehouse of knowledge about soils and land use accumulated through long-term experience and learning is crucial, especially given the accelerated rate of loss of knowledge as traditional cultures are displaced and transformed (Talawar and Rhoades, 1998). Ethnopedology can serve to document local knowledge and to help people maintain and protect their cultural knowledge base, even as modernization proceeds.

Another important reason to value indigenous soil knowledge is its role in cultural and biological diversity. Diversity is a cornerstone for function and viability in ecosystems and in human social-economic systems. Environmental and biological diversity are linked with cultural and agricultural diversity, and such diversity is critical to the conservation and protection of cultures and environmental resources (e.g., Souza et al., 1997; Prain et al., 1999; Toledo, 2001; Altieri, 2004). There is a close correspondence between indigenous homelands and remaining areas of greatest biodiversity. Traditional farmers in the humid tropics commonly grow more than 100 plant species within the same field (Altieri, 2001), which also reflects and makes for diverse soil properties and soil-crop interactions. The kinds, scale, and scope of soil knowledge covered in ethnopedological studies illustrate the diversity of soil knowledge in the world's cultures.

Much of the concern about loss of cultures and their biological and physical environments revolves around modern science and society's use of endemic or rare plants and animals in medicine. The use of these resources may involve soil; for example, soil-dwelling microorganisms such as the antibiotic-producing actinomycetes discovered by soil scientist Selman Waksman, which earned him the 1952 Nobel Prize in medicine. A number of drugs from traditional cultural areas, along with their knowledge and use among indigenous cultures, have been evaluated, and there is an intense search for others in hopes of finding cancer-fighting and other kinds of drugs. The appropriation of traditional knowledge and associated materials has also sparked much controversy regarding intellectual property rights, as companies patent genes and processes originally developed by indigenous cultures. Countries in which these indigenous groups live, such as Brazil, have taken steps to regulate the removal and use of these resources.

International agricultural development is perhaps the most currently critical area in which indigenous knowledge of soil management plays a key role (e.g., Pawluk et al., 1992; Critchley et al., 1994; Warren et al., 1995; Cools et al., 2003; Krasilnikov and Tabor, 2003). The failure of many development projects in recent decades has been attributed to top-down agricultural modernization schemes that are often seriously flawed by not carefully considering the priorities and needs of local people. There has often been a lack of two-way communication and insufficient attention paid to the culture and views of local people whose lives development is intended to improve. Even with the best of intentions on the part of development agencies, failure to increase production, alleviate poverty, and encourage continued self-sufficiency has been a common outcome and a major source of frustration and wasting of money, machinery, energy, and time. With increased recognition and incorporation of indigenous knowledge, agricultural development methods and programs can be implemented that are relevant to local people, and invite and actively involve the participation of intended recipients of aid in their own development. Ethnopedology offers a way of engaging local farmers in a dialogue about soils in a way that is more respectful of local farmers and those who seek to improve their livelihoods through development projects (e.g., on-farm agricultural experimentation). Ethnopedology can provide information and guidance on land use and management strategies for disrupted cultures that have lost knowledge or displaced peoples who are developing land use systems in new areas. Agricultural development may also, as a result of dialogue, be more productive in the long run. More recently, advancements in development strategies and results have been made, such as in the Sahel and some other regions of Africa. 2002 World Food Prize recipient Pedro Sanchez and his colleagues are using a combination of modern soil fertility principles combined with techniques such as agroforestry, which has its origins in indigenous systems, to overcome severe problems in soil fertility in eastern Sub-Saharan Africa. In agroforestry, plant species that fix atmospheric nitrogen are grown with crops and improve levels of this critical nutrient as well as other aspects of soil quality. Ameliorating soil fertility has dramatically boosted local crop yields in countries that have endured some of the worst food shortages, and work continues to expand the scale of improved soil fertility and food production.

It is important to realize that even though much of the literature on indigenous knowledge reports on traditional cultures, local knowledge exists throughout the world, in developed, as well as undeveloped regions. Substantial local knowledge of land and soil management is also found among individual farmers and others, including traditional groups (e.g., the Amish) and alternative farmers (e.g., organic "back to the land" farmers) within modern societies. This is an area needing much more inquiry, with relatively few studies to date. The knowledge of farmers in mainstream production agriculture is also an area of ethnopedology that deserves greater attention (Romig et al., 1995; Liebig and Doran, 1999; McCallister and Nowak, 1999).

A common tendency in the literature on indigenous soil knowledge is to use the term "farmer" as if this represents a homogenous individual. This is not so. Knowledge can vary considerably between male and female farmers, and also between young and old farmers (Carney, 1998). To date little research has been done to consider these differences explicitly. Studies by Carney (1991) in The Gambia and by Engel-Di Mauro (2003) in Hungary demonstrate both the need to understand the gendered nature of farming systems as well as differences in soil fertility management between men and women.

Approaches to Indigenous Knowledge Documentation and Evaluation

It is important to carefully think about methods used in scientific research because they greatly influence the collection of data, and the validity and applicability of results. It is especially important to consider methods in ethnopedology because it is a relatively new branch of science, and therefore has less well-established procedures. Also, it is at the interface of several different disciplines in both natural and social sciences, and so may employ a complex or hybrid set of methods, as well as present challenges in interdisciplinary communication. A number of approaches to ethnopedology have been used as this discipline has developed. Two general paths that have been taken to ascertain indigenous knowledge of soils involve 1) ethno-science studies that employ anthropological methodologies to directly seek to document and understand soil knowledge in cultures, and 2) inferences about soil knowledge and consequences of land use drawn from research on the soil use itself. The latter involves more indirect studies of land use and soil management in past (through archaeology and ethnohistory) and contemporary traditional cultures. For example, Vitousek et al. (2004) inferred that farmers' land selection for agriculture in precontact Hawai'i was based on climate–soil relationships. Methodologies in ethnopedology will be discussed further within the context of case studies.

Given the complex holistic nature of indigenous knowledge and cultures, the relative youth of ethnopedology, and the need for interdisciplinary approaches to study indigenous knowledge in a more integrated comprehensive manner, there is clearly a long way to go in developing effective scientifically sound methods. Relatively few ethnopedology studies have used statistical designs in collecting soil data for evaluating indigenous soil knowledge (e.g., Sandor and Furbee, 1996; Gobin et al., 2000; Pulido and Bocco, 2003). Some recent studies have quantitatively evaluated results in the context of applications such as agricultural development, with increasing use of geographic information systems (Cools et al., 2003; Payton et al., 2003). Quantitative and interdisciplinary approaches, while challenging, are essential if ethnopedology is to advance scientific understanding and apply findings to solving problems.

Case Studies to Illustrate the Scope of Indigenous Knowledge of Soil

To explore the nature and scope of knowledge about soils among traditional cultures, several case studies are presented from a range of cultures, geographic regions, and time (Fig. 2). The material presented in this chapter is by no means comprehensive, rather it is intended as an introduction to the knowledge of soils held by traditional cultures across the globe and to illustrate some main themes and the significance of this knowledge. Most of the case studies presented are from the Americas because that is where we have done our research and are most familiar, but we also include material from other regions and direct readers to world-wide studies in the reference section. Certain examples will be presented in greater depth, with other studies briefly discussed to provide cross-cultural comparisons and a global perspective. We have chosen to organize the presentation by geography, mainly emphasizing three major ecological regions where there are still many traditional cultures that have developed and applied knowledge of soil. These are the lowland humid tropics, arid and semiarid lands, and mountainous regions (Fig. 2). These regions cover a significant area of the world's land area, and have also been the focus of most ethnopedology studies because they are home to a high proportion of indigenous cultures that remain relatively more intact. The humid tropics include approximately 27% of the earth's 148 million square kilometers land area, arid and semiarid lands about 32%, and mountainous terrain about 36% (including highlands and hill country), with overlap of the mountains and the other two ecoregions (Bailey, 1998 and other sources). The humid tropics include much of the world's biological and cultural diversity. Cultures in deserts have developed much knowledge of soils in adapting subsistence strategies to cope with aridity. While only about 10% of world population lives in mountains, mountain regions have more far-reaching significance as sources of water upon which many people at lower elevations depend; for example, the major rivers originating in the Himalayas that flow through populous southern and southeastern Asia (Riley et al., 1990). Because agriculture is one of the major land uses in which people have developed and applied knowledge of soils, most of the examples concern soil knowledge within the context of agriculture.

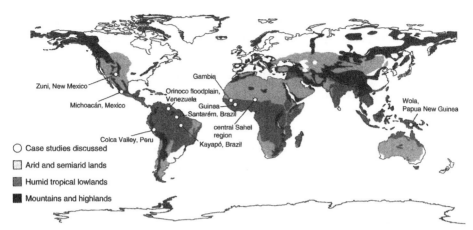

Figure 2. World ecosystem regions (humid tropical lowlands, arid and semiarid lands, and mountain regions) covered in chapter case studies, and case study locations. Cartography by Christina M. Hupy.

The different aspects of indigenous knowledge presented in the case studies are organized into three main areas: 1) concepts, religious beliefs, and attitudes about soils, 2) knowledge of soil properties and processes among cultures, and 3) the applications and use of soil knowledge. These themes parallel the conceptual framework of ethnoecology developed by Victor Toledo (2001), who frames cultural knowledge into three components: cosmos (worldview), corpus (body of knowledge), and praxis (practice, such as soil use). These three are linked in that a culture's world concept underlies its knowledge, which guides practice. Knowledge is presented from the two major investigative sources previously discussed: ethnopedology studies involving direct evaluation of soil knowledge held by local people, and knowledge inferred from observing land use and management in the past (through archaeology and ethnohistory) and contemporary traditional cultures. We hope that the material here will provide some insights into traditional knowledge of soils, and encourage readers to explore ethnopedology further.

Soil Knowledge Among Cultures in Humid Tropical Lowlands

The soil-related challenges faced by peoples of humid tropical lowland regions are multi-faceted. Many of these landscapes are relatively old geomorphologically, and experience high rainfall with concomitant high leaching conditions. This combination of conditions leaves locals with many leached Oxisols and Ultisols whose natural productivity is generally poor. For instance, one estimate states that only 6% of soils in the Amazon have no agricultural constraints (Sanchez et al., 1982). Vegetation is fast growing and abundant, with nutrients quickly recycled within the vegetation. Weeds easily overwhelm agricultural efforts, therefore the major challenge of using tropical lowland soils is one of soil fertility retention under luxurious vegetative cover.

Cultural groups occupying these regions have adapted to these conditions in their lifeways and agricultural patterns. Some of these are well known, such as variations of slash and burn agriculture and hunting and gathering strategies. Today, however, there is increasing evidence that indigenous people of these regions were also farmers, albeit a type of "forest" farmer. Substantial evidence demonstrates the existence of long-term vegetative manipulations (both forest and savanna) by indigenous groups (Martin, 1993). What to westerners look like "wild" landscapes are in fact cultural creations with centuries (if not millennia) of species management (Balée, 1989; Denevan, 2001). Other recent research demonstrates that people have also long been active agents of both soil formation and enrichment, some of which are discussed in this section. Scientists have not always been able to "see" these sorts of techniques since they are quite different from conventional agriculture. Several research projects among present-day indigenous groups as well as archaeological efforts offer insights into soil management techniques that make the lowlands quite productive without external inputs.

Concentric Ring and Mound Agriculture

Geographers and anthropologists have long been working with native people in lowland tropical regions and have come to understand that some cultural groups improve their soils over time. The Kayapó Indians of Southern Pará and Mato Grosso states of Brazil (Fig. 2) demonstrate successful long-term soil management by carefully creating patches of higher fertility within a system of "concentric ring" agriculture (Fig. 3) (Posey, 1985; Hecht, 1989; Hecht and Posey, 1989). The different zones of the rings are managed differently over space and time. Relatively higher fertility in these zones is created through the fine-tuning of the agricultural system. Careful attention is given to the placement of crops within the rings so that crops are matched carefully with their nutrient needs.

Trees are initially felled in such a way that they fall outward. The center ring is then planted with sweet potatoes. Light burning occurs after planting, so that the crop can benefit from the charcoal left over from the burn, as well as any mycorrhizal relationships already in the soil. Over time sweet potatoes thrive in the center of the rings, the driest and hottest of the zones. These conditions are ideal for sweet potato and also decrease pest problems (especially mold). Continued management includes light follow-up burns, weeding, and aeration. Maize is initially planted in the middle ring. This is a high nutrient demanding crop and utilizes the fertility flush of the initial burn. After maize there are relay plantings of manioc. The third ring has fertility pockets (especially of nitrogen) where tree canopies landed and decomposed, as well

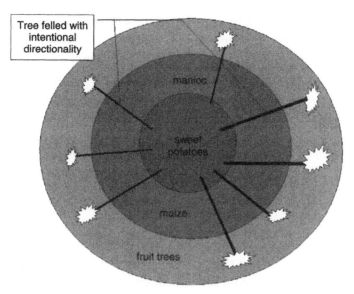

Tree felled with
intentional
directionality

manioc

sweet
potatoes

maize

fruit trees

Figure 3. Kayapó concentric ring agriculture in Eastern Amazonia.
Soils are managed differently in the different rings
so as to maximize differential crop production.
Source: Created by Antoinette WinklerPrins from text description
by Hecht and Posey, 1989.

as continual leaf litter from the adjacent forest. Piles of compost are created and into
these are planted perennials, especially fruit trees. These trees and shrubs are individually nurtured and tended over a long period of time. Some of their seedlings are
interplanted into the center and middle zones as the staple crops (sweet potato, maize,
manioc) decline in productivity. This results in a productive "fallow" patch where fruit
is collected long after the annual crops have been harvested and the rings overgrown.

Such careful soil management enables the Kayapó to obtain sustained yields of
sweet potatoes and other crops over longer periods of time than most agronomists
feel is possible on Oxisols. Additionally this method maintains and even improves the
soil over time. Kayapó farmers experience 200% higher yields than their next door
colonist farmers who do not use microvariability management on the same soils.
Therefore this type of management holds many lessons for development planners
since this system does not involve external inputs and can therefore remain affordable
to local people.

Similar directed composting activities have been documented among the Wola
of Papua New Guinea (Sillitoe, 1996, 1998). They also grow sweet potatoes and
maintain and improve soil fertility by careful soil management strategies. Hands
and digging sticks are used to build soil mounds. Plant residues are then placed
directly within these mounds, an act of careful and directed initiation of composting.

Essentially, each mound then becomes its own small compost pile. Sweet potatoes are planted directly into each mound. The benefits of the "compost" mounds are considerable, including improved infiltration and long-term fertility enhancement due to the microbial activity that the compost generates in the mounds.

Forest Islands

Another example of soil knowledge is its use in the creation of forest islands in tropical lowland areas of savanna. Several indigenous groups in Amazonia as well as people in savanna regions of West Africa have been creating forest islands in savanna landscapes for millennia. In marginal areas such as savannas, these forest "islands" offer retreat from excessive moisture or heat, and serve as places of habitation or spiritual importance. In the Amazon making the tree "islands" first requires the creation of a soil that will help support a small forest (Posey, 1985, 1990; Parker, 1992). This is done through the careful management of soil moisture and active and long term burning and composting as demonstrated in Fig. 4. Once a base has been created, forest seedlings are planted in the incipient island. Prudent

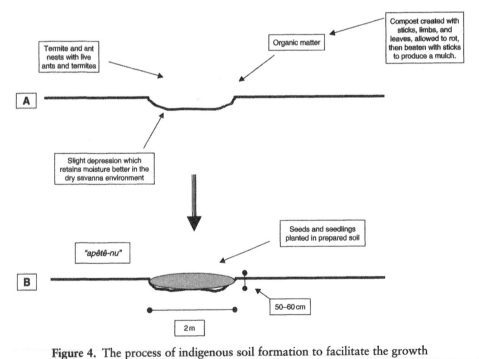

Figure 4. The process of indigenous soil formation to facilitate the growth of forest islands in the savanna region of Eastern Amazonia.

Source: Created by Antoinette WinklerPrins from text descriptions by Posey, 1985:142; Posey, 1990:18; and Parker, 1992:408, 410.

use of *Azteca* ants and termites by the Kayapó keep destructive leaf-cutter ants away during the initial period of growth for young forest trees.

In West Africa, especially in Guinea, forest islands are created through long-term use and activities that favor the growth of trees (Fairhead and Leach, 1996). This includes the creation of fire protection since fires retard forest regrowth. By clearing grasses around an area of habitation (fire-breaks) with subsequent continual planting of trees into kitchen midden soils (which are similar to the black earth discussed below), small forest islands emerge from the savanna landscape. Human-induced positive feedback loops maintain these forests over time. Termite mounds are also used to compost cropping sites.

Dark Earths

Other evidence that clearly points to human impact on soil resources are the Dark Earths found in lowland regions, especially in the Amazon Basin (Peterson et al., 2001; Lehmann et al., 2003). These organic-rich soils stand in sharp contrast to the leached Oxisols and Ultisols and are actively sought out by local farmers, both Amerindian as well as Mestizo peoples. Research shows that dark earth areas range in size from 0.5 ha to > 120 ha and are found throughout Amazonia. Some of these are called *Terra Preta do Índio* (true Indian Black Earth – usually with many ceramics), others as *Terra Mulata*, a much more subtle "brown earth" (Sombroek et al., 2002). These soils have elevated organic carbon levels that can be up to 70 times greater than those in surrounding silty-clay Oxisols and Ultisols. Much research still needs to be done on these soils, but evidence thus far points to long-term soil management practices of past indigenous groups, especially light burning and mulching probably similar to that of the Kayapó. It also appears that these soils are perpetuated in their high leaching environment due to microbial action. Stable fused aromatic ring structures in the organic matter fraction of the dark earths appear to contribute to this. It is hypothesized that the charcoal left over from light burning provides these structures. A certain threshold of biotic activity is reached at which dark earths can perpetuate and perhaps even regenerate themselves (McCann et al., 2001). This could have far reaching implications for both reconstructions of past occupancy of the Amazon as well as for future sustainable management systems throughout the tropics (Glaser et al., 2001).

Floodplain Zones

In contrast to the leached soils in upland areas of the lowland tropics, floodplain areas typically offer more naturally fertile soils. These landscapes are periodically rejuvenated by river flooding, receiving nutrients from the deposited sediments. Locals refer

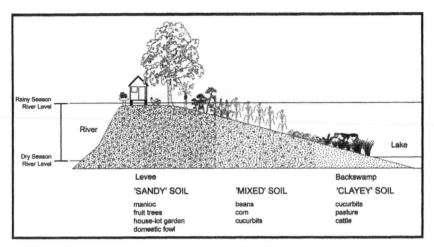

Figure 5. Generalized local soil categories and land-use along a transect of a levee
of an island in the Amazon River floodplain.
Source: Created by the Center for Remote Sensing and Geographic Information Science,
Michigan State University, based on a figure by Edinaldo Lopez.

to this annual rejuvenation as the river "refreshing" (*refrescar*) the soil. Some years,
when the river's flood is not high enough to flood commonly used agricultural land,
the soils are said to remain "tired" (*cansado*). However, periodic flooding also offers
challenges in terms of land management since land can be inundated 3–5 months
of the year (WinklerPrins, 2001). Interestingly, research from both the Amazon and
Orinoco river floodplains illustrate how similar settings can result in similar soils
knowledge systems. Figure 5 illustrates the general pattern of locally named soil types
found along the Amazon floodplain. The naming system is the same along the
Orinoco River (Barrios et al., 1994). Moisture demanding crops (watermelon) as well
as non-crops (pasture), are grown in the "clayey" (*barrento*) soils in lower lying areas;
drought tolerant crops are grown in the "sandy" (*arenoso*) soil on the highest part of
the levee (manioc). To many farmers the most useful soil is the "mixed" (*misturado*)
soils in between these extremes, where a variety of crops are planted (e.g., beans,
squash, maize). This naming and related land-use system reflects farmers' knowledge
and understanding of geomorphic process and microtopography.

Given the premium of land in the floodplain areas, locals also manage floodplain
sedimentation processes to manipulate the creation of fertile land. Along the Orinoco
locals plant an arboreal barrier along the outer edge of the levee, thereby significantly
slowing the floodwaters as they overtop the levee (Fig. 6) (Barrios et al., 1994; Barrios
and Trejo, 2003). This permits control over where sediments are deposited by the
river. Along the Amazon River there is also manipulation of slackwater sedimentation, in this case to fill in backswamp areas behind the main levees (Raffles and
WinklerPrins, 2003). Such management of geomorphic processes illustrates that local
people's knowledge of the soil extends to knowledge of soil genesis.

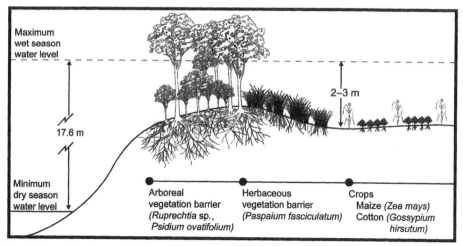

Figure 6. Deliberate tree plantings along levees in the Orinoco River to manage where the river deposits its sediments. This is an example of fluvial process manipulation and deliberate soil formation by local inhabitants.
Source: Created by the Center for Remote Sensing and Geographic Information Science, Michigan State University, based on a figure from Barrios and Trejo, 2003.

Along the floodplain of the Gambia River in West Africa numerous micro-habitats are used to cultivate rice (Carney, 1991). Taking advantage of tidal and seasonally flooded conditions, farmers manage different rice varieties over time and space. These different agro-ecological zones have different local names (Fig. 7). These names refer mostly to soil type, but really refer to the way the soil behaves in terms of water infiltration and retention. For example, in the *Bantafaro* zone rice derives its moisture

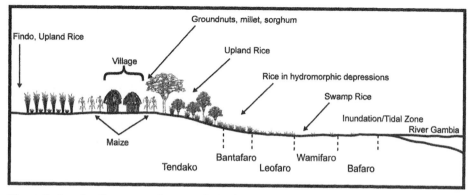

Figure 7. Local soil names and agroecological zones along the Gambia River demonstrating careful attention to microvariation in the landscape.
Source: Created by the Center for Remote Sensing and Geographic Information Science, Michigan State University, based on a figure from Carney, 1998.

from clayey soils that retain moisture from underground springs and a high water table. In contrast, rice in the *Wamifaro* zone obtains moisture only from high (lunar) tidal flooding. Farmers adjust the varieties planted to this moisture regime. This naming system demonstrates how environmental process is an integral part of indigenous soil knowledge.

Soil Knowledge Among Cultures in Arid and Semiarid Lands

Arid and semiarid lands cover about one-third of the earth's land area, with dry regions located in nearly half the world's countries. Cultures in northern Africa, the Near East, central Asia, Australia, and southwestern North America have developed substantial knowledge of soils in response to the main constraint of limited water supply. Agriculture first began about 10,000 years ago in the arid to semiarid Near East, and some of earliest methods of soil and water management, such as terracing and irrigation, also originated in this region (Hillel, 1991; Sandor, 1998). Case examples are presented of soil knowledge involved in an ancient but still important system of arid land farming known as runoff agriculture, followed by a short section highlighting the Sahel. The Sahel is a dry region in West Africa that has been a focus of global concern involving questions about land use, soil and environmental degradation (desertification), and agricultural development.

Runoff Agriculture in the North American Southwest

Runoff agriculture is a venerable farming system that has been practiced for many centuries in arid to semiarid regions of Africa, the Near East, central Asia, and the Americas, where annual precipitation alone is usually insufficient for consistent crop production (e.g., Hillel, 1991; Critchley et al., 1994; Doolittle, 2000). The common purpose that the variations of runoff agriculture share is to increase water for crops by retaining surface storm water flow generated from watersheds. Knowledge of soils and geomorphology is critical to the management of runoff agriculture. In the watershed ecosystem that is the source and pathway for runoff transport to fields, soil properties such as depth, texture, and structure influence water infiltration, hydraulic conductivity, and storage. Another aspect is the soil's role in contributing nutrients through sediment and organic debris entrained in runoff from exposed soil mineral and organic horizons. Strategic field placement on certain landscape positions and soils adjoining upland watersheds is needed to intercept and manage incoming runoff. Existing soil and slopes are modified through small dams and other techniques to enhance properties for water use, nutrient uptake, and crop physical stability and protection. Although runoff agricultures share characteristics, they also have a diverse range in geography, utilization of environmental niches, management practices, and crops. Besides the importance of runoff agriculture as a viable technique for coping with aridity and drought, much can be learned from traditional farmers and the ancient agricultural soils about successes and failures in agricultural management, land stewardship, and resource conservation.

The Zuni are among several native peoples in the North American Southwest renowned for their skills in growing maize and other crops using runoff (Fig. 8) (Doolittle, 2000; Norton et al., 2001; Muenchrath et al., 2002). Agriculture has been practiced in the Zuni region for about 3 to 4 millennia, and historic records provide evidence of productivity and attest to Zuni agricultural expertise (Ferguson and Hart, 1985). The Spanish conquistador Coronado described the Zuni as having great stores of maize, and Zuni supplied the U.S. army in the region with maize during the mid-1800s, when there were estimated to be over 4,000 hectares under mainly non-irrigated production. The Zuni also mastered other forms of agriculture, ranging from gridded gardens, to peach production in sand dunes, to floodwater farming of larger watersheds, and irrigation agriculture. The Zuni are an intensely spiritual people, and many of their religious beliefs and attitudes relate to maize agriculture, and to the water that sustains them. Their ceremonies and stories are rich in knowledge of agriculture, soil, and landscape processes.

Archaeological evidence indicates that many Zuni runoff fields have been farmed for about 1000 years and possibly longer, making them among the oldest identified fields in the United States (Fig. 8d). An important feature of runoff agriculture that relates to sustainable land use is that it functions by directly tapping into natural watershed and ecosystem processes, rather than relying on energy-intensive inputs of irrigation water and artificial fertilizers (Sandor et al., 2002). These observations formed the basis for studies of soils and crops in runoff agricultural fields, watershed geomorphology and ecology, and ethnopedology. Two approaches presented are direct study of soil knowledge through ethnoscience research, and knowledge inferred from studying the soils and agriculture.

Soil and Agricultural Knowledge Inferred from Field Location Patterns

The Zuni live in the arid to semiarid mesa country of the southeastern Colorado Plateau. They farm at relatively high elevations of about 2000 meters, higher than most modern conventional agriculture in the Southwest region. Field location with respect to elevation is likely an optimization between precipitation and temperature, which are inversely related variables with elevation. Thus, fields are at relatively

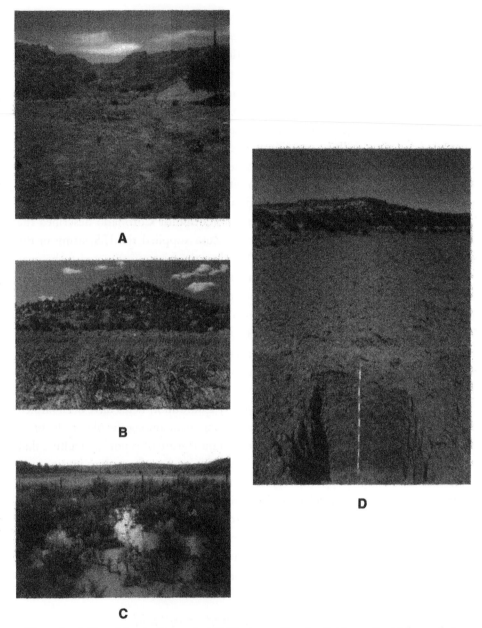

Figure 8. Arid and Semiarid Lands: (**A**) Hopi runoff maize field on alluvial fan early in growing season. Note small dams used to manage runoff. (**B**) Zuni runoff maize field on footslope, with forested watershed runoff source in background. (**C**) Runoff and associated sediment and organic debris flowing towards Zuni wheat field. Remnants of runoff control dams were observed upslope. (**D**) Zuni agricultural soil thickened by runoff deposition (note buried horizon) in same runoff field shown in Fig. 8c. Field is approximately 1000 years old or more, based on archaeological evidence. Colour version (see the colour plate section at the end of the book).

high elevations to maximize precipitation, but low enough to permit a sufficient frost-free period. Precipitation is highly variable, averaging 300–400 mm annually. May and June are usually dry, while over half the annual precipitation falls from July to September, often as intense localized thunderstorms. It is upon these monsoon rains that runoff agriculture depends.

Traditional runoff agriculture is practiced mainly on alluvial fans and other valley margin landforms where tributary ephemeral streams emerge from mesa uplands. The location of fields at valley edges directly connected to naturally vegetated watersheds is essential, because the watersheds are important sources of water, nutrients, and sediments for agricultural fields downslope (Fig. 8). Watershed vegetation is dominated by pinyon pine and juniper woodlands on mesa uplands and grasses and shrubs in valleys. Several characteristics impart effective runoff production to these watersheds, including steep hillslopes, bedrock outcrops, patchy vegetation cover, and soils with subsurface layers of clay accumulation (argillic horizons). It is interesting that while watersheds used for runoff agriculture in the Southwest are generally unmodified, alteration of watershed soils to enhance runoff has been reported in other regions. One example is from the Negev Desert, where ancient farmers successfully used runoff to augment the annual 100 mm rainfall by rearranging rock fragments on slopes, which caused soil crusts to form that increased runoff to fields (Hillel, 1991). Another example is the deliberate removal of vegetation in watersheds in Mexico to encourage sediment transport to refresh agricultural soils downslope (Doolittle, 2000).

Soil morphology and sediment transport data suggest that while subsurface soil horizons in Zuni watersheds mostly remain stable, surface horizons are more mobile, and episodically contribute organic matter and sediment to fields (Norton et al., 2003). Small frequent hillslope runoff events incrementally move sediment and organic matter downslope, where they accumulate. Larger, less frequent runoff events episodically flush accumulated materials from uplands to alluvial fans and fields. Farmers report that major runoff events can cause crop damage but recognize the long-term benefits of soil renewal. With active management of runoff that was practiced traditionally, runoff events could be controlled and were more immediately beneficial in supplying water to crops. Runoff sediments tend to be rich in organic carbon, nitrogen, and phosphorus compared to soils, supporting the idea that runoff is an important source of nutrient replenishment for agricultural soils. Besides tree litter, cryptobiotic crusts and nitrogen-fixing shrubs common in watersheds are also potential nutrient sources for fields. Ethnographic and historic reports on runoff agriculture among Southwest Native Americans discuss the importance of flood deposition in maintaining soil fertility, and farmers' knowledge of this function (Nabhan, 1984; Sandor et al., 2002).

The choice of valley margin settings for traditional agriculture is consistent from prehistoric time through the present. However, government soil surveys generally do not recognize valley margins as arable, reflecting modern emphasis on irrigated agriculture on valley floors. Nevertheless, there are good reasons for farming valley margins as a diversification strategy for managing risk and maintaining crop production stability: 1) valley margin fields could remain productive during major valley floor flooding; 2) valley floors are more freeze-prone than valley sides due to cold air drainage; 3) salt-affected and sodic soils occur on valley floors. Also, many floodplain soils are clayey, and for this reason are rated poor for maize by Zuni farmers; and 4) soil fertility is more readily maintained on alluvial fans and footslopes by more frequent deposition of runoff sediments and organic matter from adjacent uplands.

Soils in fields also have natural properties favorable for runoff agriculture, for example argillic horizons that help retain water within crop rooting zones. Research elsewhere also indicates that argillic horizons and other slowly permeable layers, such as indurated horizons cemented with silica (duripans) or calcium carbonate (petrocalcic horizons), or bedrock, are associated with many Southwest prehistoric and contemporary runoff fields (Sandor et al., 1986). Surface soil retention of runoff is greatest where sandy to loamy A horizons are underlain by strongly developed argillic horizons or analogous layers that trap moisture. Besides increased water infiltration rates, sandy surfaces also act as a mulch to reduce evaporation.

Ethnopedology Studies

Direct investigation of Zuni soil knowledge in an agricultural context indicates about 12 major soil terms, including those that emphasize knowledge about runoff processes and function in supplying water and nutrients to crops, and maintaining soil quality (Table 1) (Sandor et al., 2002). The relatively low number of Zuni soil classes is common among indigenous soil taxonomies, belying the rich knowledge they hold (Wilshusen and Stone, 1990). During interviews and in working with farmers over several years, it became clear that these terms contain meaning beyond describing discrete material. They incorporate knowledge of soil material and agronomic properties, coupled with awareness of geomorphic and ecological processes that create and distribute earth materials. In this sense the soil terms, although few, are windows into in-depth understanding of geomorphic processes connecting erosion and sedimentation from watershed to field.

Knowledge of the watershed's role in contributing sediments and organic matter to replenish agricultural soil fertility is encapsulated in the Zuni term *danaya so:we*, which translates into forest sand or soil, or simply "tree soil." Tree soil describes the

Table 1. Zuni soil and landscape concepts pertaining to runoff agriculture. Modified from Sandor et al., 2002, based on interviews with farmers by Pawluk

Danaya so:we

Translation: forest floor material (O and upper A horizons), transported in runoff to fields.

Farmer comments: "tree soil," "forest sand," "good rich soil," "the best one," "might have fertilizer" "like muje:we" (manure). "Rich soil from far away . . . forest," dark color.

Researcher comments: In Zuni dictionary, ta-na-ya means a surface collectivity of forest.

Associated terms: "ts'l'bewi:we" (organic debris) and "ummo:we" (foam in runoff).

Heyalo:we

Translation: alluvial (primarily) or eolian-derived fine sand and silt; can be loamy.

Farmer comments: "good soil," ". . . sand that comes with the water, soaks in the water and holds it," "carried in the water over the bank . . . ," "finer sand," "from up there (uplands) . . . when it rains, bring the soil down" ". . . windy soil . . ."

Sotanna

Translation: sand (coarser than Heyalo:we) from runoff and streams, some from wind.

Farmer Comments: "real sand" "out of the riverbed . . . coarse," "good rich soil" "sandy soil . . . easier and fertile to farm"; "better for catching water." "come from the topsoil from above . . . washed down here."

---

plant litter and topsoil (O and upper A horizons) transported by runoff from uplands to fields. Organic-rich materials from runoff events occur in clusters and various stages of decomposition en route to and within agricultural fields. Other Native Americans practicing runoff agriculture have parallel terms to tree soil, including terms for different components and states of organic runoff debris (Nabhan, 1984). The significance of sediment transport and deposition in this agroecosystem is also evident in two other Zuni terms: *sotanna* and *heyalo:we*. Both terms refer to runoff sediments, the first referring to coarser sediment (literally "big sand") associated with channel deposits and the latter to finer sandy or silty overbank deposits. Both were described in terms of material, origin, and depositional process, as well as reworking of fluvial deposits by wind. Farmers related that the fresh sediments and their location on alluvial fans are highly valued for enhancing water retention and fertility.

Discussions with farmers also revealed basic differences in soil concepts between Zuni and researchers (Norton et al., 2001). For example, concerning organic matter effects on soil, a farmer stated ". . . so when you add all that forest soil and manure you change the clay into sand . . . ?" To Zunis, "sand" refers not only to a coarse texture, but also to "soil," and can include other particle-sizes such as silt. Sand also connotes soil tilth and productivity, the ideal soil for maize and other crops. Conceptual differences between cultures make it challenging to accurately evaluate local soil knowledge and compare knowledge systems.

Knowledge Inferred from Soil Modification and Management

Knowledge is also expressed through traditional management practices that modify natural landscapes and soils to enhance conditions for arid land agriculture. In addition to practices such as deep planting of special maize cultivars in widely spaced clusters, small permeable dams in Zuni and other traditional Southwest runoff agriculture serve to intercept, partly retain, and distribute runoff and sediments (Fig. 8a). These dams function as agricultural terraces, decreasing slope angle and length, thereby encouraging runoff retention and sedimentation, and replenishing soils. The ephemeral dams constructed of brush and stones can be readily recon-figured, enabling farmers to respond quickly to dynamic conditions of runoff events. Traditionally, farmers prepare new fields by building dams a year or more before cropping, allowing time for soil incorporation and decomposition of fresh sediment and organic matter inputs. This illustrates the emphasis on runoff as a soil resource as well as a water resource. Although dams are not used much today (though they are in other world regions of runoff agriculture), runoff input is still important and partly managed by tillage. Other research on traditional agricultural systems and ethnopedology note the value of sedimentation in maintaining soil fertility and quality (Nabhan, 1984; Wilken, 1987; Bocco, 1991). In the analogous case of irrigation agriculture along major rivers such as the Nile, Ganges, and Amazon, farmers have long known that annual flood sediment is essential to maintaining soil (Hillel, 1991; Butzer, 1994; WinklerPrins, 2001). It is interesting to compare the perceptions of many traditional cultures that view sediment in positive terms as a resource, with those of modern science and society that usually view sediment as a negative, as a pollutant and obstacle associated with accelerated erosion; wrecking dams and reservoirs.

The research also evaluated effects of Zuni runoff agriculture on soil properties (Homburg et al., 2005). Comparisons of properties (e.g., organic matter, nutrients, microbial biomass) among cultivated and uncultivated soils suggest that tradi-tional Zuni agriculture has altered but not degraded soils. The condition of Zuni land after many centuries of farming suggests that their soil knowledge and care led to effective conservation of soil and water resources. Studies of other ancient agricultural soils in the Southwest show a range of consequences for soils ranging from relatively stable to enhanced to degraded. This illustrates how the experience of indigenous cultures can act as a resource for evaluating long-term soil quality and land use sustainability.

The Sahel

The Sahel region of West Africa is considered to be one of the world's "hotspots" in terms of land degradation and soil erosion. The Sahara Desert is said to be encroaching into what was formerly arable land, and overpopulation, regional economic change, and periodic and sustained drought are blamed for these developments. Millions of dollars of development aid have gone into soil conservation projects in the region, mostly projects instructing local farmers how to stem soil erosion. Few of these projects have had the desired effect. Recent research focusing on the local soil knowledge of farmers and herders has started to question whether land is really being degraded from the local perspective. What this research is addressing are the vast differences in perception between locals and outsiders, especially their differences in time scale. What to outsiders (primarily western scientists) appears as erosion might simply be redistribution of soil material to locals. In many cases a longer time frame is considered by locals who understand the long-term cycles of environmental change (Niemeijer and Mazzucato, 2003). In cases where erosion is acknowledged by outsiders as well as by farmers, the motivation to do anything about the process differ. Many regions are undergoing significant out-migration, leaving little incentive to conserve soil quality in the long term. The situation in the Sahel is very complex and insights from ethnopedologic research are contributing significantly to a better understanding of environmental change in this contested zone. Examples of ethnopedology studies in the Sahel, are by Critchley et al. (1994), Östberg and Reij (1998), and by several authors in Reij et al. (1996) and WinklerPrins and Sandor (2003).

Soil Knowledge Among Cultures in Mountain Regions

Mountain landscapes and ecosystems are highly diverse, as are the cultures that live in them. Peoples across the world's mountain regions have faced the special challenges of living at higher altitudes and rugged often remote and isolating topography (Jodha et al., 1990). Part of the challenge has been to develop agricultural techniques that allow production in mountainous terrain. For cultures to be successful in these environments over time requires substantial knowledge about soils and how to conserve and manage them on steep slopes, since mountain soils are commonly thin and vulnerable to erosion (Riley et al., 1990; Tamang, 1993; Östberg, 1995). Two case studies of soil knowledge, agricultural management, and guiding land stewardship concepts among indigenous communities in mountain environments are presented next.

Terrace Agriculture in the Andes of Southern Peru

In this case study, knowledge about soils in the context of agricultural terracing is emphasized. Agricultural terracing is one of the oldest and most widely used tools for soil conservation, and has been practiced in many forms across sloping lands on five continents and Oceania for millennia (Hillel, 1991; Sandor, 1998, this volume). A remarkable traditional soil knowledge system is found in the Colca Valley, located in the Andes of southern Peru, where terrace agriculture has been practiced for many centuries (Fig. 9a) (Guillet, 1992; Sandor and Furbee, 1996; Denevan, 2001). A key reason for conducting ethnopedology studies here were prior pedologic studies that indicated favorable condition of soils after at least 1500 years of agriculture (Fig. 10) (Sandor and Eash, 1991). In contrast, many soils under modern conventional agriculture show signs of degradation in such forms as erosion, compaction, decreased organic matter, and salt or sodium accumulation. Indicators of favorable tilth and fertility in Colca Valley soils included thickened A horizons, slightly lower bulk density (i.e., lack of compaction), and greater levels of organic carbon, nitrogen, and phosphorus in agricultural soils, relative to nearby uncultivated soil. The excellent condition of these soils is attributed in part to traditional soil management practices such as terracing, a form of conservation tillage, fertilization, and irrigation. On the basis of these findings, it was hypothesized that such an impressive agricultural system and management practices are supported by a substantial body of knowledge about soils. Objectives of the ethnopedology study, carried out jointly by anthropologists and soil scientists in the village of Lari, included determining which soil properties were recognized and emphasized, whether knowledge of soils was systematically organized and widespread (culturally shared), and how indigenous views of soil compared with those of soil scientists (Guillet, 1992; Sandor and Furbee, 1996).

A   **B**

Figure 9. Mountain Regions: (A) Terraced landscape of the Colca Valley, Peru. Note currently cultivated and abandoned agricultural terraces. (B) Farmer in Colca Valley sorting soil terms, resulting in a soil taxonomic tree such as that shown in Fig. 11.

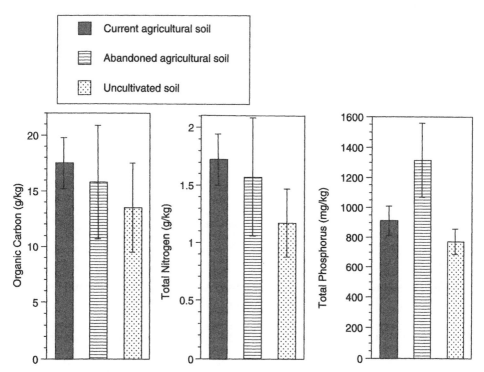

Figure 10. Comparison of soil properties in terraced agricultural and uncultivated upper A horizons in the Colca Valley, Peru. Bars are means (lines within bars show ±1 standard deviation) from 30–34 samples from each sample group. Mean differences are statistically significant minimally at the 0.05 probability level in all cases except between current and abandoned agricultural soils for organic carbon and total nitrogen.
Source: Sandor and Eash, 1991 and 1995.

The anthropological and soil science studies indicated that the terrace agriculture and careful soil management are supported by a substantial body of soil knowledge. This knowledge is in part revealed in the indigenous soil classification system, which includes nearly 50 unique soils organized in a 4-tiered hierarchical classification, with soil names in the native language Quechua as well as "borrow words" from Spanish (Figs. 9b and 11). From the soil classification, and interviews, a number of soil physical and chemical properties and landscape features are seen to be incorporated into the indigenous knowledge framework. Some properties are expressed as edaphic phrases or functional attributes such as soils that are wet and "rot roots" (certain clayey soils), that "need much water" (excessively drained, coarse-textured soils), that are "weak" or "lazy," that "need ash or fertilizer," or that do or do not "grow maize." Several morphological traits are recognized, with texture the most prominent in the classification. At the second categorical level shown in Fig. 11, soils considered suitable for agriculture form five classes on the left (left to right in

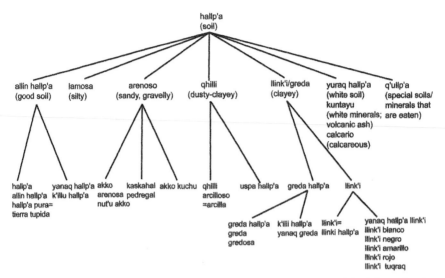

Figure 11. Soil taxonomic tree from Lari, Colca Valley, Peru. Modified from Sandor and Furbee, 1996, from the work of Furbee.

decreasing order of agricultural productivity as perceived by the farmer). Mineral materials such as volcanic ash, carbonates, and salts associated with certain soils are discerned. The far right class includes special soils or minerals eaten as condiments by people and mineral supplements fed to animals. Researchers have established that the underlying reason for this geophagy is that these earth materials act to absorb plant toxins such as the glycoalkaloid solanine found in potatoes and other Andean tubers that are important foods in the region. Soil names collected by researchers elsewhere in the Colca Valley and in the eastern Andes suggest that this indigenous knowledge system is widely shared in Andean cultures, though there is local and regional variation (e.g., Zimmerer, 1994).

Besides specific soil properties, other soil and landscape features are recognized by Lari farmers, such as horizons, spatial variability, specific pedogenic and geologic materials, and soil variation in different conditions. Although farmers didn't explain variation of soil properties with depth (e.g., by horizon differentiation or geologic stratification), subsurface soil variation not visible from the surface was recognized. For example, one farmer classified soil in his field as "akko over llink'i," which described a sandy loam A horizon overlying a clay loam argillic horizon. One activity in which farmers can directly observe soil depth variation is the construction and maintenance of agricultural terrace walls. These stone walls, commonly 2 m high, are emplaced against excavated hillslopes and anchored into subsurface horizons such as argillic horizons and duripans (silica-cemented horizons). Also, subsurface horizons affect crop production and Colca Valley farmers are acutely aware of soil properties in relation to agriculture.

Concerning recognition of soil variation and boundaries between different soils, Lari farmers are aware of lateral soil variation in their fields but use more than one soil name only when the soil variation necessitates different use or management. This is analogous to map unit concepts in U.S. soil survey, in which a map unit such as a consociation is composed of a dominant soil and inclusions of similar and dissimilar soils for practical reasons. An example is two adjacent fields in which the soil in one was named "hallp'a over peña" (hallp'a is the general name for soil and can also refer to good soil, while peña refers to a duripan) because of a shallow duripan, and soil in the other field was simply named hallp'a because the duripan was deep enough that it did not seem to affect productivity. Distinctions based on slight soil texture variation were noted by farmers, such as between sand (akko) and fine sand (nut'u akko). An instance involving anthropogenic soil properties is that farmers plant more densely near the front of terraces because they know that soils are thicker there. Awareness of the spatial distribution of soils is also indicated by a soil map of drawn by one farmer.

Certain soil and geologic materials that seem to behave similarly in terms of their effect on agriculture were differentiated by farmers during field interviews. For example, indurated soil horizons such as duripans are distinguished from bedrock, both of which are known by farmers to limit root growth if they occur at shallow depths. Differentiation of these similar-behaving materials implies knowledge of either 1) differences in origin between them, 2) practical differences in spatial distribution, or 3) subtle differences in behavior or other uses of these materials. Another discernment involves yuraq hallp'a, used to describe a light-colored, non-indurated horizon and peña, the actual duripan. A general impression from interviewing farmers is that they make subtle distinctions among soils and geologic materials. Findings from other ethnopedology studies such as in Papua New Guinea and Mexico also indicate subtle distinctions among soil and other earth materials (Ollier et al., 1971; Williams and Ortiz-Solorio, 1981). Lari farmers also recognize that properties within the same soil vary with moisture state and additions of material. They irrigate fields before tillage and planting partly to soften soil to make tillage easier and to help prepare a favorable seed bed. Another indigenous prescription for softening soil is to add ashes. Farmers distinguish between soils that are "dusty" when dry and clayey ("llink'i-like") when wet. Studies of indigenous soil knowledge world-wide have also documented differentiations of soil state.

Indigenous knowledge of soils and geomorphology is also shown in farmers' modification of soil productivity ratings according to topography and use of special landscape areas. Many aspects of agriculture in the Colca Valley represent responses to conditions imposed by geomorphic features and processes, and closely associated soil distribution. An example is the spatial pattern of crop types that conforms to the stepped topography of the valley. Barley and other cold-tolerant crops are grown on relatively level, frost-prone fluvial surfaces, whereas maize is grown on steeper slopes that separate fluvial surfaces, and down which cold air drains. Because frost hazard is a major concern in the high-altitude Colca Valley, soils considered otherwise excellent for agriculture are downgraded by farmers if such soils are located on low-lying or relatively level areas. Much of Lari's land includes inactive landslide topography and in some areas reservoirs and wetland habitats were created by constructing dams around landslide depressions. In a striking case of natural feature use, parallel ridges of volcanic ring dikes, oriented like a chute on a mountainside, were used to direct surface water to agricultural terraces constructed between these ridges.

It was also interesting to compare the indigenous classification system with U.S. soil taxonomy. Especially close similarities were found in soil textures (particle-size classes in U.S. soil taxonomy). This is not surprising, given that texture is a key soil property in physical and fertility properties relevant to crop production. Texture is an almost universally important component in indigenous soil classifications (Barrera-Bassols and Zinck, 2000, 2003a).

The integration of spiritual reverence and knowledge concerning land, common in indigenous peoples worldwide, is shown in traditional agricultural ceremonies. Geographer John Treacy observed that when preparing terraces for planting each year, Andean farmers pour chicha (a local alcoholic beverage) on the soil in a toast and offer prayers meant "to console the earth that will soon be rent by the plow." That certainly is a clear, direct statement about soil care. This land care ethic is also expressed indirectly through physical manifestation in the soils and terraced landscape themselves; the way that the terraces gracefully follow the topographic contours and fit so well within the mountain landscape (Fig. 9a), and the healthy condition of these ancient terraced soils.

Land Management in a Mesoamerican Highland Community

Land degradation in Latin American highlands is often perceived as resulting from inadequate land management practices implemented by local farmers living in fragile landscapes and exploiting marginal soils under strong climatic variability (WinklerPrins and Barrera-Bassols, 2004). In contrast, recent ethnoecological studies show that indigenous land management systems have been sustainable over long periods of time thanks to strategies that maximize land use in space and time via diversification of crops and practices, while minimizing the use of external inputs (Bocco, 1991).

An ethnoecological study carried out in San Francisco Pichátaro, a Purhépecha community in the highlands surrounding the Pátzcuaro lake, Michoacán state, in west central Mexico, reveals how land degradation, in terms of soil erosion and fertility depletion, was (and still is) handled by indigenous farmers (Barrera-Bassols and Zinck, 2003b). Their management has allowed traditional agriculture to remain sustainable over centuries thanks to co-evolution of eco- and socio-systems. The Pichátaro area is a volcanic landscape, formed by basalt cones covered with pyroclasts and separated by small valleys, between an altitude of 2300 and 3200 meters, along a bio-climatic gradient from temperate sub-humid to cold humid as elevation increases. The presence of fertile volcanic soils and permanent springs contributed to make Pichátaro an early center of maize production. Although land occupation started 2000 years ago, there are no conspicuous soil erosion features and no significant evidence of land degradation.

For indigenous people, land has a symbolic meaning based on Mesoamerican beliefs blended with practices from popular Catholicism. In this context, land is perceived as a resource, which behaves as a living being, and as a life support system for humans. Land, plants, and humans are bound by reciprocal relations, which allow perpetuation of life on earth. Land is venerated as the mother of all living beings. Cropping and crop harvesting are seen as basic activities securing people's health and survival, and thus require good land care and management. These ethical values support all local production activities. However, people consider and accept that this belief system is exposed to, and can be altered by, economic and environmental uncertainties, which means that land's behavior cannot be totally controlled by people living or working on it. Thus, humans are bound to land and have to conjure its benevolence through respect, compromise and tolerance. This is reflected in the relationships between climatic cycle, production cycle, and ritual calendar. The relationships might transcend the strict community sphere and take into account externalities, which affect the internal relationships between generations and between individuals, such as temporary out-migrations and off-farm incomes. Therefore, land care, sustainable productivity, and conservation are inherent parts of the symbolic land concept. This is then reflected in the way land is managed to meet human needs, without damaging the resource potential and thus the life support system provided by land. In a fashion very similar to the modern land concept, land is viewed as an integral whole, including water cycle, climate, relief and soils.

The word *echeri*, used by Purhépecha people to designate the soil cover, is in fact polysemic and refers at the same time to soil, land, landscape, terrain and bio-climatic zone. Thus local people perceive soil-land as a multidimensional component of the landscape. When referring to soil types and properties, the farmer conceives of soil as a three-dimensional body, similar to the technical concept of soil. When concerned with farming practices, the farmer uses *echeri* to designate a two-dimensional land surface, with variable management requirements according to local bio-climatic conditions. Beyond this practical relationship between farmer and soil-land as a resource, there is a symbolic relationship by means of which farmer's land care is rewarded by the land providing him with goods and services, including food, materials for construction and ceramics, as well as medical, ritual, and magic uses.

Local farmers recognize five major soil types: (1) dusty soils (*echeri tupuri*), (2) clayey soils (*echeri charanda*), (3) sandy soils (*echeri kutzari*), (4) gravelly soils (*echeri tzacapendini*), and (5) hard soapy soils (*echeri querekua*). Soils are further subdivided in 15 subtypes and eight varieties on the basis of textural and color differences in the upper 45 cm. Additionally, farmers distinguish composite soils at plot level as textural or color intergrades, e.g. dusty-clayey soils (*echeri tupuri-charandani*) or dusty black-yellow soils (*echeri tupuri turipiti-spambiti*). Intergrades are related to their position on the landscape, the adjacency to neighboring landscape units, the intensity of sediment transit and the volume of debris accumulation (Fig. 12).

Four principles organize the local knowledge on land management: (1) land location and soil distribution pattern on the landscape, especially the identification of ecological niches suitable for the cultivation of special crops (e.g., sacred maize landraces for religious ceremonies); (2) land behavior, with early recognition and monitoring of soil erosion and fertility depletion; (3) land resilience, which controls the application of regular practices to maintain or improve land quality and exceptional measures to rehabilitate or restore more degraded soils; and (4) land quality, which is assessed on the basis of a set of criteria including landscape position,

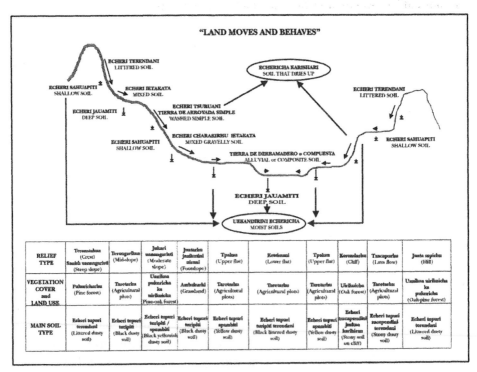

Figure 12. Purhépecha perception of soil-land distribution patterns.

micro-climatic conditions, relevant soil properties, and soil fertility (soil "strength"). On the basis of these four management principles, farmers recognize three main land classes, primarily controlled by landscape position, with identification of their qualities and limitations and prescription of the land care required: land on steep slopes, land on valley bottoms, and land in special conditions (soils on lava flows and in piedmonts, soils in homegardens and backyards).

Farmers recognize, accept, and work with the fact that land is not an immutable but a dynamic "subject." This concept is reflected in the expression: "land moves and behaves." Land behavior changes throughout the year according to seasonal rhythm, climatic variability, rainfall occurrence, and management practices. Similarly, land movement is according to its position on the landscape. The local discourse on land behavior and movement is similar to the one addressing other biological organisms. Although not explicitly stated, the farmer considers soil as a living organism. Like other living beings, soil-land can be tired, thirsty, hungry, sick or getting old. However, because soil can grow up again, be rejuvenated, recovered or rehabilitated, it is also considered fundamentally different from other living, mortal organisms.

The leaching of substances through the soil, leading to fertility depletion, and the removal, transfer and deposition of debris at the terrain surface, are perceived as "normal" processes affecting the land as a living being. The strategy adopted by the farmer to deal with these processes is to benefit from them rather than strictly control them or heavily counteract. The word "erosion" does not exist in Purhépecha language, although farmers clearly identify the process of removal and assess its severity. Soil erosion is recognized as a natural phenomenon, the severity of which might overrun the farmer's control capability, but is not perceived as a negative phenomenon leading to land degradation. It is considered as a periodical process, which depends on land management but is also an integral part of it. Soil debris eroded upslope benefit agricultural fields downslope through enrichment in mineral nutrients and organic matter. In this sense, farmers establish a difference between temporary and definitive improvement. Fields on backslopes and footslopes can only be temporarily improved, because overland-flow materials (*tsuruani*) are in transit and land requires a fallow period to fully recover from use. Instead, the constant accumulation of debris (*iorhejpiti*) on toeslopes and valley bottoms promotes permanent improvement, which allows continuous and intensive land use.

The example of San Francisco Pichátaro demonstrates that traditional agriculture does not necessarily lead to land degradation. Farmers clearly understand that land under intensive use is exposed to erosion, structural deterioration and fertility depletion, and are well prepared to identify the causes, assess the severity and apply remedies. Soil erosion is not an issue the farmer deals with because it is prevented. Soil erosion, as well as fertility depletion, is controlled and monitored the year around and year after year to take timely appropriate corrective measures. Assessment, control, and monitoring of land degradation are integral parts of land management to secure sustainable land use. Understanding land degradation and the practical experience to handle it are embodied in the knowledge system of local farmers. This indigenous knowledge is (1) shared by all members of the community, with minor variations according to age, gender and level of experience; (2) transferred from generation to generation via practical demonstrations, informal conversations and participatory meetings; (3) explained symbolically and/or logically by recognizing cause–effect relationships; and (4) conceptualized by formalizing practical experience into knowledge rules.

Farmers' knowledge body (corpus) of land degradation and sustainable land management is derived from the symbolic meaning attached to land (cosmos) and a longstanding farming experience (praxis). Altogether, it represents the social theory of land management the community has developed via centuries of co-evolution with nature. This collective knowledge is increasingly exposed to fragmentation, as the community undergoes structural changes and loses its social cohesion under the impact of externalities such as off-farm activities, out-migrations, and governmental intervention. Likewise, indigenous production systems are increasingly vulnerable to the effects of globalization and are therefore at the crossroads of sustainability, because the introduction of new farming practices often causes land degradation to increase.

Conclusions

In addition to the discipline of soil science within the modern global scientific community, there exist less familiar but highly important soil knowledge systems within indigenous cultures across the world. These encompass a wide spatial and temporal range of knowledge of soils and land use, developed in the myriad of terrestrial ecosystems. In this chapter we have presented a few examples to illustrate the scope and significance of indigenous soil knowledge. The importance of this knowledge and experience, and the ethic of land care associated with it, are echoed in Warkentin's (1999) idea of the "return of the other soil scientists," the concept of "soil care" set forth by Yaalon and Arnold (2000), in Jackson's (1994) theme of "becoming native to this place," and in the Lakota leader Luther Standing Bear's statement (1933) that "We are of the soil and the soil is of us." Ethnopedology is the interdisciplinary science that seeks to document and understand traditional soil knowledge, and to make known its importance.

Indigenous knowledge of soils deserves attention from several standpoints. Much of the knowledge and the cultures that hold it have been lost during the past few centuries. Many other traditional cultures are in the process of disappearing or being transformed in the wake of global economic development. People have a right to maintain their cultural integrity, even with technological change. The inherent rights of indigenous peoples are affirmed by the United Nations' Convention on Biological Diversity (Posey, 1999).

Indigenous knowledge is also a resource for modern society and science, providing perspectives on enhanced time and spatial scales concerning soil quality and sustainable land use. Traditional cultures possess long-term perspectives, experience, and in-depth knowledge of local environments. There is much to learn about land and environmental management from the successes and failures of cultures past and present. As the case studies illustrate, there is a strong connection between indigenous land use and sustainability. In cultures that have successfully adapted to their environments and resources over long periods of time, people have learned to use resources conservatively; in other words, to live within their means. There are also a number of examples of non-sustainable land use among cultures that resulted in soil degradation, which also provides valuable lessons from which modern soil science and society can learn. By taking careful stock of the experiences of cultures past and present which have used soils for millennia in many environments, we stand a better chance of solving challenges in food production and environmental quality.

Another area of indigenous soil knowledge relevance concerns global environmental and climate change. For example, water shortage is increasingly a problem for people and agriculture in arid lands and other regions subject to drought, and even more so with the prospect of global warming. Serious water shortages and associated land degradation already affect significant world areas, and this trend is projected to become more severe in the near future. Over 80% of the world's cultivated land is rainfed, relying solely on precipitation and runoff; these lands produce more than 60% of the global food supply (FAO, 2002). Learning from the adaptation to limited water among traditional cultures, including their soil water management strategies, may be of great help in searching for more sustainable use and conservation of water resources. For example combining the principles of traditional runoff agriculture with modern technology may result in more sustainable use of limited water resources.

Efforts to understand, respect, and use traditional knowledge of soils can play a key role in agricultural development, helping to provide a more sound basis for development that more actively involves the peoples for whom the development is intended. The relevance of ethnopedology in development is that it has the structure and perspective to support cultural viability and appropriate development, to encourage cross-cultural communication, and provide access to critical information across several disciplines. Ethnopedology is a source of information and guidance on land use and management strategies for disrupted cultures that have lost knowledge or displaced peoples who are developing land use systems in new areas. Beyond local levels, development in concert with the time-tested knowledge of cultures can help contribute to key issues world-wide in food and water security, environmental protection, resource conservation, and sustainable land use.

Indigenous knowledge, when considered in its totality, represents the aggregate experience of all humans past and present interacting with the earth's many environments. This cultural heritage, like the soils themselves and the terrestrial and biological diversity with which it is associated, is a precious resource that deserves appreciation and conservation. The integrity of the world's diverse cultures and natural resources remains in peril; there already has been too much degradation and extinction. Like indigenous cultures, indigenous soils are also at risk. Substantial areas of soils, and other land and biological resources, have become transformed and endangered during intensive use and from pressures of rapid population rise during the past century. That degradation has accelerated in many regions during modern times, despite our greater analytical and technological knowledge and tools, and best intentions. The future of life on this planet depends to a significant degree on how we conserve our collective cultural and natural resources. Indigenous knowledge of soils and its expression in a land ethic are critical to our becoming stewards of the earth upon which we live.

Acknowledgments

Ethnopedology studies rely on the patience, good will and friendship of the people who shared their soil knowledge with us. We are grateful to them.

References

Abrol, I.P., and K.K.M. Nambiar. 1997. Fertility management of Indian soils: a historical perspective. pp. 293–309. *In* D.H. Yaalon and S. Berkowicz (ed.) History of soil science: international perspectives. Advances in Geoecology 29. Catena Verlag, Reiskirchen, Germany.

Altieri, M.A. 2001. Agriculture, traditional. pp. 109–118. *In* S.A. Levin (ed.) Encyclopedia of biodiversity, Vol. 1. Academic Press, San Diego, California.

Altieri, M.A. 2004. Linking ecologists and traditional farmers in the search for sustainable agriculture. Frontiers in Ecology and the Environment 2:35–42.

Bailey, R.G. 1998. Ecoregions: the ecosystem geography of the oceans and continents. Springer-Verlag, New York.

Balée, W. 1989. The culture of Amazonian forests. Advances in Economic Botany 7:1–21.

Barrera-Bassols, N., and J.A. Zinck. 2000. Ethnopedology in a worldwide perspective: an annotated bibliography International Institute for Aerospace Survey and Earth Sciences (ITC), Enschede, The Netherlands.

Barrera-Bassols, N., and J.A. Zinck. 2003a. Ethnopedology: a worldwide view on the soil knowledge of local people. Geoderma 111:171–195.

Barrera-Bassols, N., and J.A. Zinck. 2003b. "Land moves and behaves": indigenous discourse on sustainable land management in Pichátaro, Pátzcuaro basin, Mexico. Geografiska Annaler-A 85:229–245.

Barrios, E., R. Herrera, and J.L. Valles. 1994. Tropical floodplain agroforestry systems in mid-Orinoco River Basin, Venezuela. Agroforestry Systems 28: 143–157.

Barrios, E., and M.T. Trejo. 2003. Implications of local soil knowledge for integrated soil management in Latin America. Geoderma 111:217–231.

Bocco, G. 1991. Traditional knowledge for soil conservation in central Mexico. Journal of Soil and Water Conservation 46:346–348.

Butzer, K.W. 1994. The Islamic traditions of agroecology: crosscultural experience, ideas, and innovations. Ecumene 1:7–50.

Carney, J. 1991. Indigenous soil and water management in Senegambian rice farming systems. Agriculture and Human Values 8:37–48.

Carney, J. 1998. Women's land rights in Gambian irrigated rice schemes: constraints and opportunities. Agriculture and Human Values 15:325–336.

Chadefaud, C. 1998. Anthropisation et terre arable dans l'Egypte Ancienne: la "Kemet" (terre noire), techniques d'irrigation, pratiques culturales, macrorestes végétaux. Proceedings of the 16th World Congress of Soil Science, Montpellier, France.

Cools, N., E. DePauw, and J. Deckers. 2003. Towards an integration of conventional land evaluation methods and farmers' soil suitability assessment: a case study in northwestern Syria. Agriculture, Ecosystems, and Environment 95:327–342.

Critchley, W.R.S., C. Reij, and T.J. Willcocks. 1994. Indigenous soil and water conservation: a review of the state of knowledge and prospects for building on traditions. Land Degradation and Rehabilitation 5:293–314.

Davis, S.H. (ed.) 1993. Indigenous views of land and the environment. The World Bank, Washington, D.C.

Denevan, W.M. 2001. Cultivated landscapes of native Amazonia and the Andes. Oxford University Press, New York.

Diamond, J. 2005. Collapse: how societies choose to succeed or fail. Viking, Penguin Group, Ltd., New York.

Doolittle, W.E. 2000. Cultivated landscapes of native North America. Oxford University Press, New York.

Engel-Di Mauro, S. 2003. Disaggregating local knowledge: the effects of gendered farming practices on soil fertility and soil reaction in SW Hungary. Geoderma 111:503–520.

Erickson, C.L. 1992. Prehistoric landscape management in the Andean highlands: raised field agriculture and its environmental impact. Population and Environment 13:285–300.

Fairhead, J., and M. Leach. 1996. Misreading the African landscape: society and ecology in a forest-savanna mosaic. Cambridge University Press, Cambridge, UK.

FAO. 2002. Crops and drops: making the best use of water for agriculture. Food and Agriculture Organization of the United Nations, Rome.

Ferguson, T.J., and E.R. Hart. 1985. A Zuni atlas. University of Oklahoma Press, Norman, Oklahoma.

Glaser, B., L. Haumaier, G. Guggenberger, and W. Zach. 2001. The 'Terra Preta' phenomenon: a model for sustainable agriculture in the humid tropics. Naturwissenschaften 88:37–41.

Gobin, A., P. Campling, J. Deckers, and J. Feyen. 2000. Quantifying soil morphology in tropical environments: methods and application in soil classification. Soil Science Society of America Journal 64:1423–1433.

Gong, Z., G. Zhang, and Z. Chen. 2003. Development of soil classification in China. pp. 101–125. In H. Eswaran et al. (ed.) Soil classification: a global desk reference. CRC Press, Boca Raton, Florida.

Grimes, B.F. (ed.) 1996. Ethnologue: languages of the world. 13th ed. Summer Institute of Linguistics, Dallas, Texas. http://www.sil.org/ethnologue/

Guillet, D.W. 1992. Covering ground: Communal water management and the state in the Peruvian Highlands. The University of Michigan Press, Ann Arbor.

Gupta, A. 1998. Postcolonial development. Agriculture and the making of modern India. Duke University Press, Durham, North Carolina.

Hecht, S.B. 1989. Indigenous soil management in the Amazon Basin: some implications for development. pp. 166–181. In J.O. Browder (ed.) Fragile lands of Latin America: strategies for sustainable development. Westview Press, Boulder, Colorado.

Hecht, S.B., and D.A. Posey. 1989. Preliminary results on soil management techniques of the Kayapó Indians. pp. 174–188. In D. A. Posey and W. Balée (ed.) Resources management in Amazonia: indigenous and folk strategies. The New York Botanical Garden, New York.

Hillel, D. 1991. Out of the Earth: civilization and the life of the soil. University of California Press, Berkeley.

Homburg, J.A., J.A. Sandor, and J.B. Norton. 2005. Anthropogenic influences on Zuni soils. Geoarchaeology 20:661–693.

Jackson, W. 1994. Becoming native to this place. University of Kentucky Press, Lexington.

Jodha, N.S., M. Banskota, and T. Partap (ed.) 1992. Sustainable mountain agriculture: perspectives and issues, Vol. 1. Intermediate Technology Publications, London.

Krasilnikov, P.V., and J.A. Tabor. 2003. Functional uses of ethnopedology: history and perspectives. Geoderma 111:197–215.

Lehmann, J., D.C. Kirn, B. Glaser, and W.I. Woods. 2003. Amazonian Dark Earths: origin, properties, management. Kluwer Academic Publishers, Dordrecht.

Liebig, M.A., and J.W. Doran. 1999. Evaluation of farmers' perceptions of soil quality indicators. American Journal of Alternative Agriculture 14:11–21.

Marten, G.G., and P. Vityakon. 1986. Soil management in traditional agriculture. pp. 199–225. In G.G. Marten (ed.) Traditional agriculture in southeast Asia. Westview Press, Boulder, Colorado.

Martin, G.J. 1993. Ecological classification among the Chinantec and Mixe of Oaxaca, Mexico. Etnoecológica I (2): 17–33.

McCallister, R., and P. Nowak. 1999. Whole-soil knowledge and management: a foundation of soil quality. pp. 173–193. In R. Lal (ed.) Soil quality and soil erosion. CRC Press, Boca Raton, Florida.

McCann, J.M., W.I Woods, and D.W. Meyer. 2001. Organic matter and Anthrosols in Amazonia: interpreting the Amerindian legacy. pp. 180–189. In R.M. Rees, B.C. Ball, C.D. Campbell and C.A. Watson (ed.) Sustainable management of soil organic matter. CAB International, New York.

Muenchrath, D.A., M. Kuratomi, J.A. Sandor, and J.A. Homburg. 2002. Observational study of maize production systems of Zuni farmers in semiarid New Mexico. Journal of Ethnobiology 22:1–33.

Nabhan, G.P. 1984. Soil fertility renewal and water harvesting in Sonoran Desert agriculture. Arid Lands Newsletter 20:21–28.

Needham, J., and L. Gwien-Djen. 1981. Chinese geo-botany in *Statu Nascendi*. Journal d'Agriculture Traditionelle et de Botanique Appliqee XXVIII:199–230.

Niemeijer, D., and V. Mazzucato. 2003. Moving beyond indigenous soil taxonomies: local theories of soils for sustainable development. Geoderma 111:403–424.

Norton, J.B., R.R. Pawluk, and J.A. Sandor. 2001. Farmer-scientist collaboration for research and agricultural development on the Zuni Indian Reservation, New Mexico, USA. pp. 107–120. In W.A. Payne et al. (ed.) Sustainability of agricultural systems in transition. American Society of Agronomy, Madison, Wisconsin.

Norton, J.B., J.A. Sandor, and C.S. White. 2003. Hillslope soils and organic matter dynamics within a Native American agroecosystem on the Colorado Plateau. Soil Science Society of America Journal 67:225–234.

Ollier, C.D., D.P. Drover, and M. Godelier. 1971. Soil knowledge amongst the Baruya of Wonenara, New Guinea. Oceania 42:33–41.

Östberg, W. 1995. Land is coming up: the Burunge of central Tanzania and their environments. Stockholm Studies in Social Anthropology 34. Almqvist and Wiksell International, Stockholm.

Östberg, W., and C. Reij. 1998. Culture and local knowledge – their roles in soil and water conservation. pp. 1349–1358. In H.-P. Blume et al. (ed.) Toward sustainable land use: furthering cooperation between people and institutions, Vol. 2. Catena Verlag, Reiskirchen, Germany.

Parker, E. 1992. Forest islands and Kayapó resource management in Amazonia: a reappraisal of the Apêtê. American Anthropologist 94:406–428.

Pawluk, R.R., J.A. Sandor, and J.A. Tabor. 1992. The role of indigenous soil knowledge in agricultural development. Journal of Soil and Water Conservation 47:298–302.

Payton, R.W., J.J.F. Barr, A. Martin, P. Sillitoe, J.F. Deckers, J.W. Gowing, N. Hatibu, S.B. Naseem, M. Tenywa, and M.I. Zuberi. 2003. Contrasting approaches to integrating indigenous knowledge about soils and scientific soil survey in East Africa and Bangladesh. Geoderma 111:355–386.

Petersen, J.B., E. Neves, and M.J. Heckenberger. 2001. Gift from the past: Terra Preta and prehistoric Amerindian occupation in Amazonia. pp. 86–105. In C. McEwan et al. (ed.) Unknown Amazon. The British Museum, London.

Posey, D.A. 1985. Indigenous management of tropical forest ecosystems: the case of the Kayapó Indians of the Brazilian Amazon. Agroforestry Systems 3:139–158.

Posey, D.A. 1990. The science of the Mebêngôkre. Orion Quarterly Summer:16–23.

Posey, D.A. (ed.) 1999. Cultural and spiritual values of biodiversity: a complementary contribution to the global biodiversity assessment. Intermediate Technology Publications, London.

Prain, G., S. Fujisaka, and M.D. Warren (ed.) 1999. Biological and cultural diversity: the role indigenous agricultural experimentation in development. Intermediate Technology Publications, London.

Pulido, J.S., and G. Bocco. 2003. The traditional farming system of a Mexican indigenous community: the case of Nuevo San Juan Parangaricutiro, Michoacán, Mexico. Geoderma 111:249–265.

Raffles, H., and A.M.G.A. WinklerPrins. 2003. Further reflections on Amazonian environmental history: transformations of rivers and streams. Latin American Research Review 38:165–187.

Razak, V.M. (ed.) 2003. Futures of indigenous cultures. Futures 35:907–1009.

Redman, C.L. 1999. Human impacts on ancient environments. University of Arizona Press, Tucson, Arizona.

Reij, C., I. Scoones, and C. Toulmin (ed.) 1996. Sustaining the soil: indigenous soil and water conservation in Africa. Earthscan Publications, Ltd., London.

Riley, K.W., N. Mateo, G.C. Hawtin, and R. Yadav (ed.) 1990. Mountain agriculture and crop genetic resources. Aspect Publishing, London.

Romig, D.E., M.J. Garlynd, R.R. Harris, and K. McSweeney. 1995. How farmers assess soil health and quality. Journal of Soil and Water Conservation 50:229–236.

Sanchez, P., D.E. Bandy, J.H. Villachica, and J.J. Nicholaides. 1982. Amazon Basin soils: management for continuous crop production. Science 216:821–827.

Sandor, J.A. 1998. Steps toward soil care: ancient agricultural terraces and soils. Proceedings of the 16th World Congress of Soil Science, Montpellier, France.

Sandor, J.A., P.L. Gersper, and J.W. Hawley. 1986. Soils at prehistoric agricultural terracing sites in New Mexico: I. Site placement, soil morphology, and classification. Soil Science Society of America Journal 50:166–173.

Sandor, J.A., and N.S. Eash. 1991. Significance of ancient agricultural soils for long-term agronomic studies and sustainable agriculture research. Agronomy Journal 83:29–37.

Sandor, J.A., and N.S. Eash. 1995. Ancient agricultural soils in the Andes of southern Peru. Soil Science Society of America Journal 59:170–179.

Sandor, J.A., and L. Furbee. 1996. Indigenous knowledge and classification of soils in the Andes of southern Peru. Soil Science Society of America Journal 60:1502–1512.

Sandor, J.A., J.B. Norton, R.R. Pawluk, J.A. Homburg, D.A. Muenchrath, C.S. White, S.E. Williams, C.L. Havener, and P.D. Stahl. 2002. Soil knowledge embodied in a Native American runoff agroecosystem. Transactions of the 17th World Congress of Soil Science, Bangkok, Thailand.

Sandor, J., C.L. Burras, and M. Thompson. 2005. Human impacts on soil formation pp. 520–532. In D. Hillel et al. (ed.) Encyclopedia of soils in the environment. Elsevier, Amsterdam.

Siderius, W., and H. de Bakker. 2003. Toponymy and soil nomenclature in the Netherlands. Geoderma 111:521–536.

Sillitoe, P. 1996. A place against time: land and environment in the Papua New Guinea highlands. Hardwood Academic, Amsterdam.

Sillitoe, P. 1998. Knowing the land: soil and land resource evaluation and indigenous knowledge. Soil Use and Management 14:188–193.

Sombroek, W., D. Kern, T. Rodriguez, M. da Silva Cravo, T. Jarbas Cunha, W.I. Woods, and B. Glaser. 2002. Terra Preta and Terra Mulata: Pre-Columbian Amazon kitchen middens and agricultural fields, their sustainability and their replication. Transactions of the 17th World Congress of Soil Science, Bangkok, Thailand.

Souza, V., J. Bain, C. Silva, V. Bouchet, A. Valera, E. Marquez, and L.E. Eguiarte. 1997. Ethnomicrobiology: do agricultural practices modify the population structure of the nitrogen fixing bacteria *rhizobium etli* biovar *phaseoli*? Journal of Ethnobiology 17:249–266.

Standing Bear, Chief. 1933. Land of the spotted eagle. Houghton Mifflin Co., The Riverside Press, Cambridge, Massachusetts.

Talawar, S., and R.E. Rhoades. 1998. Scientific and local classification and management of soils. Agriculture and Human Values 15:3–14.

Tamang, D. 1993. How hill farmers manage their soils. pp. 165–181. *In* D. Tamang, G.J. Gill, and G.B. Thapa (ed.) Indigenous management of natural resources. HMG Ministry of Agriculture/Winrock International, Kathmandu/London.

Toledo, V. 2001. Indigenous knowledge and biodiversity. pp. 451–463. *In* S.A. Levin (ed.) Encyclopedia of biodiversity. Academic Press, San Diego, California.

Vitousek, P.M., T.N. Ladefoged, P.V. Kirch, A.S. Hartshorn, M.W. Graves, S.C. Hotchkiss, S. Tuljapurkar, and O.A. Chadwick. 2004. Soils, agriculture, and society in precontact Hawai'i. Science 304:1665–1669.

Warkentin, B.P. 1999. The return of the "other" soil scientists. Canadian Journal of Soil Science 79:1–4.

Warren, D.M., L.J. Slikkerveer, and D. Brokensha (ed.) 1995. The cultural dimension of development: indigenous knowledge systems. Intermediate Technology Publications, London.

Wilken, G.C. 1987. Good farmers: traditional agricultural resource management in Mexico and Central America University of California Press, Berkeley, CA.

Williams, B.J. 2006. Chapter on Aztec Soil Science in this volume.

Williams, B.J., and C.A. Ortiz-Solorio. 1981. Middle American folk soil taxonomy. Annals of the Association of American Geographers 71:335–358.

Wilshusen, R.H., and G.D. Stone. 1990. An ethnoarchaeological perspective on soils. World Archaeology 22:104–114.

Winklerprins, A.M.G.A. 1999. Local soil knowledge: a tool for sustainable land management. Society and Natural Resources 12:151–161.

Winklerprins, A.M.G.A. 2001. Why context matters: local soil knowledge and management among an indigenous peasantry on the Lower Amazon floodplain, Brazil. Etnoecológia 5:6–20.

WinklerPrins, A.M.G.A., and J.A. Sandor (ed.) 2003. Ethnopedology. Geoderma 111: 165–536.

Winklerprins, A.M.G.A., and N. Barrera-Bassols. 2004. Latin American ethnopedology: a vision of its past, present, and future. Agriculture and Human Values 21:139–156.

Yaalon, D.H., and R.W. Arnold. 2000. Attitudes toward soils and their societal relevance: then and now. Soil Science 165:5–12.

Zimmerer, K.S. 1994. Local soil knowledge: answering basic questions in highland Bolivia. Journal of Soil and Water Conservation 49:29–34.

4

Some Major Scientists (Palissy, Buffon, Thaer, Darwin and Muller) Have Described Soil Profiles and Developed Soil Survey Techniques Before 1883

C. Feller[1], E. Blanchart[1] and D.H. Yaalon[2]

Abstract

Soil description, soil profiling and the use of instruments and methods for soil mapping have existed in Europe long before the admitted emergence of soil sciences at the end of the 19th century. Some little-known but definitely precursory works by major scientists of the 16th, 18th and 19th centuries are reported in the present paper. These are works by (i) Bernard Palissy (1510–1590), inventor of the auger for soil sampling (1563 & 1580); (ii) Georges-Louis Leclerc de Buffon (1707–1788), who minutely described soils and pedological traits such as ferromanganese concretions accounting for geochemical cycles (1734 & 1748); (iii) Albrecht Daniel Thaer (1752–1828), who developed a methodology to take soil samples and to describe and map soil variability in cropped plots (1809); (iv) Charles Robert Darwin (1809–1882), who published detailed drawings of pedo-architectural profiles and gave the description of a rainfall simulation (1837); (v) P.E. Müller who gave wonderful descriptions of soil profiles and soil associations with a pedogenetical approach and can be considered as the main forerunner of pedology; and (vi) some others such as Bartram (1739–1823) and Orth are evoked.

---

[1] Institut de Recherche pour le Développement (IRD), Laboratory «Matière Organique des Sols Tropicaux (MOST)», 911 Avenue Agropolis, BP 64501, 34394 MONTPELLIER Cedex 5, France.

[2] Hebrew University, Institute of Earth Sciences, Jivat Ram campus, Jerusalem, 91904 Israel.

Introduction

One of the basic activities of the pedologist consists in describing soil profiles; another is the use of a tool called the auger for sampling and for visual inspection in mapping soils. There is no doubt that these sampling and mapping tools had become of current use by the 2nd half of the 19th century, with the emergence of the applied science of agricultural geology and of the fundamental science of pedology (Yaalon, 2000). Both Dokuchaëv's publications (1883) and the comprehensive description of soils and the profile and pedogenetic processes conducted by the Swedish forester P.E. Muller in his analysis of podzolisation (1879, 1884, French transl. 1889) deserve to be mentioned here. Yet one is entitled to wonder about the possible existence of prior scientific studies on soil description, sampling and mapping.

Actually, such studies do exist, as testified by a number of little-known texts produced by some famous naturalists and agronomists of the past centuries from the 16th to the second half of the 19th century. Among those, texts by Bernard Palissy (first pub. 1563 & 1580, read in their 1880 edition); Buffon (first pub. 1783 & 1789, 1819 ed.); Bartram (1791); Thaer (1809, French transl. 1811); Darwin (1837 & 1881); Orth (1870, 1877) and Müller (1879, 1884) have been selected for the present paper. Yaalon (1989) has discussed other forerunners like W. Cobbett, J. Morton and C.S. Sprengel.

1563, 1580: The Auger and Its Use Described by Bernard Palissy

The Frenchman Bernard Palissy is best-known as a ceramist, but he is also widely acknowledged as a foundational figure in the fields of the earth sciences, geology, paleontology and the agronomic sciences – the latter thanks to the ideas he developed in his 1563 and 1580 writings on the nutrition of plants from «salts». He is not so well-known as the possible inventor of the auger as a tool for soil prospecting. He was in any case the first to describe a soil-sampling auger, and to explain its use in drawings. In the following extract from his «Admirable Speeches» taken from his complete works in their 1880 ed. p.114 (in French), he dramatizes questions and answers about the quality of soils, among other topics, through the dialogue of two characters called «Théorique» and «Practique» (T & P). The style and spelling (which would have to be in Elizabethan English for the sake of authenticity) will be somewhat modernized in the translation for practical reasons.

T. *"You have treated me with a lot of learned arguments, yet I am not wholly satisfied as to the most expedient way to discover the aforesaid marl."*

P. "*I can advise you as to no more expedient a method than that I should use myself. Should I desire to find any marl in a province where the necessary tool does not exist as yet, I should look for all and any of the augers which the potters, brick-makers and tile-makers use for their trade, and I should use them to manure a plot in my field to see if the soil had been improved in any way, and then I should seek for one that was long enough, and had a hollow socket at one end, where I would adjust a stick, and across the other end I should make sure it had a handle (. . .) in the shape of an auger, and this done, I should drill and bore away with the full length of the whole handle at all the ditches in my estate, and then I should take the said tool out and examine the concave side to tell which kind of soil had been dug out, and after cleaning the hollow, I should remove the first handle and insert one that was much longer, and I should drill the auger back deeper into the same hole in the soil with this second handle, and thus, having several handles of different lengths, I should determine the nature of the deepest layers; and not only should I dig the ditches of my estate, but I should bore all over the fields, until at length some testimony of the presence of the aforesaid marl were found at the end of my tool, and if it were, I should dig, at the very place it had been found, a sort of well.*"

T. "*Indeed. But what if there should be some hard rock beneath the surface? For such is the case in many places . . .*"

P. "*In truth that would be tiresome, though in many a place the rocks are soft and tender, especially when they are still planted in the ground.*"

In this text, Bernard Palissy mentions the auger ("la tarière") as if such a tool was already known, but he actually starts from a well-known tool created for a completely different use and adjusts it to the problem of soil prospecting. At the technological level, he provides us with the idea of the extension-pieces; he also makes it plain that this tool is designed for a double use: that of soil prospecting, and that of the detection of marl to be used as manure for soil reclaiming. It is to be noted that in the last exchange ("*in many a place the rocks are soft and tender, especially when they are still planted in the ground*") Palissy reveals to be an accurate oberver of the horizon of rock alteration.

Further investigation into the origins of the French word "tarière" (auger) seems to prove that Palissy was the first to use such a tool for soil prospecting. A complete file on the word "tarière" was kindly handed on to us (TLF, 1981) by the Institut National de la Langue Française (INALF, CNRS). This file includes about 80 quotations from a varied assortment of notes and entries taken from a number of ancient and contemporary dictionaries. According to the *Grand Larousse de la*

Langue Française (1978), the first occurrence of the word "tarière" (which takes its etymology from the low Latin *taratrum*) seems to be dated from 1212, when it had the meaning of "a tool used by carpenters, wheel wrights, etc., to bore holes into wood pieces". The *Dictionnaire de l'Académie Française* (1835) defines the word "tarière" in its zoological use, as referring to "the elongated organ situated at the end of the abdomen of some female insects which use it to deposit their eggs into the ground or plants or under the skin of animals" (that is the form of ovipositor called "terebra" in English). The Dictionnaire de Trevoux (1752) mentions the word as "a tool used by miners for drilling into the ground", which is the sense we are concerned with.

If these historical written sources are to be relied on, then Bernard Palissy was indeed the first to describe our pedological "tarière", or auger. A few years later, in 1605, Olivier de Serres describes an identical tool (which he calls "taravelle"), which he says is used as an instrument for planting (Godefroy, 1937–38). By the 18th century, the auger, also called "gauge" (in French: "sonde") had become a popular instrument for the study of soils – as testified by its frequent occurrence in writings by Buffon (1783–1788, but also the description of 1734, cf. below), Pattulo (1759, pp. 15, 16, 24), Turbilly (first anonymously published, 1760, p. 12, with a descriptive sketch p. 332), La Salle de l'Etang (1768, pp. 110–112), Delaillevault (1763, p. 64) and Thaer (1811–1816, Vol. 2, p. 1666, cf. below). However not all the authors who mention it are enthusiastic about the implement; La Salle de l'Etang, for instance, criticizes it, considering that *"this invention is neither so wonderful nor so useful to agriculture as was once imagined (. . .) good ploughmen (. . .) even regard it as superfluous."* (1768). Lastly, a fellow-researcher, Rossignol (2001) found that an auger appeared in a humourous drawing by the book illustrator Granville (1803–1847) representing a "Most Stubborn Field-Mouse" using an auger to dig his mouse-hole.

1734: Description of Soil Horizons, Erosion and Geochemical Cycles by Georges-Louis Leclerc de Buffon

Georges-Louis de Buffon (1707–1788) was a major naturalist. The publication of the 36 volumes of his *Histoire Naturelle* made him famous worldwide, but Bourde (1967, pp. 238–239) also regards him as one of the founders of rural economy. François de Neufchateau (1804, quoted by Boulaine, 1989) claims that Buffon once suggested that *"men of letters should establish a literary society concerned with land discernment"*. This is proof of Buffon's interest in agriculture in general, and soils in particular. At last, according to Kunholtz-Lordat

(1958), Buffon should even be apparently acclaimed as "the founder of Pedology".

Even without going so far, one must acknowledge Buffon's astounding skills in the matter of soil observation, as testified by the following extracts about mould from his *Histoire Naturelle des Minéraux*, in the section entitled "Of Vegetable Mould" ("De la Terre Végétale", *Oeuvres Complètes de Buffon*, 1819, Vol. 5).

Buffon occasionally calls vegetable mould ("terre végétale") a loamy soil ("terre limoneuse") when he explains that he intends to *"trace the origin and formation of loamy soil"* and its effect *"on the production of most second generation minerals"* (pp. 402–403). Buffon's formulation sounds undeniably familiar to pedogeneticists and geochemists.

This passage is followed by a long discourse on the favourable or unfavourable conditions of mould accumulation.

First, the presence of man is considered harmful for mould (p. 407, 498):

"We shall see that this layer of rich, prolific soil is always much thicker wherever Nature reigns supreme than in places inhabited by man ... (for) it cannot gain in depth in places where man and fire (this destructive minister of man's) ... annihilate ... the animal creatures and the vegetable kingdom ... Suffice it to compare long-inhabited countries with newly discovered ones: in the latter, forests, mould, alluvium everywhere; only dry sand or naked rock in the former".

Next, Buffon ponders the topographical factors in erosion (pp. 407–408):

"The layer of vegetable soil is thinner on top of the mountains than in vales and plains, because rain-waters wash it off the summits and slopes of the eminences, carrying the mould downhill."

There ensues the beginning of a horizon identification:

" ... One should follow attentively the way Nature proceeds in the production and successive formation of this vegetable mould. First entirely composed of animal and vegetable detriments, it remains for a great number of years an uncohesive, unductile and very thin, dry, blackish dust, liable to catch fire and burn like peat. The ligneous fibers and solid parts of plants can still be recognised in this mould; but as time goes by, through the intermediary agency of air and water, these dry mould particles acquire more ductibility and turn into loamy soil: this process of reduction and transformation I have observed with my own eyes."

In this description, Buffon establishes a distinction between what we would call today the organic horizons and the underlying organo-mineral and mineral horizons, even if we have learned that the latter do not directly and exclusively result from the transformation of the former (which is Buffon's thesis).

Buffon then proceeds to examining soils and the account of detailed observations:

"In 1734, I ordered a plot of about seventy acres to be probed by several auger drills, for I wanted to know how thick the good soil was in that place, where I had formerly had a number of trees planted, with satisfactory results. The ground had then been divided into several acres; and the boring being performed at all four angles of each acre, I noted the different depths of soil, the thinnest being of two feet, and the thickest, three feet and a half. I was young at the time, and was hoping to be able to observe what effect my tree plantation would have on the different depths of soil, which was good and rich soil all over. The later borings indeed enabled me to observe that everywhere the thickness of the soil layers had remained approximately the same, and I also noticed the transformation of mould into loamy soil. The said plot lies in a plain situated above the highest hills of Bourgogne; for the most part, it has always remained fallow from time immemorial; and as there is not a single eminence to tower over it, the soil is exempt from any mixture of clay or chalk, lying as it does over a horizontal layer of hard limestone.

Under the turf, or, rather, the old moss that covered the ground, a thin layer of black crumbly soil could be observed everywhere, composed of the product of the decayed grass and leaves of the preceding years. The layer just below was compounded of brown, unadhesive soil; but the deeper layers underneath the first two took by degrees more consistence and a yellowish colour, all the more so as they were deeper under the surface. The deepest layer, lying at a depth of three or three and a half feet, was a reddish orange colour; and the soil was extremely rich and ductile and stuck to the tongue like "bol"(swelling clay in French).

In this yellow soil, I noticed the presence of several grains of iron ore; at the deepest level, they were black and hard, but were only brown and still crumbly in the top layers. It is therefore obvious that animal and vegetable detriments which first constitute the mould, gradually form, through the agency of air and water, the yellow or reddish soil which is the real loamy soil we are concerned with; just as it is undoubtedly obvious that the iron contained in plants find themselves in this soil, and assemble into grains . . . As the borings (performed in 1748) *proceeded, observing carefully the different materials they brought to the surface, I concluded, without any possible doubt, that the loamy soil was carried down to great depths into the joints and cleavages of the lower layers, which were all clay, by the infiltration of waters. This loamy soil first clung to the surface of the clay clods; it then managed to infiltrate the clay layer; after which, piryte concretions of a flat and orbicular configuration could ordinarily be observed in the clay, connected by a sort of cylindrical cordon of the same pirytish substance, and the said cordon always led to a joint or cleavage filled with loamy soil. I then became convinced that this soil was a major contributor to the formation of "martial" pyrites . . . These observations convinced me that this bog-soil, produced by the complete decomposition of animals and plants, is the foremost provider of iron grains for iron ore, and that it also produces the most part of the elements neccessary to the formation of pirytes . . . The amount of iron to be found in loamy soil is so stupendous at times, that we might call this soil "a ferruginous soil" . . . .* (Buffon's underlining)

This is followed (421–423) by the descriptions and formation conditions of iron-manganese concretions. Iron specialists should take a look at these admirable pages and particularly at the passage below:

"But how does this mineral matter manage to sort itself out from the great mass of loamy soil and aggregate into such regular, tiny grains, in such large numbers, in so perfect a manner, that there is not a single one of them the surface of which fails to gleam with a metallic sheen?"

Buffon's explanation, of course, rests on hydrodynamic processes. But a more and more detailed description of pedological traits (here, concretions) ensues:

"Should the spheric grains of iron ore be divided into two hemispheres, then one would observe that they are all composed of several thin concentric layers, and that the biggest present a definite cavity, usually filled with the same ferruginous matter, which has not however acquired the same density, and which will crumble easily, like the iron grains themselves when they begin forming in the upper layers of the ferruginous soil; so that, in every grain, the outer layer, which is endowed with the metallic sheen, is the most solid and the most metallic, because it has been formed first, and, as such, has received, through infiltration, and collected, the purest ferruginous molecules, filtering the lesser pure through to produce the second layer of grain, and so on with the third and fourth layers, until the center can contain only the most soil and least metal."

These extracts show Buffon's astounding observation skills at all levels, from profile to concretions; his global approach to surficial formations with the study of materials transfer; his desire to explain the soil's successive horizons, even if, in pedogenetic terms, his interpretations have been somewhat invalidated since; and also his extraordinarily precursory reflection in the analysis of a major geochemical cycle – the cycle of iron – to which he already gives a "biological" dimension when considering the part played by organic matter.

1791: William Bartram, Traveller and Observer of Soils of the United States

William Bartram (1739–1823) was an astounding naturalist and observer of the subtropical milieu of 18th century United States. He is the author of a book entitled "Travels through North and South Carolina, Georgia, East and West Florida, the Cherokee Country, the Extensive Territories of the Muscogulges, or Creek Confederacy, and the Country of the Chactaws", first published in the USA in 1791, and republished in England in 1955 and 1988; long passages from Bartram's work were translated into French and commented on by Chatelin in 1991. The two examples given below are taken from the latter edition. In the first one Bartram relates his travels across West Florida and the part of Lousiana called the Black Bell which was to become a cotton production area; most of the soils there are "black cotton soils" (which would now be called "Vertisols" in the contemporary pedological classification). Bartram describes the area as follows:

> "The upper stratum or vegetable mould of these plains is perfectly black, soapy and rich, especially after rains, and renders the road very slippery: it lies on a deep bed of white, testaceous, limestone rocks, which in some places resemble chalk, and in other places are strata of subterrene banks of various kinds of sea shells, as ostrea, etc. These dissolving near the surface of the earth, and mixing with the superficial mould, render it extremely productive"

This passage is particularly remarkable for its precision and the pre-pedogenetic discourse on the parent material weathering. A few pages above, in the course of a development about iron ore, Bartram gives an excellent description of what is probably an iron pan (cuirasse latéritique in French):

> "the highest tops of hills provide great amounts of iron ore similar to that encountered in New Jersey and Pennsylvania which is called bog-ore; the ore occurs throughout the soil in both big separated masses and small fragments; it is heavy and appears to be rich in this especially useful metal; but a remarkable characteristic of these soil-borne stones is that they are swelled, somewhat resembling scoria, as if they had suffered from a violent fire."

One has to wait for the publication of the famous journey of Buchanan in South India (1807) to get a description of this soil horizon (iron pan) and the name of "laterite" for such a soil material.

1812: Mapping and Agricultural Soils Analysis by Daniel Albrecht Thaer

Thaer must be the most renowned agronomist of the early 19th century, owing to the fame of the 4 volumes of his "*Grundsätze der rationellen Landwirtschaft*", first published in German (1809), then translated into French in 1811–1816 "*Principes Raisonnés de l'Agriculture*" and in English under the title "*The principles of practical agriculture*" (Thaer, 1856).

In his *Rational Principles*, Thaer broaches all the subjects pertaining to agriculture (taken in its widest definition, including stock-breeding and sylviculture), in all its implications, whether socio-economic or biophysical (fertility, animal and vegetable productivities). Based on the elaboration of quantified fertility indicators, his analysis of the durability of farming and production systems is astoundingly prescient – even if the theorical premises of his system (the «humus theory») were later revealed to be erroneous. Thaer's work is remarkable for more than one good reason: he collected with minute precision a large number of quantitative data (producing over 1,600 pages of text with thousands of data); he established a typology of the contemporary farming systems, classifying them into 8 or 9 different categories, and even almost modelized the functioning of the major exploitation systems.

What will be reported here is Thaer's approach of the problems of variability, mapping and soil analysis. This aspect of soil quality and its evaluation takes up to 50 pages of Volume 2 in the section entitled: "*The different kinds of soil, their estimation, their use and properties, in correlation with the component parts of the soil.*" (Thaer, 1811–1816, French ed. pp. 115–166)

To begin with, Thaer underlines the necessity of:

"*ascertaining the composition* (of soil) *by first relying on its exterior specificities. After colour, the most positive signs of the presence of humus in a soil are its lightness, a characteristically musty smell and the presence of white growths of* <u>lichen humosus</u>. *Clay can be recognized at touch, since it is characterized by its stickiness and unctuosity; sand feels rough under the fingers when crumbled, and is clearly revealed under the scrutiny of a weak lense . . . The presence of chalk can be ascertained by the effervescence caused by its contact with acids . . .* (V.2, p. 138) *When assessing the value of a soil, the examination of its component parts must be immediately followed by the measuring of its thickness. By thickness, I mean that of the surface layer commonly called « vegetable mould », which is homogeneous and uniformly impregnated with humus. (V.5, p. 139)"*

This leads Thaer to a genuinely pedological discourse on studying soil (V.2, pp. 164–166, §575) worth quoting in its totality:

"In order to perform an accurate description of a plot of land, relying both on the nature of soil and the mixture of the component parts of this soil, a description which may at once serve as a guideline for the assessment of its value and for the choice of crops and crops rotation it is most suitable for, it is necessary to proceed by regular, carefully-determined steps. If the plot has not been divided (and thus distributed) by planks, a cross-ruling must be effected by parallel lines distant from 5, 10 or 15 perches, according to whether the nature of the soil seems to be changing or not.

At the same time, one should survey the plot to be assessed at a sufficiently large scale, that is about four times the scale used for territorial maps. On this map, the parallel lines will be reported, and crossed so as to obtain 5 to 10 perches-wide sections, or stations. These stations, extending from one line to the other, must be numbered; then the ground must be surveyed along those lines. Apart from the men in charge of measuring the ground with a surveying-chain, two more are needed, one with a spade, who will have to do the digging, and another with a basket, who must collect and carry the soil samples. The surveyor must take care both of the mapping and the procedure, unless he has an aide for the latter. The agronomist examines the nature of the soil and supervises the whole operation. Whenever he observes a change in the nature of soil, he stops and notes the station on the map; he then scrutinizes the soil more closely; and wherever he thinks it worthwhile, he orders some spadefuls of earth to be collected, if necessary as much as a pound of it, and has it put into a cornet paper bag or a small canvas bag, on which the number or letter of the station must be printed. The places where the changes in soil nature have been noticed are located and reported on the map as accurately as possible; on top of which, one will not fail to note whether the changes in the nature of soil are abrupt or gradual. All other observations pertaining to the soil poperties enumerated before will be reported on the procedure under the station number.

And thus the whole ground is surveyed following the parallel lines which have been drawn, which allows for the first step of the geological mapping.

To proceed to the latter, there are different possibilities; the most efficient being to figure on the map the various mixtures of soil compounding the ground, using different shades of water colours to indicate the lightest gradual changes; eminences and hollows will be represented by hachures, as is the common usage; the lesser or greater amounts of humus impregnating the ground are to be figured by black dots, which will be all the more concentrated as the proportion of humus is greater; so that positive signs will indicate everything worth noting. Such a map will enable one always to keep in sight an accurate representation of the available land and to take such steps as will be deemed most appropriate. The procedure will make it possible to draft a more precise description relative to the numbers on the map.

Additional information, such as uphill and downhill slopes and water-courses, may figure on that map; but for a more detailed information about such matters, levels might need to be taken with the appropriate tools. Levels can be measured in different directions, after which a profile may be established. Should there be a definite change in the lowest level of soil, worth mentioning and further analysing, this can be rendered on the relief profile by the use of colours indicating the depth of the different layers.

If that is the case, when the levels are measured, one must also use the auger or ground lead and dig it into the ground as deep and as often as necessary without much difficulty.

When the different component parts of ground are uneasy to make out through mere exterior observation, or when one feels inclined to proceed to more accurate examination, the soil must be submitted to a chemical analysis. But a mere visual comparison of the samples collected in the different stations, effected by degrees first in their damp state, then in their dry state, will make it possible to determine which are homogeneous and which are not, making in most cases the recourse to chemical analysis unneccessary.

More than any other, this operation will reward the enlightened agronomist for all his pains, for it will allow him to find answers to previously unaccountable phenomena, while enabling him to bring solutions to the problems encountered in the exploitation of his estate."

It must be acknowledged that Thaer presents us here with a real pedological mapping at the plot scale, not only for inventory and soil distribution analysis but also for the elaboration of thematic maps (depth, humus content, etc.). The only lack concerns the use and interpretation of air and satellite photographs, but this would be asking

too much from a 1809 study! The extraordinary conclusion on the primacy of observation of the soil and its variations deserves to be pondered as well.

1837: Pedological Profile Drawings by Charles Darwin

In 1881, Darwin published his last scientific writings, entitled *The Formation of
Vegetable Moulds Through the Action of Worms with Observations of their Habits*,
translated into French as *Rôle des vers de terre dans la formation de la terre végétale*.
In this work, Darwin reveals to have been an astounding precursor in the field of the
earth sciences, in such varied subjects as erosion, matter transfer at the scale of the
bigger watersheds, alteration and pedogenetic processes, ecology, soil bio-functioning
and pedo-archeology (Feller et al., 2000). But although this work was published in
1881, Darwin's first communication on the matter actually took place on 1,
November 1837 before the Geological Society of London; it was entitled "On the
Formation of Mould" (Transactions of the Geological Society of London, Vol. 5,
pp. 105–108). In writing, the communication consists of three pages, including the
profile of a soil under pasture, entitled «mould», meant to illustrate the part played
by worms in the burying of various materials, such as cinders or ashes, pieces of burnt
marl and quartz pebbles, deposited 15 years earlier at the soil surface (Figure 1 A).
The piece is most remarkable for its extremely clear identification of horizons.

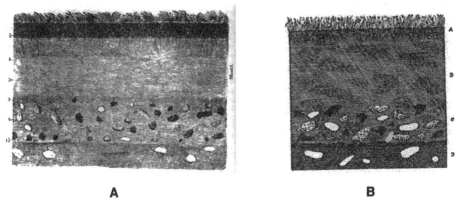

A   **B**

Figure 1. Diagrams of soil profiles presented in Darwin 1837 (**A**) and Darwin 1881 (**B**)

– *Original caption of figure 1A (Darwin, 1837): Section of field: A, cinders; B, burnt marl;
  C, quartz pebbles.*
– *Original legend of figure 1B (Darwin, 1881): Section reduced to half the natural scale,
  of the vegetable mould in a field, drained and reclaimed fifteen years previously; A, turf;
  B, vegetable mould without any stones; C, mould with fragments of burnt marl, coal-
  cinders and quartz pebbles; D, subsoil of black, peaty sand with quartz pebbles.*

On the 1837 illustration (Figure 1A), Darwin describes, from top to bottom (original measurements in inches):

– 0–0.5: the turf, or root mat

– 0.5–3.0: the vegetable mould, that is horizon A1. (on the printed reproduction of 1837, a darker sub-horizon with a considerable amount of remaining roots is perceptible in A11 between 0.5 and 1.0 inches. (It is hard to tell whether this is the rendering of an observation of Darwin's, or just an artistic liberty of the engraver's, as the subdivision is no longer apparent in the 1881 edition, Figure 1B.)

– 3.0–4.5: a layer «of fragments of burnt marl, (conspicuous from their bright red colour), of cinders, and a few quartz pebbles, mingled with earth.»

– 4.5– ?: «the original black peaty soil. We thus find, beneath the layer, nearly four inches thick, composed of fine particles of earth mixed with decayed vegetable matter, those substances which had been spread on the surface fifteen years before.»

The soil profile diagram of 1881 (Figure 1B) is definitely unlike the first wood engraving and looks different from the 1837 profile. But it is still interesting, because of the horizons denominations in A, B, C and D, even if the terms used in the nomenclature are not pedological. Darwin's caption reads: « A, turf; B, vegetable mould without any stones; C, mould with fragments of burnt marl; coal-cinders and quartz pebbles; D, subsoil of black, peaty sand with quartz pebbles.»

In the 1881 edition are presented sketchily or in great detail numerous other profiles (10 altogether) of what we would call today «archeological profiles». Figure 8 of the 1881 edition and its caption have been selected for illustration (Figure 2).

Darwin always takes pains to establish a clear distinction between undisturbed and disturbed material (cf. the comment upon FF layer on his Fig. 8).

While describing these profiles, Darwin is also one of the first to show the existence of «stone-lines». This has been abundantly commented upon by Johnson (1990, 1999).

Of course, these are not Darwin's only contributions to pedology, and many others have been reported elsewhere (Feller et al., 2000, 2003). In the passage where Darwin minutely studies the part played by worms in lands' «stripping» (that is, erosion), he has the brilliant idea of simulating rain in order to study what would be called today the stability of faunic aggregates and their transport downhill by rainfalls. As far as

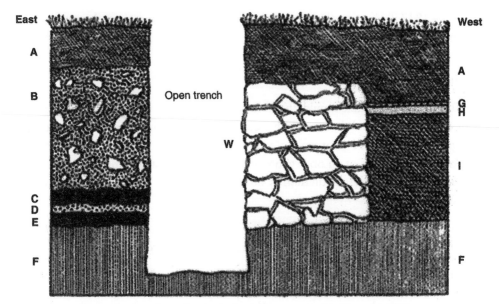

Figure 2. A pedo-archeological profile presented in Darwin 1881.

Original legend (Darwin 1881). Section through the foundations of a buried Roman villa at Abinger. AA: vegetable mould; B, dark earth full of stones, 13 in. in thickness; C, black mould; D, broken mortar; E, black mould; FF, undisturbed subsoil; G, tesserae; H, concrete; I, nature unknown; W, buried wall.

we know, Darwin's is the first rain simulator (Darwin, 1882, pp. 223–224). We shall report below an extract from this text:

"*I was led to believe that a considerable quantity of fine earth is washed quite away from castings during rain, from the surfaces of old ones being often studded with coarse particles. Accordingly a little fine precipitated chalk, moistened with saliva or gum-water, so as to be slightly viscid and of the same consistence as a fresh casting, was placed on the summits of several castings and gently mixed with them. These castings were then watered through a very fine rose, the drops from which were closer together than those of rain, but not nearly so large as those in a thunderstorm; nor did they strike the ground with nearly so much force as drops during heavy rain. A casting thus treated subsided with surprising slowness, owing as I suppose to its viscidity. It did not flow bodily down the grass-covered surface of the lawn, which was here inclined at an angle of 16° 20'; nevertheless many particles of the chalk were found three inches below the casting. The experiment was repeated on three other castings on different parts of the lawn, which sloped at 2° 30', 3° and 6°; and particles of chalk could be seen between 4 and 5 inches below the casting; and after the surface had become dry, particles were found in two cases at a distance of 5 and 6 inches."*

1870–1877: A. Orth

According to Mückenhausen (1997, p. 266), descriptions of soil profiles from the surface to parent material in soil surveying and soil mappping have also been achieved/performed by the German scientist Orth, as early as the 1870–1877 period. The complete relevant section of the Mückenhausen's paper will be quoted below.

"With the book «Die geologischen Verhältnisse des norddeutschen Schwemmlandes und die Anfertigung geognotisch-agronomischer Karten», Orth (1870) prepared the basis geologic-agronomic maps of the Prussian Geologische Landesanstalt at the scale of 1:25,000. This new presentation was so important at that time that it was displayed at the World's Fair in Vienna in 1873 (Orth 1874). While earlier soil maps either reproduced essentially the texture of the Ap-horizon or were based on the geological conditions, for Orth the soil profile down to the parent material was the basis for the map. Thus the whole soil body and its variety in the landscape was examined and represented (Orth 1873, 1875, 1877b). With the publication (Orth 1877a), "Die naturwissenschaftlichen Grundlagen der Bodenkunde", Orth contributed greatly to the development of an independent soil science."

1879, 1884: The Natural Forms of Humus and the Birth of Pedology with the Danish Forester P.E. Müller

It was Müller in his noteworthy works (1879, 1884) who laid the present day scientific bases of the study on the different forms of humus, and even of a general survey of soil genetic processes in cold and temperate climates. His book, "The natural forms of humus" (1889), which included in French the works previously quoted, can be regarded as a treatise, still vadid nowadays, on the changes of brown soils to podzols (Feller et al., 2005; Jabiol et al., 2005). A few quotations of Müller's work (1889) are provided below (translated from French):

(p. 11/12) – *"In order to study the physical properties of a forest soil . . . the soil has been prepared as it is usually done when research is carried out on*

*the structure of an organism, with the same dissection and microscopy tech-
niques. The separate parts of the soil have been examined "in situ", in the
natural conditions of deposit, before calling up the chemical analysis to
explain whenever possible what the eye could not elucidate . . . At first sight,
it looks irrational to study a complex mixture such as an alluvial soil, as if
it were a whole, organized and homogeneous . . . But the study reveals
that the various conflicting actions did not lead to any chaotic but fixed
forms, . . . but to varied types of forest soil . . . considered as an organized
entity. As usual in nature, the different types are not clearly delimitated,
they intermingle; yet each may be considered as characterized by specific
traits."*

(p. 12/13) – *"Among all the different forms that can be encountered in the
soil of beech forests, I particularly studied two, very clearly contrasted in
character and practical meaning. The distinctions relative to the superficial
layers of beech forest soil should in my opinion be based on these two . . .
I propose to name the two principal types . . . of humus . . . from beech
forest "Mull" (mould)[1] and "Torf" (peat)[1]"*

(p. 14/15) *"Terreau de hêtre (Buchenmull)"*. Following the description of
the vegetation:

*"Soil aspects. The soil . . . is covered by a more or less thick layer of isolated
(plant) remains scattered on the soil. When the cover is taken apart, the dark-
ish brown surface . . . of the soil has a gravelly and granular aspect. The line
between the cover and the soil . . . is quite marked . . . This earth is as completely loose as . . . tilled . . . garden earth"*.

This is followed by a very detailed description of the soil profile with its different layers, microscopy (of coherent elements, in particular), mechanical and
chemical analyses, and by a long study (pp. 20–28) on the litter decomposing
organisms, mycelia and earthworms, with quantitative data:

(p. 29) *"Tourbe de hêtre (Buchentorf)"*. After a description of the vegetation,
very different from that of the "Mull", the soil aspect. *"If the soil is dug with
a spade, . . . a layer of tenacious black-brown humus, is found: the peat (mor);
then, beneath, a grey-white, grey or black-grey sand . . . the sand is clearer,
the farther from the peaty layer . . . (then, is found) . . . a layer of earth having
a red-brown or brown color . . . and under it, some sandy clay, some sand"*.
The grey sand is denominated *"bleichsand"* ("sable plombifère", plumb

sand[3]) and the brown-red layer "*Ortstein*". This is followed, with drawings, by a very detailed description of the tenacious and felted peat. He noted importance of mycelia and lack of earthworms (pp. 33–44)."

Transient formations, mould-peat (mullartiger Torf) are then described (pp. 45–64); in general, they do not contain earthworms but numerous insects and mycelia.

The different forms of humus, based on the existence of two stages in the humification process 1 – "*The mechanical division of organic remains*", and 2 – "*The mixture of organic remains and mineral earth*", (according to Müller), are then discussed.

The classification scheme proposed for humus present in beech forests is:

"*Echter* (Genuine) *Mull. Perfectly divided, loose without cohesion. Does not contain more than 10% organic substance, with no free humic acids . . . well mixed with the mineral earth . . . through the work of animals . . . and the action of water*".

"*Mullartiger Torf. Perfectly divided, loose without cohesion*". (Afterwards, idem to Echter Torf).

"*Echter Torf.* (Real mor) *Imperfectly divided, firm, tenacious, coherent. Contains 30 to 60% organic substance with free and soluble humic acids; imperfectly mixed with the mineral earth, owing, almost solely due to the action of water.*"

These few quotations only provide an incomplete idea about this exceptional work which came complete with colour and black-and-white plates of soil profiles, with indications of all the changes between brown soils and podzols, and the plant–soil relationships over a few meters (Figure 3). In addition, even the very first edition of 1878 attached the greatest importance to the biological processes, and especially soil fauna. In 1878, Darwin's book, *The formation of vegetable mould through the action of worms,* had not been published yet. It was abundantly quoted in Müller's second memoir (1884).[4] Müller's study opened up the era of biological Soil Science as far as uncultivated soils were concerned.

---

[3] "Sable plombifère" was the translation of "bleichsand" given by H. Grandeau in 1889. According to an anonymous referee who reviewed this paper, Grandeau confused "blei" (= plumb) with "bleich" (= colourless). Therefore "bleichsand" would mean "colourless sand".

[4] In Müller's first memoir (1879), Darwin is only quoted through another author (Fogh). It seems that Müller was not aware of the first short publication (1837) of Darwin, *On the formation of mould.*

Figure 3. Section of an oak "Krattbusch" and the soil below. Under the leaves layer
the superficial grey mould passes little by little to the yellow sand of the subsoil; outside the
bush, where the soil carries a heath vegetation, heat peat, plumb sand and Orstein have
been formed. Oak roots are roughlky drawn. (Müller, 1889, figure 14, page 159).

Conclusion

Great naturalists have always been good observers. Where soil studies are concerned, this rule still applies, and the instances selected for the present study are all the more interesting since most of the authors reported here were not necessarily specialists in the field – which incidentally accounts for the fact that these minor aspects of their works may have gone unnoticed. This does not prevent their studies from being quite impressive, for each chosen author may be considered as a great precursor in the matter of soil description and soil mapping. After them, the earth sciences had to wait until Dokuchaev (1883) to see the rebirth of soil description, soil profiles and soil mapping methods similar to those that are in use today.

Acknowledgement

Dr. Anne-Marie Carasson, Antilles-Guyane University (France) helped with translation into English.

References

Anonyme (Marquis de Turbilly), 1760. Pratique des défrichements. 151 p., Paris.
Bartram, W., 1791. Travels through North and South Carolina, Georgia, East and West Florida, the Cherokee Country, the Extensive Territories of the Muscogulges, or Creek Confederacy, and the Country of the Chactaws. James & Johnson, Philadelphy. (Reference given by Chatelin, 1991).

Bartram, W., 1988. Travels through North and South Carolina, Georgia, East and West Florida, the Cherokee Country, the Extensive Territories of the Muscogulges, or Creek Confederacy, and the Country of the Chactaws. Introduction by James Dickey, The Penguin Nature Library. (Reference given by Chatelin, 1991).

Bourde, A., 1967. Agronomies et agronomes en France au XVIIIème siècle. SEVPEN edition, 3 vol., Paris, 1740 pages.

Buchanan, F., 1807. A journey through the countries of Mysore, Canara and Malabar. (1998, reprint, Asian Educational Services, New Delhi), 3 vol., 440, 566 and 512 p.

Buffon, G-L.L., Comte de, 1819. Oeuvres complètes de Buffon, mises en ordre par M. le Comte de Lacépède, seconde édition, 30 vol. Rapet, Impr. Plassan, Paris.

Boulaine, J., 1989. Histoire des pédologues et de la science du sol. INRA, Paris, 285 p.

Chatelin, Y., 1991. Le voyage de William Bartram (1773–1776). Découverte du paysage et invention de l'exotisme américain. KARTHALA – ORSTOM eds., Paris, 291 p.

Darwin, Ch., 1837. On the formation of mould. Proc. Geol. Soc., 2: 274–276.

Darwin, Ch., 1881. The formation of vegetable mould through the action of worms with some observations on their habits. Murray ed., London, 298 pp.

Darwin, Ch., 1882. Rôle des vers de terre dans la formation de la terre végétale. Traduit de l'anglais par M. Levêque. Préface de M. Edmond Perrier. C. Reinwald Libr.-Ed., Paris, XXVIII + 264 p.

Delaillevault, 1783. Recherches sur la houille d'engrais et les houillères. 2 vol. (161, 215 p.), La Haye.

Dokuchaev, V.V., 1883. The Russian chernozem. A report to the Free Economics Society. St Petersbourg (Reference given by Krupenikov, 1992).

Feller, C., Brown, G.G., Blanchart, E., 2000. Darwin et le biofonctionnement des sols. Etude et Gestion des Sols, 7 (4): 395–402.

Feller, C., Boulaine, J., Pedro, G., 2001. Indicateurs de fertilité et durabilité des systèmes de culture au début du 19ème siècle: l'approche de Albrecht THAER (1752–1828). Etude et Gestion des Sols, 8 (1): 33–46.

Feller, C., Thuriès, L., Manlay, R., Robin, P., Frossard, E., 2003. «The principles of rational agriculture» by A.D. Thaer (1752–1828). An approach of the sustainability of cropping systems at the beginning of the 19th Century. Journal of Plant Nutrition and Soil Science, 166: 687–698.

Feller, C., Blanchart, E., Jabiol, B., Greve, M.H., 2005. Quand l'humus est à l'origine de la pédologie. 1. Les travaux du forestier danois P.E. Müller (1840–1926). Etude et Gestion des Sols, 12(2): 101–122.

Godefroy, 1937–38. Dictionnaire de l'ancienne langue française et de tous ses dialectes du IXe au XVe siècle. Paris, Libr. des Sciences et des Arts.

Jabiol, B., Feller, C., Greve, M.H., 2005. Quand l'humus est à l'origine de la pédologie. 2. Avant et après P.E. Müller: évolution des conceptions sur la desription et la typologie des «humus». Etude et Gestion des Sols, 12(2): 123–134.

Johnson, D.L., 1990. Biomantle evolution and the redistribution of earth materials and artifacts. Soil Sci. 149: 84–102.

Johnson, D.L., 1999. Darwin the archaeologist: A lesson in unfulfilled language. Discovering Archaeology Jan.-Feb., pp. 6–7. (also can been found at Human Oasis. http://www.humanoasis.com/da%20articles/darwinthearchaeologist.html)

Krupenikov, I.A., 1992. History of soil science. From its inception to the present. Oxonian Press, New Delhi, Calcutta, 352 p.

Kuhnoltz-Lordat, G., 1958. De l'amont à l'aval. Libraire des Estudiants, Montpellier, 165 p.

La Salle de L'Etang (De), 1768. Manuel d'agriculture pour le laboureur, pour le propriétaire et pour le gouvernement. Nlle Ed., P.F. Didot, 583 p., Paris.

Mückenhausen, E., 1997. Developments in soil science in Germany in the 19th century. In: Advances in GeoEcology Vol. 29, Reiskirchen.

Müller, P.E., 1879. Studier over Skovjord, som Bidrag til Skovdyrkningens Teori. Om Bögemuld og Bögemor paa Sand og Ler. Tidsskrift for Skovbrug, t. 3, 1.

Müller, P.E., 1884. Studier over Skovjord, som Bidrag til Skovdyrkningens Teori. Om Muld og Mor i Egeskove og paa Heder. Tidsskrift for Skovbrug, t. 7, 1.

Müller, P.E., 1889. Recherches sur les formes naturelles de l'humus et leur influence sur la végétation et le sol. Berger-Levrault et Cie, Paris-Nancy.

Orth, A., 1870. Die geologischen Verhältnisse des norddeutschen Schwemmlandes mit besonderer Berücksichtigung der Mark Brandenburg und die Anfertigung geognotisch-agronomischer Karten. Halle (Pltz), 64 p.

Orth, A., 1873. Die geologische Bodenprofil nach seiner Bedeutung für den Bodenwert und die Landeskultur. Berlin.

Orth, A., 1874. Bericht über die Bodenarten, Bodenkarten und bez. Geologischen Karten auf der Weltausstellung zu Wien 1873, erstattet an das Könoglich Preussische Ministerium für die landwirthschaftlichen Angelegenheiten. Landwirthschaftl. Centrabl. 22, 641–650.

Orth, A., 1875. Die geognostisch-agronomische Kartierung, m. besond. Berücksicht. der geolog. Verhältnisse Norddeutschlands u. der Mark Brandenburg erläutert an der Aufnhme v. Rittergut Friedrichsfelde bei Berlin. Vom landwirthschaftl. Centraverein d. Reg.-Bez. Postdam gekrönte Preisschrift. Berlin (Ernst & Korn), 176 pp. 22, 641–650.

Orth, A., 1877a. Die naturwissenschaftlichen Grundlagen der Bodenkunde. Landwirthschaftl. Versuchsstationen 20, 63.

Orth, A., 1877b. Rüdersdorf un Umgegend. Auf geognostischer Grundlage agronomisch bearbeitet. Mit einer geognostisch-agronomischen Karte.

Abhandlungen zur geologischen Spezialkarte von Preussen und den Thüringischen Staaten 2, 114 p. (Berlin).

Palissy Bernard, 1880. Oeuvres complètes. Avec une notice historique et bibliographique et une table analytique par Anatole France. P. Charavay Fres Ed., 499p., Paris.

Pattulo, 1759. Essai sur l'amélioration des terres. Paris, chez Durand, 285 p.

Rossignol, J.P., 2001. Granville. In "Lettre de l'Association Française pour l'Etude des Sols" (AFES), 59 (juin 2001): 15.

Thaer, A. 1809. Grundsätze der rationnellen Landwirtschaft (1809–1812). Realschulbuch Ed., Berlin.

Thaer, A. 1811–1816. Principes raisonnés d'agriculture. Traduit de l'Allemand par E.V.B. Crud. J.J. Prechoud Ed., Paris.

Thaer, A.D., 1856. The Principles of practical agriculture. Translated by W. Shaw and C.W. Johnson. Saxton and Company Publ., New York, 551 pages.

TLF (Trésor de la Langue Française), 1981. Dictionnaire de la langue du XIXe et du XXe siècle. Editions du CNRS, Comm. Pers., Paris.

Trevoux, 1752. Dictionnaire de Trevoux, 7 volumes, Paris.

Van Doren, M., 1955. Travels of William Bartram. Dover Publ., Books on Travel and Adventures. (Citée de Chatelin, 1991).

Yaalon, D.H., 1989. Forerunners and founders of pedology as a science. Soil Science, 147: 225–226.

Yaalon, D.H., 2000. Down to earth. Why soil – and soil science – matters. Millennium essay. Nature, vol. 407, 21 September 2000, www. Nature.com 301.

5

Souls and Soils: A Survey of Worldviews

Verena Winiwarter[1] and Winfried E.H. Blum[2]

Introduction

No thinking person can live without forming some views about life and the world (Mbiti, 1996: 174). Humans perceive the world around them and try to make sense of it in terms of their daily existence as well as for their entire life and that of their community. In doing so, they have to define themselves in their relation to non-human nature. Anthropological theories have long pointed out that the impetus to define oneself in context and relation to nature yields systems of bringing order into nature's chaos (e.g. Douglas, 1966). The task to structure the world around oneself is immense. Therefore, all worldviews are essentially an attempt to reduce the complexity one has to deal with.

For the better part of human history, myth was the central tool to create order and to reduce complexity, in other words to yield a worldview which would enable humans to simplify their life by structuring it and narrating it in a way that could make sense – the latter being a central need for the complex brain humans evolved to have (Schmidt, 1997). Myth is different from history in that its time is out of the reach of humans – myth is before them, in a sense that makes myth unreachable by humans in their (historical) time.

Anthropologists have argued that the privileged theme of myth is the conflict between nature and culture and its resolution in mediation (Levi Strauss quoted in Herren, 2002: 2). In structuralist theory, the nature–culture conflict projected onto the screen of human imagination is a reflection of the very instrument that projects it. The human being belongs to nature, yet does not, and it is this dialectic which is mirrored in the nature–culture conflict played out in myth (Herren, 2002: 2). When dealing with the cultivation of the soil, we are dealing with this central fissure and its reverberations in human minds.

Myth was exchanged for other forms of interpreting the world, but this essential fissure has not been and cannot be overcome. There are several ways how this fissure

[1] APART – Fellow of the Austrian Academy of Sciences.
[2] University of Natural Resources and Applied Life Sciences (BOKU), Vienna, Peter Jordan-Str. 82, A 1190 Vienna, AUSTRIA.

can be conceptualized, among them e.g. a concept of nature and culture being like image and mirror (Latour, 2001). Worldviews are the main frame of interpretation and the main filter of perception humans use for decision making and an important context in which knowledge production happens. Worldviews are also important to determine the authority, which is granted to knowledge. Knowledge can for example be granted high authority because it is part of a holy scripture, or because it comes from a group of important people with academic degrees at important universities. What interests us here is that worldviews comprise ideas about soils and concepts of how to deal with them. As interpretative frameworks of knowledge they are of interest to the student of the history of soil sciences.

Belief systems are connected to worldviews. They are one way in which humans define their place and role in the world, part of their social universe, and offer a particular way of dealing with complexity. When it comes to religious aspects of the human relation to soils, two major kinds of religion need to be distinguished, as has been suggested by Michael Pye: "The first major type of religion is the natural religion of a particular society. As regards the social aspect, it is very clear that such religions are there for all members of the society in question. Usually they are more or less obliged to take part in it, ". . . all small-scale societies have such a religion, and for them its main features are a myth of origin, care for the ancestors, rites of transition, and calendrical economic rituals." Here the conceptual and behavioral aspects of religion are to be found. As far as the subjective aspect is concerned, the most important feeling is the sense of relation with the deities who are responsible for what happens in the world, whether beneficial or threatening. Thus, this type of religion is, in the aspect of subjectivity, correlational. This kind of religion may be called "primal religion". The expression "primal" means that historically it has a certain priority, and that the basic aspects of religion can all be found within it.

The second major type of religion can be given various names. We might call it "salvation religion", "critical religion" or "guidance religion". These religions are founded by special religious leaders who have a distinctive message to give, against the usual assumptions of the society in which they live. The major religions of the world belong to this type. But so, also, do a very large number of smaller religions which preach a special way of life to their members. During the complex history of humankind, Buddhism, Christianity and Islam established themselves as major religions. As a result, they adapted themselves to various social and political pressures and adopted some of the functions of primal religions" (Pye, 2005).

Over the long run of human history, when primal religions were dominant, soils were conceptualized as part of a world which was controlled by higher, unseen powers. Soils thus were bearing immaterial qualities, they were – to some extent – objects of religious reverence and corresponding ritual practices to ensure their sustained fertility existed.

The Current Approach: Environmental Ethics

Cultural differences in terms of soil as an object of worship and belief, and the consequences such an attitude has on behavior are an interesting facet of comparative approaches. Such approaches have quite a long intellectual history. Belief has been advocated as being of explanatory power for the actions towards nature especially by Lynn White, Jr., who argued that in order to successfully address the emerging environmental crisis, humans needed to focus on their attitudes toward nature. Ultimately, the essay concluded, our attitudes toward nature are rooted in our religious beliefs. As White expressed his conviction, "What people do about their ecology depends on what they think about themselves in relation to things around them. Human ecology is deeply conditioned by beliefs about our nature and destiny – that is, by religion." (White, 1967: 1205).

Although White's claim has been refuted since (first, and convincingly, by Yi Fu Tuan, 1968), the connection between belief systems and practices has remained an important topic of study in the field of environmental ethics. Environmental ethics can be defined, in very general terms, as efforts to articulate, systematize, and defend systems of value guiding human treatment of and behavior in the natural world. Philosophical and religious reflection on human obligations towards nature or "otherkind" has a long pedigree in human cultures, whether occidental, Asian, or indigenous. "Environmental Ethics as a distinctive sub-field within western philosophical and religious ethics, however, did not emerge until the last three decades of the twentieth century." (Taylor, 2005).

The Journal "Worldviews: Environment, Culture, Religion", which commenced publication in 1997, provides a forum for in-depth discussion of related subjects; a more practical counterpart, the "Journal of Agriculture and Human Values", offers case-studies and focussed discussions about agricultural impact on the environment and its cultural prerequisites. Other journals which have been consulted on the journey through current discussions of the interface between science, nature and spirituality in the broadest sense are the journal "Ecotheology", a specialist journal for this emerging field, and the magazine "Resurgence", which was started over 30 years ago and deals with current problems on the verge between politics, ecology and spirituality, focused on, in their own words, "the theory and practice of good living". One keyword which has to be mentioned in the context of the magazine is "Deep Ecology". This environmental movement was initiated by a Norwegian philosopher, Arne Naess, in 1972 and continues to be the focal point of much value-oriented, egalitarian environmentalism. Like most other publications referred to, the "Journal of Agricultural and Environmental Ethics", now in its 18th year, devotes only a small part of its pages to discussions of soils, often the more theoretically oriented publications deal with creation as a whole, the more practical ones increasingly deal with what seems to be perceived as the most pressing problem of today, genetically modified organisms.

Many books could be mentioned which deal with aspects of the greater subject of a spiritual engagement of humans with nature. Robert Gottlieb's anthology "This sacred earth" is particularly useful, as it contains many classical pieces of ecospirituality and a survey of texts from major religions together with scholarly discussion (Gottlieb, 1996). Some of the older articles contain outdated information, but the collection as a whole serves its purpose well enough. Gottlieb has co-published a follow-up to this collection, on the connection between Deep Ecology and world religions (Barnhill and Gottlieb, 2001). Following a series of conferences, Harvard University Press produced a series of edited volumes which cover ecological aspects of Buddhism, Christianity, Confucianism, Daoism, Hinduism, Indigenous Traditions, Jainism, Judaism, Islam, in more detail than the collection by Barnhill and Gottlieb (Tucker and Ryûken Williams, 1997; Tucker and Berthrong, 1998; Hessel and Radford Ruether, 2000; Key Chapple and Tucker, 2000; Girardot, Miller and Xiaogan, 2001; Grim, 2001; Key Chapple, 2002; Tirosh-Samuelson, 2002; Foltz, Denny and Baharuddin, 2003).

The articles in these volumes make clear among many other things, that each of the major religions consists of several denominations, which can be quite different in their approaches to nature. How these differences play out in practice has been shown in detail for the U.S.A. by environmental historian Mark Stoll, who studied the intersection between Protestantism and capitalism (Stoll, 1997).

Soil scientists' engagement with spiritual aspects is documented in an edited conference volume published first in French and more recently translated into English (Lahmar and Ribaut, 2001). The Worldwatch Institute also published a short report about the role of religion and spirituality for sustainable development (Gardner, 2002). Generally it can be said that the field of eco-spirituality has been teeming with life in recent years.

Soils in Belief Systems

Rather than talking about ecology and religion in general, we wish to concentrate on the more specific role of soil in belief systems. By describing the various ways how soil and reverence to it were and are part of worldviews over the course of history, we hope to make clear that the spiritual quality of the relation between humankind and soil is – despite the lack of a simple trajectory from attitude to behavior – pertinent to the knowledge created about them and the authority such knowledge is granted. A complete understanding of the cultural interpretation (culture taken as symbolic system in the broadest sense as one part of culture-nature interaction (see M. Fischer-Kowalski and Weisz, 1999)) of soils remains incomplete without the spiritual dimension.

While spiritual knowledge today does not carry the authority it had in primal religions, religious leaders have become increasingly aware of the present ecological crisis and try to offer spiritual guidance for a more sustainable future. For each religion, its newly defined role as an actor in the ecological realm poses a set of theological challenges. To give but one example, for Christians, this is the idea of a personal salvation – that the righteous can be saved, but the rest of creation cannot. Such a doctrine is inherently anti-ecological. But all religions can change, as they can change the current interpretation of doctrine, they even can change rituals and practice to some degree. One can assume that the interaction between ecology and religion will change both science and religions. While modern eco-theology and environmental ethics use the past, or indigenous peoples' belief systems as traditions to refer to, it is quite clear that there is no way back, a re-enchantment of the world is beyond the possible. In a functionally differentiated society, in which realms such as scholarship, politics, the economy or the health system have become incommensurable, religions, like all other subsystems, are not able to re-integrate society.

But by using traditions creatively, they can at least point to the difference between the worldviews then and now, or there and here, and thus help in developing alternatives. The currently dominant trend to commodify ever larger parts of the earth is somewhat counterbalanced by the surge to re-spiritualize our attitude to nature. In bringing together the insights of comparative studies of religion, the study of mythology, and the history of philosophy with the recent environmental ethics debate, we hope to present an interpretatory framework useful for those concerned with studying soils scientifically.

The disenchantment of the world, as the current situation is described, has never really happened fully. The famous 19th century poet Walt Whitman, to name but one example, wrote the following lines, which give testimony to a mystical, emotional bond between humans and earth, much like we assume the dominant paradigm of primal religions adhering to Mother Earth had been:

> Smile O voluptuous cool-breath'd earth!
>
> Earth of the slumbering and liquid trees!
>
> Earth of departed sunset – earth of the mountains misty-topt!
>
> Earth of the vitreous pour of the full moon just tinged with blue!
>
> Earth of shine and dark mottling the tide of the river!
>
> Earth of the limpid gray of clouds brighter and clearer for my sake!
>
> Far-swooping elbow'd earth – rich apple-blossom'd earth!
>
> Smile, for your lover comes.

(Walt Whitman, (1819–1892) excerpt from the poem 'Song of Myself', published in the collection 'Leaves of Grass', several editions and revisions from 1855 to 1882.)

Examples of poetry along similar lines could be given e.g. from Christian Morgenstern (1871–1914), whose poem "To Mother Earth" also talks about a mystical reconnection to the earth.

Gaia, the Earth Mother, has also inspired a scientific approach. James Lovelock and Lynn Margulis developed the Gaia concept, arguing that the earth as a whole has feedback-mechanisms and a tendency to homoeostasis which can only be understood if the entire planet is seen as one super-organism (Lovelock, 1979). The concept has been criticized, but it is widely acknowledged that it has allowed several crucial ecological insights, e.g. into the coupling between oceans and the atmosphere. Whether its success is really due to the name Gaia, as has been argued (Spangler, 1996: 612), is not our concern. The concept of Gaia, a planetary consciousness, is surely an important development which has been willingly connected to mythical origins, making both the myth and the concept better known, and thus might serve as an example of the mutual benefit science and myth can have from getting in touch.

On Comparing Worldviews

Scholars specializing in the comparative study of religion have developed systematic frameworks for description and interpretation of religions. Almost 50 years ago, Ernst Benz, professor of church history and historical theology and Director of Ecumenical Studies at the University of Marburg in Germany wrote an essay on the understanding of non-Christian religions (Benz, 1959). In it, he set out to explain how his experience travelling and teaching in East Asia and India had made him understand more about fundamental differences in religious practice and belief between Asian religions and Christianity. His essay is written with great respect for non-Christian religions and their spiritual practices.

Benz talks about the difficulties in translation. He especially talks about the problems of translating ontological terms into other languages, using English as an example to elucidate the much greater difficulties one will have to translate from and into more distant languages. He draws attention to the difficulty language produces, in an essay written in German and translated into English for publication.

Readers of the present chapter should be aware that it was written by two native speakers of German with a working command of English, who decided to write the essay in English rather than having a German version translated. We used material translated from a wealth of other languages as sources for this article. For each of them the same cautionary note about translation being inherently interpretation has to be made.

While religions vary in their avowed goals and in the way in which they conceive of and present the meaning of their activities, there is a basic structure which can be found in every religion. According to Gottlieb (2001:17f), religions are "organized and overlapping systems of belief, ritual, institutional life, spiritual aspiration, and ethical orientation which are premised on an understanding of human beings as other or more than simply their purely social or physical identities. ". . . Religion serves as an alternative to understandings of the human identity that center on the social successes of money, fame, political power, career achievement, or community acceptance. Religions also provide guidance that seeks to root everyday moral teachings in the ultimate nature or significance of a spiritual truth about who we really are. Religions therefore direct us toward particular ways of living with other people and with the world. Finally, religions provide rituals – acts of prayer, meditation, collective contrition, or celebration – whose goal is to awaken and reinforce an immediate and personal sense of our connection to the Sacred". To understand religions fully, we would need to consider their conceptual, behavioral, social and subjective aspects.

This paper focuses on conceptual aspects, of which one can, according to Pye, discern four types: Narrative concepts such as myth and legend, focusing concepts of the divine or "the numinous", underlying value concepts such as karma, justice and love, and interpretative concepts such as "orthodoxy", "guidance", or "consensus" (as in Islam) (Pye, 2005). In his discussion of paganism as root-religion, Michael York offers yet another typology. For him, four types can be discerned, these are Abrahamic (Judaism, Christianity and Islam), dharmic (Hinduism, Buddhism, Jainism), secular (ideologies and philosophies such as Marxism, atheism or materialism) and pagan (York, 2004). Why such frameworks are important becomes clear when trying to compare evidence: A soil goddess or god cannot be understood using the template of a personalized god, as Christ is one. Understanding has to come from an acceptance of the numinous and its personalization – accepting that the numinous goddesses and gods differ substantially from Christ, Godfather and the Holy Spirit, the Christian trinity.

Not only does one have to be aware of the difficulties in translating ontological terminology, one has also to bear in mind the conceptual differences of the entity called "god", "spirit", and, of course, "soil". In the following, we highlight a few important developments in the immaterial relations of humans and soils, rather than even trying to be encyclopedic.

The self-ascribed role of humans in mythically grounded worldviews seems more benign to us than the role humans ascribe themselves in a scientific world view. But as David Hillel has pointed out, the facts speak against such an interpretation (Hillel, 1991: 3). Human history is a history of intervention into natural systems, and neither taboos nor rites of connection had a lasting effect in mediating the long-term impact of human labor input into natural systems. If a species diverts energy from its environment, it does have an impact. By means of technology, we have greatly enhanced our power to divert ecosystem energy for our causes, but even the first agriculturists left their quintessential footprint on the soil. How they incorporated their experience with soils into their traditions and worldviews is discussed in some examples in the next section.

Earth as Part of Belief Systems

In cosmogonies the earth appears in two different forms, (a) as mother earth, deified, but as a numinous deity, with its body and organs as the different parts of nature and (b) as an element which is considered a basic constituent of natural bodies and humans alike. The mother earth goddesses do not render a cult of the soil directly, and creation myths do not refer to the soil. Tillers of the soil come late in many cosmogonies, as a second people living on the earth.

Nevertheless, earth plays a central role: The Greek Aristotelian world system knew four elements: earth, water, air and fire, its Chinese counterpart has five constituents, metal and wood, and again, earth, water, and fire. The Buddhist mahabhuds are water, soil, fire, air and space (Intrigrinova, 2004: 30). While the difference between soil and earth is principally of great importance, in the above list it is probably an artifact of translation. As food ultimately can only come from the earth (birds and fish not withstanding), the complexity of terrestrial ecosystems had somehow to be incorporated into the systems and structures in which the world was ordered. A worldview completely without earth is quite improbable, and none has come to the authors' attention.

In Greek cosmogony, the first beings, living in mythical time before agriculture, are the giants. Gaia (Gea), the earth as goddess, according to Greek mythology the one being who emerged from primeval Chaos, and Poseidon, the Greek god of the sea are the parents of the Giant Antaeus (Antaios), whose story is part of the myth of the great hero Herakles. Antaeus challenged all travelers passing his realm to a wrestling match. Prior to his encounter with the hero, he had always won these matches. He was slain by Herakles because the hero lifted the giant off the ground. Previously, whenever he had fallen to the ground in battle, the Earth, his mother, had regenerated his powers. The earth plays a powerful role in this myth, the earth as embodied in the giant's mother Gaia (Graves, 1982).

The concept of earth's healing powers has not been completely lost in European folklore over the millennia, even if Gaia was no longer worshipped as such. In German folk belief, newborn babies were ritually placed on the earth to gain strength, similarly for sick people with the same reason. Magicians and sorcerers, much like Antaios were believed to being able to escape when they were allowed to get in touch with the earth, so after capturing them, they had to be suspended above the ground. Swallowing a piece of earth was believed to heal hiccups, and after a snake bite the affected limb was put into fresh earth. A Germanic Goddess called Nerthus was considered similar to Terra mater, the Roman earth Goddess, by the Roman author Tacitus in his account of the Germanic peoples. The encyclopedia of German folk belief, from which the above examples are taken, gives also an account of soil fertility rites, mentioning e.g. the procession of naked women around the fields to ensure their fertility and refers briefly to the ritual intercourse of a chosen human pair to ensure fertility, a ritual which has been described by James George Frazer in his "Golden Bough" (Bächtold-Stäubli, 1987, Vol. 3: 895–908; the reference is to Frazer, 1922, Chapter XI, The Influence of the Sexes on Vegetation). Much of the available literature on Earth Goddesses is not of purely scholarly nature. Manfred Kurt Ehmer's small, but well illustrated book on the Goddess Earth (Ehmer, 1994) is useful for its discussion of various mother earth cults from Ancient Greece to India, and for the connections the author draws to philosophy and alchemy. Eco-feminist approaches can be found in Gottlieb's edited collection (Gottlieb, 1996), and Tikva Frymer-Kensky (1991) offers very serious treatment of the transformation of fertility goddesses into the biblical tradition.

Tilling the Soil as Serving the Deities: The Example of Hesiod's Works and Days

The practices of dealing with soil in terms of procuring a living are coupled with the spiritual dimension it is given. The history of immaterial aspects of soils in the Mediterranean shows that while this link exists, it is possible that different religious systems develop under similar agricultural regimes.

We must be careful even with historical evidence from our own past. Stephanie Nelson cautions that changes of time and place, differences of language, and the loss of assumptions shared by reader and audience can obscure a literary text (Nelson, 1998: ix). Her analysis of one of the oldest works containing agricultural practices, Hesiod's Works and Days, overcomes this difficulty very well. It is an example for the role of soils, or even more precise, of tilling the soil, in a primal religion's worldview. Nelson convincingly shows that Hesiod's poetic program is based on an immanent god by analyzing both his "Theogony" and the "Works and Days". The divine will is not discovered by reading sacred or philosophical texts, but rather in the day-to-day world in which Hesiod lives. The life of people manifests Zeus' will. What succeeds in the world is what Zeus has prescribed for humans. Farming, which is the necessary way in which humanity procures food, is, therefore, necessarily also the will of Zeus (Nelson, 1998: 63). Farming knowledge in such a belief system is knowledge necessary to perform the will of the divine beings, and has, therefore, a sacred dimension.

Hesiod does not describe soil altars or rituals, to him human life conduct is the main vehicle of worship. Tilling the soil well is a prerequisite also for spiritual fulfillment.

Many other primal religions harbor similar concepts, in some cases, especially with Nomadic people, they can be of opposite character: Touching the soil with an implement can be forbidden, and a spiritually fulfilling life implies not to touch the soil for planting (Intriginova, 2001). The case of Greek mythology is particularly interesting, as Greek philosophy of nature is the basis of the worldview which lead to the Cartesian revolution and ultimately, via technology, to the present ecological crisis. Usually these two are not considered together, but this is due to scholarly traditions rather than to them having been separated completely in Ancient Greek society. Both might well have a common ground in Pre-Socratic philosophy. G.E.R. Lloyd has drawn attention to the fact that we know the Pre-Socratic writings mostly as fragments in newer texts and not by themselves. They are rendered by those who refer to them in the light of the already developed theory and thus inconsistencies with the new system are downplayed.

But here we are interested in traces of a special treatment of the earth, or soils, as a possible connection to (spiritual or practical) agricultural experience. The philosopher Anaxagoras (ca. 500 B.C.E.) is said to have been the first to develop the scheme of the four qualities hot, cold, wet and dry, from which the qualities of the four elements are derived and which lies at the very heart of Greek nature concepts. In his Fragment 4, he specifically mentions these four opposites, along with the bright and the dark, "much earth" and "innumerable seeds" as part of the primeval mixture, when all things were together. Earth does have a special place in this cosmogony. Empedocles, in the 4th century B.C.E. gives the earth none of the four qualities, but rather identifies it with solidity, not unlike the earlier Anaximenes, who relates hot and cold to differences in density (rare/dense) (Lloyd, 1964: 92, 93, 95). Both Anaximenes and Empedocles, by describing qualities of the earth which are relevant to agriculture (density and solidity), rather than trying to abstract from it completely, to render it as one of the four elements which are the basis of later philosophy, allow us a glimpse of how an abstract worldview such as Greek natural philosophy was developed on the basis of a more practical relation with the earth. For an overview of Greek philosophy relating to the environment, see Bowler (1992: 39–65).

The various deities governing the conduct of agriculture in Greek and Roman mythology are part of the system Hesiod describes, a system in which the conduct of agriculture as the main worship to the Gods was also perceived as being at their mercy. The Greek/Roman Ceres/Demeter is but one example for a powerful divine representation, and the pantheon around agricultural operations was very differentiated in Ancient Rome. Agricultural Goddesses and Gods, while connected to the soil, are different from Earth Goddesses as they do not symbolize the entire universe or the earth, and they are also different in that they are most closely connected to the harvest and not to the fertility of the soil as such. They can be understood in the context of the primal religion of which they are part.

As has been pointed out in an overview by Kinsley, the Christian tradition has been considered as ecologically harmful in general, whereas its development into such a thought system is relatively young, and can be dated to the sixteenth to eighteenth centuries (Kinsley, 1995: 112). Kinsley also briefly refers to the medieval tradition. Only very specialized research has been done on the transmission of a spiritual engagement with the natural world in the Middle Ages (Winiwarter, 2005). Medieval monks saw it as their task to investigate the wonders of creation, to derive from that insight the singularity of the earth and all its beings for the praise and glorification of god (Hünemörder, 2003: 124). Only the female Christian mystic Hildegard von Bingen has received some recognition in this respect so far (Blume, 2003: 14f). Much is yet to be learned about the European, and also about the Jewish tradition (but see e.g. Santmire, 1985, for Christianity, or Hiebert, 1996 for Early Israel. Relevant biblical quotes and some of the Talmud, too, can be accessed in Gottlieb, 1996). Of particular interest as an ecological concept is the Sabbat, the 7th day on which nothing in nature must be changed, a day of rest for nature, so to say, and the 7th year, in which fields shall rest.

Ba'al-Worship as an Example of Place – Based Deities

Semitic religion prior to the advent of Judaism was the cult of Ba'al. The supernatural powers most obvious to the imagination of Semites were those which were supposed to supply food and drink. Gatherings and settlements were made where the soil was most inviting; that is, where it was perennially productive. Such districts were regarded as being fertilized by divine agency, and as each of them had its own divinity or demon as the "owner" of the soil, such a being was called its "Ba'al". The usage, having thus begun in agricultural settlements, was transferred to the sites of cities, all of which were in any case founded under religious auspices. There was a multiplicity of Ba'als, reflected in place names such as Baal-hazor, Baal-hermon, Baal-meon, Baal-perazim, Baal-shalisha, Baal-tamar, and Baal-zephon. Here the fertile soil is directly referred to as a divine gift (of a specific place).

Only in a later stage of development was the Ba'al of a place assigned a more abstract character as a divinity of wider functions as Baal-berith, Baalzebub. A further step was taken when the name was used absolutely of a god Ba'al without qualifications, used, for example, in antithesis to Yhwh and as the second element in names of persons, in such forms as Ish-baal ("Man of Ba'al") or Hannibal ("Favor of Ba'al") (Jastrow et al., 2004).

Place-based deities, owners of the space of a people, are common in many religions. They are responsible for fertility, but their connection to soil is indirect, they are associated with the fruit of the land rather than with land as such. The Roman *Penates* and *Lares* can also be subsumed under the category of place based deities. They are both part of ancestor worship and a place-based (House-God) idea of bringing fertility and luck. We find shrines to ancestors and the spirits of a given home in many belief systems, but their relation to soil is again an indirect one.

Chinese Soil Worship

Almost the only society which had a fully developed soil worship is traditional Chinese society. China's soils are prone to erosion today, and even in historical times erosion and other degradation was widespread. The worship of the soil did not keep people from destroying it, unintentionally, through their efforts to procure a living from the soil. Their cult is nevertheless very interesting, as it is in some respect a combination of an elemental approach and a localized fertility god. Soil altars were made on a small scale, and integrated into the greater belief system of Daoism. The best known part of this worship, and a major tourist attraction in Beijing, is the Altar of the Earth and Heavens in Sun-Yat-Sen Park, dating back to the 14th century. In it, hard-packed earth in five colors, yellow in the middle, green to the east, white to the west, red to the south and black to the north is at the center of worship. It has been remarked that the five colors chosen mirror the basic soil distribution of China. The sacrifices in the temple were held by the emperor, but earth altars were built everywhere by heaping some earth, then putting a handful of loose earth on top, adding a sorghum root and praying for good harvest. At the prefecture seats, altars similar to the imperial one were built and the sacrifices for good harvest there were made with more pomp. The God of the Earth, one of the mythical prehistoric sovereigns of China, Ju Long, is described as an expert in pedology, being able to determine what the right plants for each type of earth would be. Jens Aaberg-Jørgensen can show in his study of Hongkeng, a village in the southern Chinese province of Fujian, that the worship of the soil is still alive in rural areas (Aaberg-Jørgensen, 2000). The soil altar of Hongkeng is placed in the vicinity of a large tree under which communal activities take place.

As in all other regions of the world, conquering peoples in China forced their religion on local people. A 12th century gazetteer in the same province of Fujian tells of the previously anonymous gods of the soil, which were given anthropomorphic representations in the Tang-Song period, due to conquest. Together with their changed shapes, the places of worship changed from outdoor altars such as the one described above for present-day China, to temples. Indigenous cults themselves varied and often very local, were molded into the larger pantheons of rulers, showing both the flexibility and the stability of such belief systems (Szonyi, 2000). A short account of the role of soils on Confucianism can be found in Zhao (2001).

Chinese religion influenced Japan. In the 8th century, the mythology on which the Japanese Shinto religion is based, was compiled. There are 10 deities in Shinto which are connected to various aspects of the soil. The soil on which someone is born is considered especially relevant. Ritual practice in Shinto shrines continues until today, and reverence is also made to the 10 Kami (Motegi, 2001).

Apart from China and Japan, soil worship can be traced also in India, celebrated at least in parts of India on a special day: Dhulivandan. Its name comes from the ashes resulting from the bonfires which are part of the celebrations, and the name is derived from worshipping "the ashes" or "the dust", as a basis for fertility. In the region of Maharashtra, on this day one has to play with colors and the soil, to celebrate the new season and new crop. At some places people especially the agriculturalists worship the soil, to welcome the new season and prosperity. But like many other studies of this kind, the book by Gosling on Ecology and Religion in India and South-East Asia does not discuss soils in detail (Gosling, 2001).

Summary

In this brief survey of religious systems, different cultures through history are seen to have connected souls and soils in their worldviews. It has become clear that in the cases of earth goddesses the principle of fecundity was the focus of worship. They stood for all nature, for the rejuvenation of plant and animal life in the cycle of the seasons. Earth goddesses were not soil goddesses in particular, but embodied nature goddesses. Place-based deities such as the Ba'als are no soil gods either, again, fertility (soil fertility) was conceptualized as part of the qualities of a particular place. In Hesiod's account of farming the principal difference between primal religions and salvation religions has become clear: Worship as a way of living, in Hesiod's case, of tilling the soil, is primarily a feature of primal religions. Traditional Chinese society seems to be the only major group of people who developed a form of soil worship in which the soil itself was the object of reverence.

Soil erosion and degradation have become a theme of religious considerations (variously termed environmental ethics, deep ecology or ecotheology) only recently, in the past forty or so years. Soil scientists and their results have influenced recent world view oriented practice and theory, as can be seen by browsing through the literature mentioned at the beginning of this chapter.

In this literature, a difference is made between biocentric and ecocentric ethics, and as yet soils seem to be perceived primarily as part of ecocentric approaches. This has to do with the fact that their public perception as ecosystems teeming with life in enormous diversity is still in its infancy. For the future co-operation between scientists and those concerned with eco-theology, a focus on the soil as a living system could be a way forward.

Apart from that, looking at worldviews should also be a reflexive exercise. Like all human beings, soil scientists cannot live without forming some views about life and the world. Their work is influenced by their world-view, too.

References

Aaberg-Jørgensen, J., 2000: Hongkeng – A village in Fujian, In: Arkitekten 28, 2000, 2–9. An updated, English version is available on the www: http://www.chinad-welling.dk/hovedsider/hongkeng-tekst.htm.

Bächtold-Stäubli, H. (ed.), 1987: Handwörterbuch des deutschen Aberglaubens (10 Bände) – Berlin 1927–1942. Reprinted 1987 with an Introduction by Chistoph Daxelmüller. Walter de Gruyter, Berlin, New York.

Barnhill, D., Gottlieb, R. (eds.), 2001: Deep Ecology and World Religions: New Essays on Sacred Ground. SUNY Press, Albany.

Benz, E., 1959: On Understanding Non-Christian Religions. In: Eliade Mircea, Kitagawa Joseph M. (eds.), The History of Religions. Essays in Methodology, University of Chicago Press, 115–131.

Blume, H.-P.: Die Wurzeln der Bodenkunde. In: Handbuch der Bodenkunde 15. Ergänzungslieferung, Lieferung 5/03.

Bowler, P.J., 1992: The Fontana History of the Environmental Sciences. Fontana Press, London.

Douglas, M., 1966 (1991): Purity and Danger. An Analysis of the Concepts of Pollution and Taboo, Routledge, London.

Ehmer, M.K., 1994: Göttin Erde. Kult und Mythos der Mutter Erde. Ein Beitrag zur Ökosophie der Zukunft, Verlag Clemens Zerling, Berlin.

Fischer-Kowalski, M., Weisz, H., 1999: Society as Hybrid Between Material and Symbolic Realms. Toward a Theoretical Framework of Society-Nature Interaction. In: Human Ecology 8:215–251.

Foltz, Richard, C., Denny, F.M., Baharuddin, A. (eds.), 2003: Islam and Ecology: A Bestowed Trust. Harvard University Press, Cambridge.

Frymer-Kensky, T., 1993: In the Wake of the Goddesses. Women, Culture, and the Biblical Transformation of Pagan Myth, Ballantine Books, New York.

Gardner, G., 2002: Invoking the Spirit: Religion and Spirituality in the Quest for a Sustainable Worldwatch paper #164, ISBN: 1–878071–67–X.

Girardot, N.J., Miller, J., Liu, X. (eds.), 2001: Daoism and Ecology: Ways within a Cosmic Landscape. Harvard University Press, Cambridge, MA.

Gosling, D.L., 2001: Religion and Ecology in India and South East Asia, Routledge, London.

Gottlieb, R. (ed.), 1996: This Sacred Earth: Religion, Nature, Environment, Routledge.

Graves, R., 1982: The Greek Myths, Doubleday, Garden City, NY.

Grim, J.A. (ed.), 2001: Indigenous Traditions and Ecology: The Interbeing of Cosmology and Community, Harvard University Press, Cambridge, MA.

Herren, M.W., 2002: Nature and Culture in Mesopotamian and Greek Myths. In: Thomas M. Robinson, Laura Westra (eds.), Thinking about the Environment. Our Debt to the Classical and Medieval Past. Lanham, Boulder, New York, Oxford: Lexington, 3–15.

Hessel, D.T., Radford Ruether, R. (eds.), 2000: Christianity and Ecology: Seeking the Well-Being of Earth and Humans, Harvard University Press, Cambridge, MA.

Hiebert, T., 1996: The Yahwist's Landscape: Nature and Religion in Early Israel. Oxford University Press.

Hillel, D., 1991: Out of the Earth. Civilization and the Life of the Soil, Aurum Books, London.

Hünemörder, C., 2003: Traditionelle Naturkunde, realistische Naturbeobachtung und theologische Naturdeutung in Enzyklopädien des Hohen Mittelalters. In: Dilg, Peter (ed.), Natur im Mittelalter. Konzeptionen – Erfahrungen – Wirkungen. Akademie-Verlag, Berlin, 124–135.

Intigrinova, T., 2001: Les Bouriates et le sol: des rapports en évolution. In: Lahmar, Rabah; Ribaut, Jean-Pierre, Sols et sociétés: regards pluriculturels Paris: Charles Léopold Mayer, 57–63.

Jastrow, M., McCurdy, F.J., and McDonald, D.B.: Jewish Encyclopedia.com: http://www.jewishencyclopedia.com/view_friendly.jsp?artid=2&letter=B).

Key Chapple, C., Tucker, M.E. (eds.), 2000: Hinduism and Ecology: The Intersection of Earth, Sky, and Water, Harvard University Press, Cambridge, MA.

Key Chapple, C. (ed.), 2002: Jainism and Ecology: Nonviolence in the Web of Life.

Kinsley, D., 1995: Ecology and Religion: Ecological Spirituality in Cross-Cultural Perspective, Prentice Hall, Englewood Cliffs, N.J.

Lahmar, R., Ribaut, J.-P., 2001: Sols et sociétés: regards pluriculturels: Charles Léopold Mayer, Paris.

Latour, B., 2004: Politics of Nature – How to Bring the Sciences into Democracy (English translation by Catherine Porter), Harvard University Press, Cambridge, MA.

Lloyd, G.E.R., 1964: "The Hot and the Cold, the Dry and the Wet in Greek Philosophy." In: The Journal of Hellenic Studies, 84:92–106.

Lovelock, J.E., 1979: Gaia A New Look at Life on Earth. Oxford University Press.

Mbiti, J.S., 1996: African Views of the Universe. In: Gottlieb, Roger (ed.) This Sacred Earth: Religion, Nature, Environment, Routledge, 174–180.

Motegi, S., 2001: Le sol dans la tradition shintoïste In: Lahmar, Rabah, Ribaut, Jean-Pierre, Sols et sociétés: regards pluriculturels Paris: Charles Léopold Mayer, 161–166.

Pye, M., Patterns in comparative religion. http://philtar.ucsm.ac.uk/encyclopedia/introd.html, last viewed on June 10th, 2005.

Santmire, H.P., 1985: The Travail of Nature: The Ambiguous Ecological Promise of Christian Theology, Fortress Press, Minneapolis.

Schmidt, S.J., 1997: Geschichte beobachten. Geschichte und Geschichtswissenschaft aus konstruktivistischer Sicht. In: Österreichische Zeitschrift für Geschichtswissenschaften, 8:19–44.

Spangler, D., 1996: Imagination, Gaia, and the Sacredness of the Earth. In: Gottlieb, Roger (ed.) This Sacred Earth: Religion, Nature, Environment Routledge, London, 611–619.

Nelson, S.A., 1998: God and the Land: The Metaphysics of Farming in Hesiod and Vergil with a Translation of Hesiod's Works and Days by David Grene, Oxford University Press.

Stoll, M., 1997: Protestantism, Capitalism, and Nature in America. University of New Mexico Press, Albuquerque.

Szonyi, M., 2000: "Local cult, Lijia, and Lineage: Religious and Social Organization in Ming and Qing Fujian", In: Journal of Chinese Religions, 28:93–126.

Taylor, B. (ed.), 2005: The Encyclopedia of Religion and Nature, Continuum, London, New York.

Tirosh-Samuelson, H., 2002: Judaism and Ecology: Created World and Revealed Word, Harvard University Press, Cambridge, MA.

Tuan, Y., 1968: "Discrepancies between Environmental Attitude and Behavior: Examples from Europe and China", In: Canadian Geographer 12.3.

Tucker, M.E., Williams, D.R. (eds.), 1997: Buddhism and Ecology: The Interconnection of Dharma and Deeds, Harvard University Press, Cambridge, MA.

Tucker, M.E., Berthrong, J. (eds.), 1998: Confucianism and Ecology: The Interrelation of Heaven, Earth, and Humans, Harvard University Press, Cambridge, MA.

White, L., Jr., 1967: "The Historical Roots of Our Ecological Crisis", In: Science 155:1203–1207.

Winiwarter, V., 2005: "Über die historische Entwicklung der Bodennutzung in Europa." In: Mitteilungen der Österreichischen Bodenkundlichen Gesellschaft 72:27–33.

York, M., 2004: Paganism as Root-Religion. In: The Pomegranate 6.1, 11–18.

Zhao, Y., 2001: Le sol dans le confucianisme. In: Lahmar Rabah, Ribaut, Jean-Pierre (eds.), Sols et sociétés: regards pluriculturels, Paris, Charles Léopold Mayer, pp. 167–174.

Section II
Soil as a Natural Body

Ordering and classifying objects to understand the natural world is a common human activity. This is more difficult for soils than for plants and animals. Many of the early systems of classification were based on observable properties such as grain size composition or color, e.g. sandy soils or dark soils with high organic matter. While it is easy to see changes in soils with distance, these changes are gradual. A soil is not a discrete body. It has developed its present form as a result of the soil forming factors of subsoil or geology, climate, flora and fauna, age, and relief. This became the basis for classifying soils and understanding their genesis. This activity became one branch of soil science, applying the results of studies of soil processes discussed in Section III to soils in the field.

These ideas owe much to the concepts introduced by Dokuchaev in Russia. Catherine Evtuhov, in chapter 6, discusses his scientific and practical work in relation to the social and scientific conditions he encountered in the late 19th century. His practical work focused on describing soils for purposes such as land development planning or rating for taxation. In the course of his field studies work he observed how soils changed across the immense landscapes of Russia. There he developed his concepts of soil genesis. When his work was translated, it received ready acceptance in the rest of Europe and in the USA, where soil scientist had been facing the problems of explaining soil diversity and of interpreting what they saw.

Ronald Amundson in chapter 7 discusses an example of institutional and personal influences on the development of soil science in the USA in the early 20th century. Two dominant soil scientists with different backgrounds argued through their correspondence about whether soil fertility knowledge would be best furthered through soil chemistry or soil physics research. Each scientist had his sphere of influence. Hilgard, with interests in arid soils in the western USA and from a university position, argued for soil chemistry. Whitney, with the USDA and experience with previously forested soils, argued for soil physics research. This championship of physical vs. chemical properties was not a new phenomenon in Europe. Before the dominance of Liebig's research and his skillful advertising, physical soil conditions were seen as very important in soil fertility.

Alexander Gennadiyev and James Bockheim in chapter 8 discuss two key concepts in understanding regularities amidst the diversity of soils seen in nature. These studies in both Russia and USA aided soil classification and in understanding soils in their landscapes. In the last paper in this section, Vance Holliday discusses the importance of an understanding of the multi-disciplinary nature of soil geomorphology as a necessary background for soil classification and soil genesis. In a detailed chapter, he shows the progression of geomorphology ideas in the USA. Through a discussion of the contributions of the individuals who made the footprints, who collected and synthesized the knowledge, he shows how our present understanding developed. A complete list of references is an aid to further detailed study.

6

The Roots of Dokuchaev's Scientific Contributions: Cadastral Soil Mapping and Agro-Environmental Issues

Catherine Evtuhov[1]

Introduction

In his textbook with the terse title, *Soils* (1906), Eugene Hilgard explained the genesis of his system of soil science by the practical tasks that faced him as a professor of agriculture at U.C. Berkeley and an employee of the U.S. Geological Survey in California: how to advance farming techniques in the absence of accumulated experience, how to deal with "virgin" soils whose properties remained unknown to those who worked them (Hilgard 1906). The birth of soil science is variously attributed to Hilgard (1833–1916) and to his Russian contemporary, Vasily Dokuchaev (1846–1903).[2] In both cases, the new field of investigation arose at the juncture of science (geology and chemistry) and agricultural practice. Yet it is difficult to imagine more divergent circumstances than those confronting the German-born, Illinois-raised Hilgard, facing the challenges of taming post-Gold Rush California territories, and the no less problematic world of agricultural development in Russia following the Great Reforms of the 1860s. My task in this chapter is to look behind the scenes of the emergence of the discipline of soil science in its Russian variant (without claiming primacy for either scientist). Dokuchaev's comprehensive conception of the soil as an organic body, in constant interaction with other aspects of the natural environment, had roots in a very specific cultural and historical context in post-reform Russia. How did Dokuchaev's

---

[1] Department of History, Georgetown University, Washington, DC, USA.

[2] Soviet science was determined to claim all credit for Dokuchaev alone. As his biographer Krupenikov comments, "Only one effort was made to challenge Dokuchaev's priority in the creation of genetic soil science, attributing it to the American, Hilgard (Jenny, 1961), but it was immediately, devastatingly and conclusively disproved (Gerasimov, 1962)" (Krupenikov 1981).

"system" of soil science grow out of and interact with his practical concerns? What was the relation of his conception to those of other scientists and to the institutions with which he interacted? What does this contextualization add to our understanding of Dokuchaev's contribution to soil science?

Dokuchaev: A Biographical Note

Vasily Dokuchaev's arrival in St. Petersburg in 1867, in order to pursue his studies at the Theological Academy, coincided with an extraordinary moment in Russian cultural and intellectual life. The era of Reform was in full swing: the emancipation of the peasantry (1861), the creation of the *zemstvo* organs of local self-management (1864), the legal reforms implementing a jury system (1864) created an atmosphere of excitement and civic responsibility in educated society. Salons and dinner tables buzzed with discussion of rural poverty, land redistribution, medical care, agricultural improvement, women's rights, workers' cooperation, peasant literacy: everything, as Leo Tolstoy pointed out, took the form of a "question." Science and education, for many, held the key to the transformation and development of a new Russia. A mere three weeks spent frequenting the public lectures of St. Petersburg University professors in the shopping arcade off of Nevsky Prospect brought about an abrupt shift in Dokuchaev's intellectual orientation: the provincial priest's son and graduate of the Smolensk seminary abandoned his theological studies in favor of Natural History at the University.[3] By 1870 his successes in geology and mineralogy earned him an appointment as curator of the Geological Collection at St. Petersburg University – a post he retained for a number of years, as he progressed, with stellar success, up the academic ladder to appointments as Associate (*dotsent*) and then Professor at the University.

Dokuchaev's biographers generally perceive three or four phases in his intellectual evolution. His early work, beginning in 1869 with the summertime investigation, while he was still a student, of the riverbed of the Kachna river in his hometown of Smolensk, culminated with the controversial 1878 paper, "Formation of Riverbeds in European Russia," for which he received the degree of Magister. From rivers and the concomitant landscape structures he moved on to the study of the soil itself, yielding his seminal work, *The Russian Chernozem*, in 1883. The book gained immediate recognition, winning him not only the doctoral degree but the Makariev

---

[3] Priests' sons (*popovichi*) provided so much of the intellectual energy of the Reform era that their secularization and civic enthusiasm became a phenomenon in its own right (see Manchester 1995). The *popovichi* included the radicals Nikolai Chernyshevsky and Dobroliubov, scientists, writers, etc.

prize from the Russian Academy of Sciences and special thanks from the Imperial Free Economic Society, which had commissioned the investigation. The 1880s and 1890s were spent furthering hands-on study of the soil and soil formation, working to advance agricultural education, and founding natural history museums throughout the Russian provinces. In the final phase of his life, Dokuchaev developed the influential theory of natural zones, latitudinally traversing the Eurasian and other continents: he identified five – tundra, forest, forest-steppe, steppe, and desert. Thus, while Dokuchaev is internationally known for his contributions to soil science in particular, he might also be seen as a practitioner of environmental studies more generally: his work embodied his own early maxim that the student of the soil must take into account a variety of factors transcending a narrow analysis of the soil itself.

> For a full scientific appreciation of the chernozem, as for any other soil, one must study the following aspects of the question: the distribution of the chernozem, the flora and fauna, characteristic chemical, physical, and microscopic qualities of the soil in question and, finally, its various geological relations to other soils as well as to the underlying rock . . . (Dokuchaev 1877)

The Scientific Societies

Many of the dominant scientific paradigms in nineteenth-century Russia grew out of, or shaped, practical concerns. The primary venue for this intersection of science and practice became the scientific societies, which proliferated in Russia as in other European countries. Russia's Imperial Geographic Society, founded in 1845, was second only to the British in size and ambition, with a 20,000-ruble budget for financing scientific expeditions and publications. Much of their energy focused on the exploration of Central Asia and the Russian Far East (see Knight 1998, Bassin 1999). The Imperial Mineralogical Society (founded 1817) understood mineralogy in the broadest possible sense, bringing together physics, chemistry, geology, and paleontology, as well as practical disciplines like mining engineering and metallurgy. The Russian Technical Society (1866) sponsored cooperation between science and industry and quickly grew into one of the largest and most influential of the learned associations. The St. Petersburg Society of Naturalists (1868) operated on a high level particularly in the field of biology; it formed a bridge with the Academy of Sciences St. Petersburg in and cooperated with other learned societies based in the capital. Various medical, archeological, astronomical, mathematical and other associations blossomed in the second half of the century. In addition, periodic congresses – for example, the Congress of

Naturalists and Physicians, which met for the first time in 1867 and was modelled on the German organization of the same name – provided a forum for the intensive exchange of ideas and practical experience.

The most venerable, and one of the most influential, of these independent learned societies was the Imperial Free Economic Society, established under Catherine II (1762–1796) for the improvement of agriculture; one of its earliest and most famous undertakings was the solicitation, sanctioned by the Empress, of proposals for the abolition or amelioration of serfdom. The FES received partial funding from the Ministry of State Properties but was not subject to any government department (Vucinich 1970). The FES was a remarkable organization: from its foundation as an association of enlightened landowners (Khodnev 1865, Confino 1963), it underwent a number of transformations in the course of its 150-year history. In the post-reform era the FES became not only an engine of agricultural innovation but an informal congress where scientific information and agricultural practice intersected to mutual benefit; by the turn of the twentieth century, reflecting the progressive political radicalization of its agronomist and statistician membership, it had gradually metamorphosed into a forum for liberal and sometimes revolutionary ideas, to the point that the government felt it necessary to curb its activities and mandate a strict adherence to technical issues on the eve of revolution in 1904–07 (Evtuhov 1997).

When Dokuchaev began working with the FES in the 1870s he joined an august roster of affiliates that included fellow Petersburg University professors A.N. Beketov, A.V. Sovetov, D.I. Mendeleev, A.M. Butlerov, and others; the Society's secretary was the chemist A.I. Khodnev (Chebotareva 1961). Within the FES, there were as many ways of combining science with practical experience as there were individual members. Mendeleev (1834–1907) published the periodic table of elements in 1869, at the same time as he was completing an FES-sponsored study of the effect of chemical fertilizers on the soil; in these years, he was an organizer of the Russian Chemical Society and the First Congress of Russian Naturalists and Physicians.[4] His involvement with the world of industry was if anything more intensive: he travelled to Baku and as far as Philadelphia in his research on chemical properties of oil and natural gas; later, during the massive industrialization push of the 1890s, he became a specialist on tariffs and wrote a book on the subject. Butlerov (1828–1886), who worked out his structural theory of organic chemistry mostly in interaction with European scientists and travelled a good deal, participated in the FES primarily as an apiculturalist, contributing a number of

---

[4] Hilgard rejected the use of the "new" elements in favor of traditional titles, on the grounds that farmers would not understand the new terms.

articles on the subject to their papers; they published his book, *The Bees: Their Life and the Basic Rules of Rational Apiculture* (1871) (Vucinich 1970). The book was intended to be clear and comprehensible for a peasant readership. On the opposite end of the spectrum one can find figures like A.N. Engelgardt (1828–1893), well known to historians for his descriptions of peasant life on his estate in Smolensk (also Dokuchaev's province of origin): Engelgardt began as a landowner, but in the process developed a serious interest in chemistry, and became famous for his successful experiments with phosphate fertilizers – the object of Mendeleev's interest as well – on his own lands.[5] The man whom Soviet historians of science construed as Dokuchaev's main opponent, P.A. Kostychev, represented a type characteristic of the FES: the Brockhaus encyclopedia describes him as a "learned landowner."

In contrast to the splendid isolation in which Hilgard found himself – as a result of which, in all likelihood, he so insisted on a long voyage to Germany at the conclusion of any large project – Dokuchaev was constantly surrounded by the most active scientific minds of his time. Mendeleev, Butlerov, and others were equally passionately engaged in the chemical, geological, and ecological dimensions of agricultural transformation.

Dokuchaev's expeditions and investigations were supported and financed by a variety of organizations including, in the early days, the Petersburg Society of Naturalists, the Mineralogical Society, and later the Congress of Naturalists and Physicians and the Imperial Department of Forestry. His early work on the formation of river valleys, for example, proceeded under the joint auspices of the two former organizations. None of this involved very large sums of money, but the moral and intellectual sponsorship was essential. His most important studies, however, and the *Russian Chernozem* in particular, were commissioned by the Free Economic Society, then under the presidency of A.I. Khodnev. Dokuchaev's cooperation with the FES began in 1876 with comments on A.V. Sovetov's investigation of agriculture in the black soil region (Chebotareva 1961). The FES's endorsement of Dokuchaev's work on river valleys was followed by the full-scale sponsorship of the expedition to Central and South Russia and the Caucasus to collect materials for the *Russian Chernozem*; they contributed to his subsequent expedition to Nizhnii Novgorod and perhaps also to Central Asia. Dokuchaev's investigations of the soil became a showcase issue for the FES: they figure prominently in even the briefest of historical accounts of the Society's undertakings, as their major endeavor of the 1880s (Beketov 1890, Sudeikin 1892);

---

[5] Engelgardt's *Letters from the Countryside* have recently been translated into English by Cathy Frierson (Engelgardt 1993). Unfortunately, the passages dealing with the technical aspects of soil and agriculture have not been included.

soil science was an important enough branch of activity to warrant the foundation, on Dokuchaev's initiative, of a special Soil Committee in 1888.[6]

The regular sessions of the FES provided a forum for the presentation of data, discussion of ideas, and confrontation of conflicting views on particular scientific or agricultural problems. The Soviet version of the history of soil science isolated Kostychev, Dokuchaev, and Vil'iams as founders; in fact the discussion was much broader. The early critiques of Sovetov and Bogdanov were both presented in FES sessions; in them Dokuchaev formulated his disagreements with Ruprecht. The specific tasks of the expedition were formulated by a committee that included Khodnev, Sovetov, Bogdanov and, later, Butlerov, Mendeleev, P.A. Il'enkov, A.A. Inostrantsev, A.I. Voeikov and P.A. Kostychev. Dokuchaev submitted a report to the FES immediately upon the conclusion of the *Chernozem* investigation, presenting it orally in January 1881; these comments, "The Course and Primary Results of the FES Investigation of the Russian Black Soil," contained the kernel of the subsequent magisterial study, also published by the FES. Most importantly, Dokuchaev's own views developed in a constant exchange with other investigators interested in similar issues. The study of the soil was for Dokuchaev a means of resolving a series of intersecting debates, most significantly over the influence of climatic conditions on soil formation (with the meteorologist A.I. Voieikov) (Dokuchaev 1881); over the relative importance of the organic versus inorganic (or botanical versus geological) composition of the soil (here Kostychev was the main opponent) (Dokuchaev 1885); over the historical forestation of the steppe (Moon 2003); as well as a variety of ecological issues (to be discussed below). It is in the course of these discussions that certain crucial features of Dokuchaev's approach emerged: the independence of the soil as a natural body became his answer to those who identified it with agricultural lands on one hand, and those who saw in it a mere manifestation of underlying rock types on the other (Dokuchaev 1881); it was in dialogue with Voieikov and Kostychev in particular that Dokuchaev asserted the impossibility of isolating any one factor as determining in soil formation. A crowning presentation in 1883 officially concluded Dokuchaev's collaboration with the FES on the study of the black soil. In a sense, then, the *Chernozem* expedition, while Dokuchaev's in execution, constituted a collective undertaking in its conception.

---

[6] The Committee's original members were A.N. Beketov, P.F. Barakov, S.K. Bogushevskii, A.S. Georgievskii, V.V. Dokuchaev, A.S. Ermolov, P.A. Zemiatchenskii, V.I. Kovalevskii, V.G. Kotel'nikov, F.Iu. Levinson-Lessing, I.I. Mamontov, M.O. Musnitskii, A.V. Sovetov, G.I. Tanfil'ev, A.R. Ferkhmin, and V.M. Iakovlev. (For more on the Soil Committee and its activities see Khismatullin 2000.)

Towards the end of Dokuchaev's life, in November–December 1897, he delivered a cycle of 64 lectures at the FES on the subject of the relation of living and dead nature. He intended to publish a book on this subject but it remained unfinished; the article, "The place and role of contemporary soil science in science and in life" would have become the introduction (Dokuchaev 1897). These lectures were not reproduced in the Soviet edition of Dokuchaev's works. The published fragment, however, reflects the synthetic, ecological thinking that was a product of a lifetime of reflection upon theoretical science in its interaction with practical life.

> Mightn't our contemporary culture be too cumbersome and value-laden, particularly in light of the indisputable fact that the life of the civilized man in particular becomes more and more demanding and expensive every year? Finally, will natural resources suffice so that their *growth* corresponds to an at least marginally *significant* distribution of the benefits of civilization to the masses of humanity? Couldn't we, rather, foresee the exhaustion, in the more or less proximate future, of items so essential to civilization as coal, oil, and iron? (Dokuchaev 1897)

A distinctive feature, then, of the Russian scientific world was the intensity of scientific and practical exchange. I believe that this constant interaction influenced the manner in which Dokuchaev posed questions and the contextualization on which he always insisted in juxtaposing the study of the soil with other disciplines and with the totality of the natural environment within which soils formed, developed, and changed.

Maps and Cadasters

The U.S. Geological Survey, established in 1879, brought together under a single federal aegis a series of disparate individual surveys that sought to "conquer" the West through scientific understanding. The Congressional mandate charged the new organ, which was subject to the jurisdiction of the Department of the Interior, with "classification of the public lands, and examination of the geological structure, mineral resources, and products of the national domain." (USGS 2000) The needs of the Survey set the terms of Eugene Hilgard's work both in Mississippi and in California. Curiously, the notoriously centralized Russian state, despite its considerable antiquity, had not by the end of the nineteenth century managed to execute any such uniform and centrally-driven investigation of its own territories. Instead, Dokuchaev's detailed study of the Russian soils took shape under the auspices of local institutions: much of the hands-on work for his monumental study of the *Russian Chernozem* was executed in fulfilment of a commission from the Nizhnii Novgorod zemstvo in

1881 to complete an exhaustive investigation of the soils of Nizhnii Novgorod province for purposes of land quality assessment. A similar commission from the Poltava zemstvo (in Ukraine) followed a few years later.

Soil Maps

In the nineteenth century, soil maps became a recognized means of determining the location and distribution of natural resources and rationalizing the process of agriculture. Climate maps and hypsometric maps (for European Russia, A.A. Tillot's) found their place among scientists' efforts to conceptualize space. The first comprehensive map of the soils of European Russia was compiled in 1851. Dokuchaev's involvement in this aspect of things occurred when the statistician V.I. Chaslavsky, at work upon a much-expanded and more detailed mapping of the soils, unexpectedly died in 1878, and the task fell to his promising younger colleague. The Chaslavsky map – sponsored by the Department of Agriculture and Rural Industry – was the most advanced of its day; yet, in his conclusion to the publication, Dokuchaev noted a number of weaknesses and inadequacies. These included the lack of a consistent system of soil classification, making cross-regional comparisons very difficult, the lack of a comprehensive collection of soil types, and the need for physical and chemical analyses of the main soil types. Dokuchaev's study, "The Cartography of Russian Soils," was an important precursor to the *Chernozem*, and was entirely the result of his work on the Chaslavsky map. Most important, as Hilgard also discovered in his work for the USGS, nothing remotely resembling a total map of an entire country could even be conceived without a compilation of detailed studies of every constituent region. The whole could emerge only from the parts.

Cadasters

The significance of cadastral mapping for the study of the soil was a discovery clearly indicated in the "Cartography" essay: Dokuchaev devotes many pages to the state statistical investigations of Moscow, Iaroslavl, Nizhnii Novgorod, Vladimir, Samara, Kazan, Kostroma, Kaluga, and his native Smolensk provinces. It was these localized studies that permitted the drafting of the original 1851 map and accounted for its superiority over previous attempts. He thus fully understood the potential of the invitation he received from the Nizhnii Novgorod zemstvo in 1881 to undertake an exhaustive survey of the entire province, district by district, in order to map precisely the soil quality of each parcel of land. I would like to show here that this investigation, as well as a parallel one subsequently commissioned by the Poltava zemstvo, proved essential to the development of his own scientific paradigm, primarily because it made possible such a place-specific and detailed study. Hence, here again, the particular definition of practical goals shaped the course of scientific investigation; the subsequent interaction of scientific results with practice was in this case less felicitous, for in the end the zemstvo found that an infinitely accurate definition of soil types skirted the social and economic issues that were of paramount concern to them.

The intimate connection between cadastral mapping and the study of the soil emerged, in the 1880s, from a convergence of factors peculiar to the Russian situation. I am not proposing that all cadastral studies are conducive to the development of soil science. In fact, traditionally they are not. Cadastral mapping is a practice that in Europe dates at least to the seventeenth century. In the fifteenth- and sixteenth-century Ottoman Empire, the *tahrir*, or provincial survey, was a key instrument in the assimilation of newly-acquired territories, as well as a means of regulating existing ones, and provided a record of male population, households, crops and yields, and taxes in money and in kind (Itzkowitz 1972). The famous cadastral survey of Napoleonic France set an example of centralized information-gathering for all of Europe (see Bourguet 1988). Since the purpose is land assessment for taxation purposes, cadasters are usually concerned above all with property boundaries. The Russian ones were not. The earliest surviving *pistsovye knigi* come from fifteenth-century Novgorod; references to previous ones exist but no examples remain. These records, which a recent researcher has characterized as more sophisticated than contemporary Western European cadasters, contained a practical description of individual landholdings for tax-gathering purposes; they were instructed to "record all those lands and who lives on them and to measure accurately in obeisance to the Tsar all plowlands and fallow lands, and forests in *desiatins*, and to measure the *desiatins* as eighty *sazhens* in length and thirty in width, and to measure the plowlands and write them in the books." (Krupenikov 1981) Particularly interesting was the existence of standardized measures set in Moscow, at a time when the English feudal cadasters, for example, measured plots by simple visual observation, based on interviews. The early modern books registered land according to its quality, dividing land into "good, middling, poor, and very poor," (Krupenikov 1981)[7] based on detailed interviews with local residents.

The Catherinian Land Survey (1765–) was the first effort at a comprehensive and systematic cadaster (see German 1910). Initiated in June 1766, when parties of surveyors fanned out from Moscow, the investigation covered more then 165,000 blocs of land, comprising 144 million desiatins, in 22 provinces by 1796; it continued under Paul and had covered most of European Russian by the 1840s (De Madariaga 1981). Isabel De Madariaga, in her magisterial study of the age of

---

[7] His own references include: Veselovskii, Merzon, Bart, Rozhkov.

Catherine the Great, observes that the survey, rather than choosing individual estates as the basic unit,

> measured, described, and allotted land to individual villages, whether serf, state, peasant or odnodvortsy, just as it measured and allocated land to towns, churches, and cathedral ... The importance of labor as distinct from land is reflected in the fact that where vacant land was available, and existing allotments to villages were vague, eight desiatinas of land were to be allotted per male soul. Thus the basic principle was not the property rights of an individual landowner, but the establishment of the amount and the boundaries of the land which belonged to a given village, whether it had one or several owners or belonged to the state (De Madariaga 1981).

De Madariaga notes that the survey had the effect not only of introducing order into the landscape, but also of eliminating disputes and enlarging the sown area and intensifying agricultural production (De Madariaga 1981). The surveys of the late eighteenth century were followed by the state-sponsored local studies, mentioned above, under Nicholas I (1825–1855), in the 1840s, that resulted in reasonably complete analyses for a number of Russian provinces and, as a side-product, the 1851 soil map – further revised in 1857 and again in 1869.

The potential fruitful interaction, however, between the creation of a cadaster and a deeper understanding of Russian soils, was not fully realized until Dokuchaev's Nizhnii Novgorod study. Curiously, the most complete description of the land in Russia came not from the State, as is usual, but followed the initiative of local institutions (the zemstvo); when the "cadaster" was charted, it once again devoted its primary attention not to property divisions as in other countries but to soil quality, agricultural productivity, labor costs, and other "scientific" indicators of land value. It was Dokuchaev who, sponsored by the Nizhnii Novgorod zemstvo, created the first such "cadastral" description of the province, to be supplemented by statistician N.F. Annenskii's economic and statistical investigation, building on Dokuchaev's, of the same territory a few years later.

The Nizhnii Novgorod study came on the heels of a host of analogous local investigations over nearly two decades since the Great Reforms. The zemstvo institutions, created in 1864, were granted the right to collect their own taxes in order to manage their local needs – roads, hospitals, sanitation, lighting. The key question was *how* to collect the taxes the zemstvo was endowed with the right to gather by the 1864 legislation. Only land ownership was subject to direct taxation, and the variation in size, quality, productivity, not to mention changes in these factors over time, was such that establishing a rational and equitable system of assessment was extremely difficult. The zemstvos therefore experienced an acute need for a scientific description of the lands under their jurisdiction. Viatka province was the first to undertake such a local investigation in 1870, followed by Riazan' (1870), Tver' (1871), Kherson (1874), Moscow (1875), Chernigov (1875), Perm' (1876), Novgorod (1879), Tambov (1880), Kharkov (1880) ... etc. (Rikhter 1894; see also Stanziani 1991, Darrow 1996, and Mespoulet 1999). The techniques of description emerged on an *ad hoc* basis, with each province working out its own system; they were able to learn from each other, however, with each region providing models and experience for the next. By the 1880s two distinct methods had emerged, known by the names of the regions where they had originated – "Chernigov" and "Moscow." The "Chernigov" focused on the land itself, taking the land allotment (*mezhevaia dacha*) as the basic unit, and involving detailed on-site investigation by a team of researchers; the "Moscow" method selected people rather than land, undertaking a complete household-by-household survey and using materials from land transaction records, passport books, and taxation registers, as well as official documents.

The Nizhnii Novgorod zemstvo in particular was concerned primarily with the lack of information allowing an equitable comparative assessment of lands in different districts (*uezdy*): while each district, following the tradition of the Muscovite *pistsovye knigi*, could indicate roughly which of its lands were good, middling, or poor, there was no mechanism for comparing this information across district boundaries; thus equity among districts was a central goal. It was in this context that a special commission, established in 1880, suggested an invitation to the already well-known St. Petersburg scientist, Dokuchaev, to conduct a fully scientific, professional, and exhaustive study of the province, district by district, to evaluate the soil quality and potential productivity of absolutely every parcel of land. The zemstvo was able to allocate the significant sum of 10,000 rubles for this purpose.[8] In addition, the expedition attracted the attention of the St. Petersburg Society of Naturalists, whose president, A.N. Beketov, sent three botanists – Aggeenko, Krasnov, and Nidergefer – to the province in 1884 to supplement the expedition with a collection of 900 plant species and investigate the region from a geobotanical perspective. The head of the Society's mineralogical and geological division, A.A. Inostrantsev, deployed Dokuchaev's students Amalitskii and Sibirtsev for a geological investigation, while the respected agronomist, A.V. Sovetov, of the same society, contributed practical advice and instructions. The expedition had a high visibility in scientific circles, as evidenced by the participation of K. Shmidt (Dorpat University), and two younger researchers from St. Petersburg University – master's candidate in agronomy V.M. Iakovlev and the curator of the St. Petersburg University meteorological cabinets, A.N. Baranovskii.

---

[8] The report of the commission, from 17 January 1881, is cited in Dokuchaev 1887. N.O. Osipov gives a slightly different figure, placing the original estimated cost at 15,500 rubles, which grew to 17,800 by the time the investigation was complete, and special sums had been designated for the printing of the maps and packaging of soil samples (Osipov 1885).

Shmidt undertook a chemical analysis of the soil samples, while Iakovlev's task was physical properties, and Baranovskii's – a description of the climate (Dokuchaev 1887).[9] Thus the expedition's sponsorship, while primarily based in the zemstvo, also included financial and intellectual endorsement from St. Petersburg scientific societies.

This confluence of practical and scientific interests yielded a unique portrait of Nizhnii Novgorod province that eventually earned its own name in the annals of zemstvo statistics – the "territorial/cadastral method." Beginning in 1882, and for three summers running, Dokuchaev and a team of three students[10] arrived on the scene and spent their days combing the province, describing every stream and forested glade, every field and every ravine. They were methodical and thorough, dedicating the first summer to the southeast corner of the province (Lukoianov, Sergach, and Kniaginin districts); the second to the central province (Arzamas, Ardatov, Gorbatov, Nizhnii Novgorod); and concluding in 1884 with the densely forested districts beyond the Volga – the ancient refuge of the Old Belief (Makar'ev, Vasil', Semenov, and Balakhna). There are some interesting aspects to the procedure itself: each researcher was supposed to keep a detailed daily journal, describing every type of soil, vegetation, and rock formation he came across; they moved *volost'* by *volost'*, in every case accompanied by the village elder – who, it barely needs mentioning, in the absence of written documentation, would prove an extraordinary repository of information about land boundaries, disputes, and changes in the land contour over time. To Dokuchaev himself fell the task of coordination as well as an independent collection of samples. The finds were then catalogued and analyzed in a laboratory back in St. Petersburg, and the results painstakingly recorded on a soil map, thus yielding a full geological and topographical portrait of the entire province. The scientists originally organized their investigation along the natural orographic and hydrographic boundaries of the region – i.e. in accordance with riverbeds and geological contours – but were requested by the zemstvo board to recraft the whole study using the artificial, political district boundaries to make the information usable for practical purposes.[11]

---

[9] They were also joined by a group of students including Zaitsev, Burmachevskii, Georgievskii, Timofeev, Shaposhnikov, Mel'tsarek, and Sokolov.

[10] Eventually, the students involved were: V.P. Amalitskii, P.F. Barakov, P.A. Zemiatchenskii, F.Iu. Levinson-Lessing, N.M. Sibirtsev, and A.R. Ferkhmin. Several of them went on to become well-known scientists.

[11] The materials of the expedition were published as *Materialy k otsenke zemel' nizhegorodskoi gubernii: estestvenno-istoricheskaia chast'* (Nizhnii Novgorod, 1884–); they are reproduced in their entirety in volumes 4 & 5 of the Soviet edition of Dokuchaev's collected works (Dokuchaev 1949–1961).

By comparison with the earlier FES-commissioned study of the chernozem, the Nizhnii Novgorod investigation produced vastly more detail. It is difficult to characterize the results in a short space, because their primary value was the quantity and depth of detail that filled up fourteen volumes in the original publication. It would doubtless be of interest, for example, for the agriculturalist in Kniaginin district to learn that two patches of black soil, though they looked identical, possessed entirely different properties because they were deposited on different subsoils – quartz sands or naked rocks rather than clays, loess, or loam. The team collected several thousand soil samples, and analyzed three hundred in the laboratory. There were, however, some key insights, from the perspective of soil science, that emerged from the Nizhnii Novgorod venture: this particular investigation yielded an analysis of soil types according to geological, chemical, mechanical, and physical criteria – the classification of soils whose absence Dokuchaev lamented in his comments on the Chaslavsky map. Eight categories of soils could be ranked in terms of their thickness and percentage of organic material. The chemical dimension, which built on Mendeleev's method of evaluation, broke down the same soil in terms of their chemical composition, focusion on potassium, sodium, calcium, magnesium, phosphorus, sulfur and azote (nitrogen). Mechanical qualities refer to grain size and the proportion of small to large grain: ranging from a ratio of 1:0.7 for hilly black soil to 1:42.7 for quartz sands. Finally, physical properties could be mapped on this model. All four criteria were combined in the end to provide a generalized evaluation of the soil (Margulis 19–). The results, from Dokuchaev's perspective and that of soil science, were spectacular. The cadastral expedition yielded an extraordinarily detailed analysis of the chernozem and other things, and created the ideal soil map – because of its regional orientation – that had eluded Chaslavsky.

Dokuchaev believed firmly in the nexus of soil science and cadastral assessment; he was perfectly clear and self-conscious concerning the principles and methodology of his expedition. Dokuchaev outlined his goals and procedures, post factum, in a presentation to the Free Economic Society, "O normal'noi otsenke pochv evropeiskoi Rossii," from 19 February 1887. Dokuchaev's most basic proposition was that a scientific study of the soil must of necessity underlie any attempt at a rational and equitable assessment of land value, and that, to this end, the natural qualities of the soil, abstracted from any human activity or intervention, must form the object of independent and primary investigation. As he points out, he formulated this basic supposition already in 1879:

Soils which, together with the climate, and with the growth and productivity of various cultured plants, must naturally influence the well-being and way of life of residents. Further, because they are of varying thickness, composition and structure, and supported by a great variety of subsoils, fertile lands *must, of necessity*, give rise to different systems of economy, to different methods of working the land, etc. etc. In a word, *soil maps* together with *climatic maps must form the basis of any agricultural statistics*. (Dokuchaev 1887)

Soil quality is the first and most fundamental determinant of land value; it is the most constant and most measurable; it must form the object of accurate scientific study, which in turn can positively influence agricultural production. On occasion, it provides the only accurately measurable index of land value; and no program of statistical investigation can be adequate unless it is sufficiently grounded in the study of the soil (Dokuchaev 1884–87). Dokuchaev did not deny the need for further statistical investigation of economic conditions; indeed, he initially proposed that such a study should follow his own. But he did argue against such "pure" statisticians as Kablukov, in Moscow, who tried to play down the role of natural factors, while, in contrast, endorsing the efforts of the Riazan, Chernigov, Viatka, Vladimir, Tver, Saratov, Ufa, Samara, and Kazan zemstvos, which understood the necessity of placing a study of the soil at the foundation of their assessment strategy (Dokuchaev 1884–87).[12]

Parting of the Ways

Ultimately, Dokuchaev was looking for both more and less than the zemstvo needed. Complaints sounded that the level of detail in the investigation of the soil was unnecessary for practical application – whether for agricultural improvement or accurate land assessment. The zemstvo turned – as had been originally projected – to a representative of a different field, statistics, to complement the strictly "territorial" or soil-scientific investigation conducted by Dokuchaev. This story in itself is quite interesting for the history of science and the history of ideas, and resulted in the creation of a specifically Nizhegorodian method: the above-mentioned territorial-cadastral method, in which Dokuchaev's detailed study of the soil was supplemented by an equally exhaustive study of social and economic conditions. This "economic" part of the investigation was carried out by the statistician N.F. Annenskii, who outlined its principles in the seminal article, "The zemstvo cadaster and zemstvo statistics" (Annenskii 1894). Here, Annenskii presents a sort of agrarian calculus that factors in labor, grain prices, and a variety of other factors. The maps produced by the Annenskii expedition built heavily on Dokuchaev's; their primary interest, however, was not the macro-picture of soil types, but a division of the land into "assessment categories" that factored in not just soil type but also all of the above economic considerations. [Fig. 1, Gorbatov map]. Here, however, soil science and zemstvo interests

---

[12] Dokuchaev's most vehement critic was N.O. Osipov, who called his investigation superficial and useless for any practical purposes. See ch. 4, "Raboty Nizhegorodskogo zemstva" (Osipov 1885).

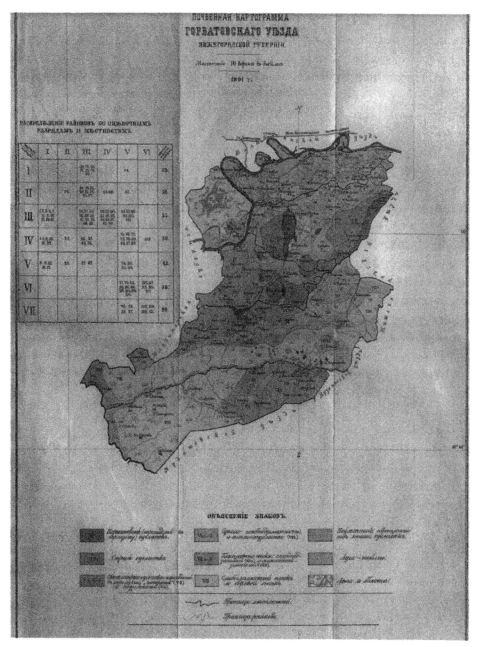

Figure 1. The Gorbatov Map. Color version (see the color plate section at the end of the book).

parted ways, as Dokuchaev went on with his laboratory studies and, ultimately, his investigation of natural zones, while the zemstvo pursued its practical ends in the realm of social science.

Grain Yields and Ecological Issues

The twentieth century arguably wreaked greater and more lasting damage upon the natural environment than any preceding period in human history (McNeill 2000). The demographer Murray Feshbach attributed the collapse of the Soviet Union not to politics or economics, but to the destruction of nature (Feshbach and Friendly 1992). Yet, neither damage to the environment nor awareness of the potential dangers of such damage are twentieth-century inventions. The Icelandic forests disappeared as early as the fifteenth century; all nineteenth-century efforts to restore them ended in naught, resulting in the eerie landscape that now seems emblematic of the country. In Russia, potash extraction wiped out large swaths of forested land in the seventeenth century; the process, which involves multiple stages of reduction of timber to ash, inflicts permanent damage on the forest. The eighteenth-century traveller Pallas commented on the pollution of streams and rivers from leather production, one of the "dirtiest" of proto-industrial activities (Pallas 1771–76). The most damaging of human activities before the twentieth century, however, was agriculture; as one researcher has commented, the beginning of settled agriculture created the conditions for population growth and marked the incipient imbalance between humanity and nature (Herrmann 1989).

The emancipation of the serfs in 1861 raised a host of issues that, in the half-century that followed, came to be known as the "agrarian question." The peasants were freed with land – a decision contained in the Emancipation decree itself.[13] But the size and distribution of allotments, the type of ownership (communal v. individual), peasant-landlord contracts and schedules for redemption payments, remained subject to negotiation. In this atmosphere of rural transformation, the conduct of agriculture itself became a topic of intensive discussion: the improvement of farming methods, more efficient use of available land, and an understanding of specific environmental conditions that might affect the process of agriculture were high on the agenda. At stake was not only the issue of negative human impact upon nature, but "ecological" concerns broadly defined as any active relation of human beings and the natural

---

[13] The emancipation affected 20 million serfs belonging to individual landlords; a milder reform was implemented for 50 million state peasants. Thus the numbers involved ultimately reached close to 80% of the total population.

environment. Although today, Dokuchaev's name appears in the annals of the history of science almost exclusively in connection with his work on the soil and the chernozem in particular, these studies were at the time imbedded in a whole spectrum of environmental questions whose investigation yielded a better understanding of the conditions underlying Russian agriculture.

The landscape of European Russia, like that of any geographical region, has certain peculiarities, with implications for methods of human exploitation. Once covered by the immense Scandinavian-Russian glacier, the Eurasian plain is characterized by dense forests; slowly meandering rivers with a hilly right bank; a "continental" climate – a sharp contrast between summer heat and winter cold, together with massive spring flooding; and soft, loose soils and gently sloping hills. Dokuchaev's scholarly concerns extended to the forests, the steppes, and particularly the rivers. These "macro" issues, which involved the action upon the soil of a variety of other factors, included deforestation, the apparent progressive shallowing of rivers (due to increased traffic?), erosion, and the possible encroachment of the steppe upon central Russian forest and agricultural land. Dokuchaev devoted a good deal of attention to such questions as the formation of riverbeds and particularly of ravines – an extremely common (and from the human point of view frequently destructive) feature of the Eurasian loose topsoils and flat or sloping landscapes. In fact the study of the soil interacted with and grew out of these broader ecological concerns.

Grain Yields

The first issue was, how might one extract more from the land, or intensive versus extensive agriculture. Peasant land allotments were not small, and exceeded peasant land holdings in most European countries. Yet, traditional agricultural techniques resulted in shockingly low grain yields; in non-black-soil regions yields could be as low as 1:2. It was here that soil science, as well as chemistry (fertilizers) could make a contribution. The study of soil properties, in conjunction with the environment, would permit the agriculturalist, whether peasant or landlord, to realize not only the real, but the potential value of his land. As Stebut put it, "The proprietor [*khoziain*] certainly does know what the soil produces, but he does not know what it might produce in certain given conditions. It's also fair to say that the significance of the soil changes; for once we know the soil in its actual state and know the changes it brings about and are brought about in it, we can also trace what it will become in time, [in 1, 20, 100 years] . . ." (Dokuchaev 1879) Dokuchaev's farmer is not, from a scientific perspective, that different from Hilgard's: years of agricultural practice do not necessarily bring enlightenment, and only scientific investigation can bring awareness of potentials for exploitation.

The issue of increased productivity, and the necessity of analyzing soil conditions for that purpose, became a motivation for Dokuchaev's development of his system of soil classification and soil rating.

Drought and Famine

Agrarian reform did nothing to stop the periodic natural catastrophes that ravaged large parts of the Russian countryside. Major droughts occurred, regionally, in 1864, 1865, 1867, 1875, 1885, 1889, 1890 (Kahan 1989). Famine, caused by drought, struck in 1891–92, affecting the whole blacksoil region, all of the Volga, southern Ukraine and the North Caucasus. The plight of the peasantry was a concern that animated Russia's educated elite throughout the decades following the reform. Social reform, medical improvement, education, charity, credit institutions, dissolution of the peasant commune, were all among measures proposed to remedy rural poverty, while radicals advocated insurrection and revolution. Dokuchaev, and fellow scientists and FES members, turned to in-depth scientific study of the natural conditions underlying Russian agriculture, in the hope that better management of the natural environment could prevent the devastation of drought and famine.

An ecological passion – a deep concern for the possible negative changes wreaked by humans upon nature – runs through much of Dokuchaev's work. During his on-site investigations in Nizhnii Novgorod and Poltava, his team repeatedly paid attention to the much-discussed problem of deforestation (Costlow 2003). The Dokuchaev expedition reported complete lack of forests in the northern parts of Lukoianov and severe deforestation in Sergach – though they attributed the latter to the last 30–40 years only; they assessed Kniaginin as completely deforested, with only tiny groves remaining on individual owners' land (Dokuchaev 1884–87). The team noted that the only extensive groves preserved in Kniaginin district were those belonging to the largest landowner, a Mr. Baryshnikov, who evidently appreciated their value (Dokuchaev 1884–87). Human industry – potash production, shipbuilding, wooden spoon manufacture, consumption for firewood joined natural disaster (fires) as factors contributing to the decline of the forests; an 1888 Statute on the Preservation of Forests forbade landowners to cut down forests, and instructed them to construct plans for their conservation and replacement (Arnol'd 1890).

A second problem was the perceived shallowing of rivers. Shipping companies complained about the shallowness of rivers, which they attributed to increased commercial activity. Dokuchaev, as a specialist on rivers and their formation, directed his attention to this problem quite early in his career. The application of scientific methods to this particular issue led to the conclusion that, in fact, the riverbeds were not becoming shallower; instead, their configuration constantly shifted, thus also displacing the fairway. On this issue, then, the ecological situation turned out not to be as dire as it at first appeared (Dokuchaev 1876).

The primary concern for Dokuchaev remained erosion, particularly the formation of ravines so characteristic of the Eurasian topsoil. Ravines can form when a stream of water meets even the smallest crack or indentation in the earth's surface; once the initial path has been carved the process continues almost automatically as new streams of water are attracted to the existing bed (Dokuchaev 1878, Ototskii 1897). Scientists agreed that Russia provided a remarkable convergence of factors favorable to the formation and development of ravines: in addition to the soft, loose soil itself, there was a tendency for the top layer of diluvial soil to consist of clay, and the layer just below it of sand – which contributed to the ravines' rapid expansion. Hot summers, cold winters and massive spring flooding provided an additional stimulus (Dokuchaev 1878).[14] Human cultivation of the soil became an integral part of the natural drama of ravine formation. So far as nature was concerned, there was no difference between a spontaneously-appearing crack in the earth and, for example, a furrow plowed by human hands. Thus scientists hypothesized that the southern steppes had indeed reached an equilibrium in which the formation of gullies and ravines had come to a standstill – until intensive cultivation opened the process up anew (Ototskii 1897). Among the types of human activity tending to facilitate the expansion of ravines, E.E. Kern cited: cutting down forests and removing stumps along the banks of ravines for cultivation; plowing any steep slopes; tracing dividing furrows between parcels of land; digging ditches; pasturing cattle on slopes; plowing up small "saucer-like" pits that sometimes occurred in cultivated fields; laying rails; creating inclines for harvesting logs; and general carelessness attributed to peasant agriculturalists (Kern 1913). The ravines, in turn, washed off fertile topsoil, drained off rainwater necessary for irrigation as well as underground water deposits, increased the earth surface susceptible to evaporation,[15] dumped sand onto arable land, and cut across and destroyed roads (Ototskii 1897). Dokuchaev's earliest work attributed to ravines a positive function as well: they were in fact responsible for the very shape of the Russian countryside, for redistributing soils and minerals, and even, he argued, for creating beds in which rivers would later flow. They were an essential part of the continually shifting landscape, carving contours in the earth only to be covered up once more a decade later, or deepening and growing branches, occasionally even joining erstwhile separate rivers or streams together (Dokuchaev 1878).

---

[14] Dokuchaev's explanations are adduced in all the subsequent literature (e.g. Kern 1913), and so far as I can tell have not been replaced. See also Murchison 1845.

[15] Dokuchaev spoke of ravines as "natural drainage pipes," sucking off waters and desiccating the steppe (Dokuchaev 1892).

These environmental issues entered the public consciousness in 1891, as famine struck large portions of the Russian countryside. The dangers of erosion by the 1890s were great enough to prompt the philosopher Vladimir Soloviev to speak of the "Enemy from the East" – meaning not Mongol invaders but the encroaching arid steppe (Soloviev 1891).[16] Dokuchaev's 1892 pamphlet, *Our Steppes Before and Now*, was one of a plethora of publications that placed environmental questions at the center of discussion.

The solutions Dokuchaev advocated for these environmental problems complicating the process of agriculture – some of them humanly initiated and others the result of an interaction between human beings and nature – varied according to the particular problem. To augment grain yields he advocated the founding of agricultural stations (precursors to the *zapovedniki* or nature reserves). In the United States as well these stations became important for agricultural experimentation. Like Hilgard, Dokuchaev devoted a good deal of effort to agricultural education, becoming closely involved with the Agricultural Institute. An important part of this educative process was the establishment of soil and natural history museums, the first of which he established in Nizhnii Novgorod. For Dokuchaev, it was not sufficient to study the soil as a scientific problem: the information needed to be disseminated to those who actually farmed the land (again, the nexus of science and practice). He proposed the regulation of rivers, artificially reshaping their beds and limiting spring flooding. Finally he, like Kern and other specialists, advocated the planting of trees and other measures to combat erosion and ravine formation. All these measures, however, whether involving chemical improvement, conservation, or reforestation, were premised on a careful and respectful symbiosis with the natural environment, never disrupting completely the millenial patterns of interaction between human beings and nature.

Conclusion

Russkii chernozem is a book rich in ideas and practical materials. Still, it is possible to isolate a basic theme. This fundamental principle, articulated repeatedly, saw the soil "as an *independent natural-historical body*, resulting from the collective influence of a) subsoils, b) climate, c) flora and fauna, d) geological age, and e) relief of the locality." (Dokuchaev 1890) The soil, in other words, lay at the intersection of a variety of apparently external factors; studying it seriously required bringing into play insights from mineralogy, geology, chemistry, physics, meteorology, biology,

---

[16] The title, "Vrag s Vostoka," is a pun on the Russian word for ravine, "ovrag," which folk pronunciation often shortened to "vrag" (= enemy).

and geography. If it is possible to speak of a single result of this many-layered and complex approach, it would be: multiplicity and variability. The rich examples of *Russkii chernozem* would convince the reader "that *each* of the above-mentioned *soil-fashioners* (a-e) represents an endless series of *degrees, variations, peculiarities* etc., and in addition all of these differences and degress are related to each other (separately of course for *each soil-fashioner*) through exceedingly refined and barely perceptible transitions . . ." Since all of these different factors – what Dokuchaev calls soil-fashioners – operated not in isolation but simultaneously, the resulting spectrum would be virtually infinite. Ultimately, a perfected science of the soil would be capable of arranging all existing soils in a continuous series, broken down by these five factors.

It is this theory of soil-fashioners that constitutes the originality of Dokuchaev's contribution. In this paper, I have tried to show that the peculiar attention to the soil in its relation to a multiplicity of other factors grew out of the particular set of concerns that Dokuchaev sought to address within the Russian context of the last third of the nineteenth century. The context in which Dokuchaev worked included first an unusually intensive interrelation of scientific and practical concerns – a common project that awakened the intellectual and creative energies of scientists like Dmitrii Mendeleev and enlightened landowners like Alexander Engelgardt, within the framework of the Russian learned societies. Second, like Hilgard's participation in the U.S. Geological Survey in Mississippi and then California, Dokuchaev's involvement in the Chaslavsky soil map and subsequently in the zemstvo-commissioned cadastral surveys made possible a depth of scientific investigation necessary to his theoretical formulation of the basic principles of soil science. Finally, the study of the soil for Dokuchaev stood at the crux of a much broader set of environmental concerns, ranging from river and ravine formation which preoccupied him from the earliest days of his career, to the classification of natural zones or belts traversing the globe and enabling a juxtaposition, for example, of the Russian central black soil region and the Kansas prairies. By the end of his life, Dokuchaev's researches brought him to the deepest of spiritual questions: the coexistence and struggle of man and nature, after all, evokes the fundamental mysteries of our existence on earth.

References

Annenskii, N.F. (1894): "Zemskii kadastr i zemskaia statistika." *Trudy podsektsii statistiki 9-go s"ezda russkikh estestvoispytatelei i vrachei*. Chernigov, pp. 17–44.

Arnol'd, F.K. (1890): *Russkii les*. St. Petersburg, 3 vols.

Bassin, Mark (1999): *Imperial Visions: Nationalist Imagination and Geographical Expansion in the Russian Far East, 1840–1865*. Cambridge UP, 329 p.

Beketov, A.N. (1890): *Istoricheskii ocherk 25-letnei deiatel'nosti IVEO s 1865 do 1890 goda*. St. Petersburg.

Bourguet, Marie-Noëlle (1988): *Déchiffrer la France: la statistique départementale à l'époque napoléonienne*. Paris, 476 p.

Chebotareva, L.A. (1961): "Vasilii Vasil'evich Dokuchaev (1846–1903). Biograficheskii ocherk," in V.V. Dokuchaev, *Sochineniia*, v. IX. Moscow, pp. 49–152.

Confino, Michael (1963): *Domaines et seigneurs en Russie vers la fin du XVIIIe siècle: étude de structures agraires et de mentalités économiques*. Paris, 306 p.

Costlow, Jane (2003): "Imaginations of Destruction: The 'Forest Question' in Nineteenth-century Russian Culture," *The Russian Review* 62:91–118.

Darrow, David (1996): *The Politics of Numbers: Statistics and the Search for a Theory of the Peasant Economy in Russia, 1860–1917*. PhD diss., University of Iowa, 532 p.

De Madariaga, Isabel (1981): *Russia in the Age of Catherine the Great*. Yale UP, 698 p.

Dokuchaev, V.V. (1876): *Predpolagaemoe obmelenie rek Evropeiskoi Rossii*. St. Petersburg, 16 p.

Dokuchaev, V.V. (1877): "Itogi o russkom chernozeme." *Trudy IVEO*, SPb I, part 4, pp. 415–432.

Dokuchaev, V.V. (1878): *Sposoby obrazovaniia rechnykh dolin*. SPb, 222 p.

Dokuchaev, V.V. (1879): *Kartografiia russkikh pochv*. SPb.

Dokuchaev, V.V. (1881): "Otvet na vozrazheniia A.I. Voeikova po povodu doklada o zakonnosti izvestnogo geograficheskogo raspredeleniia nazemno-rastitel'nykh pochv na territorii evropeiskoi Rossii." *Trudy IVEO*. SPb XII part 1, pp. 87–97.

Dokuchaev, V.V. (1881): "Khod i glavneishie rezul'taty predpriniatogo IVEO issledovaniia russkogo chernozema." SPb, 68 p.

Dokuchaev, V.V. (1884–87): *Materialy k otsenke zemel' nizhegorodskoi gubernii: estestvennoistoricheskaia chast'*. Nizhnii Novgorod.

Dokuchaev, V.V. (1885): "K voprosu o russkom chernozeme." *Trudy IVEO*, SPb II, part 4: pp. 444–469; III, part 1: pp. 22–44.

Dokuchaev, V.V. (1887): "O normal'noi otsenke pochv evropeiskoi Rossii." *Trudy IVEO*, SPb, III, No. 9: pp. 1–47.

Dokuchaev, V.V. (1890): "O glavneishikh rezul'tatakh pochvennykh issledovanii v Rossii za poslednee vremia." *VIII s"ezd russkikh estestvoispytatelei i vrachei v Sankt-Peterburge ot 28 dekabria 1889 g. do 7 ianvaria 1890 g*. SPb, pp. 9–10, section 9.

Dokuchaev, V.V. (1892): *Nashi stepi prezhde i teper'*. SPb.

Dokuchaev, V.V. (1898): "K voprosu o sootnosheniiakh mezhdu zhivoi i mertvoi prirodoi." *Sankt-Peterburgskie vedomosti* 11 February, No. 41.

Engelgardt, A.N. (1993): *Letters from the Country, 1872–1887*. Trans. & ed. Cathy Frierson. Oxford UP, 272 p.

Evtuhov, Catherine (1997): *The Cross & the Sickle: Sergei Bulgakov and the Fate of Russian Religious Philosophy, 1890–1920*. Cornell UP, 278 p.

Feshbach, Murray and Alfred Friendly (1992): *Ecocide in the USSR: Health and Nature under Siege*. Basic Books, 376 p.

German, I.E. (1910): *Istoriie russkogo mezhevaniia*. Moscow.

Herrmann, Bernd, ed. (1989): *Umwelt in der Geschichte: Beiträge zur umweltgeschichte*. 152 p.

Hilgard, Eugene (1906): *Soils*. Macmillan Co., 593 p.

Itzkowitz, Norman (1972): *Ottoman Empire and Islamic Tradition*. Chicago UP, 117 p.

Kahan, Arcadius (1989): *Russian Economic History*. Chicago UP, 244 p.

Kern, E.E. (1913): *Ovragi, ikh zakreplenie, oblesenie i zapruzhivanie*. 5th ed., SPb.

Khismatullin, Sh.D. (2000): "Activities of the Soil Committee of the Free Economic Society," *Eurasian Soil Science* 33:4, 441–446.

Khodnev, A.I. (1890): *Istoriia Imperatorskogo Vol'nogo Ekonomicheskogo Obshchestva s 1865 do 1890 g*. St. Petersburg.

Knight, Nathaniel (1998): "Science, Empire, and Nationality: Ethnography in the Russian Geographical Society, 1845–1855," in Jane Burbank & David Ransel (eds.), *Imperial Russia: new histories for the empire*. Indiana UP, pp. 108–141.

Krupenikov, Igor (1981): *Istoriia pochvovedeniia ot vremeni ego zarozhdeniia do nashikh dnei*. Moscow.

Manchester, Laurie (1995): *Secular Ascetics: The Mentality of Orthodox Clergymen's Sons in Late Imperial Russia*. PhD diss., Columbia University, 634 p.

Margulis, H. (19–): *Aux sources de la pédologie (Dokoutchaïev – Sibirtzev)*. Toulouse, pp. 37–46.

McNeill, J.R. (2000): *Something New under the Sun: An Environmental History of the Twentieth-century World*. W.W. Norton & Co., 421 p.

Mespoulet, Martine (1999): *Personnel et production du bureau statistique de la province de Saratov: Histoire d'une professionnalisation interrompue (1880–1930)*. Thèse de doctorat, EHESS, Paris.

Moon, David (2003): "Were the Steppes Ever Forested? Science, Economic Development, and Identity in the Russian Empire in the Nineteenth Century," in Laos Jelecek et al., (eds.), *Dealing with Diversity: 2nd International conference of the European Society for Environmental History: Proceedings*. Prague, pp. 206–209.

Murchison, Sir Roderick Impey (1845): *The Geology of Russia in Europe and the Ural Mountains*. London.

Osipov, N.O. (1885): *Kratkii ocherk zemskikh statisticheskikh issledovanii*. Kazan.

Ototskii, N. (1897): "Ovrag," *Entsiklopediia Brokgauz-Efrona*. SPb.

Pallas, Peter Simon (1771–76): *Reise durch die verschiedene Provinzen des russischen Reichs.* SPb, 3 v.

Rikhter, D. (1894): "Zemskaia statistika." *Entsiklopediia Brokgauz-Efrona.* SPb.

Soloviev, Vladimir (1891): "Vrag s Vostoka." *Severnyi vestnik*, SPb, No. 7.

Stanziani, Alessandro. "Statisticiens, *zemstva* et état dans la Russie des années 1880," *Cahiers du Monde russe et soviétique*, XXXII (4), oct.-déc. 1991, pp. 445–468.

Sudeikin, V. "Vol'noe ekonomicheskoe obshchestvo." *Entsiklopediia Brokgauz-Efrona.* SPb.

USGS (2000): http://pubs.usgs.gov/circular/c1050/intro.htm.

Vucinich, Alexander (1970): *Science in Russian Culture, 1861–1917.* Stanford UP, 575 p.

7

Philosophical Developments in Pedology in the United States: Eugene Hilgard and Milton Whitney

Ronald Amundson[1]

Introduction

The development of Soil Science in the United States began to occur rapidly late in the 19th century. A large contributor to this success was an influx of funding from the Federal (and state) governments to land grant universities as a result of the Hatch Act of 1887 and the establishment of Agricultural Experiment stations throughout the country. A second impact was the growing presence of soil science within the Federal Bureaucracy[2] within the USDA. During the past 100 years the USDA/Academy relationship in pedology has remained strong and tightly interwoven. Numerous university faculty have left their academic posts for appointments in the Soil Survey Program in its various forms. The US Soil Classification System, instigated and published by the USDA, casts a large shadow over the organization of pedology text books and course contents in most universities. USDA-organized research programs have lent considerable knowledge to our understanding of soil and geomorphic relationships (Holliday et al., 2002). This situation is implicitly recognized by the numerous authors who have summarized the history of pedology (e.g. Holliday et al., 2002), but this unique relationship and its positive or negative effects on science have yet to be fully examined and discussed.

From this present vantage point, it is instructive and inherently interesting to look back to the late 19th century, when soil science and pedological research was in its truly formative stages, and was nurtured mainly within the walls of

---

[1] Division of Ecosystem Sciences, University of California, Berkeley, CA 94720.

[2] I use the term bureaucracy in the non-pejorative sense: "a body of nonelective government officials", "government characterized by specialization of functions, adherence to fixed rules, and a hierarchy of authority" (Webster's Seventh New Collegiate Dictionary).

academia and state experiment stations.[3] As Jenny (1961) and I (Amundson and Yaalon, 1995; Amundson, 2004) have outlined, the key intellectual leader in this development was Eugene Hilgard, Professor at the University of California from 1875 until his death in 1916. Hilgard was a towering figure in American science, education, agriculture, and politics. His voluminous and impeccably organized correspondence contains letters to and from the giants of science and politics in the late 19th century.[4] By the late 1880s, despite being located in a virtual academic outpost on the far edge of the continent, Hilgard had achieved a stature and influence that was remarkable for any man of his time. His soil chemical research was the model for the period, and this chemistry, combined with Hilgard's international reputation as a geologist, led to numerous ground-breaking concepts and publications beginning with his landmark work: *Report on the Geology and Agriculture of the State of Mississippi* (1860). In that report, Hilgard set out with remarkable clarity to discuss the origin, character-istics, and uses of soils. This document is additionally noteworthy in that Hilgard also produced an "Agricultural Survey" of Mississippi, a document and accompanying map that might be viewed as a reconnaissance soil survey and a legitimate precursor to our modern soil survey program.

During the 1880s, much of Hilgard's considerable energy and talent was devoted to inducing the newly formed USGS, under J.W. Powell, to incorporate "agricultural survey" into its mandated activities. At various points, Hilgard was offered a position within the USGS to develop such a program, an offer which he turned down, apparently viewing it as a reduction of his standing within the aca-demic community. Finally, in 1888, the Agriculture Department was elevated to a cabinet-level Department of Agriculture. Two key events occurred during this period that likely have had a continuing impact on the subsequent development of geology and pedology in the United States. First, J.W. Powell sought vigor-ously to transfer the USGS from the Department of Interior to the new Department of Agriculture (Amundson and Yaalon, 1995; Worster, 2001), an effort which collapsed due to numerous skirmishes that Powell had with Congress and administrators (Worster, 2001). Second, Hilgard was put forward (and approved by Congress) as the first Assistant Secretary of Agriculture, a per-son who would be "the man who is to be the organizing and controlling agent of the scientific work of the Department around or in which, is soon to centre,

---

[3] Notable exceptions to this rule were the occasional efforts directed to soils by the early USGS (e.g. Shaler, 1891).

[4] Please see the Appendix for a discussion of the Hilgard papers and the documentation methods used in this article.

the scientific affairs of the U.S. Govt." (Alvord, March 14, 1889 letter to Hilgard). To the dismay and embarrassment of everyone involved (except Hilgard it seems), Hilgard turned down the offer when the Regents of the University of California met and offered him a raise in pay, reduction in teaching, and additional laboratory assistance. Viewed from Hilgard's perspective, the Regents attention to his demands was a balm to his ego, and Hilgard seemed to truly enjoy his prominent standing within California and on the Berkeley campus. However, it is likely, in hindsight, that no single decision in Hilgard's career had a greater impact on his scientific legacy or on the very future of soil science (Amundson and Yaalon, 1995; Helms et al., 2002).

The purpose of this paper is to begin at this stage in Hilgard's life (1889), and to begin to examine (primarily from his perspective), the sweeping changes in soil science that were to occur over the next 20 years, changes in which he and some other academic scientists were reduced to vocal bystanders. The period was first examined in careful detail by Jenny (1961). The goal here is to illuminate more fully the start of the conflict between Hilgard and Milton Whitney, and to present an expanded version of this truly interesting struggle in American soil science. My method here is to freely quote from Hilgard's correspondence, allowing the reader to sense, firsthand, the personalities and opinions of some of the founders of our science, and to watch the drama unfold in their own language. The conflict was long and complex, and due to space limitations, I will focus on the first few years, when Hilgard and Whitney first meet and clashed scientifically and ideologically – the time when Whitney began developing a physics-based view of soil fertility that he would use to dominate Federal Soil Science for the next 35 years.

The Calm Before the Storm

In 1890 Hilgard was in his late fifties, secure in his stature in the growing University of California and politically and socially connected with a who's who of scientists, administrators and politicians from around the country. Yet, instead of comfortably seeing his legacy grow and spread in his later years, he was soon to be thrust into a battle over the future of soil survey, and soil science philosophy, in America. Hilgard's nemesis in this struggle would be Milton Whitney, the leader and originator of the US Soil Survey, and the leader of Federal soil investigations from 1894 until 1924, when C.F. Marbut assumed the role of the Chief of the Division of Soil Survey. However vigorous and personal the battle became, it began innocuously with a letter dated June 10 to Hilgard, from the then relatively unknown thirty year old Whitney, requesting that Hilgard review and comment on his enclosed manuscript.[5] In October, Hilgard replied (October 9, 1890):

> *"I owe you an apology for – or rather and explanation of – my failure to answer soon yours of June 10, with Ms. Inclosure. I was under such heavy strain near the end of the session that it was all I could do to stand up under it; and then when vacation came it proved no vacation at all, but as busy a time as the session, . . . Moreover, my eyes are so weak that I found the Ms. too much for them. Now that I have seen your article in print, I hasten to say at least a few worlds about your work and results, without, in the brief space of a letter, being able to do scientific justice"*[6]

Hilgard went into some detail to discuss the important differences between particle size analyses and "flocculated vs. single grain" states of the particles (due to soil chemical conditions), and sent Whitney his paper on "flocculation" (Hilgard, 1873). He concluded:

> *"I have so much of this sort of material* (physical data) *on my hands for elaboration that it seems a great pity not be enabled to work it up; but I am so loaded down with other work I can not get at it, much less pursue the investigation further in a systematic manner. I wish I could get a gentleman of your ability to work with me right here where I have the largest soil-collection now extant, probably: – some 1600 specimens from Cal. And from the Northwest – But not men enough to handle more than the current demand for work!"*

The manuscript that Whitney had sent Hilgard, *Soil Investigations* (Whitney, 1891), set out in a provocative and bold manner the thesis that most soils have adequate nutrients for plant growth, but that the key difference between soil's abilities to produce crop *"is to be found in the study of the physical structure of soil and the physical relation to meteorology and plant growth"* (Whitney, 1891). The mild, and paternal, reply made by Hilgard in this letter was to become a rare

---

[5] This letter does not exist in Hilgard's files, and is known from Hilgard's later reply. While the breadth and organization of Hilgard's correspondence in the Bancroft Library at Berkeley is remarkable, notable gaps or missing letters exist, due (according to communication with the late Hans Jenny) to materials carelessly discarded at the time of Hilgard's daughters' deaths, and the sale of their home.

[6] The tone of this introduction, and Hilgard's references to his health and work load, become very familiar to the reader of his voluminous correspondence.

occurrence in the correspondence between the two men. The difference of opinion, and the tenor of the conflict, soon escalated, and consumed much of Hilgard's attention for the remainder of his life. Before I begin this narrative, I review some brief details of Whitney's background, and his as yet unexplained conversion to a doctrine of the dominating importance of soil physical, as opposed to chemical, characteristics on crop growth.

Helms (2002) provides an important biographical overview of Whitney. As Jenny (1961) noted, Whitney was an ambitious and strong-willed adversary for Hilgard, who himself appeared to relish academic and political skirmishes. Whitney (1892a) boasted that he *"was brought up on a farm myself, and think I have had considerable experience in the methods and practice of farming"*. Biographical sketches note the constant presence of a cigar about him during his adult years, a habit which added even more to the impression of a vigorous and ambitious nature. Yet, somewhat incompatible with these impressions are details gleaned from his application to Johns Hopkins University in 1879, in which he stated he had been largely educated at home *"my health not having permitted close confinement"*. Additionally, he was not admitted to Hopkins in a regular degree program, but to a three year program of special study: *"My object in wishing to come to the University is to perfect myself in the study of analytical chemistry desiring to take that up as a profession, not being physically able to go through an entire college or university course . . ."*.

Following his course work at Johns Hopkins (work which focused on chemistry), Whitney became an assistant chemist (1883) with the Connecticut Agricultural Experiment Station (Helms, 2002). Following subsequent positions in North and South Carolina, Whitney obtained employment in the Maryland Agricultural Experiment Station, where he acquired the title "Professor", a title that followed him the remainder of his career. The Station was directed by Major Henry E. Alvord, president of the Maryland Agricultural College, and a longtime friend and confidant of Eugene Hilgard. Alvord made soil investigations a priority, and Whitney proved to be productive in this effort (Helms, 2002). It was thus, under the tutelage and protection of Alvord, that Whitney developed his ideas on soil physical properties and published his ideas in the Annual Report of the Maryland Experiment Station in 1891.

Stormy Skies

Jenny (1961), in his characteristically precise and concise manner, notes that in 1892, Hilgard published a review of Whitney's (1891) paper, a review "almost fatherly in tone". Yet, the backdrop that led to this review is illuminating, and one might certainly disagree with Jenny that it was "fatherly" either in tone or intent. The events leading to the 1892 review began subtly. Hilgard (April 14, 1891) wrote a letter to his friend Alvord, which included the following comments about Whitney:

> *"I think you have secured a good worker in Whitney; soil physics needs a great deal of additional work and he has started well, although I think he will have to fall back a few steps and lean on what has been done before him, before he will get a definite result. But we can not have too many at work on this important subject of soils."*

However, as the year went on, it becomes clear that Whitney's paper, and its effect on soils work, was beginning to gnaw at Hilgard. In a letter to H.P. Armsby (Feb. 22, 1892):

> *"By the way, there seems to be some stirring up of old topics in the matter of soil study. Arizona has sent a delegate to study up our methods here; Oregon, Washington, and Colorado, and lately Nebraska, have come for the same by letters and otherwise ... However, ... some of their methods, are signs of the times for which I have hoped for 35 years; a curious outgrowth of the same is Whitney's latest, in which all our researchers in vegetable development ... are knocked sky-high."*

What knocked Hilgard's staff "sky-high" was Whitney's focus on particle size and soil volume as keys to soil water relations and plant fertility. Whitney's (1891) report makes interesting reading, and some of his approaches to soil physical properties (and interpretations of data) will be familiar to many today. However, some aspects that severely weaken his thesis, and particularly galled Hilgard and others, were the lack of data, lack of literature review and citations, and the array of assumptions made. At the outset, Whitney assumed an exceedingly paternalistic tone with respect to Maryland farmers, and their naive knowledge about soils – a tone that it must have struck Hilgard (who had spent decades elevating the study of soils to a science) wrong:

> *"It takes really very little experience for one to judge at a glance whether a soil is suited to grass or wheat or tobacco or watermelons, and he has but to turn up a small handful of earth to see if the soil is in good condition ... And yet, agricultural chemists have worked over this problem for years, arguing points from minute differences in chemical composition of the soils or plants ... overlooking the fact that the farmer can tell from the appearance whether a soil is*

in "good heart" . . . Those of us who are engaged in agricultural investigations,
even in soil studies, are not as far advanced as the farmer in our knowledge of
the soil."

Whitney spent much of the remainder of his report focusing on particle size
analyses of selected agricultural soils in the state, a discussion which leads to an
agricultural classification of soil *based on the number of mineral grains per gram*
of soil:

> *"For example, no crop can be successfully grown, except under highly arti-*
> *ficial conditions of manuring with organic matter, or by irrigation, on a*
> *soil having so few as* <u>*one thousand seven hundred million*</u> *grains per*
> *gram."*

Based on particle size analyses, Whitney calculated (but did not measure) soil
hydraulic conductivities and discussed the abilities of soil to supply water to crops
via the upward movement of moisture by capillary action. Many of these calculations, as Whitney himself noted, involved assumptions that had not been experimentally validated. Despite these uncertainties, the tone of the report is such that
a reader unaware of the work of Hilgard and others at the time, would be led to
view Whitney's work as some of the first truly scientific studies on soils, and certainly a paradigm shift in soil-plant relations.

In 1892, Professor William Frear of Pennsylvania State College was editor of
Agricultural Science, and among the Associate Editors of the journal were Hilgard
and Whitney. Frear contacted Hilgard to write a review of Whitney's (1891) paper
for the journal. Hilgard replied (March 12, 1892):

> *"Your suggestion to take Whitney's work in hand is emphasized in a sim-*
> *ilar request from Alvord, who I am afraid will be disappointed in what I*
> *shall say . . .".*

In a letter to Alvord, Hilgard wrote (March 19, 1892)

> *"I have been so extremely busy that I have not yet had time to go quite in*
> *detail over Whitney's report on soils; I will do so because as co-editor of*
> *'Agr. Science' I am called upon to review it, and I do not yet see a way to*
> *do so without rubbing him too hard the wrong way . . . Candidly, I fail to*
> *see in his writings and work any true induction; he has conceived a certain*
> *notion and he pursues that through thick and thin . . . He starts out, as did*
> *the old scholastics, with a thesis which he sets out to maintain, and does*

maintain regardless of consequences . . . he can no more make his crops live on soil flocculation than he himself can live on the movements of water in his body."

Jenny (1961) writes that the conflict between Hilgard and Whitney became, for Hilgard, particularly personal and heated early in the next century, though this and other letters clearly establish the first phases of this development. Despite his growing frustration with Whitney and his scientific approach, Hilgard was willing to defer to his friend Alvord, and withdraw his review of Whitney's work from *Agricultural Science* (April 5, 1892):

"Your's of the 28ᵗʰ is to hand. What you say is not new to me – I know that Whitney has claimed to be in possession of just such experimental proof, in fact has done so toward me last summer. But why in the name of sense and modern scientific methods don't he give these first of all instead of rushing into print with a lot of unsupported assertions? Quite independently of the actual facts of the case, I think he deserves a severe overhauling for knowing no better than to same as he does – the whole tone of the paper is of a kind that we have ceased to see for a number of years, I do not think it desirable that the old style should be revived."

Hilgard then offered to withdraw his review and send it to Alvord, who could then pass it to Whitney, so he could see what type of trouble he was possibly headed for. Frear agreed to this, and in a reply to this affirmation, Hilgard wrote (April 18, 1892):

"You are quite right about Whitney's deserving richly to be scored at least as much as my review would do . . . I only wish, on account of my personal relations to Alvord . . . that he ought to have stood in the way of the stultification of American sciences by such a pronunciation."

Alvord replied to Hilgard, apparently assuming some of the responsibility for the style of Whitney's paper, to which Hilgard replied (April 22, 1892):

"I cannot omit demurring pointedly to your assumption of the responsibility for the criticism to which I think Prof. Whitney has made himself liable in this publication . . . I leave it to you and Frear to determine about the publication of my review . . . I hope you have shown it to him (Whitney); let him say what he has to reply at once and publish them both together."

In a letter to F.H. King (whose censoring and firing later by Whitney was to play prominently in Hilgard's dislike of Whitney), Hilgard wrote (April 23, 1892):

> *"I find it* (King's enclosed paper on soil moisture) *refreshing reading after my unsavory task of trying vainly to find something in the same spirit in Whitney's treatise. He ought to have lived a century ago . . ."*

Awaiting a reply from Alvord and Whitney, Hilgard wrote Frear (May 16, 1892): "*I hear nothing . . . I am not especially anxious about publication – but I am getting 'mad'.*"

Finally, in a letter to Frear dated May 30, 1892, Hilgard gladly notes that Alvord "has had the good sense" to send Hilgard's review back to Agricultural Science for publication.[7] The review appeared in the July issue (Hilgard, 1892a).

The review was detailed and biting. Hilgard begins by again stating that it is good that soil physics is again being revived, but notes "*few probably will agree with the author in the deductions made from these premises, until more experimental proof shall be forthcoming in their support*". Hilgard criticized the lack of acknowledgments of previous work on soil water and physical properties, saying "*anyone not acquainted with the history and literature of agricultural science would be led to infer that up to this time, little or no attention had been paid to the investigation of water movement in soils*".

But what must have particularly chaffed Hilgard was Whitney's suggestion that "*it really takes very little experience for one to judge at a glance whether a soil is suited to grass, or wheat, or tobacco or watermelons*". The tone of this statement essentially flew in the face of all of Hilgard's career work on soil fertility and the management and remediation of salt-affected soils. Hilgard attacked this in several ways, first noting that while Maryland farmers might know their own soils, they certainly wouldn't be able to do the same in California, and only a scientist and a scientific approach would be able to accomplish this feat. Hilgard then showed numerous exceptions to Whitney's particle size classificaion of soils and their capabilities, indicating that such classifications were neither universal or even practiced by farmers ("*In other words, the grounds for selection* (of a soil for a given crop type) *are financial, not intrinsic*").

Hilgard criticized Whitney's lack of recognition that the chemical composition of soil particles (not just size) is also critical to soil physical behavior.

---

[7] In July of that year, Hilgard and his family traveled to Europe for a year of vacation and study. There is an apparent gap in the correspondence record at the Bancroft Library for this time period.

Hilgard gave an erudite discussion of the differences in physical behavior of clay sized quartz vs. soil clay (which upon *"wetting enlarges to twenty or more times its original size"*). He also faulted Whitney for ignoring humus, iron hydroxides, etc.

Hilgard particularly attacked Whitney's premise that *"deterioration of soils by cultivation is not due to loss of plant food so much as to a change in the physical structure of the soil"*, noting no data were provided to support such a claim. Hilgard concluded: *"That the relations of soils and plants to the water of the soil are of primary importance, has been denied by no one, either in the past or present. But the movements of soil water should possess the almost exclusive importance attributed to it by Whitney ... will have to be substantiated by more cogent experiments and reasoning than, thus far, has been shown by the published record of the author's work."*

One might expect that such a scalding and detailed review by such a senior scientist might cause a young scientist to pause, and if submitting a reply, it would be of a reflective nature. However, Whitney showed no such inclinations, and published a reply that only elevated the rhetoric and tensions (Whitney, 1892a):

> *"Prof. Hilgard has spoken as a very able exponent of the view of the 'agricultural old school,' and I think ... my work marks a distinct advance in agricultural science ... The work was not undertaken with a view to combating the prevailing ideas of agricultural chemists[8]. It seems to me that it can be left largely to others to prove or disprove the generalizations which have been made, although I shall by no means leave it to others to do so. There has been neither time nor opportunity to sum up the work of others, and this has not seemed necessary ... since the work is largely based upon that 'empiric' knowledge the farmer has of local soils ..."*

In what we now know to be an entirely incorrect interpretation of soil mineralogy, Whitney went on:

> *"I cannot agree with Prof. Hilgard in regard to the essential properties of clay. I have pointed out ... that these essential properties of clay, as recognized in mass, can probably all be explained on purely physical principles, being a function principally of the size of the grains ... I think that if quartz were powdered as fine as the limits I have assigned to the clay groups ..., it would have all the essential properties of 'clay' in mass."*

---

[8] This conflicts with Whitney's later statements that took a much stronger stance.

As a final blow to Hilgard and his laboratory expertise:

"I would take this opportunity to urge that the best place to carry on these investigations is in the field and in close touch and sympathy with the farmer, rather than exclusively in the laboratory."

This comment was obliquely directed to one of the premier field geologists in America, and a man who spent much of his energies on soil investigations for the farmers of California!

While Hilgard's personal correspondence (and his reaction to these statements) is missing, Hilgard wasted little time in submitting a reply to *Agricultural Science* (even though he was then on vacation in Europe). It was scathing and direct, brusquely correcting Whitney's errors, pointing out his experimental flaws, and critiquing his many assumptions and lack of scientific protocol (*"It is a sound article in the code of scientific investigation, too often violated in these days, that a report of work should be preceded or accompanied by a brief reference, at least, to the existing status of the subject investigated . . ."*). Hilgard strongly criticized Whitney's assertion that all clay-sized particles have the same properties, giving examples of the difference between the behavior of kaolinite and quartz. Hilgard challenged Whitney's unsubstantiated claims that differences in soil's abilities to produce crops is due to "a change in the physical structure of the soil." by concluding *"The least one can say in examining his record, is 'not proven'."*

It is worth noting that Whitney, in these initial stages, appears to have backed off, at least a bit, in the tenor of his original statements regarding the utility of soil chemical analyses, and tried to make the point that his new line of work on soil physical properties would "supplement" the work of soil chemistry, not conflict with it (Whitney, 1892a). Though, it's fair to say he still made critical remarks regarding the status and utility of soil chemistry, and these critical remarks continued to attract Hilgard's attention and likely contributed to the polarization of their positions.

While the public debate over Whitney's (1891) first report was being waged, Hilgard and Whitney published back-to-back papers in the USDA Weather Bureau Bulletin. Hilgard's (1892b) paper, *The Relation of Soils to Climate*, has been hailed by Jenny (1961) as Hilgard's *non plus ultra* report, one that gained Hilgard considerable notoriety in Europe and among select scientists in the American soils world. The report begins with Hilgard's advanced concepts of soil and soil formation, then proceeds to present chemical and physical data of humid and arid regions to support the theoretical hypotheses forwarded in the work. To the modern scientist, Hilgard's discussion of clay, humus, and carbonate in relation to climate is particularly interesting. As Jenny discusses, many of these concepts emerged again, years later, many times by authors unaware of Hilgard's ground-breaking work.

Jenny (1961) argues that Hilgard's obscurity was due to the rise of Whitney, who published the paper, that immediately followed Hilgard's, entitled *Some Physical Properties of Soils in Their Relation to Moisture and Crop Distribution* (Whitney, 1892b). The contrast to Hilgard's paper is sharp. Whereas Hilgard presents a virtual *tour de force* on soil formation processes and soil chemistry, Whitney article begins with an extended lament that "*It has not been many years since it was generally believed that the chemical analysis of a soil would show what class of plants the soil is best adapted to produce, and what elements of plant food are lacking in the soil for the best development of other crops.*" Whitney proceeds on what will become, for him, a mantra for his work: soil chemistry research has shown little tangible benefit to farming, soil physical properties have "recently" (e.g. Whitney) become recognized as important – but little is known about them, and Whitney's goal is to expand knowledge on this front. This report ("*Bulletin 4*"), contains considerably more data and experimental background than the 1891 report, though many of Hilgard's (and others) objections to the interpretations and extensions of the data to agriculture are not addressed or altered. This volume of the Weather Bureau Report sets in sharp contrast the philosophical schism between what would become the "Whitney school", and the "old school" as Whitney referred to Hilgard (Whitney, 1892a).

A Deluge Begins

At this point, the dispute between the two men was between two scientists in academic and experiment station appointments. However, this was soon to change. When Hilgard turned down the offer to become Assistant Secretary of Agriculture in 1889, Edwin Willets was subsequently selected. With the change of the presidential administration in 1892, Willets was replaced by Charles Dabney (Helms et al., 2002). Hilgard took great interest in these political changes (and their effects on the Experiment Station), and corresponded with his colleagues about the fate of the USDA. In a letter to Frear (Dec. 27, 1893), Hilgard discussed the future of the experiment stations and the support of the government. With respect to the change in leadership, he said: "*I think we are very fortunate indeed in having him* (Dabney) *in the position at this somewhat critical time*", and he told Frear that he would write a "private' letter to Dabney. Hilgard wrote to Dabney (Dec. 28, 1893):

> "*Permit me to congratulate you . . . upon your appointment as Ass't Sec., 'y, which I understand is now a settled fact*". Hilgard noted that the new

Secretary (Morton) had made a number of unsettling comments regarding the Agricultural Experiment Stations, and vaguely threatened *"it may come to the stations making a common cause and going on the warpath in self-defense. I trust (you can) avert any such unpleasant necessity."*

While Hilgard may have had ample reason to have faith in Dabney's leadership – which proved to be strong and innovative (Helms et al., 2002), Dabney moved quickly to achieve his vision of elevating the importance of soils to agriculture, and appointed Milton Whitney as chief of a new Division of Agricultural Soils in 1894 (Helms et al., 2002). Thus Dabney, in the position that Hilgard himself had unceremoniously declined only 5 years earlier, had appointed Hilgard's new nemesis as the person in charge of Federal soil investigations! Dabney apparently was greatly influenced by Whitney's pronouncements, and told a meeting of representatives of state experiment stations in 1894 (Helms et al., 2002):

"I speak this morning to many agricultural chemists, so that I need not take time to explain the disappointments that we have all felt with regard to the results of the chemical analyses of soils."

It seems that despite their large differences in opinion, Hilgard took the developments in stride and with a sense of humor. In a letter to Smith (Feb. 26, 1894), Hilgard wrote:

"There is a lively revival of this favorite pastime (soils) *just now, so I have had to put it in shape and in order to put a quietus on a lot of silly propositions made by youngsters who are just starting out on their first soil work and think they have found me napping . . . So I have been giving it out hot and heavy of late, and there is more to come."*

This last phrase proved to be prophetic. Nonetheless, Hilgard was magnanimous upon Whitney's appointment. In a letter to Frear (April 23, 1894), Hilgard wrote:

"Whitney sent me special announcement of his establishment at the dep.t of Agriculture; whereon I wrote him a letter of congratulation, with a little emphasis upon the fact that he would now have a wider field from which to draw conclusions . . . I hope he may now become really useful, even though he still proclaims 'it is now coming to be thought' that soil exhaustion etc."

"There is More to Come"

In a mere five years after Hilgard turned down the Assistant Secretary of Agriculture post, soil investigations had indeed come to the USDA. From Hilgard's perspective, the fact that the man in charge of this program was a "youngster" with "silly propositions" was probably not lost on him, though he was certainly unaware of the magnitude of the difficulties to come between them. It is here, at the end of the prologue of their drama that I choose to pause in this narrative, with the main battle between them in the early 1900s still ahead. I hope that this introduction to Hilgard and Whitney gives some personal insight into their characters, and provides some rationale for the events that would later unfold, events that would become part of a very public and acrimonious debate (Jenny, 1961).

Acknowledgments

I thank Professor Dan Yaalon for his encouragement to pursue this topic, and his patience during the process.

Appendix

The historical data used in this article are taken from documents contained in the *Hilgard Family Papers*, in the Bancroft Library of the University of California at Berkeley. The Bancroft Library officially was established in 1905, but has existed in its present location with its expanded mission since 1973. The library includes a number of "special collections", including a Rare Books Collection, the Mark Twain Papers and Project, the Regional Oral History Office, the University of California Archives, the History of Science and Technology Program, and the Pictorial Collection. It is one of the largest special collections in the United States.

The Hilgard papers (call number: BANC MSS C-B 972) consist of 32 boxes, 33 volumes, 7 cartons, 1 package of glass slides, 5 oversize folders. Part of the collection is stored off-site, and advanced notice is required for use. The collection consists primarily of the papers of Eugene Hilgard relating to his work on the Northern Transcontinental Survey, the U.S. Census in 1880, and as professor in the University of California College of Agriculture. Included are letters, manuscripts of his writings, subject files, lecture notes, accounts, personalia, clippings, etc. Also in the collection are a few papers of his father, Theodor E. Hilgard; papers of his brother Julius E. Hilgard concerning his work with the U.S. Coast and Geodetic Survey; papers of his brother Theodore C. Hilgard, some of which relate to the magnetic survey; and papers of his daughters, Alice and Louise, primarily letters of condolence on the death of their father.

A finding aid to assist scholars is available. Most of Hilgard's out-going mail exists as volumes of letterpress copies – essentially carbon copies – of typed letters (with occasional handwritten notes or corrections to typed documents). A few boxes of other out-going letters exist in boxes arranged by year. In-coming mail is arranged by author. In this chapter, the dates (and direction) of the correspondence should allow scholars to quickly locate the cited letters within the Hilgard collection.

References

Amundson, R. 2004. History of Soil Science: Hilgard. Encyclopedia of Soils in the Environment. D. Hillel (ed.). *In press.*

Amundson, R. and D.H. Yaalon. 1995. E.W. Hilgard and John Wesley Powell: Efforts for a joint agricultural and geological survey. Soil Sci. Soc. of Am. J. 59:4–13.

Helms, D., A.B.W. Effland, and S.E. Phillips. 2002. Founding the USDA's Division of Agricultural Soils: Charles Dabney, Milton Whitney, and the State Agricultural Experiment Stations. pp. 1–18, In: D. Helms, A.B.W. Effland, and P.J. Durana (eds), Profiles in the History of the U.S. Soil Survey. Iowa State Press, Ames, IA.

Helms, D. 2002. Early Leaders of the Soil Survey. pp. 19–64. In: D. Helms, A.B.W. Effland, and P.J. Durana (eds), Profiles in the History of the U.S. Soil Survey. Iowa State Press, Ames, IA.

Hilgard, E.W. 1860. Report on the Geology and Agriculture of the State of Mississippi. Jackson, MS.

Hilgard, E.W. 1873. On the silt analysis of soils and clays. Am. J. of Sci. & Arts 106:288–296, 333–339.

Hilgard, E.W. 1892a. Soil investigations. Agricultural Science 6:321–328.

Hilgard, E.W. 1892b. A report on the relations of soil to climate. Bulletin No. 3. US Department of Agriculture Weather Bureau, Washington, D.C.

Hilgard, E.W. 1892c. Soil investigation and soil physics. Agricultural Science 6:566–570.

Holliday, V.T., L.D. McFadden, E.A. Betttis, and P.W. Birkeland. 2002. Soil Survey and Soil-Geomorphology. pp. 233–274. In: D. Helms, A.B.W. Effland, and P.J. 1Durana (eds), Profiles in the History of the U.S. Soil Survey. Iowa State Press, Ames, IA.

Jenny, H. 1961. E.W. Hilgard and the Birth of Modern Soil Science. Collana della Rivista "Agrochimica", Pisa, Italy.

Shaler, N.S. 1891. The origin and nature of soils. pp. 213–352. In: J.W. Powell (ed.) Annual Report of the US Geological Survey to the Secretary of Interior, 12th. 1890–91. Part I – Geology. US Government Printing Office, Washington, DC.

Whitney, M. 1891. Soil Investigations. Fourth Annual Report of the Maryland Agricultural Experiment Station and Agricultural College. C.H. Baughman and Co., Annapolis.

Whitney, M. 1892a. Soil physics and crop production. Agricultural Science 6:427–431.

Whitney, M. 1892b. Some physical properties of soils in their relation to moisture and crop production. Bulletin No. 4. US Department of Agriculture Weather Bureau, Washington, D.C.

Worster, D. 2001. A River Running West: The Life of John Wesley Powell. Oxford University Press, New York.

Inset

Eugene Woldemar Hilgard was born January 5, 1833 in Zweibrücken, Germany and died January 8, 1916 in Berkeley, California. In 1836, due to political reasons, his family moved to Illinois where, because of the state of the local schools, the Hilgard children were home-schooled by their father Theodore, focusing especially on mathematics and languages. Hilgard's only formal education took place in Europe, culminating with a PhD in Chemistry from the University of Heidelberg in 1853. Hilgard returned to the United States looking for employment, and to the surprise of many, accepted a position as assistant to the State Geologist of Mississippi. As Hilgard would do many times in his life, he turned this seemingly "uninteresting" position to his advantage. It was in Mississippi that Hilgard formally turned his attention to soils, and his final report on the *Geology and Agriculture of the State of Mississippi* (1860) marked a conceptual breakthrough in the models of soil formation and chemistry, and was the beginning of Hilgard's long career that continuously focused on the utility of soil chemistry to agriculture, and the importance of agricultural surveys to society (e.g. soil surveys).

Following stints as Professor at the Universities of Mississippi and Michigan, Hilgard accepted the post of Professor and Dean of the College of Agriculture at the fledgling University of California at Berkeley. Once again, beset by physical isolation, inadequate facilities, and skepticism by both colleagues and constituents, Hilgard rose to the challenge and arguably established the first agricultural experiment station in the USA, and made the study of soils a bona fide academic discipline. Some of Hilgard's major contributions and accomplishments include the enormous *Report on the Cotton Production of the United*

States, Vols. V to VI (1882) (which included maps of soil regions for the southern states and California), work done as part of the Northern Transcontinental Rail Survey, and landmark papers on the remediation (and chemistry) of salt affected soils and the relation of soils to climate. All during this period, Hilgard directed the California Agricultural Experiment Station, writing voluminous reports, directing research on wide ranging problems (including plant pathology and viticulture/enology), and personally corresponding with constituents throughout the state. Though much of Hilgard's biographical researchers have focused on his soil and agricultural research, Hilgard was a prominent geologist, who corresponded frequently and widely among the key figures of his time, including Sir Charles Lyell. His research on the Cenozoic formations of Mississippi, Louisiana, and Texas led to him later being called the "father of American Tertiary stratigraphy".

Eugene Hilgard received wide acclaim and recognition in his long lifetime for his varied contributions to science. Hilgard was elected to the National Academy of Sciences early in his career. In later years, he received honorary Doctorate degrees from Columbia, Michigan, Mississippi and Berkeley. He was only the second graduate of Heidelberg to receive its "Golden Degree", commemorating the 50th anniversary of his dissertation. Mountains, streets, and University buildings in California now bear his name, and mark his legacy.

8

Development of the Soil Cover Pattern and Soil Catena Concepts

A.N. Gennadiyev[1] and J.G. Bockheim[2]

Introduction

At the beginning of the 20th century, two groups of pedogeographical regularities were clearly distinguished. One group, controlling the variations of soils across continental expanses, is contingent on climatic and resultant vegetational variations. Dokuchaev (1899) clarified general principles of global soil distribution and developed the doctrine of natural soil zones, including the laws of horizontal and vertical bio-climatic zonation of soils. These changes were reflected by him in the first schematic soil map of the northern hemisphere, presented at the World Exhibition in Paris in 1900 (see Glazovskaya, 1981). Prasolov (1922) divided the soil bio-climatic zones and subzones into provinces and facies. Fundamental doctrine of zonation and provinciality was the theoretical basis for the world soil maps compiled by Glinka in 1927, and Prasolov in 1937 (see Dobrovolsky, 1994).

The other group of regularities, those controlling changes of soils within small areas, is contingent on relief. Although Dokuchaev (1879, 1883) developed mainly the ideas on bioclimatic zonation of soils, he stressed the importance of "the topography of soils", that was much later included in the pedological concepts of "soil cover pattern" and "soil catena". The initial accumulation of facts and regularities related to the concepts was begun mainly due to development of some large-scale soil surveys (Gennadiyev et al., 1995).

The objective of this review is to examine the history of the soil cover pattern and soil catena concepts.

[1] Faculty of Geography, Moscow State University, Moscow, Russia.

[2] Department of Soil Science, University of Wisconsin, Madison, USA.

Early Soil Surveys and Initial Accumulation of Soil Cover Pattern Data

The approaches to understanding of soil topography began to develop in Russia during the Dokuchaev and Post-Dokuchaev periods by his disciples and followers (see Gennadiyev and Olson, 1998).

In 1900 Sibirtsev (Sibirtsev, 1900a, 1900b), summarizing the materials of the Dokuchaev's Nizhni-Novgorod and Poltava expeditions, introduced the term of *soil combinations* in order to characterize "constant changes" of components of soil cover. He wrote: "Schematic soil zones that appear homogenous on small-scale maps are in fact complex . . . When soils are highly variable, the concept of cropland type or topographic type for a given region is useful. Cropland may be complex, but if this complexity is similar for a group of landholdings and can be portrayed by a single scheme, then such a scheme . . . can be considered a unit of comparison of the region . . . If a certain area, as represented on the map, includes many different soil types (i.e. if there is a frequent alternation of loam and sandy loam or of chernozem and eroded or coarse soils), a special symbol may be used" (p. 455). Sibirtsev (1900b) provided the first classification of soil cover patterns by distinguishing seven "plowland types" for the podzol zone, six for the forest-steppe zone and nine for the chernozem zone. This was the first contribution to the general concept of soil cover pattern (Fridland, 1972).

Vysotskii (1906) gave the typical schemes of the dependence between the soil pattern and relief at small distances for several natural zones – forest, forest-steppe, and others. These schemes were developed on Dokuchaev's idea of topographic soil chains and were very similar to the soil catenas established by Milne in the 1930s. As early as 1900 to 1903, Sibirtsev (1900b) and D.N. Vikhman studied selected "keys" in Pskov region in order to explain in detail peculiarities of the soil cover structure and to account for them in smaller scale mapping of extensive areas (Shvergunova, 1974).

The Russian soil surveys of that period almost always contained some data on the pattern of soil cover. Bogdan (1900) described in detail the arid steppe soil complexes of the soil cover developed within the Valniki Experimental Station area. Dimo and Keller (1907) furnished a painstaking description of semidesert soil complexes of the Caspian lowland. They explained the appearance of the complexes by the redistribution of moisture and salts between elements of microrelief.

Neustruev (1915) examined soil combinations of plains and mountains within the Trans-Volga region and Soviet Central Asia. He suggested the term "mesorelief" and described two groups of soil combinations: complexes related to the microrelief, and combinations related to the meso- and microrelief. Neustuev suggested replacing the concept of zonal soils by the concept of zonal soil complexes. He first began to develop the idea of the degree of contrast in soils within combinations. Following these approaches he proposed that "any map of soils is a map of soil complexes" (p. 3). Neustruev presented the map legend as a system of soil combinations.

Existence and diversity of soil combinations was confirmed by using aerial photographs, that were widely applied to soil surveys in Russia in the 1930s. This technology allowed characterizing soil complexes by quantitative and statistical methods. Statistical analysis was used for the first time by Serdobol'skii (1937) to describe quantitatively the solonets soil complexes within the Trans-Volga region.

Prasolov's paper presented in 1927 at the First International Congress of Soil Science in Washington, D.C., indicated Russian achievements in the field of soil mapping and stressed the great importance of investigations of soil complexes and combinations (see Gennadiyev and Olson, 1998). A manual on soil surveying (Krasyuk, 1931) contained a special chapter on mapping methods of soil complexes.

Soil surveying began in the USA in 1899. By 1912 a large portion (520,234 mi^2) of the USA had been mapped, commonly at a scale of 1:63,360 (Marbut et al., 1913). The primary soil map unit was the soil type, which was based on texture of the topsoil. At this time, very little was mentioned about the complexity of the soil cover, though some leaders of American soil science of that period were close to understanding such phenomena as soil combinations. Hilgard (1906) emphasized that soils occur on landscapes not only as spatially homogenous bodies but also as transitional.

In the 1930s Kellogg (1933), as a head of the U.S. Soil Survey, supported introduction of the complexity in the soil survey. Under his leadership U.S. soil cartography began to make extensive use of the complex cartographic units known as associations. However, these soil associations were not the same as the Russian soil complexes. They were not interrelated genetically and didn't consist of regularly repeating soil cover patterns. The soil associations that were showed on the U.S. maps were mostly formed by the different soil series of significant genetic and morphological diversity. According to the opinion of some American soil scientists (Simonson, 1997), as the U.S. soil surveys developed through the 20th century genetic relationships among soil series were expressly de-emphasized, because soil survey work focused mainly on describing, classifying and interpreting soil series. The soil associations were used as an instrument of generalization in small-scale mapping. As the scale of the map is decreased, the number of soil associations is increased. The associations were extensively shown on the map of the USA (Soil Associations of the United States) in the 1938 USDA *Yearbook of Agriculture* (Soil Survey Division, Bureau of Chemistry and Soils, 1938) and on maps of many states including Georgia, Indiana, Iowa, and Oklahoma.

In 1951 the U.S.A. Soil Survey Staff (1951) adopted the ideas of the complex structure of soil cover and the necessity for cartographic delineation of contours with heterogeneous soil covers. Soils were described as "dynamic three-dimensional land-scapes" as well as profiles (pp. 5–8). More specifically, "influence of soil behavior on any one characteristic, or on a variation in any one, depends on the others in the combination" (p. 7). The *Soil Survey Manual* (Soil Survey Staff, 1951) emphasized the differences between soil taxonomic and soil map units. The soil complex was defined as a "combined" soil taxonomic unit containing two or more recognized units.

The concept of a soil combination, i.e. interrelated soils, was more correctly reflected in the soil-geomorphological research program of the U.S. Soil Conservation Service, created under the impetus of C.E. Kellogg and G.D. Smith in 1953. They considered the program as a methodological support of soil survey work which has to be based on a fully developed multi-site series of soil-geomorphology studies (Effland and Effland, 1992). By this time the Milne's and Bushnell's soil catena concepts were well-known in the U.S.A.

Soil Catena Concept as a Component of a Soil Geography Paradigm

Milne (1935, 1936a, 1936b) introduced the concept of the catena in his work in East Africa. A catena was defined as "a unit of mapping convenience . . . a grouping of soils which while they fall wide apart in a natural system of classification on account of fundamental and morphological differences, are yet linked in their occurrence by conditions of topography and are repeated in the same relationships to each other wherever the same conditions are met with" (Milne, 1935, p. 197) (Figure 1). According to Milne, a catena is a "chain" of soils "hanging" between two summits. Milne considered erosion as one of the factors of soil catena formation. He suggested that a soil sequence within a catena varies with the maturity of the relief, with the underlying parent materials and with new cycles of erosion. Milne arrived at the conclusion, previously deduced by Neustruev (1915), that the soil-climatic zones con-sist of zonal complexes rather than zonal soil types. Milne (1935) not only recognized the relation between soils and landforms, but he also saw the catena as a soil map unit and a soil classification unit. He argued that soils along catenas in East Africa are influenced by erosion and deposition of sediments, a precursor of the K-cycle later introduced by Butler (1959).

Bushnell (1942) elaborated on the concept of catenas and redefined the catena as a group of soils having close geographic relationship and association but having different profile characteristics owing to the influence of differences in relief, drainage, and time exposed to soil-forming processes. It is interesting that Bushnell (1942)

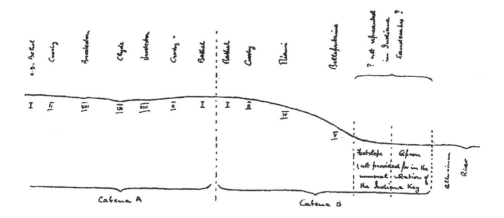

Figure 1. The original soil catena concept as depicted by Milne in a letter sent in November 1938 to T. Bushnell (Bushnell, 1942, p. 467).

corresponded directly with Milne prior to Milne's death. Bushnell restricted the catena concept to taxonomic usage, i.e., "taxonomic catenas". The simplest components, i.e., soil units, were called "major profiles". "Catenary complexes" were certain associations of soil types. He also identified chrono-catenas, flor-catenas, etc. After Bushnell, the terms *well drained, moderately well drained, imperfectly drained,* and *poorly drained* were recommended by the Soil Survey Staff (1951) to describe individual soils associated in toposequences where drainage was related to the water table.

From 1953 to 1980 the USDA Soil Conservation Service (now the Natural Resources Conservation Service) sponsored several soil-geomorphology projects in representative areas of the country. Originally proposed by C.E. Kellogg and G.D. Smith, the purposes of the projects were to understand the processes responsible for the soil cover pattern, to improve the quality of the soil mapping program, and to make more informed land-use interpretations. Holliday discusses these projects in Chapter 9.

According to Holliday and others (2002), the soil-geomorphology projects were the single greatest contribution of the soil survey in the USA to an understanding of factors and processes controlling soil cover patterns at a variety of scales. Not only did these studies enable the training of leaders in pedology and geomorphology, but also they enhanced our understanding of the soil-forming factors, soil-forming processes, and soil geography and mapping.

In 1953 Hole (1953) suggested terminology for describing soils as three-dimensional bodies rather than strictly as two-dimensional soil profiles. He emphasized "Bushnell natural drainage" or toposequence designations, as well as landscape position.

Polynov (1953) grouped all the varieties of elementary landscape on the earth's surface and the associated soils into three groups: eluvial, superaqual, and subaqual landscapes. The main focus in studying them is on radial fluxes of matter and on interrelationships in soil-plant, soil-parent material, and plant-soil-groundwater systems. Elementary landscapes and the associated soils located at different elevations in the relief and linked by lateral fluxes of matter were called "geochemical landscapes" (Figure 2). Butler (1959) identified two "frames" in the study of soils: (1) groundsurface status, which included elements of soil stratigraphy and soil history, and (2) the "component of the groundsurface", which reflects the soil cover pattern.

Ruhe (1960, 1969) developed a soil-geomorphic model that integrated soil characteristics with hillslope models. He studied hillslopes in three dimensions, subdividing them into three units: summit, shoulder, and backslope. This approach was an extension of Milne's (1935) catena concept. Ruhe (1960) alluded to Butler's (1959) K-cycle model. Later in his book *Quaternary Landscapes of Iowa*, Ruhe (1969) emphasized a quantitative approach to soil-landscape relations. He introduced the concept of "open" and "closed" drainage systems. Ruhe and Walker (1968) later presented a five-unit hillslope classification, including the flat

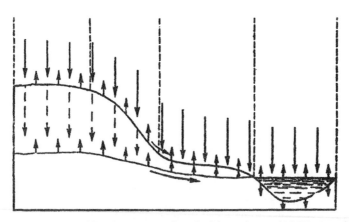

Figure 2. Geochemical landscapes as identified by Polynov (1953): 1 – eluvial, 2 – trans-eluvial, 3 – superaqual, 4 – aqual. Arrows show direction of additions and transfers of matter.

summit, convex portion of the shoulder, the more or less uniform slope of the backslope, and the concave portion of the base, which was further divided into upper footslope and lower toeslope position.

Yaalon and his colleagues (Dan and Yaalon, 1968; Yaalon, 1971, 1975) introduced the concepts of pedomorphic forms and pedomorphic surfaces. A pedomorphic surface was considered to be a landform in which soils and relief are genetically and evolutionarily inter-dependent. The various members of a catena or pedomorphic surface that contain a definite horizon sequence are termed a pedomorphic form. Pedomorphic forms were described in terms of degree of development, slope position, and drainage class. Pedomorphic surfaces were equated to the "ground surfaces" of Butler (1959). They were recognized as often being polygenetic, i.e., containing buried, relict or exhumed soils as defined by Ruhe (1956).

Conacher and Dalrymple (1977) provided a nine-unit landscape model based on geomorphic position, pedogeomorphic process, and slope shape and steepness. A "landsurface catena" was defined as a three-dimensional slope showing "landsurface units" including soils. Their ideas were also an extension of Milne's (1935) catena concept.

Gerrard's (1981) book *Soils and Landforms* began with the concept of the catena by Milne (1935) and continued with the work of Butler (1959), Ruhe and his associates, and Conacher and Dalrymple (1977). The catena concept was considered by Gerrard in detail. "The real significance of catenas lies in the recognition of the essential processes involved in catenary differentiation" (p. 79). Gerrard (1990) studied soil variations on hillslopes in humid temperate climates. He questioned the validity of employing the traditional catena concept in soil-landform relations of humid-temperate regions by challenging the idea that many slopes are integrated along their entire length. Individual components of slopes possess soil characteristics that appear to be related to the morphologic nature of these components, but the slopes did not possess integrated soil systems. Different parts of the slopes appeared to act independently. The catena concept may also be invalid in areas containing "inverted relief" (Pain and Ollier, 1995). Inversion of relief may occur when materials on valley floors become more resistant to erosion than the adjacent valley slopes.

Sommer and Schlichting (1997) identified three "archetypes" of catenas, including (1) transformation catenas showing no gains or losses of constituents (i.e., only transformations), (2) leaching catenas with losses occurring in part of the catena and no gains elsewhere, and (3) accumulation catenas showing only gains in part of the catena and no losses elsewhere. They illustrated how these archetypes of catenas could be used to extrapolate from the landscape level to greater scales. Birkeland (1974, 1999) presented topography-soil relations with time in different climatic settings. He reviewed the soil catena concept in detail and summarized soil-forming processes in catenas of different bioclimatic and geographic regions.

Gennadiyev (1990) studied the chronosequences of soil catenas developed on various-aged earthy mounds at archeological sites within different natural zones, and generalized the results of investigations regarding the rate, stages and trends of soil catena formation.

According to Gerrard (1981) the major processes of soil catenary differentiation are surface wash, soil creep, rapid mass movements and solution transport. The downslope movement of soil solution is of great significance in soil catena formation as well. Thus, soils of lower slope positions are affected by the flows of solutions from soils of higher slope positions and this lateral transport and redeposition of matter is in a way analogous to the processes linked the eluvial A horizon and the illuvial B horizon of vertical soil profile. Greene (1947) investigated soil catenary differentiation affected by lateral movement of soluble materials within humid tropical regions.

Perel'man (1961), Glazovskaya (1964), and Kasimov and Perel'man (1992) expanded on Polynov's (1953) ideas about geochemical landscapes, which considered the soil as an integrative result of vertical and horizontal fluxes of energy and matter. Glazovskaya (1964) introduced the concept of a soil-geochemical catena as paragenetic associations of soils belonging a single geochemical landscape. Within the soil-geochemical catenas, the sequences of autonomic and subordinate (heteronomic) soils are related to one another by lateral migration of matter. Her concept included the migrational and geochemical structure of the catena, the leaching zones, soil-geochemical barriers, etc. Glazovskaya suggested that the set of catenas bounded by a common drainage basin should be called "a landscape-geochemical arena".

Kasimov and Perel'man (1992) classified soil-geochemical catenas. The highest level of the classification is catena *type*. The most important criterion for this level is the degree of geochemical association of autonomic and subordinate soils. Two types of catenas are *autochthonous* and *allochthonous*. In the former the geochemistry of subordinate soils is governed entirely by the lateral fluxes from local autonomic soils. They occur primarily in basins of the first or second order, with no inputs of material from large distances. Geochemical differentiation of allochthonous catenas is affected not only by local lateral redistribution of material, but also longitudinal transport from higher-level basins. Catena types are divided into two subtypes: *monolithic* and *heterolithic*.

The classification of *families* is based on the migrational structure of geochemical landscapes: catenas with predominantly aqueous migration (1 – water-surface-soil, 2 – water-subsoil; 3 – water-surface-soil-potuscular, 4 – water-surface-soil-subsoil) and catenas with mechanical migration (5 – water-soil-soliflution, 6 – water-soil-erosion, 7 – gravitational-seepage, 8 – deflution).

Distinguishing the *classes*, *genera* and *species* of soil-geochemical catenas is based on substantive features which indicate the variability of the main geochemical properties of soils within a catena, their prominence, kinds of soil-geochemical barriers, and the relationship between the lithochemical and lateral-migrational differentiation of the catenas.

Soil Landscape Models and the Soil Cover Pattern Concept

Soil-landscape models address issues of the functional boundaries of soil systems. Smeck et al. (1983) subdivided soil models into: (1) state-factor analysis, i.e., capitalizing on original concepts of Dokuchaev (1879, 1883); (2) energy models; (3) residua or haplosoil models that emphasized independence of the soil-forming factors; (4) generalized process models, i.e., Simonson's (1959) model of inputs, outputs, transfers, and transformation of matter and energy; and (5) soil-landscape models.

A key example of a soil-landscape model was Huggett's (1975) use of the soil-landscape system as the basic three-dimensional unit. He tried to simulate the plasmic material flows within an ideal valley basin. Huggett (1975) viewed the soil-landscape as an "open system" similar to that envisioned by Ruhe and Walker (1968) and emphasized the effect of inputs, outputs, transfers, and transformations of material and energy on pedogenesis and landscape evolution. Huggett's (1975) model was a precursor to a set of quantitative soil-landscape models that have appeared in the last decade, e.g., Gessler et al. (2000). These models integrate short- and long-term pedogeomorphologic processes. Statistics are used to correlate measured and predicted soil-landscape variables. These models often use digital terrain analysis, intensive field sampling, remote sensing, statistical modeling, and geographic information systems. Gessler et al. (2000) modeled soil-landscape and ecosystem properties using terrain attributes. They were able to account for from 52 to 88% of soil property variance using terrain variables such as slope and flow accumulation. The development of Geographic Information Systems (GIS) and digital elevation models (DEMs) will continue to enhance our understanding of soil development along hillslopes.

Hole (1953, Fig. 3) emphasized "Bushnell natural drainage" or toposequence designations and suggested terminology for describing soils as three-demensional bodies including soil body patterns and landscape positions, as well as drainage conditions. Hole (1953) constructed indices of patterns of soil bodies similar to those later provided by Fridland (1972).

Fridland (1965, 1972, 1984) advanced the study of soil cover patterns. He understood "soil cover pattern" as regularly repeating spatial inhomogenities of soil geography. "A certain pattern of soil cover is characterized by set of soil areas which repeats multiply and rhythmically in a space, resulting in a consistent composition and structure of soil cover and consistent geochemical and geophysical linkages

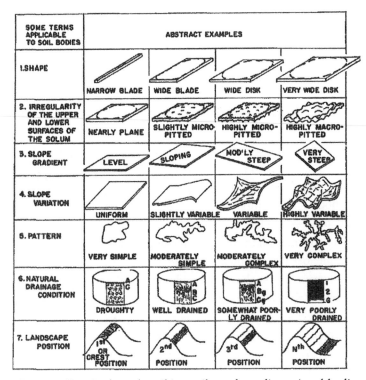

Figure 3. Terminology describing soils as three-dimensional bodies
(Hole, 1953, p. 132).

between soils within the sets. Each soil cover pattern has the same history of development" (1984, p. 4).

An important component of the Fridland concept was definition of basic soil-geographic units. Literature on pedology contained different approaches to this problem. Arnold (1983) found and discussed several proposals concerning basic soil-geographic units. Some of them were: soil body (Hole, 1953), soil-area unit (Muir, 1962), artificial soil body (van Wambeke, 1966), and pedomorphic forms (Dan and Yaalon, 1968).

Hole (1953, Fig. 3), proposed that the landscape contains a continuum of soil bodies. Using examples of general soil-landscape relations in portions of Wisconsin, USA, he depicted the interconnected soil bodies in terms of their shape, irregularity of the upper and lower surfaces of the solum, slope gradient, pattern, natural drainage condition, and landscape position. He classified soil body patterns into six classes that range from "very simple" to "very complex". The landscape position classes were based on toposequences, or catenas. Soil types were recognized as multi-factor units.

Fridland (1972) introduced the concept of an *elementary soil areal (ESA)*, defining it as a soil body without internal pedogeographic boundaries. Fridland

Box 1

Vladimir M. Fridland (1919–1983) – internationally acknowledged as a classical scholar of Russian pedology. He completed an academic program at the Faculty of Geology and Soil Science of the Moscow State University in 1941. He received the degree of Doctor of Science in 1964 at the Dokuchaev Soil Research Institute, in which he started his scientific career and worked until his very last day.

For many years, Fridland headed the Division of the Genesis, Geography, and Classification of Soils at the Institute, and developed the traditions of such outstanding representatives of the Russian school of pedology as L.I. Prasolov, I.P. Gerasimov, and E.N. Ivanova. His concepts of soil genesis and evolution, and of classification were based on the results of soil geographical investigations performed in regions with different natural conditions, such as the Caucasus, Trans-Carpathians, the Urals, the Caspian Lowland, Vietnam, Pakistan, Indonesia, Romania, France, and others.

Fridland laid the basis for a new understanding of the structure of the soil mantle. His main ideas in this field are presented in two monographs "Pattern of the Soil Cover" (1972) and "Patterns of the Soil Cover of the World" (1984). He devoted his principal attention to compiling highly informative soil maps. The most famous of these works is the "Soil Map of the Russian Federation, scale 1: 2 500 000, the first map of this scale depicting the soil cover structure.

identified three kinds of discrete elementary soil areals: (1) homogeneous ESAs, (2) heterogeneous biopatterned ESAs, and (3) regularly cyclic ESAs. Homogeneous ESAs are uniform, have small contrast due to patchiness of organisms, and vary considerably in scale. In contrast biopatterned ESAs have organism-induced variations on a small scale, due to the role of individual tree species, tree-tips, canopy gaps, and other biopatterns. Regularly cyclic ESAs are soil bodies that occur in regular patterns as in the case of patterned ground in areas underlain by permafrost.

Fridland (1972) also defined combinational soil areals (CSAs) as those consisting of two or more elementary soil bodies, with or without associated bodies of non-soil. The soil cover pattern as defined by Fridland (1972) pertains to the detailed arrangement of ESAs, CSAs, and associated bodies of non-soil. Topographic sequences of soils, or catenas, constitute CSAs that occur in repetitive patterns.

Fridland (1972) suggested indices to describe and classify ESAs and soil cover patterns. They are the taxonomic soil names, the size and composition, spatial-geometry characteristics of ESA and its links to the environmental conditions. The soil cover pattern is described by two groups of characteristics: (1) the taxonomic nomenclature of the soil pattern's components and the degree of contrast between their properties; (2) characteristics of spatial proportion between ESAs – their morphology and relative participation of ESAs in the soil cover pattern. The latter characterized the complexity of soil cover pattern.

Fridland (1972) developed the typology of soil combinations grouped according to the soil forming factors. Soil combinations included genetically linked components related to microrelief, and were divided into strongly contrasting *complexes* (e.g. complex of solonchaks, meadow-steppe solonetzes, and light chestnut meadow-like soils, and meadow chestnut soils within Caspian lowland) and weakly contrasting *spottiness* (i.e. spottiness of typical and leached chernozems of the Middle-Russian Upland). Soil combinations including the genetically linked components resulting from mesotopography were divided into strongly contrasting *combinations, or catenas* (e.g. catena-variation of soddy-podzolic, boggy-podzolic and boggy soils of hilly morainic plain), and weakly contrasting *variations* (e.g. variation of soddy slightly- and medium-podzolic sandy soils within the Meshchera Lowland). Soil combinations where components consisted of genetically in linked components were divided into strongly contrasting *mozaics* (e.g. soil mozaics of South Australia related to repeating spatial changes of parent materials) and weakly contrasting *tachets* (e.g. a tachet of slightly solonetzic ordinary chernozems in the interlake areas of Trans-Urals abrasian plain, where the only differences between soils were colour and structure).

Fridland (1972) was the first to examine all the major aspects of the soil cover pattern problem. Topics included: soil cover pattern as a system, elementary soil area as the starting unit of pedogeographical taxonomy, causes giving rise to soil combinations, natural and anthropogenic evolution of soil combinations, typology of soil combinations, main features of soil cover patterns in the different regions of the world, research methods for soil cover pattern, and solution of practical problems by means of information on soil cover pattern. A method for studying multivariant natural phenomena with modern mathematical statistics and cybernetic analysis was also proposed.

Rozanov (1977) differentiated hierarchical levels of the world's soil cover organization: mega-, macro-, meso-, micro-, and nanostructures. The megastructure is the genetic-morphostructural regions (soils of mountains, plains, lowlands, etc.), produced by the geologic peculiarities of the continents. Macrostructures consist of bioclimatic zones, facies, and districts. The mesostructures are soil cover areas consisting of soil combinations related to mesotopographical forms. The soil cover microstructures related to microrelief and nanostructures are the elementary soil areals.

Fridland (1984) did not support the idea of a zonal-provincial structure of soil cover as the upper levels of an overall hierarchical system of soil-geographical organization. He proposed six levels of soil cover pattern: elementary soil area, elementary soil cover pattern (microstructure), mesostructure, soil region, soil district, and soil country. And, as separate system, he recognized several levels in zonal-provincial (bio-climatic) structure of soil cover – province, facies, subzone, zone, and region.

The book *Soil Landscape Analysis* by Hole and Campbell (1985) was a landmark study in soil-landscape studies. Dedicated to Fridland, the book included an analysis of soil cover patterns and catenas. Hole and Campbell provided a perspective for soil-landscape studies (Figure 4). Although Hole (1978) had recommended use of the term "soilscapes", he considered all elements of the soil landscape, including elementary and combinational soil bodies as defined by Fridland (1972). Hole and Campbell (1985) denoted a point of departure: "Just as an elementary soil body is a natural, relatively distinct cluster of pedons, so a combinational soil body is a distinct unit in the soil cover" (p. 4). They concluded the chapter with nine fundamental concepts that describe the factors controlling soil cover patterns.

Hole and Campbell (1985) described the nature of the soil cover, including the spatial distribution of soil bodies at different scales. They used Fridland's "elementary

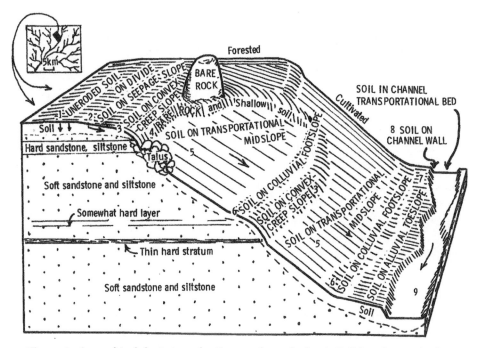

Figure 4. A graphical depiction of soilscapes from the book *Soil Landscape Analysis* by Hole and Campbell (1985, p. 33).

Box 2

Dr. Francis D. Hole (1913–2002), was a major U.S. contributor to the ideas of soil cover patterns and catenas. He graduated in 1933, and earned a Ph.D. degree in geology and soil science from the University of Wisconsin-Madison in 1944. He joined this university in 1946, where he worked until 1983, during the last decades as a professor of geography and soil science. His research interests have been primarily in the areas of soil characterization, soil mapping, and soil-landscape relationships.

Among his many publications were "Soils of Wisconsin" and "Soil Genesis and Classification", co-authored with McCracken and Buol. He created the journal "Soil Survey Horizons", and compiled the map "Soils of Wisconsin". He emphasized the pedon concept and in his classic article in 1953 suggested terminology for describing soils as three-dimensional units. He recognized the importance of Fridland's ideas about soil cover patterns and built upon these ideas in his book co-authored with J.B. Campbell, "Soil Landscape Analysis "(1985). In the epilogue of their book, Hole wrote the poem "Final Comment on Soil Landscape Dynamics":

Five things influence soil landscapes;
Biota, climate, terrain shapes,
Initial stuff and human kind.
A blend of these may loose or bind
The land skin of hills and dales:
Here turns soil dark, and elsewhere pales;
Leaches it poor or makes it rich,
Defining each natural niche,
Erodes or catches soil debris,
Changing the landscape endlessly.

After his retirement, Dr. Hole continued his activities, giving lectures and courses on the importance of soil.

soil bodies" (ESBs) and "combination soil bodies" (CSBs) and gave graphic depictions of soilscapes. They emphasize the soil catena concept and discuss "bonding regimes" in CSBs, i.e., the flow of water and matter across and through the landscape.

Methods of soil landscape analysis are described by Hole and Campbell, including field mapping, laboratory characterization, and graphic techniques for displaying soils data. They provided numerous indices for showing contrasts in the soil cover pattern, and discuss the principles of landscape mapping, including spatial order and map representation.

According to Hole and Campbell (1985), there are two contrasting views of soil geography: (1) the traditional view employed for soil surveying and mapping and (2) Fridland's soil cover patterns approach (i.e., "the other soil geography"). They pointed out that the latter approach is seldom explicitly acknowledged in textbooks and professional journals. "The 'geography' of this soil geography denotes those properties that pertain to soil bodies specifically as *areal* entities possessing characteristic sizes, shapes, volumes, slopes, and internal variability" (p. 97).

Hole and Campbell (1985) distinguished among taxonomic, genetic, and soil map units. They provided a detailed discussion of specific pedological units including the soil series, the pedon, the polypedon, consociations, complexes and associations, and taxadjuncts. They suggest that the term "soil cover" defines the entire population of soils in a region. The soil cover is composed of elementary soil areals (ESAs), defined as soils belonging to a single taxon of the lowest rank and occupying a space that is bounded on all sides by other ESAs (or non-soil formations). They further suggest that ESAs are analogous to polypedons. ESAs occur in soil combinations (i.e., CSAs) that are linked by specific pedogenetic processes. The last chapter discusses the generalizations of soil maps, including precision and accuracy of detailed soil maps.

In the revised *Soil Survey Manual* (Soil Survey Division Staff, 1993, Ch. 1), there is a special subchapter concerning soil-landscape relations. Arnold (1994) wrote that "working models of soil-landscape relations – that is regressions of soil properties related to specific landscape positions and kinds of materials within the survey area – must be visualized, described, documented and tested in order to adequately form the predictive hypotheses with which the detailed soil survey can be conducted." Several kinds of "identification catena" or "toposequence keys" were developed to assist field mappers in properly associating materials and soil properties (Arnold, 1990).

Soil map units defined in the *Soil Survey Manual* (1993) include: (1) consociations, (2) complexes, (3) associations, and (4) undifferentiated groups. Consociations are delineated areas dominated (≥50%) by a single taxon. Complexes and associations consist of two or more dissimilar components occurring in a regularly repeating pattern. The main difference between complexes and associations is that the components of a complex cannot be mapped separately at a scale of about 1:24,000, whereas the components of a soil association can be separated at a scale of 1:24,000. Undifferentiated groups consist of two or more taxa components consistently associated geographically.

Summary

The history of the soil cover pattern and soil catena concepts is traced from the initial data accumulation for the period of early soil surveys (the end of the 19th century) to contemporary approaches using the mathematical modeling, digital terrain analysis, and geographic information systems.

The concepts of soil cover pattern and soil catenas were enunciated at the end of the 19th century. Dokuchaev and later Sibirtsev identified spatial units of the soil cover and stressed the importance of "the topography of soils" and "soil combinations". The ideas of soil cover patterns were synthesized by Fridland into a single doctrine in the 1970s. In 1985 Hole and Campbell adopted the concept of soil cover patterns in their book "Soil Landscape Analysis".

Russian soil-geographical approaches were precursors of the catena concept introduced by Milne in 1935. During the 20th century important contributions to developing both soil cover pattern and catena concepts were made by Neustruev, Visotsky, Kellogg, Bushnell, Ruhe, Hole, Polynov, Perelman, Yaalon, Glazovskaya, Gerrard, Birkland and others scientists, who were involved in investigations of soil cover complexity, typology of soil combinations, basic soil-geographic units, pedomorphic surfaces and hillslope soil processes, soil-landscape relations, soil catenary differentiation, and soil catena taxonomy. Both the concepts of soil cover pattern and soil catena provided a methodological basis for much basic research. Its fundamental postulates assume that a researcher is obliged to take into account the spatial heterogeneity of the soil cover while planning an experiment and interpreting the results obtained. These concepts were utilized in various countries from 1930s to 1990s to study: (1) lateral genetic relations between soils; (2) spatial variability of soils and soil properties; (3) the "purity" of soil map units; (4) the topographic and lithological factors in pedogenesis; (5) processes of soil matter transport and erosion, etc. At the same time, the concepts of soil cover pattern and soil catena have found recognition in many areas of applied activities such as soil survey, land use, land inventory, soil monitoring, environmental land management and soil conservation.

References Cited

Arnold, R.W. 1983. Concepts of soils and pedology. In: L.P. Wilding, N.E. Smeck, and G.F. Hall (eds.) Pedogenesis and Soil Taxonomy. 1. Concepts and Interactions. New York: Elsevier. pp. 1–21.

Arnold, R.W. 1990. Detailed soil survey in the United States (in Russian). Pochvovedenie (Soil Science). 2:21–29.

Arnold, R.W. 1994. Soil geography and factor functionality: Interacting concepts. In: R. Amundson et al. (ed.) Factors of Soil Formation: A Fiftieth Anniversary Retrospective. SSSA Spec. Publ. 33. SSSA, Madison, WI. pp. 99–109.

Birkeland, P.W. 1974. Pedology, Weathering, and Geomorphological Research. Oxford University Press. New York-London-Toronto. 286 pp.

Birkeland, P.W. 1999. Soils and Geomorphology. 3rd edit. Oxford University Press, New York. 430 pp.

Bogdan, V.S. 1900. Report of the Valuiki Agricultural Experimental Station. 1895–1896 (in Russian). St. Petersburg.

Bushnell, T.M. 1942. Some aspects of the catena concept. Soil Sci. Soc. Am. Proc. 7:466–476.

Butler, B.E. 1959. Periodic phenomena in landscapes as a basis for soil studies. CRIRO Australia Soil Publ. No. 14.

Conacher, A.J. and J.B. Dalrymple. 1977. The nine unit landsurface model: an approach to pedogeomorphic research. Geoderma 18:1–154.

Dan, J. and D.H. Yaalon. 1968. Pedomorphic forms and pedomorphic surfaces. Trans. 9th Internat. Cong. Soil Sci. IV:577–584.

Dimo, N.A. and B.A. Keller. 1907. In the semi-desert region (in Russian). Saratov.

Dobrovolsky, G.V. 1994. The Development of Studies on the Structure of the Soil Cover as a Branch of Soil Geography//Eurasian Soil Sience, 26(4):1–9.

Dokuchaev, V.V. 1879. Mapping the Russian Soils (in Russian). St. Petersburg.

Dokuchaev, V.V. 1883. The Russian Chernozem Report to the Imperial Free Economic Society (in Russian). St. Petersburg.

Dokuchaev, V.V. 1899. A Contribution to the Theory of Natural Zones: Horizontal and Vertical Soil Zones (in Russian). St. Petersburg.

Effland, A.B.W. and W.R. Effland. 1992. The Soil Geomorphology Studies in the U.S. Soil Survey Program. Agricultural History. 66:189–212.

Fridland, V.M. 1965. Make-up of the soil cover. Sov. Soil Sci. 4:343–354. (Transl. from Pochvovedeniye, 19654, 4:15–27).

Fridland, V.M. 1972. Pattern of the Soil Cover. Transl. From Russian by N. Kaner, Israel Program for Scientific Translations, Jerusalem. U.S. Dep. Commerce, Nat. Tech. Infor. Serv., Springfield, VA.

Fridland, V.M. 1984. Patterns of the soil cover of the World (in Russian). Moscow. Mysl'. 234 pp.

Gennadiyev A.N. 1990. Soils and Time: Models of Development. Moscow University Press. 240 pp.

Gennadiyev, A., M. Gerasimova, and R. Arnold. 1995. Evolving approaches to soil classification in Russia and the United States of America: their divergence and convergence. Soil Surv. Horizons 36:104–111.

Gennadiyev, A.N. and K.R. Olson. 1998. Pedological cooperation between Russia and the USA: Past to Present. Soil Sci. Soc. Am. J. 62(5):1153–1161.

Gerrard, A.J. 1981. Soils and Landforms; an Integration of Geomorphology and Pedology. London. George Allen & Unwin. 220 p.

Gerrard, A.J. 1990. Soil variations on hillslopes in humid temperate climates. Geomorphology 3:225–244.

Gessler, P.E., O.A. Chadwick, F. Chamran, L. Althouse, and K. Holmes. 2000. Modeling soil-landscape and ecosystem properties using terrain analysis. Soil Sci. Soc. Am. J. 64:2046–2056.

Glazovskaya, M.A. 1964. Geochemical foundations of the typology and methodology of investigating natural landscapes (in Russian). Moscow Univ. Press. Moscow. 400 p. (in Russian).

Glazovskaya, M.A. 1981. General Soil Science and Soil Geography. Moscow University Press.

Greene, H. 1947. Soil Formation and Water Movement in Tropics. Soils and Fertilizer. 10:253–256.

Hilgard, E.G. 1906. Soils: Their formation, properties, composition and relation to climate and plant growth. MacMillan, New York.

Hole, F.D. 1953. Suggested terminology for describing soils as three-dimensional bodies. Soil Sci. Soc. Am. Proc. 17:131–135.

Hole, F.D. 1978. An approach to landscape analysis with special emphasis on soils. Geoderma 21:1–23.

Hole, F.D. and J.B. Campbell. 1985. Soil Landscape Analysis. Rowman and Allanheld, Totowa, NJ. 196 pp.

Holliday, V.T., L.D. McFadden, E.A.Bettis, and P.W. Birkland. 2002. Soil Survey and Soil Geomorphology. Profiles in the History of the U.S. Soil Survey. Eds. D. Helms, A.B.W. Effland, and P.J. Durana. Iowa State Press. A Blackwell Publ. Co.

Huggett, R.J. 1975. Soil landscape systems: a model of soil genesis. Geoderma 13:1–22.

Kasimov, N.S. and A.N. Perel'man. 1992. The Geochemistry of Soils. Eurasian Soil Science. 24(4):59–76.

Kellogg, C.E. 1933. A method for the classification of rural lands for assessment in North Dakota. J. Land Public Utility Econ. 9:10–14.

Krasyuk, A.A. 1931. Soils and their investigations in nature (in Russian). Gosizdat. Moscow-Leningrad.

Marbut, C.F., H.H. Bennett, J.E. Lapham, and M.H. Lapham. 1913. Soils of the United States. U.S. Bur. Soils Bull. 96:1–791.

Milne, G. 1935. Some suggested units of classification and mapping particularly for East African soils. Soil Res. 4:183–198.

Milne, G.A. 1936a. Provisional Soil Map of East Africa//Amani Memories, # 28. East African Agr. Res. Stn., Tanganyika Territory; 34.

Milne, G. 1936b. Normal Erosion as a Factor in Soil Profile Development. Nature. 138:548.

Muir, J.W. 1962. The General Principles of Classification with Reference to Soils. J. Soil Sci. 13:22–30.

Neustruev, S.S. 1915. On soil combinations of plains and uplands (in Russian). Pochvovedenie (Soil Science). 1:62–73.

Pain, C.F. and C.D. Ollier. 1995. Inversion of relief; a component of landscape evolution. Geomorphology 12:151–165.

Perel'man, A.I. 1961. Geochemistry of Landscapes. Geographic Literature Press. Moscow. 496 pp. (in Russian).

Polynov, B.B. 1953. Doctrine of landscapes. (in Russian). Geogr. Proc. 33. USSR Acad. Sci. Press. Moscow.

Prasolov, L.I. 1922. Soil Regions of European Russia (in Russian). Petrograd.

Rozanov, B.G. 1977. Soil Cover of the Earth (in Russian). Moscow Univ. Press.

Ruhe, R.V. 1956. Geomorphic Surfaces and Nature of Soils. Soil Science. 82:441–455.

Ruhe, R.V. 1960. Elements of Soil Landscapes. Trans. Int. Cong. Soil. Sci. 7th. 4:165–170.

Ruhe, R.V. 1969. Quaternary Landscapes in Iowa. Iowa State Univ. Press, Ames.

Ruhe, R.V. and P.H. Walker. 1968. Hillslope models and soil formation. I. Open systems. Trans. 9th Internat. Cong. Soil Sci. 4:551–560.

Serdobol'skii, I.P. 1937. Soil Property Variation within Solonets Soil Complexes. Proc. Commision on Irrigation AS USSR. Vol. 9. Moscow-Leningrad.

Shvergunova, G.A. 1974. Stages of development of large scale mapping in USSR//Large-scale soil mapping. (in Russian). Nauka Publ. Moscow. pp. 144–166.

Sibirtsev, N.M. 1900a. Soil Sciece (in Russian) Y.N. Skorokhodva, St. Petersburg, Russia.

Sibirtsev, N.M. 1900b. Soils of the upper reaches of Velikaya river basin (in Russian). Pskov.

Simonson, R.W. 1959. Outline of a generalized theory of soil genesis. Soil Sci. Soc. Am. Proc. 23:152–156.

Simonson, R.W. 1997. Evolution of Soil Series and Type Concepts in the United States. In: D.H. Yaalon and S. Berkowisz. Reiskirchen (eds.) History of Soil Science. Catena Verlag. pp. 79–108.

Smeck, N.E., E.C.A. Runge, and E.E. Macintosh. 1980. Dynamics and genetic modeling of soil systems, pp. 51–81. In: L.P. Wilding et al. (eds.) Pedogenesis and Soil Taxonomy. I. Concepts and Interactions. Elsevier, Amsterdam.

Soil Survey Division. Bureau of Chemistry and Soils. 1938. Soils of the United States. Soils and men. Washington D.C. pp. 1019–1161.

Soil Survey Division Staff. 1993. Soil Survey Manual. Agric. Handbook No. 9, U.S. Dep. Agric., U.S. Govt. Printing Office, Washington, D.C.

Soil Survey Staff. 1951. Soil Survey Manual. USDA Agric. Handbook # 18. Washington, D.C.

Sommer, M. and E. Schlichting. 1997. Archetypes of catenas in respect to matter; a concept for restructuring and grouping catenas. Geoderma 76:1–33.

U.S. Department of Agriculture. 1938. Soils and Men. Yearbook of Agriculture, 1938. USDA, Washington, D.C.

van Wambeke, A. 1966. Soil Bodies and Soil Classifications. Soils and Fert. 29:507–510.

Vysotskii, G.N. 1906. Oro-climatological principles of soil classification (in Russian). Pochvovedenie (Soil Science). 1:14–26.

Yaalon, D.H. 1971. Soil-Forming Processes in Time and Space. In: Paleopedology: Origin, Nature and Dating of Paleosols. Israel University Press, Jerusalem, Israel. pp. 29–39.

Yaalon, D.H. 1975. Conceptual models in pedogenesis: can soil-forming functions be solved? Geoderma 14:189–205.

9
A History of Soil Geomorphology in the United States

Vance T. Holliday[1]

Introduction

Soil geomorphology has a long and rich history in the United States and has had a significant impact on soil-geomorphic research worldwide. It has been particularly important in unraveling the ages and evolution of landscapes. The history of soil geomorphology is very poorly documented, however. What little is written is confined to very specific components of the story (e.g., Birkeland, 1989; Olson, 1989, 1997; Effland and Effland, 1992; Johnson and Hole, 1994; Holliday et al., 2002) or is tangential to it (e.g., Tandarich and Sprecher, 1994; Tandarich, 1998a,b; Bockheim et al., 2005). This is probably because: 1) the history of this subdiscipline is very diverse, with roots in Quaternary geology, pedology, and physical geography; none of which are well-documented; and 2) because soil geomorphology was not formally recognized as a distinct subdiscipline until late in the 20th century.

Outlining the history of soil geomorphology is important precisely because of its interdisciplinary roots and outlook. Although pedology has its intellectual roots in geology (Tandarich et al., 1988; Tandarich, 1998a,b), most pedologic training and research in both academic and governmental settings in the U.S. has tended to focus on soil description and classification, mapping, contemporary land use, soil quality, and plant productivity, and not on reconstructing the past (Daniels, 1988; McFadden and Kneupfer, 1990; Daniels and Hammer, 1992; Swanson, 1993; Tandarich and Sprecher, 1994; Bronger and Catt, 1998; McFadden and McDonald, 1998; Holliday et al., 2002). While important in characterizing soils, describing their spatial distribution, and understanding their potential for supporting crops, these subfields tell us little about how soils (and associated landscapes and environments)

---

[1] Departments of Anthropology & Geosciences, University of Arizona, Tucson, USA.

evolved and why soils are where they are on the landscape. Such characteristics of soils are crucial in their conservation. Moreover, soil geomorphology is an important tool for reconstructing the past. Soil genesis is dependent on a variety of environmental factors. Chemical, physical, and biological characteristics of soils uniquely reflect the integration of these factors. Soils, therefore, under the right circumstances, can help us understand these environmental factors in the past and how they interacted (e.g., Gerrard, 1992; Birkeland, 1999).

Research focusing on the relationship of soils and landscapes has rarely been a priority in either pedology or geomorphology, however. This situation largely evolved from the intellectual and physical separation of soil science from geoscience. Soil science became grounded in agricultural colleges, often on campuses physically separated from those with major research programs in geology and geography. This division mirrored the situation in the Federal government, where soil science research was a component of the U.S. Department of Agriculture (USDA) and generally not part of the U.S. Geological Survey (USGS) in the Department of Interior. Thus, the possibilities for "cross fertilization" and training among researchers, faculty, and students was and continues to be minimized. A goal of this chapter is to not only outline these unfortunate trends, but to highlight individuals, institutions, and programs that pursued and promoted research cooperation that became soil geomorphology.

This chapter is a first approximation of the development of soil geomorphology in the U.S.[2] It follows several threads that evolved out of soil science and geology in the 19th century and went on to include physical geography in the years preceding WWII. The post-war explosion in scientific research in the U.S. is mirrored in the several lines of research, focused in Quaternary geology, pedology, and physical geography in both academic and governmental settings that led to the field of soil geomorphology as it stood at the end of the 20th century. The chapter is divided into five time blocks, based on general, convenient (though not altogether arbitrary) groupings of significant events in the history of the discipline: 19th century to the 1930s; 1930s to 1941; the war years; 1945 to 1974, and 1974 to 2004. Several individuals not from the U.S. and not working in the U.S. are mentioned in the discussion because their research and writing significantly influenced soil geomorphology in the U.S.

For this paper a broad view of what constitutes soil geomorphology is taken. At its most fundamental level, soil geomorphology is the study of genetic relationships between soils and landscapes (e.g., Ruhe, 1956a, 1965a; McFadden and Knuepfer, 1990; Daniels and Hammer, 1992; Gerrard, 1992, 1993). The focus is on pedogenic processes and geomorphic processes, and sometimes a strong component of hydrology, to understand the distribution of soils in the present (contemporary soil geography) and in the past. In a much broader sense, however, soil geomorphology includes the investigation of soils as a means of studying and reconstructing the past, with a particular focus on soils as age indicators and soils as clues to past environments (especially vegetation and climate). These aspects of soil geomorphology evolved through both the "state factor approach" and through soil stratigraphy.

19th Century to the 1930s

There are two key threads in the early development of soil geomorphology. Both began in the 19th century, but include very different perspectives on soils. One thread is in pedology in Europe, studying soils at the surface. The other thread is in Quaternary geology in North America, looking at soils buried beneath the surface.

Russian Roots

An important step in the development of pedology and soil geomorphology was also one of the first steps in formulation of the "state factor concept" of soil genesis. This approach to soil genesis began with Russian scientists in the 1870s through 1890s, most notably V. V. Dokuchaev, but also N. M. Sibertsev and K. D. Glinka (Johnson and Hole, 1994; Tandarich and Sprecher, 1994; Tandarich, 1998a; Eutuhov, this volume). A noteworthy aspect of the five-factor approach as originally devised by Dokuchaev was its grounding in geology (Tandarich and Sprecher, 1994). Dokuchaev (1883/1967) outlined a framework for studying soils from a "geologic and geographic perspective." His writing anticipated aspects of soil geomorphology a century later. For example, in his book on the *Russian Chernozem* (1883, translated in 1967) a chapter on the age of chernozems (pp. 373–386) includes discussion of the age of burial mounds and ancient fortifications as a means of dating underlying soils and soils on the artificial features. This approach to dating soils and estimating rates of pedogenesis continues to be important in soil geomorphology (e.g., Bettis, 1988; Alexandrovskiy, 2000).

The five-factor paradigm for soil genesis took root in the U.S. early in the 20th century, most widely promoted by C.F. Marbut, Chief of the Soil Survey Division in the Bureau of Soils (USDA), but also endorsed and espoused by E.W. Hilgard (1892) and G.N. Coffey (1909a,b; 1912) (see also Amundson, this volume). These soil scientists viewed the factors driving soil genesis in different ways, however, and the original Russian concepts evolved significantly as they were passed along. Coffey (1912) strongly emphasized the importance of climate in soil formation, but still recognized the importance of parent material. Hilgard expressly recognized the importance of late Quaternary geology in "agricultural survey" (soil survey) (cited in Amundson and Yaalon, 1995) and, indeed, lobbied to have a division of agricultural geology (as soil science was then known) established in the USGS. After that effort failed, Hilgard worked with J. W. Powell, the eminent geologist who directed the USGS, to have that agency transferred to the USDA (Amundson and Yaalon, 1995). That attempt also failed, resulting in the institutional separation of pedology from geology.

Succeeding pedologists clearly tried to distance pedology from its geologic roots, although research demonstrating the significance of geology, especially geomorphology, to soils continued in the USDA. For example, E.E. Free (1911), with the Bureau of Soils, first demonstrated the significance of dust inputs in soil genesis. Glinka (1914) and Marbut apparently embraced the basic concept of the soil-forming factors, but most of Marbut's lectures and writings, and those of his successor C.E. Kellog, focused on the morphology and classification of soils. Marbut thus further distanced pedology from geology.

Even though the five factors of soil genesis became a leading conceptual framework in pedology, especially in the USDA, they almost disappeared from field work in the U.S. As the soil surveys developed through the 20th century, genetic and factorial relationships among series as well as geologic and other "physiographic" aspects, were expressly de-emphasized (Simonson, 1987, 1997a). Some pedologists apparently were eager to distance pedology from its geological roots (Wilding, 1994). Marbut (1928) himself indicated that soil characteristics and geologic characteristics should be considered completely independent of one another.

Midwestern Soil Stratigraphy

Another starting point for what is now soil geomorphology can be found in the early work on glacial stratigraphy in the Midwestern U.S. (Totten and White, 1985). Some of the earliest attempts at stratigraphic subdivision and correlation of glacial drift include reference to "buried forests" and buried plant remains (e.g., Whittlesey, 1848, 1866; Newberry, 1862; Worthen, 1868; Orton, 1870), although the full significance of these zones as indicators of past landscapes apparently went unrecognized. That quickly changed, however. Worthen (1870, 1873), Orton (1873), and Winchell (1873), among others, all recognized and described buried forest beds or peats, and believed that they represented interglacial intervals, and thus evidence of multiple glaciation. Newberry (1878, p. 38) later agreed that a buried forest bed represented an interglacial period and also "an interval of mild climate in the ice period," an early allusion to the environmental significance of associated buried soils.

The most influential and widely known early work on soil stratigraphy in Midwestern glacial studies is the monumental body of research by T.C. Chamberlin and Frank Leverett. Chamberlin is best known for his field studies that resulted in

Table 1. Origin of glacial and interglacial stage names in North America
(from Totten and White, 1985, Table 1)

Glacial	Interglacial	First reference
Wisconsinan		Chamberlin, 1894, 1895
	Sangamonian	Leverett, 1898a
Illinoian		Leverett, 1899
	Yarmouthian	Leverett, 1898b
Kansan		Chamberlin, 1894, 1895
	Aftonian	Chamberlin, 1894
Nebraskan		Shimek, 1909

naming some of the classic interglacial stage names (Table 1). Less well known, however, is his training in soil science (or "agricultural geology") (Tandarich et al., 1988; Tandarich, 1998a), and his use of degree of weathering of till to differentiate younger (Wisconsin) drift inside the classic Wisconsin moraine from older drift outside the moraine (Chamberlin, 1878), presaging similar soil-geomorphic work in the western U.S. almost a century later. Leverett's work more directly related to soil stratigraphy. He recognized several buried soils and used them to establish several interglacial stages (Table 1). The Midwestern glacial and interglacial terminology quickly took hold and was used to identify glacial and interglacial stages. Thus, the Sangamon Interglacial was represented in the field by the Sangamon Soil, first named by Leverett (1898a) at its type section in central Illinois, developed in the Illinoian till sheet (Follmer, 1979). This soil was the first formally-named soil-stratigraphic unit, and is the best known and probably most intensively studied soil-stratigraphic unit in the world.

The Chamberlin-Leverett scheme, modified by Shimek (1909) (Table 1) became widely used in the early decades of the 20th century and helped focus geologic attention on the significance of buried soils in the glacial stratigraphic record. For example, poorly drained, gleyed deposits buried in till attracted attention and research, particularly the Sangamon Soil (Leighton and MacClintock, 1962; Follmer, 1979). This lead to identification of "gumbotil" (Kay, 1916), which was the poorly drained facies of the Sangamon Soil (though at the time, better-drained upland facies were not explicitly recognized). Subsequent research (Kay and Pearce, 1920) provided insights into the geochemical evolution of gumbotil that "was a major contribution toward understanding soil genesis" (Follmer, 1979, p. 82).

During the early work on the stratigraphy and character of the soils buried in Midwestern tills, little attention was paid to subtle variations in soil morphology that we would now recognize as soil horizons (Follmer, 1979). In any case, most of the work tended to focus on gleyed, organic-rich "gumbo" soils. The field geologists who applied the soil-stratigraphic nomenclature typically knew little about soils or soil morphology (Follmer, 1979). This approach to soils began to change in 1923 when C.F. Marbut presented a series of lectures at the University of Illinois, making his case for the Russian approach to soil genesis and also presenting the A-B-C scheme of soil horizonation (Follmer, 1979). Soil scientists E. A. Norton and R. S. Smith (1928) were the first to describe the buried soils in terms of soil horizons, and geologists Leighton and MacClintock (1930) first recognized a relationship between well-drained and poorly-drained variants of the Sangamon Soil across a buried landscape, subsequently recognized as a buried catena (Follmer, 1979). By the 1930s, therefore, some geologists were beginning to apply the newly emerging concepts of pedology in unraveling the evolution of soils and landscapes.

1930s to 1941

In the 10 years or so before the U.S. became engaged in the Second World War, there were a series of developments in research and publication that had profound effects on the subareas within pedology, Quaternary geology, and physical geography that would emerge as soil geomorphology later in the century. In retrospect, that time, more broadly characterized in the history of the United States by economic turmoil (the Great Depression), environmental catastrophe (the Dust Bowl), and looming war, was something of a Golden Age in attempts at understanding the relationship of soils to landscapes, both modern and ancient. Indeed, some of these otherwise adverse conditions helped to drive what would become soil-geomorphic research. The research threads in pedology and Quaternary geology outlined above (promotion of the five-factor view of soil genesis and recognition of buried soils in the Midwest glacial sequence) continued, but several new research agendas and key players, and resulting publications, appeared on the scene.

USDA, The Soil Conservation Service and Soil Erosion Studies

In pedology, Marbut and then Kellog continued to promote the Russian approach to soil genesis both within the USDA and in other research and academic settings (Simonson, 1997b). In the USDA, the culmination of the Russian/Marbut influence can be seen in the official soil classification system published in 1938 (Baldwin et al., 1938). The classification was developed by Marbut in the years before his death in 1935 (Marbut, 1935; Kizer, 1985) and the importance of the factors can be seen at the higher levels of classification, with broad groupings based on physiographic characteristics (e.g., "Tundra soils," "Desert soils," and "Prairie soils") (Simonson, 1987). Further evidence of Marbut's geologic view of soils can be seen in the emphasis on "mature soils" (or "zonal soils"). Notions of soil maturity and soil genesis undoubtedly stemmed from the concepts of landscape evolution espoused by one of Marbut's professors, the eminent geomorphologist W.M. Davis (Cline, 1961; Johnson, 1985).

Another important event in the institutional study of soils in the 1930s was the establishment of the Soil Conservation Service (SCS) in the USDA. The roots of the SCS were in the Soil Erosion Service (SES), established in 1933 within the Department of the Interior (Effland and Effland, 1992). In 1935, due to severe wind erosion on the Great Plains, the SES moved to the USDA and became the SCS.

Within the SCS, a new approach to soils research, with significant geomorphic implications, was launched, driven by the advent of the New Deal and the severe environmental impact of the 1930s Dust Bowl. The following summary of the SCS soil erosion studies is from Effland and Effland (1992). In 1933 the "Science Advisory Board" (established during the New Deal) identified "land use" as an important problem that could be addressed by Federally-funded research. The Board contacted Carl Sauer (University of California-Berkeley), one of the leading geographers of his day, and he agreed with their assessment. He proposed developing a science of "surface and soil" and suggested that pedologists, geologists, and climatologists work together to study landscapes. Of particular interest at this time was soil erosion.

While these events were transpiring, the SCS established a Division of Climate and Physiographic Studies and hired climatologist C. Warren Thornthwaite, also an eminent researcher (with a PhD under Sauer), as chief of the Division. Thornthwaite took the recommendations of the Science Advisory Board, and he and Sauer established a series of soil erosion research projects at sites around the U.S. They made plans for research on the relationships of "surface and soil," but most of the field scientists hired were geologists and geographers because few pedologists of the time were trained in erosion studies. The projects produced good data on erosion, but were ultimately limited by staff sizes, funding, and base-line data on soils and geology. This project, however, laid the groundwork for a remarkable series of soil-geomorphic studies several decades later. The work also brought geographers into the soil-geomorphic picture.

Hans Jenny and *The Factors*

The factorial paradigm of soil development culminated with publication of Hans Jenny's *Factors of Soil Formation* (1941). Jenny (BioBox 1) prepared and published the book shortly after joining the University of California-Berkeley (following in the footsteps of Hilgard). Much of his thinking evolved while at the University of Missouri, however, including his "rediscovery" of the soil-forming factors of Hilgard and the Russian school (Jenny, 1980, xi). Jenny presented the most comprehensive, detailed, and integrated statement up to that time on the

BioBox 1 Hans Jenny, photographer unknown

HANS JENNY
b. 1899 in Zürich,
Switzerland
d. 1992 in Oakland,
California, USA

B.S. in Agriculture (1923)
and Ph.D. in Agricultural
Chemistry (1927), both from
the Swiss Federal Technical
Institute, Zürich. His teaching
career began in Soil Science at
the University of Missouri,
Columbia in 1928. He moved
to the Department of Plant
Nutrition at the University of
California, Berkeley, in 1936,
then transferred to the newly
created Department of Soils
(1940) where he remained
until he retired in 1967.
Jenny's initial interests were in

soil chemistry and fertility, but exposure to soil mapping and classification in Missouri, and soil tours across the central and eastern U.S. sparked an interest in the factors responsible for pedogenesis. His focus was first on the factors of time and climate, but eventually grew into a desire to develop a "logical theory of soil-forming factors." The result was *Factors of Soil Formation* (1941). His subsequent research and writing was diverse, including more on the factors (and an updated version of his book in 1980), and on soil chemistry and fertility, but also broader issues of ecology and conservation.

Amundson, 1994; Jenny, 1980, vii–xiii

influence of the soil forming factors (climate, organisms, relief, parent material, and time; the "clorpt factors"). The immediate impact of this volume was muted by the U.S. entry into Second World War (Arnold, 1994), but it went on to become one of the most influential books in pedology, and had a profound effect on soil geomorphology. Johnson and Hole (1994, p. 113) provide a succinct summary of the attractiveness of Jenny's approach: ". . . Jenny's distinctive contribution was to theoretically and methodologically *showcase* the formational-factorial approach by: using clear and simple language, using many excellent illustrations, bringing together under one cover numerous examples of soil forming situations, and calling attention to potential quantitative applications in pedology" (italics in original). Certainly the two most influential components of Jenny's thinking are formally outlining the "clorpt" formula and proposing a means of "solving" the formula by defining climofunctions, biofunctions, topofunctions, lithofunctions, and chronofunctions. The latter was an aspect of the factorial approach that was a significant step beyond the Russian/Marbut view of the factors.

In and of itself, however, *Factors* is not a volume on soil geomorphology nor is it a means to reconstructing the past. Jenny's training was in Agricultural Chemistry and his interest in writing his book was focused on systematizing soil data in ways other than classification (Jenny, 1941, xi) which by the 1930s was dominating much soil science research (p. xi). He wanted to organize soils data by means of "laws and theories" and "assemble soil data into a comprehensive scheme based on numerical relationships" (p. xi). The ultimate goal of the book was to "assist in the under-standing of soil differentiations and . . . help to explain the geographical distribution of soil types" (p. xi). Jenny did not apply the factors as a means of reconstructing the past, and *Factors* was not intended as a textbook on soil geomorphology (though it inspired such work, discussed below). It was designed to "formulate a conceptual scheme" or "logical theory of soil-forming factors" (Jenny, 1980, xi). Applying the factorial approach to reconstructing the past had to wait for several more decades.

James Thorp

In the same year that *Factors* was published, James Thorp published a paper on the significance of the environment on soil formation (Thorp, 1941). Thorp (BioBox 2) was the leading figure within USDA, promoting continued attention to the geologic aspects of soils and to the importance of soils in understanding the past. Thorp had been a member of the Bureau of Soils under Marbut and, follow-ing Marbut's death, had gone on to further Marbut's views on soil genesis (and was a co-author of the 1938 classification). He also had training in and a contin-uing interest in geology (as did most soil scientists of the day) (Thorp, 1985), and this clearly comes through in the 1941 paper. Further, Thorp spent three years

(1933–1936) studying soils in China (Tandarich et al., 1985; Thorp, 1936, 1985)
and became interested in the buried soils that are so prominent in the massive loess
deposits (Thorp, 1935). His 1935 paper includes a review of the Russian approach
to soil genesis, discussion of basic soil forming processes (e.g., laterization), and
the role of soils in interpretations of Quaternary stratigraphy. This work was one
of the first attempts to link the five factors, soil forming processes, buried soils, and
reconstructions of past landscapes.

Midwestern Soil Stratigraphy

In Quaternary geology, soil stratigraphy continued to be applied for correlation and
environmental reconstruction (see references in Scholtes et al., 1951; Thorp et al.,
1951; Simonson, 1954). Thorp's 1935 paper on his work in China (and published
in China), was, as noted earlier, a landmark in this regard, but is little-known.
Among those investigating buried soils, perhaps the best-known practitioner in the
Midwest at the time was Morris M. Leighton of the Illinois Geological Survey.
He was interested in using soils to determine the relative ages of glacial deposits
(Ray, 1974). Leighton was strongly influenced by the Russian/Marbut approach to
soils (indeed, he apparently spent time in the field with Glinka, according to Ray,
1974, p. 136), and attended lectures by Marbut at the University of Illinois in 1923

(Leighton and MacClintock, 1930, 30). Leighton also apparently worked closely with soil scientists at the University of Illinois (Tandarich et al., 2002). The result was the promotion of the "weathering profile concept" (e.g., Leighton and MacClintock, 1930), which was essentially the geologic counterpart of the soil profile in pedology (Leighton, 1958). Tandarich et al. (2002) see this dichotomy in the geologic vs pedologic views of profiles as an important and unfortunate step in the distancing of pedology from its geologic roots (noted further below). Leighton (1934) also used relative degree of soil profile development to estimate the ages of Indian mounds in Illinois, an early application of soil development as an age indicator and one of the first applications of soil geomorphology in archaeological research in North America. A 1937 paper further carried the "weathering profile" concept to archaeological stratigraphy.

Kirk Bryan

Kirk Bryan is best known for his work on arid-land geomorphology and on archaeological sites (e.g., Haynes, 1990), but also published several influential papers dealing with soils and their significance in reconstructing the past (see BioBox 3). This research evolved from Bryan's geological and archaeological investigations in the Big Bend and Davis Mountains regions of far western Texas and in the Sandia Mountains of central New Mexico in the 1930s (Albritton and Bryan, 1939; Bryan, 1941; Bryan and Albritton, 1943). Bryan's work with his graduate student Claude Albritton in western Texas had the most far-reaching impact on soil stratigraphy. They were clearly influenced by the Russian view of soil genesis and by Jenny's then new volume (Bryan and Albritton, 1943, pp. 470–471), and also by a manuscript review from Thorp (Bryan and Albritton, 1943, p. 473). They formulated the concepts of monogenetic vs polygenetic soils and composite soils. As Johnson and Hole (1994, p. 115) note, "Their paper had a tremendous impact on pedology and geomorphology. The concept of polygenesis in particular has had a dramatic and lasting affect on these fields." Their concept of monogenetic soils was also important because it complemented Jenny's (1941) notion of a soil chronosequence whereby a set of soils varies only as a function of time (i.e., climate and the rest of the clorpt variables, except time, remained constant) (Johnson and Hole, 1994, p. 115).

Bryan was also involved in archaeological research at Sandia Cave, New Mexico. Though the archaeological work was in a cave (Hibben, 1941), most of Bryan's interests seemed to be drawn to the soils on the slopes outside (Bryan, 1941). He attempted to estimate the age of some enigmatic deposits and artifacts in the cave by comparing weathering characteristics (secondary iron content) of the cave deposits. He concluded that iron in the cave sediment must have been deposited when waters moving from the surface soil and through the rock carried more iron, i.e., the source of the iron was from pedalfer soils. The current soils contain pedogenic carbonate, indicating and environment in which iron is not very mobile, so Bryan surmised that the iron in the cave originated during a past, wetter climate. Further, he correlated the cave stratigraphy and inferences about soils with the loess-, till-, and soil stratigraphy of the Midwest to estimate the antiquity of the cave inhabitants. The specifics of his reasoning and conclusions clearly do not hold up today, but this study was a milestone in applying principals of pedology and weathering to issues of Quaternary climate and archaeology.

International Influence: Milne and the Catena Concept

Through the 1930s Geoffrey Milne (East African Agricultural Research Station) and fellow soil chemists were attempting to come to grips with complex soil associations during the mapping of soils in Uganda.[3] In one of the earliest statements on what would become known as the "catena concept," Milne (1932, p. 5) wrote:

"over large areas where local variation in topography were regularly repeated, a given colour on any soil map . . . would have to be interpreted as

indicating the occurrence of not a single soil but of a sequence of soils occurring generally over the area, to be worked out on the actual ground in each
instance according to topography and other local influences."

Milne's first formal publication on the topic, where he coined the term "catena"
(1935a) essentially followed this original concept (see Gennadiyev and Bockheim,
this volume). In a subsequent paper for the Third International Congress of Soil
Science, Milne (1935b) expanded on his original ideas by noting that catenary associations can be due to differences in drainage conditions combined with the effects
of erosion and redeposition along a slope. In a later paper, Milne (1936) argued that
if erosion (including anthropogenic erosion) and redeposition kept pace with pedogenesis, then those processes should be considered pedogenic processes.

Milne was not the first to realize the importance of the soil-topography relation
(these were long employed in the U.S. soil survey; Bushnell, 1927). However,
Milne's unique contribution was linking catenary patterns to specific processes:
differential drainage, solute leaching downslope, and erosion/deposition (Milne,
1935b). A significant aspect of Milne's research was that it focused entirely on
mapping soils. Milne was not a geomorphologist and apparently had no interest in
the evolution of landscapes in a geologic sense. His concept of the catena was
described in terms of contemporary landscapes and processes.

The concept of the catena quickly caught on among U.S. pedologists (e.g.,
Bushnell, 1942) and was incorporated in the 1938 soil classification (Baldwin et al.,
1938, p. 989). But here the basic idea was changed, in spite of objections by pedologists working in the tropics (ap Griffith, 1952). The catena was defined as a
drainage sequence on uniform parent material. This was apparently done to aid
classification of soil series into "geographic groups," i.e., to classify different series
by slope and varying drainage conditions, but otherwise formed in similar parent
material (Thorp, 1941, p. 42). The notion of uniform parent material was never
considered by Milne, and his key concept dealing with transport/depositional
processes was eliminated. Thus, the catena concept entered U.S. pedology in a form
significantly altered from that originally proposed.

The War Years

Scientific research generally came to a halt during the Second World War. One
significant exception deemed important to the war effort was joining geology and
soil studies to create the Military Geology Unit (MGU) in the U.S. Geological Survey.
Their mission was to provide geologic and other environmental data to answer
military questions (e.g., evaluating, mapping, and illustrating landscapes to provide

indications of basic terrain characteristics, relief, and hydrology; suitability of areas for trafficability and construction, and availability of construction materials; and water supply). The individuals in the MGU had a dramatic influence on post-war soil geomorphology. During the years of the MGU (1942–1945), which operated under contract to the U.S. Army Corps of Engineers, 88 geologists, 11 soil scientists, and 15 other specialists were brought together to produce detailed strategic planning studies (USGS and USACE, 1945; Hunt, 1950; Terman, 1998). The leadership included Charles B. Hunt, who was first Assistant Chief and later Chief. As described below, he and several others in the MGU went on to become key players in combining soil studies with research on Quaternary geology and geomorphology within the U.S. Geological Survey (Table 2). Hunt (1972, v) clearly states that his interest in surface deposits and soils was stimulated in part by his participation in the MGU and "the opportunity to work with teams of soil scientists, plant geographers, hydrologists, and engineers, as well as geologists." Others in the USGS, including Gerald M. Richmond (GMR to VTH, Feb 14, 1992, April 2, 1992) and Roger Morrison (RM to VTH, Jan 24, 1992) echoed those sentiments. James Thorp was also in the MGU, and his interests and background in both geology and soils provided additional stimulus for developing an appreciation of soils among the geologists (GMR to VTH, Feb 14, 1992, April 2, 1992). Hans Jenny also apparently had some connection to the MGU because both Richmond and Morrison first met him through the MGU, but no published information on the MGU or on Jenny note the nature of this relationship.

Table 2. A selection of geomorphologists and pedologists (and institutional affiliations) in the Military Geology Unit of the USGS, 1942–1945[1]

Geologists			
Albritton, Claude C., Jr	SMU	Moss, John	Harvard (student)
Denny, Charles S.	Wesleyan U	Ray, Louis L., Jr	Michigan State
Hack, John T.	Hofstra College	Richmond, Gerald M.	Corps of Engineers
Hunt, Charles B.	USGS	Smith, H.T.U.	U of Kansas
Morrison, Roger B.	USGS		
Pedologists[2]			
Baldwin, Mark	USDA	Sokoloff, Vladimir P.	UCLA
Cady, John G.	U of Idaho	Thorp, James	USDA

[1] From USGS and USACE (1945) and Terman (1998, table 1). According to G. Richmond (GMR to VTH, Feb 14, 1992, April 2, 1992) and R. Morrison (personal comm., 2003), Hans Jenny was part of the MGU, but histories of the unit do not mention this. He may have had an informal, unofficial or temporary attachment.

[2] Roy Simonson was "loaned" to the MGU for one year (Simonson, 1987, p. 1).

As indicated earlier and also discussed below, pedology and geology diverged in orientation and outlook through the 20th century. Relatively few individuals in either field were interested in research that combined the two field fields. Pedologists in particular became more focused on mapping and description, as well as other more agricultural applications. The MGU literally put geologists, including geomorphologists and Quaternary stratigraphers, together with pedologists. Their duties and goals clearly were not focused on soil geomorphology, but their daily close contact fostered an appreciation for shared interests in soils, landscapes, and environments both past and present that flowered in the post-war years and decades.

1945–1974

The rapid expansion and pace of scientific research in the U.S. following the Second World War is mirrored in the many developments in pedology, geomorphology, and Quaternary geology that grew into the field of soil geomorphology. As in the above discussion, these developments can be seen institutionally as well as in the activities of a few key individuals.

Institutional and Academic Pedology

Pedology in the USDA and in academic settings seems to have followed two very different paths in research agendas in the decades following the end of the war. Much of the USDA effort was on expansion of the soil survey program, which further distanced pedology from geology, but also included starting a separate research program in soil geomorphology that put pedologists and geologists together. The efforts of the soil survey program were most visible in the production of the *Soil Survey Manual* (1951) and in the development of a new system of soil classification, which became Soil Taxonomy (Simonson, 1987). Both of these efforts resulted in standardized, systematic, unambiguous nomenclature for describing and classifying soils. Because of the size and scope of the soil survey program, adoption of the description nomenclature and the new classification system quickly spread to academic soil science and, therefore, became the *lingua franca* of pedologists in the U.S. This standardized nomenclature for field descriptions was commonly applied in soil-geomorphic studies, sometimes with "unofficial" modifications to suit particular situations (Holliday et al., 2002; Holliday, 2004). Modifications to Soil Taxonomy to accommodate thick sequences of buried soils were proposed, including some from the USDA (e.g., Thorp, 1949; Ruhe and Daniels, 1958), but were never adopted. The *Soil Survey Manual* also re-emphasized the concept of the catena as a set of "strongly-contrasting soil series" along a slope, developed in similar parent material (Soil Survey Staff, 1951, p. 160);

different from the original concept and devoid of any notion of the significance of erosion or deposition. Indeed, by this time, several pedologists equated the catena with a toposequence (Bushnell, 1942; Jenny, 1946). In any case, because of the influence of the manual, the dramatically altered concept of the catena among U.S. pedologists became more entrenched (e.g., Buol et al., 1997, pp. 154–155).

A few pedologists in both governmental and academic settings continued to press for better integration of geologic and pedologic concepts, in sometimes eloquent essays. James Thorp continued his work along these lines. For example, he presented an exposition on the relationship of Pleistocene geology to soil science, reflecting his years of interest and work on the topic (Thorp, 1949). The paper outlines the importance of Pleistocene geology to mapping and understanding soils, and the importance of understanding soils for reconstructing Pleistocene landscapes. Thorp makes a forceful argument for cooperation between geologists and soil scientists, reflecting, in part, his association with geologists in the MGU. In this paper, Thorp also introduced the term "paleosol" to the English-language literature, equating it with "buried soil" (Johnson and Hole, 1994, 117). Later, Thorp et al. (1951) presented a state-of-the-art inventory of Quaternary soil stratigraphy of the central U.S. that still remains a standard reference. Other important contributions to soils and geomorphology include a monumental map compilation of eolian deposits in the U.S. and Canada (Thorp and Smith, 1952) and one of the first attempts at summarizing pedogenesis through the Quaternary (Thorp, 1965).

C.C. Nikiforoff, a Russian pedologist who worked for USDA, also published a series of papers that, like Thorp's, strongly espoused a reintegration of geology with pedology. In particular, he urged development of the nascent field of paleopedology (Nikiforoff, 1943). This paper was followed by one emphasizing that an important implication of the factorial paradigm is that soils evolve, both downward and upward, depending on the erosional/depositional setting (i.e., the landscape) (Nikiforoff, 1949).

Roy Simonson, a pedologist who spent most of his career with USDA and eventually became Director of Soil Classification and Correlation in the SCS, also made several significant contributions to pedology that had ramifications in soil geomorphology. Trained in soil science, Simonson maintained an active interest in buried soils (Simonson, 1941, 1954) and dealt with issues of polygenesis in buried soils in till (Simonson, 1941). His most far-reaching contribution to soil geomorphology, however, was his "generalized theory" or "multiple process model" of soil formation (Simonson, 1959, 1978). Simonson described soil genesis more broadly than it was usually portrayed, as a set of individual processes such as podzolization or calcification. He grouped the wide array of known soil-forming processes into the now familiar four categories of additions, removals, translocations, and transformations, arguing that soil formation results from the interaction of processes within and among these broad sets of processes. This approach was not unique nor necessarily original, but was well explained in a highly visible soil science journal, and was widely accepted (e.g., Buol et al., 1973, 1980, 1989, 1997). His groupings provided an excellent complement to Jenny's factors; the factors are external or environmental controls that drive the internal pedogenic processes. This has been an important concept, particularly in teaching (e.g., Buol et al., 1973, 1997).

Marlin G. Cline, soil scientist at Cornell University, also presented a strong argument for maintaining links between pedology and geology, especially geomorphology. In a presentation designed to highlight the role of pedology as a component of soil science (in honor of the 25th anniversary of the Soil Science Society of America), Cline (1961, p. 443) succinctly argued that pedology ". . . must incorporate the basic concepts of geology; rocks and minerals and their transformations are keystones of our concepts; and geomorphology is the foundation of our interpretations of past events." Cline (1961, p. 444) goes on to argue that two of the three most important concepts in pedology to emerge in the previous 25 years were the importance of geomorphology and the importance of time (the third important concept being better understanding of the processes of soil formation).

The words and examples of Thorp, Cline, and a few others attempting to maintain links between pedology and geomorphology were largely unheeded, however. With development and publication of the revolutionary new system of soil classification, Soil Taxonomy (Soil Survey Staff, 1960, 1975, 1999), much of the emphasis in both governmental and academic pedology shifted to description and classification, and understanding site-specific (and profile specific) soil-forming processes, as well as mapping. This trend further separated a significant amount of soils research from geologic or geomorphologic research.

The USDA Soil-Geomorphology Projects

One significant exception to the trend within governmental soil science away from research on soil-landscape relations was the USDA soil geomorphology program of the 1950s–1970s. In this venture, the Soil Survey program of the SCS supported the most intensive and extensive systematic investigations of soil geomorphology attempted in North America. The following discussion is based on the history of the soil geomorphology projects prepared by Effland and Effland (1992, n.d.; see also Holliday et al., 2002). The projects evolved from the USDA soil erosion studies of the 1930s. The USDA was reorganized in 1952, resulting in placement of a "more research-oriented staff" in the soil survey program (Effland and Effland, 1992, p. 203). In 1952 and 1953, Charles D. Kellogg, head of the Soil Survey program, and Guy D. Smith, chief of Soil Survey Investigation, ". . . advocated a fully developed multi-site series of soil geomorphology studies as the basis of a research program in support of the soil survey" (Effland and Effland, 1992, 204). In 1953, Robert V. Ruhe (BioBox 4, a geologist with interests and training in pedology, was hired to begin the first of these studies and to direct the entire soil geomorphology research program, which he did until leaving the SCS in 1970. The individual field studies were carried out by pedologists and geomorphologists working together.

BioBox 4 Robert Ruhe, photograph by P.W. Birkeland

ROBERT V. RUHE
b. 1918 in Chicago Heights, Illinois, USA
d. 1993 in Bloomington, Indiana, USA

BA Carleton College, MA Iowa State university, PhD, University of Iowa. He worked for the U.S. State Department in the Belgian Congo, 1951–1952, then 1953–1970 the USDA Soil Conservation Service, directing the soil-geomorphology projects 1953–1970, and spent the remainder of his career at Indiana University in Bloomington as Professor of Geology and Director of the Indiana University Water Resources Institute until retirement in 1985. Ruhe's research career began in the Belgian Congo, but became firmly established with the USDA soil-geomorphology projects. He set up and directed the Iowa Project and then initiated the Desert Project before returning to Iowa. After leaving USDA, his work focused on soils and loess in the Midcontinental US.

Wright, 1995

Seven USDA soil geomorphology projects were authorized. The study areas were selected to represent a wide range of climatic and topographic features. Four of the planned seven projects were completed: 1) the humid, glaciated landscape of southwestern Iowa; 2) the arid and semiarid basin-and-range country of south-central New Mexico; 3) the humid Coastal Plain of North Carolina; and 4) the humid, maritime Pacific coast environment of western Oregon (Table 3). Two of the other three projects were partially completed (Table 3). Most of these investigations were in cooperation with local universities, agricultural experiment stations, and state geological surveys. They were landmark studies of the relationship of soils, geology, and landscapes and demonstrated the benefits of detailed soil geomorphology studies for interpreting landscape history and facilitating soil survey. The research had a variety of impacts on the evolution of soil geomorphology as a subdiscipline, at specific, local scales as well as more broadly in understanding and conceptualizing regional soil-landscape relations. Some of the studies became the standard for academic research in North America and the world, though they appear to have had minimal impact in the Soil Survey.

The Iowa Project and the New Mexico "Desert Project" were two of the most influential soil-geomorphic studies in the history of the subdiscipline. The

Table 3. Selected publications resulting from the USDA and USGS soil geomorphology projects

Project *Principal Investigators*	References
USDA Iowa Project *R.B. Daniels, R.V. Ruhe*	Ruhe, 1954b, 1956a, 1960, 1969, 1970, 1975; Ruhe et al., 1967, 1968; Ruhe & Daniels, 1958; Vreeken, 1975a; Walker, 1966; Walker & Ruhe, 1968
USDA Desert Project *L.H. Gile, F.F. Peterson,* *J.W. Hawley, R.V. Ruhe*	Gile, 1975a,b, 1977; Gile & Grossman, 1979; Gile et al., 1965, 1966, 1981, 1995; Ruhe, 1962, 1964a, 1967
USDA Coastal Plain Project *R.B. Daniels, E.E. Gamble*	Daniels & Gamble, 1978; Daniels et al., 1967, 1970, 1973, 1978; Gamble et al., 1970
USDA Oregon Project *C.A. Balster, R.B. Parsons*	Balster & Parsons, 1968; Parsons, 1978; Parsons et al., 1968, 1970; Parsons & Balster, 1967
USDA Hawaii *R.V. Ruhe*	Ruhe, 1964b, 1965; Ruhe et al., 1965a,b,c
USDA Southern Plains *L.H. Gile, J.W. Hawley*	Gile, 1979, 1985

impacts of these investigations are more fully enumerated by Holliday et al. (2002), but several examples provide illustrations. The Iowa Project was the first of the studies to get under way and therefore had some of the broadest impact. This was the first large-scale, long-term research that explicitly linked pedology, geomorphology, and Quaternary stratigraphy. As a result of the early stages of the work, Ruhe (1960) argued that because weathering takes place from the land surface downward, the soils and the landscape elements on which they occur are intimately linked, thus laying the conceptual groundwork for modern soil geomorphology. Another important concept introduced during the Iowa Project was the "hillslope profile," a commonly used system describing the components of a slope profile, which include summit, shoulder, backslope, footslope, and toeslope (Ruhe, 1960). Ruhe and co-workers also related the Holocene history of erosion and deposition to slope evolution and basin infilling and to vegetation and inferred climatic changes, thus establishing linkages among vegetation cover, soil erosion, and sedimentation. These studies strongly influenced Holocene landscape evolution concepts in recently glaciated humid to subhumid continental regions (Gerrard, 1992).

The Desert Project was arguably the most influential study of arid-land soils in the history of pedology and soil geomorphology. The results of the research are prominent in some of the most widely read volumes on soils, geomorphology, and Quaternary geology (Birkeland, 1984, 1999; Bull, 1991; Gerrard, 1992) and in the major geomorphology textbooks (e.g., Ritter et al., 1995; Easterbrook, 1993). One especially significant study was on sequences of carbonate accumulation in desert soils by Gile et al. (1966), which first presented the now classic "stage" concept for carbonate accumulation in desert soils and their dependence on time. Recognition of this time dependency provided a powerful means to estimate the age of alluvial fan and terrace deposits throughout desert regions (e.g., Machette, 1985) because they seldom contain materials suitable for the numerical dating techniques available in the 1960s to 1980s. In the year preceding publication of this important paper, the Desert Project team introduced the K horizon as a new master horizon (Gile et al., 1965), a massive zone of carbonate accumulation that had a profound influence on subsequent soil development and landscape evolution. Though never adopted by the USDA, the concept of the K horizon and the morphologic stages of carbonate accumulation horizon were quickly adopted by many Earth scientists working in deserts, especially North Americans.

Another very significant result of Desert Project research was demonstrating the importance of dust and other external additions to soils in arid region soil genesis. Other researchers, notably Israeli scientists (e.g., Yaalon and Ganor, 1973), had recognized the potentially large influence of dust on the genesis of calcareous and/or clay-rich horizons of desert soils (see also Free, 1911, noted above). The

Desert Project, by linking the atmospheric with the soil data, provided the data to critically test and ultimately prove the hypothesis.

The Coastal Plain and Oregon projects were somewhat more limited in scope, but both were important contributions to broader themes in soil geomorphology. The Coastal Plain project produced one of the first comprehensive studies of Ultisol genesis (Daniels et al., 1970, 1978; Daniels and Gamble, 1978). This project also showed that soil variability decreased from young to old surfaces due to the convergence of soil profile characteristics over extended periods of weathering and soil development (Gamble et al., 1970). The "edge effect" concept was developed during an investigation of soil variability within geomorphic surfaces in the project area. This concept explains variations in soil color, E horizon thickness, and gibbsite and free iron oxide content with respect to water-table flux in different landscape positions (Daniels et al., 1967). Of particular interest in the Oregon Project studies were investigations of soil variability as a function the buried soil-stratigraphic record, which exerted considerable influence on the geography and interpretation of surface soils (Parsons and Balster, 1967; Parsons et al., 1968).

Robert V. Ruhe

The USDA soil geomorphology projects provided an important venue for Robert Ruhe (BioBox 4) to develop many basic concepts of soil geomorphology that go beyond any one of the regional studies, and established Ruhe as one of the key figures in the development of modern soil geomorphology. Ruhe pioneered the integration of process-oriented quantitative geomorphology and hydrology with modern pedology (Olson, 1997, 415; see also Olson, 1989). Ruhe's view of landscape evolution began to gel during geomorphological research in the Belgian Congo from 1951 to 1952 (Ruhe, 1954a, 1956b) and was heavily influenced by the pedimentation (slope back wearing) concepts of Lester King (1949, 1950, 1953) and especially the catena concept of Milne (1935a,b).

The papers by Ruhe (1956a) and Ruhe and Scholtes (1956), resulting from the Iowa Project, were two of the first papers in what we now recognize as soil geomorphology, documenting (rather than simply asserting or casually noting) the effects of parent material, landscape position, and landscape age on soil geography, and also using both surface and buried soils to reconstruct the evolution of the landscape. Ruhe's chapter on "Quaternary Paleopedology" (Ruhe, 1965a) became one of the more influential papers on the topic, detailing his "tripartite" perspective on types of "paleosols" (buried, exhumed, and relict), although his definitions of the terms were somewhat at odds with the priority of their usage (Johnson and Hole, 1994, p. 118). Finally, Ruhe (1974) introduced the term "soil geomorphology," though without explicitly defining it. In that paper he also describes the "soil geomorphic unit" which is the repetitive occurrence of a soil-hillslope system that "illustrates the fit of specific soils with specific properties to the erosional and depositional parts of the hillslope" (Ruhe, 1974, pp. 493–494). He effectively reintroduced the catena concept as originally defined by Milne, though without citing him.

Several methodological approaches, seen as routine today, also emerged from the soil geomorphology projects and reflected Ruhe's influence. One was the application of radiocarbon dating to determine the ages of landscapes and soils, particularly in Iowa and New Mexico. This use of radiocarbon dating was the first on such a large and regional scale, coming within a decade or so after the invention of the technique. The other methodological advance (summarized by Olsen, 1997) was in the use of widespread coring and trenching to work out stratigraphic and landscape relationships. This again reflected the post-war advances in technological (and financial) support available for such large-scale efforts.

U.S. Geological Survey

In the years following WWII, research on what would now be recognized as soil geomorphology also became an important part of Quaternary studies in the USGS. This was a direct outgrowth of the close collaboration between geologists and pedologists in the MGU during the war, and is best seen in the work of Charles B. Hunt, Gerald M. Richmond, and Roger Morrison. Hunt, who became Chief of the MGU, is emphatic on this point in the preface to his book *Geology of Soils* (Hunt, 1972, v). After the Second World War, Hunt stayed with the USGS for a time, then went on to The Johns Hopkins University and New Mexico State University, and ended his career back in the USGS. He published several stratigraphic studies that incorporated soils, including alluvial stratigraphy along the South Platte River (Hunt and Sokoloff, 1950; Hunt, 1954) and shoreline stratigraphy in the Bonneville Basin (Hunt and Sokoloff, 1950). Hunt's work with soils in these settings inspired the research of co-workers Gerald Richmond in the La Sal Mountains and Roger Morrison in the Bonneville Basin (see below) and led to the classic soil-geomorphic work of Glen Scott (1963) in the Kassler Quadrangle of the South Platte, one of the first applications of surface soils both for correlation of terraces and for relative age estimates. Work on the coastal plain of Florida (Hunt and Hunt, 1957) incorporated a catenary approach (though not described as such) that also had archaeological implications (Holliday, 2004). His final publication dealing with soils was *Geology of Soils* (Hunt, 1972), the first English-language book that explicitly linked soils and geology in an integrated manner.

Gerald Richmond (BioBox 5) was also clearly influenced by his work in the MGU and association with C.B. Hunt. His initial exposure to soils came from Kirk Bryan (apparently before WWII), but the MGU provided a longer and more continuous apprenticeship. James Thorp was especially prominent in this role during and after the war (GMR to VTH, Feb 14, 1992). This association can be most prominently seen in Richmond's (1962) study of glacial stratigraphy in the La Sal

BioBox 5 Gerald Richmond, photographer unknown

GERALD M. RICHMOND b. 1914 in Providence, Rhode Island, USA d. 2001 in Denver, Colorado, USA

AB in biology and psychology, Brown University (1936), MA in geology, Harvard University (1936), and PhD. in geology, University of Colorado (1954). He joined the USGS in 1941. He stayed with the survey for his entire career, over 50 years, including work for the MGU (1943–1945) where his appreciation for and interest in soils began. Most of his work focused on Quaternary geology, particularly the glacial history of the Rocky Mountains and other parts of the mountain west. He was also a key figure in the 1965 INQUA Congress in Boulder, CO, where much U.S. soil-geomorphic and soil-stratigraphic work was showcased to the international Quaternary community.

Letters to VTH, February 14, 1992; April 2, 1992

Mountains, Utah. Soils were the most important tool for correlating deposits locally and regionally. Richmond took concepts from Thorp's Midwestern soils studies and applied them to the mountains. More importantly, changes in soil morphology with altitude (soil facies) were recognized; probably a first in this sort of geologic mapping. This work showed a direct influence of Thorp, based on his work in Wyoming before WWII (Thorp, 1930). Field consultations with Hunt also benefitted Richmond's soils applications in the La Sals (GMR to VTH, Feb 14, 1992). The La Sal work was the basis for his PhD dissertation in geology at the University of Colorado (the first in Quaternary geology in that department). Richmond's influence extended beyond field work. He was a leader in incorporating soils as stratigraphic units in the Code of Stratigraphic Nomenclature (Richmond and Frye, 1957).

Roger Morrison's work (BioBox 6) with soils began largely after WWII, when he worked in the Lahontan Basin of Nevada for the USGS (RM to VTH, Jan 24, 1992). Though exposed to the geologic significance of soils via the MGU, most of his initial soils training was largely in the field from USGS (and former MGU) colleagues Charles Hunt and Gerald Richmond. This background was important in Morrison's research on the history of Pleistocene Lakes Lahontan and Bonneville (Morrison, 1964, 1965). Soils were an important part of the stratigraphic sequence he recognized and also were an important tool in mapping the surficial geology, which set the standard for others to follow (P. W. Birkeland, pers. communication, 2003). He came to recognize buried soils as important components of the stratigraphic record. His interest in soils resulted in a PhD dissertation (from the University of Nevada-Reno; also the first in Quaternary geology there) were he presented his views on Quaternary soil stratigraphy, subsequently published (1967) and revised (1978). This was perhaps the first systematic review of soils as stratigraphic units. Morrison's dissertation and the 1967 paper are where the Geosol concept first appeared, later to be adopted (with some modification) as the official pedostratigraphic unit in the North American Code of Stratigraphic Nomenclature (NACOSN, 1983).

Midwestern Soil Science

Important soil-stratigraphic and related soil-geomorphic research continued in the stratified glacial and eolian deosits in the Midwest throughout the post-war years. Beyond the work of the USDA and R.V. Ruhe in Iowa, much of this research was in state geological surveys. M.M. Leighton, as Director of the Illinois Geological Survey, continued his work on the Quaternary evolution of Illinois and neighboring areas through the rest of his career (e.g., Leighton and Willman, 1950; Leighton, 1960, 1965; Leighton and MacClintock, 1962).

BioBox 6 Roger Morrison, photography by P.W. Birkeland

ROGER MORRISON
b. 1914 in Madison, Wisconsin, USA

A.B. in Geology (1933), M.A. in Economic Geology (1934), both from Cornell University. Morrison began working for the USGS in 1942, and served with the MGU until 1947, largely working on water supply (ground and surface water). In 1947 he moved out of Military Geology and into the General Geology Branch where he started working on the history of the Lahontan Basin. This is where his work on soils began, with encouragement from Charles B. Hunt and Gerald M. Richmond, and was followed by similar work in the Bonneville Basin. He received a PhD from the Mackay School of Mines at the University of Nevada-Reno. Much of his later work was for the EROS program of the USGS and focused on geomorphic mapping in the southwestern U.S. He retired from USGS in 1976.

Letter to VTH, January 24, 1992; personal communication, 2003

Another important figure was John C. Frye. He began his career in Kansas, working for the Kansas Geologic Survey from the war years to 1954. His work on Pleistocene stratigraphy and geomorphology included the use of soils for correlation and environmental reconstruction (e.g., Frye, 1949, 1951; Frye and Leonard, 1951, 1952; Frye et al., 1948). He then moved to the Illinois Geological Survey, succeeding Leighton as Director. Frye continued his own and Leighton's work with soils as key stratigraphic markers (Frye and Leonard, 1955a; Richmond and Frye, 1957), furthered studies of the "gumbotil dilemma" that reformulated concepts of the Sangamon Soil (Follmer, 1979; Frye et al., 1960; Frye and Willman, 1963, 1970), and pioneered the use of petrography and mineralogy to study Cenozoic soils (Frye and Leonard, 1955b, 1957a,b, 1967; Frye et al., 1963, 1966). Frye also collaborated with Roger Morrison to propose long distance correlations, based in large measure on soils, of Quaternary stratigraphy between the Midwest, southern Great Plains, Wasatch Range, Bonnevile Basin, and Lahontan Basin (Morrison and Frye, 1965).

In Wisconsin, soil mapping and subsequent research by Francis Hole were significant steps in understanding soil-landscape relationships. Soil survey in Wisconsin was part of the Wisconsin Geological and Natural History Survey, beginning in 1909 (Hole, 1962). Hole was a product of this unusual collaboration of soil science and geology. He was trained in geology, like so many pre-War pedologists, but went to work for the University of Wisconsin-Madison as a soil scientist. He spent much of his career in the decades after the war teaching in the Department of Soil Science and mapping soils in Wisconsin for the Geological and Natural History Survey. This work honed his insights on soil-landscape and soil-environment relations. In 1967 he gained a joint-appointment between the Department of Soil Science and the Department of Geography at the University of Wisconsin. He further developed his teaching and writing on soils within a geographic approach. His principal contributions in soil geomorphology were his views on soils as dynamic entities; he was an early proponent of the concept of "pedoturbation," emphasizing the often subtle and microscopic, but nevertheless important movements within soils due to a wide range of processes during pedogenesis (Hole, 1961, 1981). His thoughts on the "vibrant" nature of soils are perhaps best captured in his last scientific publication, detailing "terra vibrata," which he defined as the dynamics of soil landscapes (Hole, 1988). Hole and O. W. Bidwell (Bidwell and Hole, 1965) also anticipated the late 20th century interest and concern about human impacts on soils (e.g., Amundson and Jenny, 1991). Francis Hole left a significant teaching legacy for pedologists and soil geomorphologists as co-author of *Soil Genesis and Classification* (Buol et al., 1973, 1980, 1989, 1997), the most widely used and disseminated pedology text in the U.S. and perhaps the world.

Kirk Bryan Legacy

Kirk Bryan continued his work on geomorphology, soils, and archaeology in the years immediately after WWII. In soils, he published a short piece on "cryopedology" (permafrost soils) (Bryan, 1949) and a paper on the relationship of buried soils to climate change (Bryan, 1948). The latter remains almost wholly unknown because it was published in Spanish. Bryan died in the field in 1951, ending a remarkable career. He also guided a group of graduate students who exerted significant influences in geomorphology, Quaternary geology, and paleoecology (Haynes, 1990). Oddly, though none of these students continued Bryan's interests in soils, the geomorphic significance of buried soils was recognized and incorporated into their geologic interpretations (e.g., Hopkins and Giddings, 1953; Judson, 1953; Leopold and Miller, 1954). His principal intellectual descendant in this regard is geoarchaeologist C. Vance Haynes, Jr., who further developed Bryan's late Quaternary "alluvial chronology" by, among other things, utilizing buried soils as stratigraphic markers and as indicators of geomorphic stability (Haynes, 1968). More broadly, the work of Bryan, particularly the paper by Bryan and Albritton (1943), had long lasting results in geomorphology. "Polygenesis soon became the theoretical core of the geomorphic concept of 'polygenetic landscapes'" (Johnson and Hole, 1994, p. 115).

INQUA 1965

In the years from 1965 to 1974, there were a series of events that culminated in the appearance of modern (late 20th/early 21st century) soil geomorphology. These events are built around several key publications and individuals. In 1965 the International Union for Quaternary Research (INQUA) held its VIIth Congress in Boulder, CO. Gerald Richmond was a key player in getting the meetings to Boulder and was the Secretary General of the Congress. Soils in geologic contexts were an important component of the themes, papers, and field trips of the Congress and in subsequent INQUA publications. In many ways the Congress and papers represent the initial integration of decades of interests in Quaternary soil stratigraphy among many geologists, the influence of the MGU, and the data emerging from the USDA soil geomorphology projects. For example, in the monumental review volume for the Congress, *The Quaternary of the United States* (Wright and Frey, 1965), soils are prominent components of the stratigraphic summaries that make up close to half of the book. The impact of the soil geomorphology projects can be seen in several chapters (Kottlowski et al., 1965; Wright and Ruhe, 1965) and, in particular, in Robert Ruhe's "Quaternary Paleopedology" (Ruhe, 1965a), noted above.

Beyond that center-piece volume for the Congress, several other books evolved from the conference that featured soil stratigraphy and soil geomorphology, largely through Roger Morrison's efforts. Most notable among these books is *Quaternary Soils* (Morrison and Wright, 1967), the first edited volume to focus exclusively on the significance of soils in Quaternary research, and *Means of Correlation of Quaternary Successions* (Morrison and Wright, 1968), which includes discussion of soils as important correlation tools. *Quaternary Soils* is also the venue for Morrison's views on Quaternary soil stratigraphy, from his PhD dissertation (Morrison, 1967).

International Influence: Butler K-Cycles; Yaalon and *Paleopedology*

Several post-war developments in pedology outside of the U.S. had significant influences on pedology and soil geomorphology in the states. In the 1950s B. E. Butler in Australia developed the K-cycle concept. A key tenet in soil geomorphology (and soil stratigraphy in particular) is the relationship between soils and landscape stability. Soils in the stratigraphic record are uniquely suited to identifying phases of stability vs phases of instability (Gerrard, 1993). Put simply, soils form on stable landscapes, i.e., landscapes with no or little erosion or aggradation. There are exceptions to this concept (e.g., slowly aggrading floodplains or eolian landscapes or regions of continuous, intense weathering, e.g., tropical settings), but the basic idea has proven useful in the interpretation of buried soils. This view of soils and landscapes as "periodic phenomena" was formalized by Butler (1959, 1982) in his "K-cycle" concept (the K referring to the Greek *khronos* for time). This approach to soils evolved from the initial stages of soil-stratigraphic and soil-geomorphic research in Australia in the 1950s "when pedology was concerned mainly with soil classification and mapping" (Walker, 1989, 589). Recognition of a complex Quaternary stratigraphic and geomorphic record by Walker (who would work with Ruhe on the Iowa Project; Table 3) and other Australian geoscientists led to recognition of cycles of stability and instability on the landscape (e.g., Tonkin and Basher, 1990).

Each K-cycle includes an unstable phase of erosion and deposition, followed by a stable phase with concomitant pedogenesis. A clear implication of this formalized approach to describing the events represented by buried soils is that they represent a sequence of landscapes, and recurrent cycles of landscape stability and instability. Butler (1959, 1982) also emphasizes the lateral variability of soils on both modern and buried soilscapes and thus, without saying so, describes catenas and paleocatenas. Though not particularly well known by its name in the U.S., the K-cycle concept has been widely applied (e.g., in the USDA projects) and has proven useful for understanding the relationship between pedogenesis and other earth surface processes (Brewer, 1972; Huggett, 1975; Catt, 1986, p. 166; Gerrard, 1992, pp. 216–220). The K-cycle concept is also important because it emphasizes the relationship of stratigraphy to landscapes, and also emphasizes soils, buried and unburied, as three- and four-dimensional entities.

On the other side of the globe, in 1971, Dan Yaalon edited and published a series of papers on "the origin, nature and dating of paleosols" in the volume *Paleopedology* (Yaalon, 1971a). Yaalon is an Israeli pedologist who worked on issues of soil-landscape and soil-stratigraphic relationships (e.g., Yaalon, 1967; Dan et al., 1968). The book was the result of the "Symposium on the Age of Parent Materials and Soils" held in Amsterdam in 1970 and co-sponsored by INQUA and the International Society of Soil Science. The symposium brought together "pedologists, geomorphologists, sedimentologists, stratigraphers, radiocarbon chemists, micromorphologists and clay mineralogists interested in paleosols" (Yaalon, 1971a, x). This was probably the first international gathering of such a diverse group of scientists in a symposium devoted exclusively to the topic of paleopedology. The volume was also the first attempt to systematically deal with the stratigraphic, paleoenvironmental, and geochronologic aspects of buried soils.

There were several influential papers in the volume. Yaalon (1971b) presented a summary discussion and table indicating relative rates of development of various pedogenic features, and the direct relationship between these rates and the persistence of the features in buried soils (i.e., features that take longer to form, tend to persist longer in the geologic record). This become a standard guide in the study of rates of soil formation and buried soils (e.g., Birkeland, 1999, table 1.5). Most significant, however, are the papers that deal with the issue of radiocarbon dating of soils; the first collection of papers on this topic. The papers by Geyh et al. (1971), Polach and Costin (1971), and Scharpenseel (1971), for example, became classic overviews of the problems and potentials of dating soil organic matter and, in particular, the issues of "mean residence time" and the mobility of humic and fulvic acids.

University of California, Berkeley

Following the Second World War, Hans Jenny continued his work on the soil-forming factors (Jenny, 1946, 1958, 1961; Jenny et al., 1968) as well as many other topics (Amundson, 1994). This work culminated in what was essentially an update of *Factors* (Jenny, 1980), but with more quantitative data. Quantification of the state-factor equation was a continuing interest of Jenny and his students. One of these, R. J. Arkley, eventually joined the Soil and Plant Nutrition faculty at Berkeley. Arkley was a pedologist but used geomorphic and Quaternary stratigraphic relationships to map soils in California (e.g., Arkley, 1964). He also published a pioneering study in the quantification of soil formation and using soils data to get at past environment and to explain the distribution of soil. Arkley (1963) compared depth to the top of the calcic horizon to climate in a transect across the Sierra Nevada. Depths that seem discordant with present climate could be due to past climate. Arkley (1967) also looked at small scale patterns of soil distribution and showed that they could be linked to water-balance climatic parameters, based on soil-moisture storage capacity and mean annual temperature. These studies were several decades ahead of research on modeling climate and calcic horizon development, discussed below.

Another development in soil geomorphology at Berkeley was a decision to hire a geologist in Soils and Plant nutrition, a rare if not unique step for a soil science program. That hire (in 1962) was Peter W. Birkeland (BioBox 7), a Quaternary geologist and geomorphologist interested in using soils to work out Quaternary stratigraphic relationships in the Sierra Nevada. He credits Jenny and Arkley with his formal training in soils. This combined with working and teaching alongside them led Birkeland to systematically combine soils and geomorphology in both teaching and research (Birkeland, 1989). He also came to recognize the importance of soil stratigraphy when he continued his Quaternary stratigraphic work in the Sierra. Tracing units down the Truckee River he came in contact with Roger Morrison and his work in the Lahontan Basin (Birkeland, 1989, 2001). An immediate result of these collaborations was using soils for mapping Quaternary geology. For example, Birkeland worked with geologist R. J. Janda, and, using the careful geomorphically-informed soil mapping by colleague R. J. Arkley, prepared detailed maps of Quaternary geology in the San Joaquin Valley of California (Birkeland, 1974, fig. 2–3, 1999, fig. 2–3).

Peter W. Birkeland

In 1974, Peter W. Birkeland (BioBox 7) published *Pedology, Weathering and Geomorphological Research*, which, along with later editions, influenced much late 20th century soil geomorphology in the U.S. In 1967 Birkeland left Berkeley and moved to the University of Colorado where Richmond's work in the Rocky Mountains made a strong impression (Birkeland, 1989; pers. communication, 2003). In a later edition of his book (Birkeland, 1999), he specifically recognizes Richmond, Morrison, Jenny, Frye, and Ruhe for their influence on him and on soil geomorphology.

In his book, Birkeland took Jenny's "clorpt paradigm" for describing and quantifying the impacts of the five factors on the genesis and distribution of soils and recast it to establish a research paradigm for using soils to help reconstruct the past. A significant advantage of Birkeland's approach is the availability of a wide

BioBox 7 Peter Birkeland, photograph by V.T. Holliday

PETER W. BIRKELAND b. 1934 in Seattle, Washington, USA

B.A. in Geology, University of Washington (1958); PhD in Geology, Stanford University (1962). At the invitation of Hans Jenny, he joined the Department of Soils and Plant Nutrition at the University of California-Berkeley as a geomorphologist in 1962, then moved onto Geological Sciences at the University of Colorado-Boulder in 1967. He retired in 1997. Birkeland was a Quaternary geologist who initially became interested in soils as a means of dating moraines in the mountains of the western U.S. Under the influence and tutelage of Hans Jenny, Roger Morrison, Gerald Richmond, Robert Ruhe, and Dwight Crandell (USGS) Birkeland recast the *clorpt* formula to study the past. He further combined the factorial approach with stratigraphic and geomorphic approaches to pedology evolved from the MGU (via Richmond and Morrison) and the USDA soil-geomorphology projects (particularly the Iowa and Desert projects). His own field work was in the glaciated terrain of the western U.S., largely in the Rockies and Sierra Nevada, but by the late 1970s he started working internationally, including field research in New Zealand, the Peruvian Andes, Israel, and Baffin Island. Most of this work focused on soil chronosequences and topo-chronosequences.

Birkeland, 1989, 2001; Hawley, 1989

variety of analytical methods, especially numerical dating techniques, to truly establish quantitative soil-forming functions. The "clorpt equation" combined with Birkeland's quantitative approach provides a powerful means of establishing research questions regarding soil-landscape relationships. *Pedology, Weathering and Geomorphological Research* (Birkeland, 1974) and subsequent editions titled *Soils and Geomorphology* (Birkeland, 1984, 1999) became the standard textbook for soil geomorphology. As described by John Hawley (1989) (geomorphologist on the Desert Project), Birkeland's book is ". . . by far the best and most comprehensive presentation of soil-geomorphic relations from a broad-based geological perspective."

1974–2004

The last quarter of the 20th century saw soil geomorphology fully formed, insofar as we see it in the early years of the 21st century. In the same year that Birkeland published the first edition of his influential book, Ruhe (1974) coined the term "soil geomorphology" (ironically, also about the time the USDA closed the soil geomorphology projects). The importance of using soils to understand the evolution of landscapes and other aspects of the geologic past are well illustrated by use of the term "soil geomorphology." In addition to subsequent editions of Birkeland's volume, the term is applied to other author-written, systematic treatments (Daniels and Hammer, 1992; Gerrard, 1992), and several edited volumes (Richards et al., 1985; Knuepfer and McFadden, 1990). Through these publications as well as many others, most of the historical threads outlined above can be seen.

Birkeland's approach to soil geomorphology broadened somewhat in subsequent editions of his book (1984, 1999). The catena concept and the effects of erosion and deposition across soil landscapes in different climatic regimes were added, including a full chapter in the 3rd edition (Birkeland, 1999). He and his students also began investigating topo-chronosequences, where soils within a catena are compared to those related to older deposits of a chronosequence (e.g., across moraines) (Table 4). This approach provided additional insight into soil variability within chronosequences and identified settings where age-related pedologic features might be well expressed or poorly expressed.

Birkeland's influence can be seen in a variety of studies, both by direct academic descendants and by indirect intellectual descendants (e.g., Table 4). In particular, the study of chronosequences (Table 4) and the identification of paleoclimatic indicators in soils has been widely embraced. "There is no doubt that soil chronosequences are immensely powerful tools for probing the rate and direction of soil

Table 4. Examples of chronosequence and topo-chronosequence studies

Location	Soil variables[1]	Reference
Chronosequences		
Coastal Oregon	overall profile morphology, Fe & Al, podzolization, clay illuviation	Langley-Turnbaugh & Bockheim, 1997
Metolius River, OR	alteration & translocation of Fe, Al, & P	Scott, 1977
Rock Creek Basin, MT	argillic & calcic horizon formation	Reheis[2], 1987c
Rocky Mountain, CO, Sierra Nevada, CA	cambic & argillic horizon formation; overall profile morphology & thickness; rubification; alteration & neoformation of clay minerals	Shroba[2] & Birkeland, 1983
South Platte River, CO	argillic & calcic horizon formation; overall profile morphology & thickness	Machette[2], 1975; Holliday[2], 1987; McFaul et al., 1994
Mojave Desert, CA & AZ	cambic, argillic & calcic horizon formation	Shlemon[2], 1978; Bischoff et al., 1981; Shlemon & Budinger, 1990
Eastern Mojave Desert, CA	argillic & calcic horizon formation; overall profile morphology; rubification; alteration or translocation of Fe	McFadden et al., 1986
Ventura Basin, CA	argillic horizon formation; overall profile morphology & thickness; rubification	Rockwell et al., 1985
Transverse Ranges, CA	alteration of Fe & clay minerals	McFadden & Hendricks, 1985
Transverse Ranges, CA	argillic horizon formation; rubification; overall profile morphology; alteration or translocation of Fe	McFadden & Weldon, 1987
Sacramento Valley, CA	cambic & argillic horizon formation; overall profile thickness & morphology; rubification	Busacca, 1987
Silver Lake Playa, CA	calcic horizon formation; overall profile morphology; rubification	Reheis[1] et al., 1989
San Clemente Is, CA	argillic horizon, rubification, Vertisols	Muhs, 1982

(*Continued*)

Table 4. Examples of chronosequence and topo-chronosequence
studies—*Cont'd*

Location	Soil variables[1]	Reference
Black Mesa, AZ	cambic, argillic, natric & calcic horizon formation, argillic horizon thickness; overall profile morphology & thickness; rubification	Karlstrom, 1988
Lubbock Lake site, TX	cambic, argillic & calcic horizon formation; rubification	Holliday[2], 1985, 1988
Ridge & Valley area, PA	argillic & cambic horizon & fragipan formation; overall profile morphology; neoformation of clay minerals	Bilzi & Ciolkosz, 1977a
Susquehanna River, NY	cambic horizon formation; overall profile morphology; rubification	Scully & Arnold, 1981
Susquehanna River, PA	clay illuviation and argillic horizon thickness; overall profile morphology and thickness; rubification; alteration or translocation of Fe	Engel et al., 1996
Des Moines River, IA	cambic & argillic horizon formation; rubification; over profile morphology	Bettis, 1992
Northern Michigan	Overall profile morphology; podzolization	Barrett & Schaetzl, 1992
Blue Ridge Mtns, GA	clay illuviation & argillic horizon thickness; overall profile morphology & thickness; rubification; alteration or translocation of Fe	Leigh, 1996; Leigh & Cable, 1997
Southeast U.S.	argillic horizon formation; alteration and translocation of Fe	Markewich & Pavich, 1991
Cordillera Blanca, Peru	overall profile morphology & thickness; rubification	Rodbell[2], 1993
Topo-chronosequences		
Salmon River Mtns, ID	clay illuviation & argillic horizon thickness; overall profile morphology & thickness; rubification; alteration or translocation of Fe	Berry[2], 1987; Birkeland et al., 1991

Table 4. Examples of chronosequence and topo-chronosequence studies—*Cont'd*

Location	Soil variables[1]	Reference
Wind River Mtns, WY	clay illuviation & argillic horizon thickness; overall profile morphology & thickness; rubification; alteration or translocation of Fe	Swanson[2], 1985; Birkeland et al., 1991; Dahms, 1994
Whiskey Basin, WY	clay illuviation & argillic horizon thickness; overall profile morphology & thickness; rubification	Applegarth & Dahms[2], 2001
Eastern Sierra Nevada, CA	clay illuviation & argillic horizon thickness; overall profile morphology & thickness; rubification; alteration or translocation of Fe	Birkeland & Burke[2], 1988; Berry[2], 1994
Southern Israel	gypsum, other salts & argillic horizon formation; overall profile morphology	Birkeland & Gerson, 1991
South Island, New Zealand	overall profile morphology & thickness; rubification; leaching; alteration or translocation of Fe	Birkeland, 1994
Peruvian Andes	clay illuviation & argillic horizon thickness; overall profile morphology & thickness; rubification; alteration or translocation of Fe	Miller[2] & Birkeland, 1992

[1] Rate of argillic horizon formation can include data on rates of clay illuviation; Fe alteration can refer to Fe oxidation or alteration of Fe oxides.
[2] Student of P.W. Birkeland.

evolution ... Well dated chronosequences are therefore a boon to pedologists. They are also invaluable to geomorphologists, for, once a soil chronosequence is established, it may be used to investigate other landscape processes" (Huggett, 1998, p. 159). Willem Vreeken (1975b), a student of Ruhe's, also defined subcategories of chronosequences (noted below). Birkeland furthered his influence on the practice of soil geomorphology through his own research. For example, the Utah Geological and Mineral Survey brought him and two former students, to train their geologic mappers and produce a manual on applying the five-factor approach in Quaternary geology and the mapping of Quaternary units (Birkeland, Machette, and Haller, 1991).

The importance of chronosequences was recognized by geomorphologists in the USGS who sponsored a series of soil geomorphology investigations in the eastern and western U.S. (Table 5). Developed by Dennis E. Marchand and continued by Jennifer W. Harden, Michael N. Machette, Helene Markewich, and Milan Pavich, the work was aimed at understanding age relationships between soils, landscapes, and various weathering criteria (i.e., the work focused on soil chronosequences). The field work was supported by an array of physical, chemical, and mineralogical data. The studies in the east were cooperative projects of the USGS and the USDA, thus representing a melding of sorts of several historic predecessors.

Evolving out of both the Jenny/Birkeland factorial approach to soil geomorphology and quantitative geomorphology were attempts at modeling soil forming processes. Much of this work has focused on modeling carbonate accumulation (e.g., McFadden and Tinsley, 1985; Mayer et al., 1988; McFadden et al., 1991, 1998), following in the footsteps of Arkley (1963), but also includes more general attempts at modeling pedogenesis (Chadwick et al., 1990). A very different approach to quantifying soil geomorphic studies was taken by Jennifer Harden, a USGS geologist who studied with Arkley and Jenny at Berkeley. She developed a "semi-quantitative" assessment of soil profile morphology as a means of objectively comparing soil profile development (Harden, 1982), following the work of Bilzi and Ciolkosz (1977b). The "profile development index" (or more informally the

Table 5. USGS/USDA chronosequence and landscape studies, 1970s–1980s

Project Duration	Setting	References
USGS Western U.S. 1978–1983	Merced River, CA	Harden, 1987
	Ventura River & Ventura Coast, CA	Harden et al., 1986
	Kane fans, Cottonwood Creek, WY	Reheis, 1987a
	Rock Creek, MT	Reheis, 1987b
	Alpine soils, Front Range, CO	Birkeland et al., 1987
	Cowlitz River, WA	Dethier, 1988
USGS/USDA Eastern U.S. 1979–1984	Coastal Plain, MD, VA	Markewich et al., 1987
	Coastal Plain, SC	Markewich et al., 1986
	Uphapee Creek & Tallapoosa River, AL	Markewich et al., 1988
	Guide to soil & weathering profile data, Coastal Plain	Markewich et al., 1989

"Harden index") has proven useful in soil development studies in a wide variety of settings (Harden and Taylor, 1983; Busacca, 1987; Reheis et al., 1989; Birkeland, Berry, and Swanson, 1991; Vidic and Lobnik, 1997) but it is not necessarily universally applicable (Birkeland, 1999, 293). In a related development, Schaetzl and Mokma (1988) developed the "POD index" for semi-quantitative evaluation of podzolization.

The five-factor approach advocated by Birkeland has been criticized by some workers (see summaries in Birkeland, 1999, pp. 144–145; Johnson and Watson-Stegner, 1987; and Gerrard, 1992, 3–7). Ruhe (1975b, p. 177) viewed Birkeland's factorial approach as simply "compartmentalization . . . useful for instruction" but otherwise "disappointing." Daniels and Hammer (1992, pp. 195–202) explicitly and strongly criticize the use of soils as age indicators. They correctly note the many variables that can complicate the use of soils to date landscapes, but nowhere do they explicitly deal with the many successful chronosequence studies that illustrate the rate-dependent nature of a wide array of pedologic features (e.g., Table 4 and references in Huggett, 1998, and in chapters 8 and 9 of Birkeland, 1999).

Paton et al. (1995) attempted to overturn what they see as the "clorpt paradigm." In their "new global view" of soils, they essentially reject the idea that the distribution of soil zones is determined by the five factors, particularly climate and vegetation, and the corollary concept that the factors drive soil-forming processes vertically down through the soil resulting in A-B-C profiles. Paton and his coauthors propose to replace the clorpt concept with one emphasizing surficial and biological processes (such as bioturbation) acting upon a weathered and mobile mantle of sediment. Their model appears to be most effective in explaining the spatial pattern of soils on tropical landscapes (Schaetzl, 2000, pp. 772–773). The "new global view" has come under heavy criticism, however (e.g., Catt, 1996). In particular, Beatty (2000), Courchenese (2000), Johnson (2000), and Schaetzl (2000) forcefully argue that the traditional "clorpt, A-B-C" model does successfully account for the genesis and distribution of soils on younger landscapes in the middle latitudes. Johnson (2000, p. 780) is probably correct in his view that the old clorpt, A-B-C model "formulated for soils in the plainlands of European Russia and North America, conceptually promotes an atypical (but not aberrant) style of pedogenesis relative to most of the nonglaciated and loess-free rest of the world, an atypical style that has been applied, unfortunately, as the world standard." Most soil-geomorphologic research, however, has been carried out in the Quaternary glacial, alluvial, eolian, and desert regions of the middle latitudes, where the traditional model works, and generally works well.

Certainly the five factors do not deal with soils and pedogenic process, but with external factors that affect the soil. In Quaternary research, however, the external factors often are the object of concern, and soils can be a means of reconstructing them; the soils are environmental proxies. The five-factor approach cannot describe landscapes, however, and soils are selected on a landscape for their ability to solve the equation. The five-factor approach also tends to treat the factors individually as independent variables, although they often act together, such as climate and biota. This point was raised by Jenny in his original study as well as in his own revision in 1980. Indeed, in the final version of his book, Birkeland (1999) combined discussion of climate and vegetation because they are in many ways inseparable. The time factor is the only truly independent variable, but the passage of time in and of itself does not form a soil; it simply allows significant time for the other factors and processes to operate.

For the most part the general validity of the state factor approach has been upheld (e.g., Yaalon, 1975; Bockheim, 1980; Huggett, 1998), especially in soil-geomorphic research in the middle latitudes, where, historically most pedologic and Quaternary geologic work has been conducted. The application of the factorial equation has been especially useful in chronosequence studies on moraines and alluvial terraces, and in assessing the influence of parent materials on soils (i.e., in lithosequences) (e.g., Tables 4, 5). As an example of the significance and impact of chronosequence research, the topic has generated its own body of theoretical literature (Schaetzl et al., 1994; Rabenhorst, 1997; Huggett, 1998).

Ruhe did not leave a legacy in the form of a book focusing on soil geomorphology, and he produced a relatively small number of graduate students because much of his career was with USDA, but his influence is evident. His approach of combining pedology, geomorphology, and hydrology, as well as soil stratigraphy, evolving from his Iowa project, had a lasting influence on geologists and pedologists in the Midwestern U.S. (Olson, 1989, 1987). Ruhe's student Raymond Daniels also went on to co-author a volume on *Soil Geomorphology* (Daniels and Hammer, 1992). It presents an approach very much in the Ruhe tradition that applies process pedology and geomorphology with hydrology to interpret past soil-forming environments. As indicated above, their approach contrasts strongly with Birkeland's. Ironically, nowhere do they use the term "catena" (a criticism of Birkeland's first edition in Ruhe's 1975 review), though they describe a variety of catenary settings and soil associations. Other students worked with Ruhe or continued his work on topics such as landscape evolution in Iowa (Vreeken, 1975a), "welded soils" (Ruhe and Olson, 1980; Olson and Nettleton, 1998) and the Sangamon Soil (Ruhe et al., 1974), and mapping soils to better understand landscape evolution (Brevik and Fenton, 1999). Several of his students melded Ruhe's approach to soil geomorphology, using stratigraphy and hydrology, with the clorpt factorial approach, including chronosequences (Vreeken, 1975b; Hall, 1999; Hall and Anderson, 2000).

Although Kirk Bryan left no academic legacy insofar as his students of geomorphology interested in soils are concerned, his name is now attached to several soils-related studies. In 1958, as a result of an "outpouring of affection and funds . . . upon his untimely death" (Sharp, 1993, p. 190) the Quaternary Geology and Geomorphology Division of the Geological Society of America established the annual Kirk Bryan Award for an outstanding publication. As of 2004, five of the awards went to research directly or indirectly relating to soil geomorphology. The first was Richmond's 1962 study of the La Sal Mountains (1965 award), which relied on soils. Two of the awards were for research that resulted directly from the USDA soil geomorphology projects: Ruhe's 1969 book *Quaternary Landscapes in Iowa* (1974 award), and Gile, Hawley, and Grossman's 1981 *Guidebook to the Desert Project* (1983 award). The other two awards, for Birkeland (1984) (1988 award) and Holliday (1995) (1998 award), were heavily influenced by the other awardees.

Other trends in research related to soil geomorphology noted above continued through the last quarter of the 20th century. The Midwestern tradition of research on genesis of soil-stratigraphic units to understand the evolution of glacial and eolian landscapes continued (e.g., Norton et al., 1988; Jacobs and Knox, 1994; Leigh and Knox, 1994; Jacobs et al., 1997; Jacobs, 1998) and was also applied to alluvial systems (e.g., Bettis, 1992; Mandel, 1994). A key player has been Leon Follmer of the Illinois Geological Survey, following in the footsteps of Leighton and Frye (Follmer, 1978, 1979, 1982, 1983; Follmer et al., 1998). Studies in the tradition of the USDA soil geomorphology projects, using soils to reconstruct landscape evolution, have also continued (e.g., Busacca, 1989; Holliday, 1990; McDonald and Busacca, 1990, 1992; Snyder and Bryant, 1992; McFadden and McAuliffe, 1997; Schaetzl et al., 2000; Birkeland et al., 2003; Gaylord et al., 2003), though typically at significantly reduced scales of time and funding. One particular line of research inspired by the Desert Project is in the importance of dust infiltration as a factor in pedogenesis (e.g., McFadden et al., 1986, 1987, 1998; Litaor, 1987; Dahms, 1993; Reheis et al., 1995).

Some pedologists have maintained an interest in soil-geomorphic relations, even though research in pedology through the last quarter of the 20th century tended to focus on description and classification. In particular, pedologists from UC Berkeley have followed the path of Jenny and Arkley regarding processes of soil genesis, landscape relationships, and the state factors (e.g., Amundson et al., 1989; Harden, 1990; Amundson and Jenny, 1991). Soil-geomorphic research by other pedologists in the U.S. include work on rates of development and other factor-related studies (e.g., Bockheim, 1979, 1982, 1990; Foss and Segovia, 1984; Eash and Sandor, 1995; Langley-Turnbaugh and Bockheim, 1997), landscape evolution (e.g., Ciolcosz et al., 1990), including work in Uganda at Milne's "type area" for the catena (Brown et al., 2004), and studies of buried soils (e.g., Busacca, 1989; Carter et al., 1990; Busacca and Cremaschi, 1998; Ward and Carter, 1998).

As soil geomorphology became established as a subdiscipline in its own right, it became applied in allied disciplines. A particularly important area of research is in using soil-stratigraphy, chronosequences, and other age relationships to determine the timing and recurrence rates of faults (e.g., Machette, 1978, 1988; Douglas, 1980; Harden and Matti, 1989; Berry, 1990; McCalpin and Berry, 1996; Keller et al., 1998). In archaeological research, soils were recognized as important stratigraphic and environmental indicators by such pioneers in soil geomorphology as Leighton (1937) and Bryan (1941). But not until the 1970s, probably due to Ruhe's work in the Midwest and the first edition of Birkeland's book, did soils research become an important component of geoarchaeological research, including work by pedologists (e.g., Foss, 1977; Foss et al., 1995; Cremeens, 1995; Cremeens and Hart, 1995; Morris, 2002) as well as geoscientists (e.g., Holliday, 1989, 1994, 2004; Reider, 1990; Bettis, 1992; Mandel, 1994, 1995; Bettis and Hajic, 1995; Mandel and Bettis, 2001).

The most significant conceptual advances in soil geomorphology since 1974 resulted from Francis Hole's work, especially on the dynamic nature of soils (e.g., Hole, 1981), and are seen most prominently in the work of geographers influenced by Hole's teaching and writing. For example, Curtis J. Sorenson (University of Kansas), who took courses from Hole at the University of Wisconsin, and his students (including Randall Schaetzl, who also took courses from Hole) produced an impressive body of literature dealing with soils as clues to climate change and landscape evolution (e.g., Sorenson, 1977; Barrett and Schaetzl, 1992; Dahms, 1994, 2002; Mandel, 1994; Schaetzl et al., 2000) and applying the factorial approach to understanding the distribution of soils (Schaetzl and Isard, 1996). The intellectual thread of Hole's views on soil dynamism can be seen in the work of Donald L. Johnson and his students at the University of Illinois. Johnson and Donna Watson-Stegner (1987) formulated the "Soil Evolution Model," which incorporates aspects of several other models of pedogenesis, including Simonson's and Jenny's approaches, the thermodynamics and energy-transfer concepts of Nikiforoff (1959) and Runge (1973), the notion of soil evolution espoused by Nikiforoff (1949), and the concept of landscape evolution (though not expressly the idea of K-cycles) (see also Johnson et al., 1987, 1990). The authors recognize that soil formation is not a linear, unidirectional, process. Soils do not simply exhibit more horizons and get thicker over time, for example. They are affected by processes (external and internal) that promote or inhibit horizonation (or both), or promote or inhibit profile thickening (or both). At any given site or in any given region a subset of these "progressive" and "regressive" processes will operate and the resulting soil usually reflects the dominance of one group of processes over the other. The dominance of one group could shift to the other group over time as either internal or external processes or factors change. Such changes could include shifts in climate, vegetation, animal communities, human activity or geologic events and processes.

The soil evolution model in its holistic view of soil formation is the most sophisticated of the several models that have been proposed (and as a result, Johnson was awarded the 1990 G.K. Gilbert Award of the Association of American Geographers). It also provides considerable explanatory power in discussing soils once they have been studied. The model does not provide the *a priori* means of investigating soil development that the Jenny factorial approach holds, but it is a useful conceptual approach that links the factors to the soil itself.

Perhaps a fitting culmination of the research and influence of Hole, Johnson, and Sorenson, and the work of many others linking pedology and geomorphology is publication of the volume *Soils: Genesis and Geomorphology* by R. Schaetzl (a student of Hole, Johnson, and Sorenson) and S. Anderson (2005). The authors produced the most comprehensive English-language volume integrating traditional pedology with soil geography and soil geomorphology. The book treats soils and landscapes as dynamic components of one another, evolving together through space and time. Publication of this volume is a fitting and encouraging end-point to this history.

Summary and Conclusions

Soil geomorphology in the U.S. has disparate roots in pedology, Quaternary geology, and physical geography going back to the late 19th century. Several historical threads leading to modern soil geomorphology can be traced through the early 20th century, but what most of us would recognize as soil geomorphology did not appear until after the Second World War. The beginnings, however, were in Russian pedology, with its emphasis on the landscape and the factors of soil formation, and in American Quaternary stratigraphy, with a focus on past environments and the Pleistocene time scale. These emphases remain important components of soil geomorphology today. One aspect of the field focuses on the relationships between soils and landscapes, emphasizing pedogenic processes, geomorphic processes, and hydrology to understand the distribution of soils in the present and in the past. This approach evolved more directly from pedology and process geomorphology as well as geography, largely under the influence of R.V. Ruhe and F. D. Hole. Institutionally, this approach derived from traditional soil mapping and from the USDA-sponsored soil geomorphology projects of the 1950s to 1970s. Soil stratigraphy, particularly Midwestern Quaternary stratigraphy, also played a role in this work.

A significant amount of soil-geomorphic research and writing also focuses on soils as a means of studying and reconstructing the past, with a particular focus on soils as age indicators and soils as clues to past environments (especially vegetation and climate). These aspects of soil geomorphology evolved more directly from study of the relationship between soils and the soil-forming factors that began with the

Russians, was most famously espoused by H. Jenny, and then refocused on the past by P. W. Birkeland. Institutionally, this research approach also evolved from some of the USDA soil geomorphology projects, but also from the U.S. Geological Survey, beginning with the Military Geology Unit in Second World War and continuing with a variety of Quaternary geologic studies, largely in the western U.S., first by C. B. Hunt, R. Morrison, and G. R. Richmond, and later by workers encouraged by D. E. Marchand and J. Harden.

A result of the evolution of soil geomorphology in the U.S. is a geographic distinction between what can be over simplified as soil geomorphology in the Midwest and soil geomorphology in the mountain and desert West. The Midwestern approach, following Ruhe and his many predecessors, emphasizes hillslopes, hydrology, and stratigraphy. Much of the Midwestern landscape today is undergoing erosion, which is superimposed on till and loess sheets with multiple buried soils (forming the "time-transgressive chronosequences without historical overlap" of Vreeken, 1975b) traceable over vast (multi-state) regions. Given the humid climate, overland and subsurface flow of water are also ubiquitous processes. The emphasis on soil geography as a component of soil geomorphology in the Midwest also is related to the close cooperation among pedologists and Quaternary geologists and geomorphologists probably due to the long tradition of Quaternary geology in the region and because of the agricultural significance of the regional soils. The tradition of field conferences that bring Quaternary geoscientists and pedologists together, such as the Friends of the Pleistocene, is also an old one in the Midwest, and likely contributed to the cross-fertilization between geologists, geographers, and pedologists.

In the West, soil geomorphology includes soil stratigraphy, but buried soils are restricted to specific settings such as paleo-lake basins or alluvial valleys. More commonly, soil geomorphic work has focused on chronosequences and topochronosequences on moraines, alluvial terraces, and paleo-lake shorelines (Table 4). These Quaternary landform assemblages are ubiquitous throughout the western U.S., and their associated chronosequences (the "post-incisive" variety of Vreeken, 1975b) provide a unique opportunity to study rates of soil development in a variety of settings and to use soils to correlate landforms. Soil geography has rarely been emphasized, probably because there were fewer instances of pedologists working with Quaternary geologists in the west. This situation, in turn, is likely due to the much more limited nature of farmable land in the west.

Since the very beginnings of research into the geologic aspects of soil genesis and the development of the field of pedology, pedologists have increasingly distanced themselves from geologists and geology. To a large extent this was an institutional accident resulting from the evolution of schools of agriculture separate from schools of natural science, the housing of soil science in those agricultural colleges, and most government-funded soil science research coming from the USDA. The emergence of soil geomorphology after Second World War represents, in many ways, the most substantive re-integration of pedology and geology. This was due in large measure to the USDA soil geomorphology projects. Since they ended, however, there have been few comparable investigations based in soil science beyond the exceptions noted above. Otherwise, most substantive soil-geomorphic research has been based in the USGS (e.g., Table 5) and in departments of geology and geography. The broader implications of this history in soil science are unclear, but certainly the decline in hirings of pedologists in soil science departments and in pedologic research into the relationship of soils to contemporary and past landscapes and environments is an unsettling trend. More positive developments can be seen in the soil-oriented research of geologists and geographers, represented by the books from Birkeland (1999) and Schaetzl and Anderson (2005). Ultimately, however, maintaining soil-geomorphic research in soil science as well as geology and geography is the healthiest direction.

Notes

2. Because soil geomorphology is a young field with roots in several disciplines, few will agree on the relative importance of specific individuals, events, and agencies in its evolution. Our individual views on these matters are strongly colored by our training and our research and (in some cases) academic experience. In my own case, I was trained in pedology at Texas Tech University under B.L. Allen, and received a PhD in Geosciences at the University of Colorado under Peter Birkeland. I then spent a total of 17 years teaching Geography and Geology at the University of Wisconsin-Madison. This experience exposed me to two very different but widely applied approaches to soil geomorphology and soil stratigraphy (the western Morrison/Richmond/Birkeland tradition and the midwestern Ruhe/Hole tradition). This paper resulted from my curiosity about the origins of these approaches. My years in the Midwest also alerted to me to the broader issues of soil geomorphology coming out of physical geography and pedology.
3. The discussion and comments on the catena concept throughout this paper is based on conversations with, and unpublished papers from, David Brown (2004), now at Montana State University.

Acknowledgments

My thanks to Benno Warkentin and Dan Yaalon for organizing and shepard-ing this volume. My interest in this subject began when I was on the Geography faculty at the University of Wisconsin-Madison following my graduate training and research in soils. As I became exposed to Midwestern soil-stratigraphy and soil-geomorphology I quickly recognized the very different approaches to the topics there vs what I was exposed to in my training "out west." So I casually and unsystematically began talking with and corresponding with others interested in the history of science and those who made the history. Among the former was John Tandarich, who has become the leading historian of pedology in the U.S. In the latter category, Roger Morrison and the late Gerry Richmond kindly answered my many questions and shared their professional development with me. Richmond alerted me to the very important role of the MGU. Francis Hole also enthusiastically talked about his many years of soil mapping in Wisconsin and the evolution of his thinking about soils. John Hawley and B.L. Allen shared their recollections of the Desert Project and the other USDA soil-geomorphology projects. David Brown was the source of my information on the history of the catena concept, when he was a graduate student in Soil Science at Wisconsin. David also helped me crystalize my thinking on many aspects of the history of pedology and soil-geomorphology via long, enjoyable conversations. As the manuscript came together a number of colleagues provided comments on various drafts and I thank them. Pete Birkeland read three drafts (!), Erik Brevik, Jeff Homburg, Bill Farrand, Tom Fenton, and Alan Busacca all read through it once. Erik also alerted me to the work of E.E. Free and G.N. Coffey.

Finally, an anonymous reviewer prodded me to look further into the prominent role of U.C. Berkeley in this story. This paper is dedicated to Roger Morrison, Gerry Richmond, Lee Gile, John Hawley, and all of the men and women of the MGU, USGS, and USDA who worked hard to maintain the link between pedology and geomorphology.

References Cited

ap Griffith, G. 1952. Catena. Soils and fertilizers 15:169–170.

Albritton, C. C., Jr., and K. Bryan. 1939. Quaternary stratigraphy in the Davis Mountains, trans-Pecos, Texas. Geol. Soc. Am. Bull. 50:1423–1474.

Alexandrovskiy, A. L. 2000. Holocene development of soils in response to environmental changes: The Novosvobodnaya archaeological site, North Caucasus. Catena 41:237–248.

Amundson, R. 1994. Appendix 3: Brief highlights of Han Jenny's life. pp. 153–160. *In* R. Amundson (ed.) Factors of soil formation: A fiftieth anniversary retrospective. SSSA Spec. Publ. 33. SSSA, Madison, WI.

Amundson, R., and H. Jenny. 1991. The place of humans in the state factor theory of ecosystems and their soils. Soil Sci. 151:99–109.

Amundson, R., and D. H. Yaalon. 1995. E. W. Hilgard and John Wesley Powell: Efforts for a joint Agricultural and Geological Survey. SSSA Jour. 59:4–13.

Amundson, R. G., O. A. Chadwick, J. M. Sowers, and H. E. Doner. 1989. Soil evolution along an altitudinal transect in the eastern Mojave Desert of Nevada, U.S.A. Geoderma 43:349–371.

Applegarth, M. and D. E. Dahms. 2001. Soil catenas of calcareous tills, Whiskey Basin, Wyoming, USA. Catena 42:17–38.

Arkley, R. J. 1963. Calculation of carbonate and water movement in soil from climatic data. Soil Sci. 96:239–248.

Arkley, R. J. 1964. Soil survey of the eastern Stanislaus area, California. USDA Soil Series 1957, no. 20, U.S. Gov. Print. Office, Washington, D.C.

Arkley, R. J. 1967. Climates of some Great Soil Groups of the western United States. Soil Sci. 103:389–400.

Arnold, R. W. 1994. Soil geography and factor functionality: Interacting concepts. pp. 99–109. In R. Amundson (ed.) Factors of soil formation: A fiftieth anniversary retrospective. SSSA Spec. Publ. 33. SSSA, Madison, WI.

Baldwin, M., C. E. Kellogg, and J. Thorp. 1938. Soil classification. In Soils and men, Yearbook of Agriculture 1938, pp. 979–1001. USDA, U.S. Gov. Print. Office, Washington, D.C.

Balster, C. A. and R. B. Parsons. 1968. Geomorphology and soils, Willamette Valley, Oregon. Oregon Agric. Exp. Stn., Oregon State U., Spec. Rpt 265. Corvallis, OR.

Barrett, L. R., and R. J. Schaetzl. 1992. An examination of podzolization near Lake Michigan using chronofunctions. Canadian J. Soil Sci. 72:527–541.

Beatty, S. W. 2000. On the rocks; shaken not stirred. Book review forum on "Soils: A New Global View." Annals Assoc. Am. Geog. 90:785–787.

Berry, M. E. 1987. Morphological and chemical characteristics of soil catenas on Pinedale and Bull Lake moraine slopes in the Salmon River Mountains, Idaho. Quaternary Res. 28:210–225.

Berry, M. E. 1990. Soil catena development on fault scarps of different ages, eastern escarpment of the Sierra Nevada, California. In P. L. K. Knuepfer and L. D. McFadden (eds.) Soils and landscape evolution. pp. 333–350. Geomorphology (special issue) 3.

Berry, M. E. 1994. Soil-geomorphic analysis of late-Pleistocene glacial sequences in the McGee, Pine, and Bishop Creek drainages, east-central Sierra Nevada, California. Quaternary Res. 41:160–175.

Bettis, E. A., III. 1988. Pedogenesis in late Prehistoric indian mounds, upper Mississippi valley. Phys. Geog. 9:263–279.

Bettis, E. A., III. 1992. Soil morphologic properties and weathering zone characteristics as age indicators in Holocene alluvium in the Upper Midwest. pp. 119–144. *In* V. T. Holliday (ed.) Soils and landscape evolution. Smithsonian Inst. Press, Washington, D.C.

Bettis, E. A., III, and E. R. Hajic. 1995. Landscape development and the location of evidence of Archaic cultures in the Upper Midwest. pp. 87–113. *In* E. A. Bettis, III, Archaeological geology of the Archaic period in North America. Geol. Soc. of Am. Spec. Paper 297, Boulder, CO.

Bidwell, O. W., and F. D. Hole. 1965. Man as a factor of soil formation. Soil Sci. 99:65–72.

Bilzi, A. F., and E. J. Ciolkosz. 1977a. Time as a factor in the genesis of four soils developed in recent alluvium in Pennsylvania. SSSA Jour. 41:122–127.

Bilzi, A. F., and E. J. Ciolkosz. 1977b. A field morphology rating scale for calculating pedogenic development. Soil Sci. 124:45–48.

Birkeland, P. W. 1974. Pedology, weathering, and geomorphological research. Oxford University Press, New York, NY.

Birkeland, P. W. 1984. Soils and geomorphology (2nd ed.). Oxford University Press, New York, NY.

Birkeland, P. W. 1989. Response (to presentation of the Kirk Bryan Award). Geol. Soc. Am. Bull. 101:1007–1008.

Birkeland, P. W. 1994. Variation in soil-catena characteristics of moraines with time and climate, South Island, New Zealand. Quaternary Res. 42:49–59.

Birkeland, P. W. 1999. Soils and geomorphology (3rd ed.). Oxford University Press, New York, NY.

Birkeland, P. W. 2001. Response (to presentation of the Quaternary Geology and Geomorphology Distinguished Career Award). Quaternary Geol. Geom. (Newsletter of the Quaternary Geology and Geomorphology Division of the Geol. Soc. Am.), 42(1):3–6.

Birkeland, P. W., and R. M. Burke. 1988. Soil catena chronosequences on eastern Sierra Nevada moraines, California, U.S.A. Arctic and Alpine Res. 20:473–484.

Birkeland, P. W., and R. Gerson. 1991. Soil-catena development with time in a hot desert, southern Israel – Field data and salt distribution. J. Arid Env. 21:267–281.

Birkeland, P. W., M. E. Berry, and D. K. Swanson. 1991. Use of soil catena field data for estimating relative ages of moraines. Geology 19:281–283.

Birkeland, P. W., R. M. Burke, and R. R. Shroba. 1987. Holocene alpine soils in gneissic cirque deposits, Colorado Front Range. USGS Bull. 1590-E. Washington, D.C.

Birkeland, P. W., M. N. Machette, and K. M. Haller. 1991. Soils as a tool for applied Quaternary geology. Utah Geol. & Min. Survey, Misc. Pub. 91–3, Salt Lake City.

Birkeland, P. W., R. R. Shroba, S. F. Burns, A. B. Price, and P. J. Tonkin. 2003. Integrating soils and geomorphology in mountains – an example from the Front Range of Colorado. Geomorphology 55:329–344.

Bischoff, J. L., R. J. Shlemon, T. L. Ku, R. D. Simpson, R. J. Rosenbauer, and F. E. Budinger. 1981. Uranium-series and soil-geomorphic dating of the Calico archaeological site, California. Geology 9:576–582.

Bockheim, J. G. 1979. Properties and relative age of soils of southwestern Cumberland Peninsula, Baffin Island, N.W.T., Canada. Arctic & Alp. Res. 11:289–306.

Bockheim, J. G. 1980. Solution and use of chronofunctions in studying soil development. Geoderma 24:71–85.

Bockheim, J. G. 1982. Properties of a chronosequence of ultraxerous soils in the Trans-antarctic Mountains. Geoderma 28:239–255.

Bockheim, J. G. 1990. Soil development rates in the Transantarctic Mountains. Geoderma 47: 59–77.

Bockheim, J. G. 2005. Historical development of key concepts in pedology. Geoderma 124:23–36.

Brewer, R. 1972. Use of macro- and micromorphological data in soil stratigraphy to elucidate surficial geology and soil genesis. J. Geol. Soc. Australia 19:331–344.

Brevik, E. C., and T. E. Fenton. 1999. Improved mapping of the Lake Agassiz Herman strandline by integrating geological and soil maps. J. Paleolimnology 22:253–257.

Bronger, A., and J. A. Catt. 1998. The position of paleopedology in geosciences and agricultural sciences. Quaternary International 51/52:87–93.

Brown, D. J., K. McSweeney, and P. A. Helmke. 2004. Statistical, geochemical, and morphological analyses of stone line formation in Uganda. Geomorphology 62:217–237.

Bryan, K. 1941. Correlation of the deposits of Sandia Cave, New Mexico, with the glacial chronology (appendix). pp. 45–64. In F. C. Hibben, Evidences of early occupation in Sandia Cave, New Mexico, and other sites in the Sandia-Manzano region. Smithsonian Misc. Coll. 99.

Bryan, K. 1948. Los suelos complejos y fosiles de la altiplanicie de Mexico en relacion a los cambios climaticos. Sociedad Geológia, Mexicana, Boletin 13:1–20.

Bryan, K. 1949. The geologic implications of cryopedology. J. Geol. 57:101–104.

Bryan, K., and C. C. Albritton, Jr., 1943. Soil phenomena as evidence of climatic changes. Am. J. Sci. 241:469–490.

Bull, W. B. 1991. Geomorphic responses to climatic change. Oxford University Press, New York, NY.

Buol, S. W., F. D. Hole, and R. J. McCracken. 1973. Soil genesis and classification. Iowa State U. Press, Ames.

Buol, S. W., F. D. Hole, and R. J. McCracken. 1980. Soil genesis and classification, 2rd Ed. Iowa State U. Press, Ames.

Buol, S. W., F. D. Hole, and R. J. McCracken. 1989. Soil genesis and classification, 3rd Ed. Iowa State U. Press, Ames.

Buol, S. W., F. D. Hole, R. J. McCracken, and R. J. Southard. 1997. Soil genesis and classification, 4th Ed. Iowa State U. Press, Ames.

Busacca, A. J. 1987. Pedogenesis of a chronosequence in the Sacramento Valley, California, U.S.A., I. Application of a soil development index. Geoderma 41:123–148.

Busacca, A. J. 1989. Long Quaternary record in eastern Washington USA interpreted from multiple buried paleosols. Geoderma 45:105–122.

Busacca, A. J., and M. Cremaschi. 1998. The role of time versus climate in the formation of deep soils of the Apennine fringe of the Po Valley, Italy. Quaternary International 51/52:95–107.

Bushnell, T. M. 1927. To what extent should location, topography or physiography constitute a basis for differentiating soil into units or groups. Paper read at First Intl. Congress of Soil Sci. June 13–22, Washington, D.C.

Bushnell, T. M. 1942. Some aspects of the soil catena concept. Proc. SSSA 7:158–163.

Butler, B. E. 1959. Periodic phenomena in landscapes as a basis for soil studies. CSIRO, Soil Publ. 14, Canberra, Australia.

Butler, B. E. 1982. A new system for soil studies. *Journal of Soil Science.* v.33, pp. 581–595.

Carter, B. J., P. A. Ward, III, and J. T. Shannon. 1990. Soil and geomorphic evolution within the Rolling Red Plains using Pleistocene volcanic ash deposits. pp. 471–488. *In* P. L. K. Knuepfer and L. D. McFadden (eds.) Soils and landscape evolution, Geomorphology (special issue) 3.

Catt, J. A. 1986. Soils and Quaternary geology: A handbook for field scientists. Oxford University Press, London.

Catt, J. A. 1996. Review of "Soils: A new global view." Quaternary Sci. Rev. 15:92–93.

Chadwick, O. A., G. H. Brimhall, and D. M. Hendricks. 1990. From a black to a gray box – A mass balance interpretation of pedogenesis. pp. 369–390. P. L. K. Knuepfer and L. D. McFadden (eds.) Soils and landscape evolution, Geomorphology (special issue) 3.

Chamberlin, T. C. 1878. On the extent and significance of the Wisconsin Kettle Moraine. Wisconsin Academy of Science, Arts, and Letters Trans. 4:201–234.

Chamberlin, T. C. 1894. Glacial phenomena of North America. pp. 724–775. *In* J. Geike (ed.) The great Ice Age and its relation to the antiquity of Man, 3rd Ed. Stanford, London.

Chamberlin, T. C. 1895. The classification of American glacial deposits. J Geol. 3:270–277.

Ciolcosz, E. J., B. J. Carter, M. T. Hoover, R. C. Cronce, W. J. Waltman, and R. R. Dobos. 1990. Genesis of soils and landscapes in the Ridge and Valley province of central Pennsylvania. pp. 245–262. *In* P. L. K. Knuepfer and L. D. McFadden (eds.) Soils and landscape evolution. Geomorphology (special issue) 3.

Cline, M. G. 1961. The changing model of soil. SSSA Proc. 25:442–446.

Coffey, G. N. 1909a. Physical principles of soil classification. Proc. Am. Soc. Agron. 1:175–185.

Coffey, G. N. 1909b. Value of the field study of soils. Proc. Am. Soc. Agron. 1:168–175.

Coffey, G. N. 1912. A study of the soils of the United States USDA Bureau of Soils Bull. 85. U.S. Gov. Print. Office, Washington, D.C.

Courchesne, F. 2000. Breaking the barrier of conceptual locks. Book review forum on "Soils: A New Global View." Annals Assoc. Am. Geographers 90:782–785.

Cremeens, D. L. 1995. Pedogenesis of Cotiga Mound, a 2100-year-old Woodland mound in southwest Virginia. SSSA Jour. 59:1377–1388.

Cremeens, D. L., and J. P. Hart. 1995. On chronostratigraphy, pedostratigraphy, and archaeological context. *In* M. E. Collins, B. J. Carter, B. G. Gladfelter, and R. J. Southard (eds.) Pedological perspectives in archaeological research. pp. 15–33. SSSA Spec. Pub 44. Madison, WI.

Dahms, D. E. 1993. Mineralogical evidence for eolian contribution to soils of late Quaternary moraines, Wind River Mountains, Wyoming. Geoderma 59:175–196.

Dahms, D. E. 1994. Mid-Holocene erosion of soil catenas on moraines near the type Pinedale till, Wind River Range, Wyoming. Quaternary Res. 42:41–48.

Dahms, D. E. 2002. Glacial stratigraphy of Stough Creek Basin, Wind River Range, Wyoming. Geomorphology 42:59–83.

Dan, J., D. H. Yaalon, and H. Koyumdjinsky. 1968. Catenary soil relationships in Israel. A. The Netanya catena on coastal dunes of the Sharon. Geoderma 2:95–120.

Daniels, R. B. 1988. Pedology, a field or laboratory science? SSSA Jour. 52:1518–1519.

Daniels, R. B., and R. D. Hammer. 1992. Soil geomorphology. John Wiley and Sons, New York.

Daniels, R. B., and E. E. Gamble. 1978. Relations between stratigraphy, geomorphology, and soils in Coastal Plain areas of southeastern U.S.A. Geoderma 21:41–65.

Daniels, R. B., E. E. Gamble, and S. W. Buol. 1973. Oxygen content in the groundwater of some North Carolina Aquults and Udults. pp. 153–166. *In* Field water regime. SSSA Spec Pub. 5. SSSA, Madison, WI.

Daniels, R. B., E. E. Gamble, and J. G. Cady. 1970. Some relationships among Coastal Plain soils and geomorphic surfaces in North Carolina. SSSA Proc. 34:648–653.

Daniels, R. B., E. E. Gamble, and L. A. Nelson. 1967. Relations between A2 horizon characteristics and drainage in some fine loamy Ultisols. Soil Sci. 104:364–369.

Daniels, R. B., E. E. Gamble, and W. H. Wheeler. 1978. Age of soil landscapes in the Coastal Plain of North Carolina. SSSA Jour. 42:98–105.

Dethier, D. P., 1988. The soil chronosequence along the Cowlitz River, Washington. USGS Bull. 1590-F. Washington, D.C.

Dokuchaev, V. V. 1967. Russian Chernozem (translated from Russian): Selected works of V.V. Dokuchaev, vol 1. Israel Program for Scientific Translation for the USDA and NSF, Jerusalem.

Douglas, L. A. 1980. The use of soils in estimating the time of last movement of faults. Soil Sci. 129(6):345–352.

Eash, N. S., and J. A. Sandor. 1995. Soil chronosequence and geomorphology in a semi-arid valley in the Andes of southern Peru. Geoderma 65:59–79.

Easterbrook, D. J. 1993. Surface processes and landforms. Macmillan Publishing Co., New York.

Effland, A. B. W., and W. R. Effland. 1992. Soil geomorphology studies in the U.S. soil survey program. Agri. History 66:189–212.

Effland, W. R., and A. B. W. Effland. n.d. The soil geomorphology studies of the U.S. Soil Conservation Service: Description and applications. Unpublished manuscript.

Engel, S. A., T. W. Gardner, and E. J. Ciolkosz. 1996. Quaternary soil chronosequences on terraces of the Susquehanna River, Pennsylvania. Geomorphology 17:273–294.

Follmer, L. R. 1978. The Sangamon Soil in its type area – a review. pp. 125–165. In W. C. Mahaney (ed.) Quaternary soils Geo Abstracts, Norwich, UK.

Follmer, L. R. 1979. A historical review of the Sangamon Soil. pp. 79–91. In L. R. Follmer, E. D. McKay, J. A. Lineback, and D. L. Gross, (field trip leaders), Wisconsinan, Sangamonian, and Illinoian stratigraphy in Central Illinois. Ill. State Geol. Survey Guidebook 13 (Midwest Friends of the Pleistocene 26th Field Conference), Urbana-Champaign.

Follmer, L. R. 1982. The geomorphology of the Sangamon surfaces: Its spatial and temporal attributes. pp. 117–146. In C. E. Thorn (ed.) Space and time in geomorphology. Allen and Unwin, London.

Follmer, L. R. 1983. Sangamonian and Wisconsinan pedogenesis in the midwestern United States. pp. 138–144. In S. C. Porter (ed.) Late Quaternary environments of the United States, vol. 1: The late Pleistocene. University of Minnesota Press, Minneapolis.

Follmer, L. R., D. L. Johnson, and J. A. Catt (eds.). 1998. Revisitation of concepts of paleopedology. Quaternary International 51/52:1–221.

Foss, J. E. 1977. The pedological record at several Paleoindian sites in the Northeast. pp. 234–244. *In* W. S. Newman and B. Salwen (eds.) Amerinds and their paleoenvironments in northeastern North America. Annals of the NY Acad. Sci. 228.

Foss, J. E., and A. V. Segovia. 1984. Rates of soil formation. pp. 1–17. *In* R. G. LaFleur (ed.) Groundwater as a geomorphic agent. Binghamton Symposium in Geomorphology 13. Allen and Unwin, Boston.

Foss, J. E., R. J. Lewis, and M. E. Timpson. 1995. Soils in alluvial sequences: Some archaeological implications. pp. 1–14. *In* M. E. Collins, B. J. Carter, B. G. Gladfelter, and R. J. Southard (eds.) Pedological perspectives in archaeological research. SSSA Spec. Pub. 44, SSSA, Madison, WI.

Free, E. E. 1911. The movement of soil material by the wind. USDA Bureau of Soils Bull. No. 68. Washington DC.

Frye, J. C. 1949. Use of fossil soils in Kansas Pleistocene stratigraphy. Kansas Acad. Sci. Trans. 52:478–482.

Frye, J. C. 1951. Soil-forming intervals evidenced in the Kansas Pleistocene. Soil Sci. 71:403–408.

Frye, J. C., and A. B. Leonard. 1951. Stratigraphy of the late Pleistocene loesses of Kansas. Jour. Geol. 59:287–305.

Frye, J. C., and A. B. Leonard. 1952. Pleistocene geology of Kansas. Kansas Geol. Survey, Bull. 99, Lawrence.

Frye, J. C., and A. B. Leonard. 1955a. The Brady soil and subdivision of post-Sangamonian time in the Midcontinent region. Am. Jour. Sci. 253:358–364.

Frye, J. C., and A. B. Leonard. 1955b. Petrographic comparison of some loess samples from western Europe with Kansas loess. Jour. Sed. Petrology 25:3–23.

Frye, J. C., and A. B. Leonard. 1957a. Ecological interpretations of Pliocene and Pleistocene stratigraphy in the Great Plains region. Am. J. Sci. 255:1–11.

Frye, J. C., and A. B. Leonard. 1957b. Studies of Cenozoic geology along eastern margin of Texas High Plains, Armstrong and Howard counties. Univ. Texas, Bureau of Econ. Geol., Rept. Inv. 32, Austin.

Frye, J. C., and A. B. Leonard. 1967. Buried soils, fossil molluscs, and late Cenozoic paleoenvironment. pp. 429–444. *In* C. Teichert and E. L. Yochelson (eds.). Paleontology and stratigraphy. Univ. Kansas Press, Lawrence.

Frye, J. C., and H. B. Willman. 1963. Loess stratigraphy, Wisconsinan classification and accretion-gleys in central western Illinois. Ill. State Geol. Surv. Guidebook Series 5, Champaign-Urbana.

Frye, J. C., and H. B. Willman. 1970. Pleistocene stratigraphy of Illinois. Ill. State Geol. Surv. Bulletin 94, Champaign-Urbana.

Frye, J. C., R. R. Shaffer, H. B. Willman, and G. E. Ekblaw. 1960. Accretion-gley and the gumbotil dilemma. Am. Jour. Sci. 258:185–190.

Frye, J. C., A. Swineford, and A. B. Leonard. 1948. Correlation of Pleistocene deposits of the Central Great Plains with the glacial section. Jour. Geol. 56:505–525.

Frye, J. C., H. B. Willman, and H. D. Glass. 1963. Mineralogy of glacial tills and their weathering profiles in Illinois – Part I. Glacial tills. Ill. State Geol. Surv. Circ. 347, Champaign-Urbana.

Frye, J. C., H. B. Willman, and H. D. Glass. 1966. Mineralogy of glacial tills and their weathering profiles in Illinois – Part II. Weathering profiles. Ill. State Geol. Surv. Circ. 400, Champaign-Urbana.

Gamble, E. E., R. B. Daniels, and W. D. Nettleton. 1970. Geomorphic surfaces and soils in the Black Creek Valley, Johnston County, North Carolina. SSSA Proc. 34:276–281.

Gaylord, D. R, A. J. Busacca, M. R. Sweeney. 2003. The Palouse Loess and the Channeled Scabland; a paired ice-age geologic system. pp. 123–134. In D. J. Easterbrook (ed.) Quaternary geology of the United States. INQUA field guide volume. Desert Research Institute. Reno, NV.

Gerrard, J. 1992. Soils and geomorphology. Chapman and Hall, London.

Gerrard, J. 1993. Soil geomorphology – Present dilemmas and future challenges. Geomorphology 7:61–84.

Geyh, M. A., J.-H. Benzler, and G. Roeschmann. 1971. Problems of dating Pleistocene and Holocene soils by radiometric methods. pp. 63–76. In D. H. Yaalon (ed.) Paleopedology: Origin, nature and dating of paleosols. University of Israel Press, Jerusalem.

Gile, L. H. 1975a. Causes of soil boundaries in an arid region. SSSA Proc. 39:316–330.

Gile, L. H. 1975b. Holocene soils and soil-geomorphic relations in an arid region of southern New Mexico. Quaternary Res. 5:321–360.

Gile, L. H. 1977. Holocene soils and soil-geomorphic relations in a semi-arid region of southern New Mexico. Quaternary Res. 7:112–132.

Gile, L. H. 1979. Holocene soils in eolian sediments of Bailey County, Texas. SSSA Jour. 43:994–1003.

Gile, L. H. 1985. The Sandhills Project soil monograph. New Mexico State University, Rio Grande Historical Collections, Las Cruces.

Gile, L. H., and R. B. Grossman. 1979. The Desert Project soil monograph. USDA-SCS, U.S. Gov Print. Office, Washington, D.C.

Gile, L. H., F. F. Peterson, and R. B. Grossman. 1965. The K horizon – A master soil horizon of carbonate accumulation. Soil Science 99:74–82.

Gile, L. H., F. F. Peterson, and R. B. Grossman. 1966. Morphological and genetic sequences of carbonate accumulation in desert soils. Soil Science 101:347–360.

Gile, L. H., J. W. Hawley, and R. B. Grossman. 1981. Soils and geomorphology in the Basin and Range area of southern New Mexico: Guidebook to the Desert Project. New Mexico Bureau of Mines and Mineral Resources, Mem. 39.

Gile L. H., J. W. Hawley, R. B. Grossman, H. C. Monger, and G. H. Mack. 1995. Supplement to the Desert Project Guidebook, with Emphasis on Soil Micromorphology. New Mexico Bureau of Mines and Mineral Resources, Bull. 142.

Glinka, K. D. 1914. Die typen der bodenbildung. Gebruder Borntrager, Berlin, 365 p.

Hall, R. D. 1999. A comparison of surface soils and buried soils: Factors of soil development. Soil Science 164:264–287.

Hall, R. D., and A. K. Anderson. 2000. Comparative soil development of Quaternary paleosols of the central United States. Palaeogeography, Palaeoclimatology, Palaeoecology 158:109–145.

Harden, J. W. 1982. A quantitative index of soil development from field descriptions: Examples from a chronosequence in central California. Geoderma 28:1–28.

Harden, J. W. 1987. Soils developed in granitic alluvium near Merced, California. USGS Bull. 1590-A, Washington, D.C., 65 p.

Harden, J. W. 1990. Soil development on stable landforms and implications for landscape studies. pp. 391–398. In P. L. K. Knuepfer and L. D. McFadden (eds.) Soils and landscape evolution. Geomorphology (special issue) 3.

Harden, J. W., and J. C. Matti. 1989. Holocene and late Pleistocene slip rates on the San Andreas fault in Yucaipa, California using displaced alluvial-fan deposits and soil chronology. Geol. Soc. Am. Bull. 101:1107–1117.

Harden, J. W., and E. M. Taylor. 1983. A quantitative comparison of soil development in four climatic regimes. Quaternary Res. 20:342–359.

Harden, J. W., A. M. Sarna-Wojcicki, and G. R. Dembroff. 1986. Soils developed on coastal and fluvial terraces near Ventura, California. USGS Bull. 1590-B, Washington, D.C., 34 p.

Hawley, J. W. 1989. Citation (presentation of the Kirk Bryan Award). Geol. Soc. Am. Bull. 101:1006–1007

Haynes, C. V., Jr. 1968. Geochronology of late-Quaternary alluvium. In R. B. Morrison and H. E. Wright, Jr. (eds.) Means of correlation of Quaternary successions. pp. 591–631. University of Utah Press, Salt Lake City.

Haynes, C. V., Jr. 1990. The Antevs-Bryan years and the legacy for Paleoindian geochronology. In L. F. Laporte (ed.) Establishment of a geologic framework for paleoanthropology. pp. 55–68. Geol. Soc. Am. Special Paper 242, Boulder, CO.

Hibben, F. C. 1941. Evidences of early occupation in Sandia Cave, New Mexico, and other sites in the Sandia-Manzano Region. Smithsonian Miscellaneous Collections 99, 64 p. Washington, D.C.

Hilgard, E. W. 1892. A report on the relations of soil to climate. USDS Weather Bull. 3:1–59.

Hole, F. D. 1961. A classification of pedoturbations and some other processes and factors of soil formation in relation to isotropism and anisotropism. Soil Science 91:375–377.

Hole, F. D. 1962. History and progress in soil survey in Wisconsin. Wisconsin Academy Review 9:167–169.

Hole, F. D. 1981. Effects of animals on soils. Geoderma 25:75–112.

Hole, F. D. 1988. *Terra Vibrata*: Some observations on the dynamics of soil landscapes. Physical Geog. 9:175–185.

Holliday, V. T. 1985. Morphology of late Holocene soils at the Lubbock Lake site, Texas. SSSA Jour. 49:938–946.

Holliday, V. T. 1987. Geoarchaeology and late Quaternary geomorphology of the middle South Platte River, northeastern Colorado. Geoarchaeology 2:317–329.

Holliday, V. T. 1988. Genesis of a late Holocene soil chronosequence at the Lubbock Lake archaeological site, Texas. Annals Assoc. Am Geog. 78:594–610.

Holliday, V. T. 1989. Paleopedology in archeology. *In* A. Bronger and J. Catt (eds.) Paleopedology: Nature and applications of paleosols. pp. 187–206. Catena Supplement 16, Cremlingen, Germany.

Holliday, V. T. 1990. Soils and landscape evolution of eolian plains: the Southern High Plains of Texas and New Mexico. *In* P. L. K. Knuepfer and L. D. McFadden (eds.) Soils and landscape evolution. pp. 489–515. Geomorphology (special issue) 3.

Holliday, V. T. 1994. The "state factor" approach in geoarchaeology. *In* R. Amundson (ed.) Factors of soil formation: A fiftieth anniversary retrospective. pp. 65–86. SSSA Spec Pub. 33. SSSA, Madison, WI.

Holliday, V. T. 1995. Stratigraphy and paleoenvironments of late Quaternary valley fills on the Southern High Plains. Geol. Soc. Am. Mem. 186. GSA, Boulder, CO.

Holliday, V. T. 2004. Soils in archaeological research. Oxford University Press, New York.

Holliday, V. T., L. D. McFadden, E. A. Bettis, and P. W. Birkeland. 2002. The soil survey and soil geomorphology. *In* D. Helms, A. B. W. Effland, P. J. Durana (eds.) Profiles in the history of the U.S. Soil Survey. pp. 233–274. Iowa State Press and Blackwell Publishers, Ames, IA.

Hopkins, D. M., and J. L. Giddings. 1953. Geological background of the Iyatayet archeological site, Cape Denbigh, Alaska. Smithsonian Misc. Coll. 121. Washington, D.C., 33 p.

Huggett, R. J. 1975. Soil landscape systems: A model of soil genesis. Geoderma 13:1–22.

Huggett, R. J. 1998. Soil chronosequences, soil development, and soil evolution: A critical review. Catena 32:155–172.

Hunt, C. B. 1950. Military geology. *In* S. Paige (ed.) Application of geology to engineering practice. pp. 295–327. Geol. Soc. Am. Berkey Volume, New York.

Hunt, C. B. 1954. Pleistocene and Recent deposits in the Denver area, Colorado. USGS Bull. 966-C:91–140. Washington, D.C.

Hunt, C. B. 1972. Geology of soils: Their evolution, classification, and uses. W. H. Freeman and Co., San Francisco, 344 p.

Hunt, C. B., and V. P. Sokoloff. 1950. Pre-Wisconsin soil in the Rocky Mountain region. USGS Prof. Paper 221-G:109–121. Washington, D.C.

Hunt, C. B., and A. P. Hunt. 1957. Stratigraphy and archeology of some Florida soils. Geol. Soc. Am. Bull. 68:797–806.

Jacobs, P. M. 1998. Influence of parent material grain size on genesis of the Sangamon Geosol in south-central Indiana. Quaternary International 51/52: 127–133.

Jacobs, P. M., and J. C. Knox. 1994. Provenance and pedology of a long-term Pleistocene depositional sequence in Wisconsin's Driftless Area. Catena 22:49–68.

Jacobs, P. M., J. C. Knox, and J. A. Mason. 1997. Preservation and recognition of middle and early Pleistocene loess in the Driftless Area, Wisconsin. Quaternary Res. 47:147–154.

Jenny, H. 1941. Factors of soil formation: A system of quantitative pedology. McGraw-Hill Book Company, New York, NY.

Jenny, H. 1946. Arrangement of soil series and types according to functions of soil-forming factors. Soil Science 61:375–391.

Jenny, H. 1958. Role of the plant factor in the pedogenic functions. Ecology 39:5–16.

Jenny, H. 1961. Derivation of state factor equations of soils and ecosystems. SSSA Proc. 25:385–388.

Jenny, H. 1980. The soil resource. Springer-Verlag, New York, 377 p.

Jenny, H., A. E. Salem, and J. R. Wallis. 1968. Interplay of soil organic matter and soil fertility with state factors and soil properties. Pontif. Acad. Sci. Scr. Varia 32:1–44.

Johnson, C. H. 1985. Dr. Marbut: The geologist. Soil Survey Horizons 26(1):10–13.

Johnson, D. L. 2000. Soils and soil-geomorphology theories and models: The Macquarie connection. Book review forum on "Soils: A new global view." Annals Assoc. Am. Geographers 90:775–782.

Johnson, D. L., and F. D. Hole. 1994. Soil formation theory: A summary of its principal impacts on geography, geomorphology, soil-geomorphology, Quaternary geology, and paleopedology. *In* R. Amundson (ed.) Factors of soil formation: A fiftieth anniversary retrospective. pp. 111–126. SSSA Spec. Publ. 33. SSSA, Madison, WI.

Johnson D. L., and D. Watson-Stegner. 1987. Evolution model of pedogenesis. Soil Science 143:349–366.

Johnson, D. L., D. Watson-Stegner, D. N. Johnson, and R. J. Schaetzl. 1987. Proisotropic and proanisotropic processes of pedoturbation. Soil Science 143:278–292.

Johnson, D. L., E. A. Keller, and T. K. Rockwell. 1990. Dynamic pedogenesis: New views on some key soil concepts, and a model for interpreting Quaternary soils. Quaternary Res. 33:306–319.

Judson, S. 1953. Geology of the San Jon site, eastern New Mexico. Smithsonian Miscellaneous Collection 121. Washington, D.C.

Karlstrom, E. T. 1988. Rates of soil formation on Black Mesa, northeast Arizona: A chronosequence in late Quaternary alluvium. Phys. Geog. 9:301–327.

Kay, G. F. 1916. Gumbotil, a new term in Pleistocene geology. Science, new series 44:637–638.

Kay, G. F., and J. N. Pearce. 1920. The origin of gumbotil. Jour. Geol. 28:89–125.

Keller, E. A., R. L. Zepeda, T. K. Rockwell, T. L. Ku, and W. S. Dinklage. 1998. Active tectonics at Wheeler Ridge, southern San Joaquin Valley, California. Geol. Soc. Am. Bul. 110(3):298–310.

King, L. C. 1949. The pediment landform: some current problems. Geological Magazine 86:245–250.

King, L. C. 1950. The world's plainlands: a new approach in geomorphology. Geol. Soc. London, Quarterly Jour. 106:101–131.

King, L. C. 1953. Cannons of landscape evolution. Geo. Soc. Am. Bull. 64:721–751.

Kizer, S. 1985. Dr. Marbut: The soil scientist. Soil Survey Horizons 26(1):16–19.

Knuepfer, P. L. K., and L. D. McFadden (eds.). 1990. Soils and landscape evolution. Geomorphology (special issue) 3:489–515.

Kottlowski, F. E., M. E. Cooley, and R. V. Ruhe. 1965. Quaternary geology of the Southwest. In H. E. Wright, Jr. and D. G. Frey (eds.) Quaternary of the United States. pp. 287–298. Princeton University Press.

Langley-Turnbaugh, S. J., and J. G. Bockheim. 1997. Time-dependent changes in pedogenic processes on marine terraces in coastal Oregon. SSSA Jour. 61:1428–1440.

Larsen, E. S. 1951. Memorial to Kirk Bryan. Proc Geol. Soc. Am., Annual Report for 1950, 91–96.

Leigh, D. S. 1996. Soil chronosequence of Brasstown Creek, Blue Ridge Mountains, USA. Catena 26:99–114.

Leigh, D. S., and J. S. Cable, 1997. Stratigraphy and soil chronosequence of the Brasstown sites: A model for age assessment of alluvium in the southern Blue Ridge, U.S.A. In A. C. Goodyear, J. E. Foss, and K. E. Sassaman (eds.) Proc. 2nd Intl. Conf. on Pedo-Archaeology. pp. 1–9. Occasional Papers of the South

Carolina Institute of Archaeology and Anthropology, University of South Carolina, Columbia, SC.

Leigh, D. S., and J. C. Knox. 1994. Loess of the Upper Mississippi Valley Driftless Area. Quaternary Res. 42:30–40

Leighton, M. M. 1934. Some observations on the Antiquity of Man in Illinois. Trans. Ill. State Acad. Sci. 25, 83 p.

Leighton, M. M. 1937. The significance of profiles of weathering in stratigraphic archaeology. *In* G. G. MacCurdy (ed.) Early Man. pp. 163–172. Lippincott, New York.

Leighton, M. M. 1958. Principles and viewpoints in formulating the stratigraphic classifications of the Pleistocene. Jour. Geol. 66:700–709.

Leighton, M. M. 1960. The classification of the Wisconsin glacial stage of north central United States. Jour. Geol. 68:529–552.

Leighton, M. M. 1965. The stratigraphic succession of Wisconsin loesses in the Upper Mississippi River Valley. Jour. Geol. 73:323–345.

Leighton, M. M., and P. MacClintock. 1930. Weathered zones of the drift-sheets of Illinois. Jour. Geol. 38:28–53.

Leighton, M. M., and P. MacClintock. 1962. The weathered mantle of glacial tills beneath original surfaces in north-central United States. Jour. Geol. 70:267–293.

Leighton, M. M., and H. B. Willman. 1950. Loess formations of the Mississippi Valley. Jour. Geol. 58:599–623.

Leopold, L. B., and J. P. Miller. 1954. A postglacial chronology for some alluvial valleys in Wyoming. USGS Water Supply Paper 1261. Washington, D.C., 90 p.

Leverett, F. 1898a. The weathered zone (Sangamon) between the Iowan loess and Illinoian till sheet. Jour. Geol. 6:171–181.

Leverett, F. 1898b. The weathered zone (Yarmouth) between the Illinoian and Kansan till sheet. Jour. Geol. 6:238–243.

Leverett, F. 1899. The Illinois glacial lobe. USGS Monograph 38. Washington, D.C., 818 p.

Litaor, M. I. 1987. The influence of eolian dust on the genesis of alpine soils in the Front Range, Colorado. SSSA Jour. 51:142–147.

Machette, M. N. 1975. Geologic map of the Lafayette Quadrangle, Adams, Boulder, and Jefferson Counties, Colorado. USGS Map MF-656. Washington, D.C.

Machette, M. N. 1978. Dating Quaternary faults in the southwestern United States by using buried calcic paleosols. USGS Jour. Res. 6:369–381.

Machette, M. N. 1985. Calcic soils of the southwestern United States. *In* D. L. Weide (ed.) Soils and Quaternary geology of the Southwestern United States. pp. 1–21. by Geol. Soc. Am. Special Paper 203. GSA, Boulder, CO.

Machette, M. N. 1988. Quaternary movement along the La Jencia fault, central New Mexico. USGS Prof. Paper 1440. Washington, D.C.

Mandel, R. D. 1994. Holocene landscape evolution in the Pawnee River Valley, southwestern Kansas. Kansas Geol. Surv. Bull. 236. Lawrence, KS, 117 p.

Mandel, R. D. 1995. Geomorphic controls of the Archaic record in the Central Plains of the United States. *In* E. A. Bettis III (ed.) Archaeological geology of the Archaic Period in North America. pp. 37–66. Geol. Soc. Am. Special Paper 297. GSA, Boulder, CO.

Mandel, R. D., and E. A. Bettis III. 2001. Use and analysis of soils by archaeologists and geoscientists: A North American perspective. *In* P. Goldberg, V. T. Holliday and C. R. Ferring (eds.) Earth sciences and archaeology. pp. 173–204. Kluwer Academic/Plenum Publishers, New York.

Marbut, C. F. 1928. A scheme for soil classification. Proc. 1st Intl. Cong. Soil Science 4:1–31.

Marbut, C. F. 1935. Soils of the United States, Part III. pp. 1–98. USDA Atlas of American Agriculture. Washington, D.C.

Markewich, H. W., and M. J. Pavich. 1991. Soil chronosequence studies in temperate to subtropical, low-latitude, low-relief terrain with data from the eastern United States. *In* M. J. Pavich (ed.) Weathering and soils, pp. 213–239. Geoderma (special issue) 51.

Markewich, H. W., W. C. Lynn, M. J. Pavich, R. G. Johnson, and J. C. Meetz. 1988. Analyses of four Inceptisols of Holocene age, east-central Alabama. USGS Bull. 1589-C. Washington, D.C.

Markewich, H. W., M. J. Pavich, M. J. Mausbach, R. L. Hall, R. G. Johnson, and P. P. Hearn. 1987. Age relations between soils and geology in the Coastal Plain of Maryland and Virginia. USGS Bull. 1589-A. Washington, D.C.

Markewich, H. W., M. J. Pavich, M. J. Mausbach, R. G. Johnson, and V. M. Gonzales, 1989. A guide for using soil and weathering profile data in chronosequence studies on the Coastal Plain of the eastern United States. USGS Bull. 1589-D. Washington, D.C.

Markewich, H. W., M. J. Pavich, M. J. Mausbach, B. N. Stuckey, R. G. Johnson, and V. Gonzales. 1986. Soil development and its relation to the ages of morphostratigraphic units in Horry County, South Carolina. USGS Bull.1589-B. Washington, D.C.

Mayer, L., L. D. McFadden, and J. W. Harden. 1988. Distribution of calcium carbonate in desert soils: A model. Geology 16:303–306.

McCalpin, J. P., and M. E. Berry. 1996. Soil catenas to estimate ages of movement on normal fault scarps, with an example from the Wasatch fault zone, Utah, USA. Catena, 27:265–286.

McDonald, E. V., and A. Busacca. 1990. Interaction between aggrading geomorphic surfaces and the formation of a late Pleistocene paleosol in the Palouse loess of eastern Washington. *In* P. L. K. Knuepfer and L. D. McFadden (eds.) Soils and landscape evolution. pp. 449–470. Geomorphology (special issue) 3.

McDonald, E. V., and A. Busacca. 1992. Late Quaternary stratigraphy of loess in the channeled scabland and Palouse regions of Washington State. Quaternary Res. 38:141–156.

McFadden, L. D., and D. M. Hendricks. 1985. Changes in the content and composition of pedogenic iron oxyhydroxides in a chronosequence of soils in southern California. Quaternary Res. 23:189–204.

McFadden, L. D., and R. J. Weldon. 1987. Rates and processes of soil development on Quaternary terraces in Cajon Pass, California. Geol. Soc. Am. Bull. 98:280–293.

McFadden, L. D., and P. L. K. Knuepfer. 1990. Soil geomorphology: The linkage of pedology and surficial processes. In P. L. K. Knuepfer and L. D. McFadden (eds.) Soils and landscape evolution. pp. 197–205. Geomorphology (special issue) 3.

McFadden, L. D., and E. V. McDonald. 1998. Paleopedology, paradigms, and paleosols: A comment on the issue of disciplinary status. Quaternary Intl. 51/52:23–26.

McFadden, L. D., and J. C. Tinsley. 1985. The rate and depth of accumulation of pedogenic carbonate accumulation in soils: Formulation and testing of a compartment model. In D. L. Weide (ed.) Soils and Quaternary geology of the southwestern United States. pp. 23–42. Geol. Soc. Am. Spec. Paper 203. GSA, Boulder, CO.

McFadden, L. D., and J. R. McAuliffe. 1997. Lithologically influenced geomorphic responses to Holocene climatic changes in the Southern Colorado Plateau, Arizona: A soil-geomorphic and ecologic perspective. Geomorphology 19:303–332.

McFadden, L. D., S. G. Wells, and J. C. Dohrenwend. 1986. Influences of Quaternary climatic changes on processes of soil development on desert loess deposits of the Cima volcanic field, California. Catena 13:361–389.

McFadden, L. D., S. G. Wells, and M. J. Jercinovic. 1987. Influences of eolian and pedogenic processes on the evolution and origin of desert pavements. Geology 15:504–508.

McFadden, L. D., R. G. Amundson, and O. A. Chadwick. 1991. Numerical modeling, chemical, and isotopic studies of carbonate accumulation in soils of arid regions. In W. D. Nettleton (ed.) Occurrence, characteristics, and genesis of carbonate, gypsum, and silica accumulations in soils. pp. 17–36. SSSA Spec. Publ. 26. Madison, Wisconsin.

McFadden, L. D., E. V. McDonald, S. G. Wells, K. Anderson, J. Quade, and S. L. Forman. 1998. The vesicular layer and carbonate collars of desert soils and pavements: formation, age and relation to climatic change. Geomorphology 24:101–145.

McFaul, M., K. L. Traugh, G. D. Smith, and W. Doering. 1994. Geoarchaeologic analysis of South Platte River terraces: Kersey, Colorado. Geoarchaeology 9:345–374.

Miller, C. D., and P. W. Birkeland. 1992. Soil catena variation along an alpine climate transect, northern Peruvian Andes. Geoderma 55:211–223.

Milne, G. 1932. Conference of East African soil chemists. Technical Conferences of the East African Dependencies. Amani, Tanganyika Territory, Government Printer, Nairobi.

Milne, G. 1935a. Some suggested units for classification and mapping, particularly for East African soils. Soils Research – Supplements to the Proceedings of the International Society of Soil Scienmce (Berlin) 4:183–198.

Milne, G. 1935b. Composite units for the mapping of complex soil associations. Transactions of the Third International Congress of Soil Science 1:345–347.

Milne, G. 1936. Normal erosion as a factor in soil profile development. Nature 138:548–549.

Morris, M. W. 2002. Soil science and archaeology: Three test cases from Minoan Crete. The Institute for Aegean Prehistory Academic Press, Prehistory Monographs 4, Philadelphia, 141 p.

Morrison, R. B. 1964. Lake Lahontan: Geology of Southern Desert, Nevada. USGS Prof. Paper 424-D. Washington, D.C., 156 p.

Morrison, R. B. 1965. Lake Bonneville: Quaternary stratigraphy of eastern Jordan Valley, Utah. USGS Prof. Paper 477. Washington, D.C., 80 p.

Morrison, R. B. 1967. Principles of Quaternary soil stratigraphy. In R. B. Morrison and H. E. Wright, Jr. (eds.) Quaternary soils. pp. 1–69. Desert Research Institute, University of Nevada, Reno.

Morrison, R. B. 1978. Quaternary soil stratigraphy-concepts, methods, and problems. In W. C. Mahaney (eds.) Quaternary soils. pp. 77–108. Geo Abstracts, Norwich, UK.

Morrison, R. B., and J. C. Frye. 1965. Correlation of the middle and late Quaternary successions of the Lake Lahontan, Lake Bonneville, Rocky Mountain (Wasatch Range), southern Great Plains, and eastern Midwest areas. Nevada Bur. Mines Rpt 9, 45 p., Reno.

Morrison, R. B., and H. E. Wright, Jr. (eds.). 1967. Quaternary soils. Desert Research Institute, University of Nevada, Reno.

Morrison, R. B., and H. E. Wright, Jr. (eds.). 1968. Means of correlation of Quaternary successions. University of Utah Press, Salt Lake City.

Muhs, D. R. 1982. A soil chronosequence on Quaternary marine terraces, San Clemente Island, California. Geoderma 28:257–283.

NACOSN (North American Commission on Stratigraphic Nomenclature). 1983. North American stratigraphic code. Am. Assoc. Petroleum Geologists Bull. 67:841–875.

Newberry, J. S. 1862. Notes on the surface geology of the basins of the Great Lakes. Boston Soc. Natural History Proc. 9:42–46.

Newberry, J. S. 1878. Review of geological structure of Ohio. Report Geol. Surv. Ohio 3:1–51.

Nikiforoff, C. C. 1943. Introduction to paleopedology. Am. Jour. Sci. 241:194–200.

Nikiforoff, C. C. 1949. Weathering and soil evolution. Soil Science 67:219–230.

Nikiforoff, C. C. 1959. Reappraisal of the soil. Science 129:186–196.

Norton, E. A., and R. S. Smith. 1928. Horizon designation. Am. Soil Surv. Assoc. Bull. 9:83–86.

Norton, L. D., L. T. West, and K. McSweeney. 1988, Soil development and loess stratigraphy of the midcontinental USA. In D. N. Eden and R. J. Furkert (eds.) Loess: Its distribution, geology, and soils. pp. 145–159. A. A. Balkema, Rotterdam.

Olson, C. G. 1989. Soil geomorphic research and the importance of paleosol stratigraphy to Quaternary investigations, Midwestern USA. In A. Bronger and J. Catt (eds.) Paleopedology: Nature and applications of paleosols. pp. 129–142. Catena Verlag, Catena Supplement 16. Cremlingen, Germany.

Olson, C. G. 1997. Systematic soil-geomorphic investigations – Contributions of R. V. Ruhe to pedologic interpretation. In D. H. Yaalon and S. Berkowicz (eds.) History of soil science. pp. 415–438. Catena Verlag, Advances in GeoEcology 29. Reiskirchen, Germany.

Olson, C. G., and W. D. Nettleton. 1998. Paleosols and the effects of alteration. Quaternary International 51/52:185–194.

Orton, E. 1870. On the occurrence of a peat bed beneath deposits of drift in south-western Ohio. Am. Jour. Sci. 50:54–57, 293.

Orton, E. 1873. Report on the third geological district; Geology of the Cincinnati group: Hamilton, Clermont, Clarke Cos. Ohio Geol. Surv. Report 1:365–480.

Parsons, R. B. 1978. Soil-geomorphology relationships in mountains of Oregon, U.S.A. Geoderma 21:25–39.

Parsons, R. B., and C. A. Balster. 1967. Dayton – a depositional Planosol, Willamette Valley, Oregon. SSSA Proc. 31:255–258.

Parsons, R. B., C. A. Balster, and O. A. Ness. 1970. Soil Development and geomorphic surfaces, Willamette Valley, Oregon. SSSA Proc. 34:485–491.

Parsons, R. B., G. H. Simonson, and C. A. Balster. 1968. Pedogenic and geomorphic relationships of associated Aqualfs, Albolls, and Xerolls in western Oregon. SSSA Proc. 32:556–563.

Paton, T. R., G. S. Humphreys, and P. B. Mitchell. 1995. Soils: A new global view. Yale University Press. New Haven, CT.

Polach, H. A., and A. B. Costin. 1971. Validity of soil organic matter radiocarbon dating: Buried soils in Snowy Mountains, southeastern Australia as example. In D. H. Yaalon (ed.) Paleopedology: Origin, nature and dating of paleosols. pp. 89–96. University of Israel Press, Jerusalem.

Rabenhorst, M. C. 1997. The chrono-continuum: An approach to modeling pedogenesis in marsh soils along transgressive coastlines. Soil Science 162:2–9.

Ray, L. L. 1951. Obituary – Kirk Bryan. The Geographical Review 41:165–166.

Ray, L. L. 1974. Memorial to Morris Morgan Leighton 1887–1971. Geol. Soc. Am. Memorials 3:133–153.

Reheis, M. C. 1987a. Gypsic soils on the Kane alluvial fans, Big Horn County, Wyoming. USGS Bull. 1590-C. Washington, D.C., 39 p.

Reheis, M. C. 1987b. Soils in granitic alluvium in humid and semiarid climates along Rock Creek, Carbon County, Montana. USGS Bull. 1590-D. Washington, D.C., 71 p.

Reheis, M. C. 1987c. Climatic implications of alternating clay and carbonate formation in semiarid soils of south-central Montana. Quaternary Res. 27:270–282.

Reheis, M. C., J. W. Harden, L. D. McFadden, and R. R. Shroba. 1989. Development rates of late Quaternary soils, Silver Lake Playa, California. SSSA Jour. 53:1127–1140.

Reheis, M. C., J. C. Goodmacher, J. W. Harden, L. D. McFadden, T. K. Rockwell, R. R. Shroba, J. M. Sowers, and E. M. Taylor. 1995. Quaternary soils and dust deposition in southern Nevada and California. Geol. Soc. Am. Bull. 107:1003–1022.

Reider, R. G. 1990. Late Pleistocene and Holocene pedogenic and environmental trends at archaeological sites in plains and mountain areas of Colorado and Wyoming. In N. P. Lasca and J. Donahue (eds.) Archaeological geology of North America. pp. 335–360. Centennial Volume 4. Geological Society of America, Boulder, CO.

Richards, K. S., R. R. Arnett and J. Ellis (eds.). 1985. Geomorphology and soils. George Allen and Unwin, London, 441 p.

Richmond, G. M. 1962. Quaternary stratigraphy of the La Sal Mountains, Utah. USGS Prof. Paper 324. Washington, D.C., 135 p.

Richmond, G. M., and J. C. Frye. 1957. Status of soils in stratigraphic nomenclature. Bull. Assoc. Petroleum Geologists 41:758–763.

Ritter, D. F., R. C. Kochel, and J. R. Miller. 1995. Process geomorphology (3rd ed.). W. C. Brown Publishing. Dubuque, Iowa, 538 p.

Rockwell, T. K., D. L. Johnson, E. A. Keller, and G. R. Dembroff. 1985. A late Pleistocene-Holocene soil chronosequence in the Ventura Basin, southern California, USA. In K. S. Richards, R. R. Arnett and J. Ellis (eds.) Geomorphology and soils. pp. 309–327. George Allen and Unwin, London.

Rodbell, D. T. 1993. Subdivision of late Pleistocene moraines in the Cordillera Blanca, Peru, based on rock-weathering features, soils, and radiocarbon dates. Quaternary Res. 39:133–143.

Ruhe, R. V. 1954a. Erosion surfaces of central African interior high plateaus. Publications de L'Institut National Pour L'Etude Agronomique du Congo Belge (I.N.E.A.C.), Bruxelles, Ser. Sci. 59.

Ruhe, R. V. 1954b. Pleistocene soils along the Rock Island relocation in southwestern Iowa. American Railway Engineering Association Bulletin 514:639–645.

Ruhe, R. V. 1956a. Geomorphic surfaces and the nature of soils. Soil Science 82:441–455.

Ruhe, R. V. 1956b. Landscape evolution in the High Ituri, Belgian Congo. Publications de L' I.N.E.A.C., Bruxelles, Ser. Sci. 66.

Ruhe, R. V. 1960. Elements of the soil landscape. 7th Intl. Congress Soil Sci. Trans. 4:165–170. Madison, Wisconsin.

Ruhe, R. V. 1962. Age of the Rio Grande Valley in southern New Mexico. Jour. Geol. 70:151–167.

Ruhe, R. V. 1964a. Landscape morphology and alluvial deposits in southern New Mexico. Annals Assoc. Am. Geographers 54:147–159.

Ruhe, R. V. 1964b. An estimate of paleoclimate in Oahu, Hawaii. Am. Jour. Sci. 262:1098–1115.

Ruhe, R. V. 1965a. Quaternary paleopedology. In H.E. Wright, Jr. and D. G. Frey (eds.) Quaternary of the United States. pp. 755–764. Princeton University Press.

Ruhe, R. V. 1965b. Relation of fluctuations of sea level to soil genesis in the Quaternary. Soil Science 99:23–29.

Ruhe, R. V. 1967. Geomorphic surfaces and surficial deposits in southern New Mexico. N. M. Bureau Mines and Mineral Resources, Memoir 18. Socorro.

Ruhe, R. V. 1969. Quaternary landscapes in Iowa. Iowa State University Press. Ames.

Ruhe, R. V. 1970. Soil-geomorphology studies 1953–1970. Manuscript on file, Iowa State University Agronomy Library, Ames, 12 p.

Ruhe, R. V. 1974. Holocene environments and soil geomorphology in the Midwestern United States. Quaternary Res. 4:487–495.

Ruhe, R. V. 1975a. Geomorphology. Houghton Mifflin. Boston.

Ruhe, R. V. 1975b. Review of "Pedology, weathering, and geomorphological research." Geoderma 14:176–177.

Ruhe, R. V., and R. B. Daniels. 1958. Soils, paleosols, and soil-horizon nomenclature. SSSA Proc. 22:66–69.

Ruhe, R. V., and C. G. Olson. 1980. Soil welding. Soil Science 130:132–139.

Ruhe, R. V. and W. H. Scholtes. 1956. Ages and development of soil landscapes in relation to climatic and vegetational changes in Iowa. SSSA Proc. 20:264–273.

Ruhe, R. A., R. B. Daniels, and J. G. Cady, 1965a. Landscape evolution and soil development in southwestern Iowa. USDA Tech. Bull. 1349. Washington, D.C.

Ruhe, R. V., J. M. Williams, and E. L. Hill. 1965b. Shorelines and submarine shelves, Oahu, Hawaii. Jour. Geol. 73:485–497.

Ruhe, R. V., J. M. Williams, R. C. Shuman, and E. L. Hill. 1965c. Nature of soil parent materials in Ewa-Waipahu area, Oahu. SSSA Proc. 29:282–287.

Ruhe, R. V., W. P. Dietz, T. E. Fenton and G. F. Hall. 1968. Iowan drift problem northeastern Iowa. Iowa Geol. Survey Rept. Investigations 7. Iowa City.

Ruhe, R. V., R. D. Hall, and A. P. Canepa. 1974. Sangamon paleosols of south-western Indiana. Geoderma 12:191–200.

Runge, E. C. A. 1973. Soil development sequences and energy models. Soil Science 115:183–193.

Schaetzl, R. J. 2000. Shock the world (and then some). Book review forum on "Soils: A New Global View." Annals Assoc. Am. Geographers 90:772–774.

Schaetzl, R. J., and S. Anderson. 2005. Soils: Genesis and geomorphology. Cambridge University Press, New York, 817 p.

Schaetzl, R. J., and S. A. Isard. 1996. Regional-scale relationships between climate and strength of podzolization in the Great Lakes region, North America. Catena 28:47–69.

Schaetzl, R. J., and D. L. Mokma. 1988. A numerical index of podzol and podzolic soil development. Physical Geography 9:232–246.

Schaetzl, R. J., L. R. Barrett, and J. A. Winkler. 1994. Choosing models for soil chronofunctions and fitting them to data. European Jour. Soil Sci. 45:219–232.

Schaetzl, R. J., F. J. Krist, Jr., P. R. Rindfleisch, J. Liebens, and T. E. Williams. 2000. Postglacial landscape evolution of northeastern Lower Michigan, inter-preted from soils and sediments. Annals Assoc. Am. Geographers 90:443–466.

Scharpenseel, H. W. 1971. Radiocarbon dating of soils – Problems, troubles, hopes. In D. H. Yaalon (ed.) Paleopedology: Origin, nature and dating of paleosols. pp. 77–87. University of Israel Press, Jerusalem.

Scholtes, W. H., R. V. Ruhe, and F. F. Reicken. 1951. Use of the morphology of buried soil profiles in the Pleistocene of Iowa. Iowa Academy of Science 58:295–306.

Scott, G. R. 1963. Quaternary geology and geomorphic history of the Kassler Quadrangle, Colorado. USGS Prof. Paper 421-A. Washington, D.C., 70 p.

Scott, W. E. 1977. Quaternary glaciation and volcanism, Metolius River area, Oregon. Geol. Soc. Am. Bull. 88:113–124.

Scully, R. W., and R. W. Arnold. 1981. Holocene alluvial stratigraphy in the upper Susquehanna River basin, New York. Quaternary Res. 15:327–344.

Sharp, R. P. 1993. Recollections of Kirk Bryan: A biographical sketch. Geomorphology 6:189–205.

Shimek, B. 1909. Aftonian sands and gravels in western Iowa. Geol. Soc. Am. Bull. 20:399–408.

Shlemon, R. J. 1978. Quaternary soil-geomorphic relationships, southeastern Mojave desert, California and Arizona. *In* W. C. Mahaney (ed.) Quaternary soils. pp. 187–207. Geo Abstracts, Norwich, UK.

Shlemon, R. J. and F. E. Budinger. 1990. The archaeological geology of the Calico site, Mojave Desert, California. *In* N. P. Lasca and J. Donahue (eds.) Archaeological geology of North America. pp. 301–314. Centennial Volume 4. Geological Society of America, Boulder, CO.

Shroba, R. R., and P. W. Birkeland. 1983. Trends in late-Quaternary soil development in the Rocky Mountains and Sierra Nevada of the western United States. *In* H. E. Wright, Jr. (ed.) Late-Quaternary environments of the United States, vol. 1. pp. 145–146. University of Minnesota Press, Minneapolis.

Simonson, R. W. 1941. Studies of buried soils formed from till in Iowa. SSSA Proc. 6:373–381.

Simonson, R. W. 1954. Identification and interpretation of buried soils. Am. Jour. Sci. 252:705–732.

Simonson, R. W. 1959. Outline of a generalized theory of soil genesis. SSSA Proc. 23:152–156.

Simonson, R. W. 1978. Multiple process model of soil genesis. *In* W. C. Mahaney (ed.) Quaternary soils. pp. 1–25. Geo Abstracts, Norwich.

Simonson, R. W. 1987. Historical aspects of Soil Survey and Soil Classification. SSSA (reprinted from Soil Survey Horizons). Madison, WI, 30 p.

Simonson, R. W. 1997a. Evolution of soil series and type concepts in the United States. *In* D. H. Yaalon and S. Berkowicz (eds.) History of soil science. Advances in GeoEcology 29:79–108.

Simonson, R. W. 1997b. Early teaching in USA of Dokuchaiev factors of soil formation. SSSA Jour. 61:11–16.

Snyder, K. E., and R. B. Bryant. 1992. Late Pleistocene surficial stratigraphy and landscape development in the Salamanca re-entrant, southwestern New York. Geol. Soc. Am. Bull. 104:242–251.

Soil Survey Staff. 1951. Soil survey manual. USDA Agric. Handbook 18. Washington, D.C.

Soil Survey Staff. 1960. Soil classification, A comprehensive system, 7th approximation. USDA. Washington, D.C., 503 p.

Soil Survey Staff. 1975. Soil taxonomy. USDA Agric. Handbook 436, Washington, D.C., 754 p.

Soil Survey Staff. 1999. Soil taxonomy, 2nd Ed. USDA Agric. Handbook 436, Washington, D.C., 869 p.

Sorenson, C. J. 1977. Reconstructed Holocene bioclimates. Annals Assoc. Am. Geographers 67:214–222.

Swanson, D. K. 1985. Soil catenas on Pinedale and Bull Lake moraines, Willow Lake, Wind River Mountains, Wyoming. Catena 12:329–342.

Swanson, D. K. 1993. Comments on "The soil survey as paradigm-based science." SSSA Jour. 57:1164.

Tandarich, J. P. 1998a. Agricultural geology: Disciplinary history. *In* G. A. Good (ed.) Science of the earth: An encyclopedia of events, people, and phenomena. pp. 23–29. Garland Publishing, Inc. New York.

Tandarich, J. P. 1998b. Pedology: Disciplinary history. *In* G. A. Good (ed.) Science of the earth: An encyclopedia of events, people, and phenomena. pp. 666–670. Garland Publishing, Inc. New York.

Tandarich, J. P., and S. W. Sprecher. 1994. The intellectual background for the factors of soil formation. *In* R. Amundson (ed.) Factors of soil formation: A fiftieth anniversary retrospective. pp. 1–13. SSSA Spec. Publ. 33. Madison, WI.

Tandarich, J. P., C. J. Johannsen, and W. W. Wildman. 1985. James Thorp talks about Soil Survey, C. F. Marbut, and China. Soil Survey Horizons 26(2):5–13

Tandarich, J. P., R. G. Darmody, and L. R. Follmer. 1988. The development of pedological thought: Some people involved. Physical Geography 9:162–174.

Tandarich, J. P., R. G. Darmody, L. R. Follmer, and D. L. Johnson, 2002. The historical development of soil and weathering profile concepts. SSSA Jour. 66:1–14.

Terman, M. J. 1998. Military Geology Unit of the U.S. Geological Survey during World War II. *In* J. R. Underwood, Jr., and P. L. Guth (eds.) Military geology in war and peace. pp. 49–54. Geol. Soc. Am. Reviews in Engineering XIII, Boulder, CO.

Thorp, J. 1930. The effects of vegetation and climate upon soil profiles in northern and northwestern Wyoming. Am. Soil Survey Assoc. Bull. 11:45–52.

Thorp, J. 1935. Soil profile studies as an aid to understanding recent geology. Bull. Geol. Soc. China 14:360–393.

Thorp, J. 1936. Geography of the soils of China. National Geological Survey of China, Nanking, 552 p.

Thorp, J. 1941. The influence of environment on soil formation. SSSA Proc. 6:39–46.

Thorp, J. 1949. Interrelations of Pleistocene geology and soil science. Geol. Soc. Am. Bull. 60(9):1517–1526.

Thorp, J. 1965. The nature of the pedological record in the Quaternary. Soil Science 99:1–8.

Thorp, J. 1985. Impressions of Dr. Curtis Fletcher Marbut, 1921–1935. Soil Survey Horizons 26(1): 26–30.

Thorp, J., and H. T. U. Smith. 1952. Pleistocene eolian deposits of the United States, Alaska, and parts of Canada, 1:2,500,000. Geol. Soc. Am. Map, Boulder, CO.

Thorp, J., W. M. Johnson, and E. C. Reed. 1951. Some post-Pliocene buried soils of central United States. Jour. Soil Sci. 2:1–19.

Tonkin, P. J., and L. R. Basher. 1990. Soil-stratigraphic techniques in the study of soil and landform evolution across the Southern Alps, New Zealand. *In* P. L. K. Knuepfer and L. D. McFadden (eds.) Soils and landscape evolution. pp. 547–575. Geomorphology (special issue) 3.

Totten, S. M., and G. W. White. 1985. Glacial geology and the North American Craton: Significant concepts and contributions of the nineteenth century. *In* E. T. Drake and W. M. Jordan (eds.) Geologists and ideas: A history of North American geology. pp. 125–141. Centennial Special Volume 1. Geological Society of America, Boulder, CO.

USGS (U. S. Geological Survey) and USACE (U. S. Army Corps of Engineers), 1945. The Military Geology Unit. Geol. Soc. Am. Pittsburgh Meeting (pamphlet), 22 p.

Vidic, N. J., and F. Lobnik, 1997. Rates of soil development of the chronosequence in the Ljubljana Basin, Slovenia. Geoderma 76:35–64.

Vreeken, W. J. 1975a. Quaternary evolution in Tama County, Iowa. Annals Assoc. Am. Geographers 65:283–296.

Vreeken, W. J. 1975b. Principal kinds of chronosequences and their significance in soil history. Jour. Soil Sci. 26:378–394.

Walker, P. H. 1966. Postglacial environments in relation to landscape and soils. Iowa Agric. Res. Sta. Bull. 549:838–875.

Walker, P. H. 1989. Contributions to the understanding of soil and landscape relationships. Australian Jour. Soil Res. 27:589–605.

Walker, P. H., and R. V. Ruhe. 1968. Hillslope models and soil formation: Closed systems. Trans. the 9th Intl. Congress Soil Sci. 4:561–568. Adelaide, Australia.

Ward, P. A., III, and B. J. Carter. 1998. Paleopedologic interpretations of soils buried by Tertiary and Pleistocene-age volcanic ashes; southcentral Kansas, western Oklahoma, and northwestern Texas, U.S.A. Quaternary Intl. 51/52:213–221.

Whittlesey, C. 1848. Notes upon the drift and alluvium of Ohio and the West. Am. Jour. Sci. 5:205–217.

Whittlesey, C. 1866. On the fresh-water glacial drift of the northwestern states. Smithsonian Contributions to Knowledge 197, 32 p.

Whittlesey, D. 1951. Kirk Bryan, 1888–1950. Annals Assoc. Am. Geographers 41:88–94.

Wildman, W. E. 1985. Dr. James Thorp, a close Marbut associate – Memorial tribute. Soil Survey Horizons 26(1):25–26.

Wilding, L. P. 1994. Factors of soil formation: Contributions to pedology. *In* R. Amundson (ed.) Factors of soil formation: A fiftieth anniversary retrospective. pp. 15–30. SSSA Spec. Pub. 33. Madison, WI.

Winchell, N. H. 1873. The surface geology. Minnesota Geol. Natural Hist. Survey, 3rd Annual Report, pp. 61–62. Minneapolis.

Worthen, A. H. 1868. Geology and paleontology. Geol. Surv. Illinois 5, 574 p. Springfield.

Worthen, A. H. 1870. Geology and paleontology. Geol. Surv. Illinois 4, 508 p. Springfield.

Worthen, A. H. 1873. Geology of Peoria County, McDonough County, Monroe County, Macoupin County, Sangamon County, Illinois. Geol. Surv. Illinois 5, 235–319.

Wright, H. E., Jr. 1995. Memorial to Robert V. Ruhe 1918–1993. Geol. Soc. Am. Memorials 25:41–44.

Wright, H. E., Jr., and D. G. Frey (eds.) 1965. Quaternary of the United States. Princeton University Press, 922 p.

Wright, H. E., Jr., and R. V. Ruhe. 1965. Glaciation of Minnesota and Iowa. In H. E. Wright, Jr., and D. G. Frey (eds.) Quaternary of the United States. pp. 29–42. Princeton University Press.

Yaalon, D. H. 1967. Factors affecting the lithification of aeolianite and interpretation of its environmental significance in the coastal plain of Israel. Jour. Sed. Petrology 37:1189–1199.

Yaalon, D. H. (ed.). 1971a. Paleopedology: Origin, nature and dating of paleosols. University of Israel Press, Jerusalem, 350 p.

Yaalon, Dan H., 1971b. Soil-forming processes in time and space. In D. H. Yaalon (ed.) Paleopedology: Origin, nature and dating of paleosols. pp. 29–40. University of Israel Press, Jerusalem.

Yaalon, D. H. 1975. Conceptual models in pedogenesis: Can soil-forming functions be solved? Geoderma 14:189–205.

Yaalon, D. H., and E. Ganor. 1973. The influence of dust on soils in the Quaternary. Soil Science 116:146–155.

Section III
Soil Properties and Processes

About 200 years ago the footsteps of the analytical scientists began to show in the soil. After 1850 these scientists took over from the practitioners to become the main source of knowledge about soils. They introduced divisions of soils study based on the experimental methods they used. These divisions of soil chemistry, soil physics, soil biology, soil ecology, etc. have largely held in the ensuing 150 years in spite of urgent recent calls for integrated research and education in the solution of current earth science problems. The divisions form the basis for specialization in university post-graduate study in soil science, for organization of the literature, and in defining the content of textbooks. The study of soils as natural bodies, soil description, soil genesis and soil classification, became a separate division that drew its methods from several sciences and from a tradition of close observation of the natural world. These studies were examined in Section II.

In this section the development of soils knowledge, or soil science, is examined in four research areas. Binkley discusses the interaction of ecology and soil science, how they have influenced each other. This is a relatively recent activity for soil scientists. Ecologists looking at ecosystem functions realized the vital part that soils play in controlling water distribution, recycling of nutrients, storage and release, etc. Soil scientists in the last four decades have turned their attention to these ideas, in both undisturbed and in agro-ecosystems. The interaction has been most prominent in examining soils under forests and also in application of ideas from ecology to explain observations in soil formation and to aid in soil taxonomy. Binkley argues that the full use of ideas from one discipline in the other has been slow and spotty. Ecology has broadened interests from plant ecology and soil science has broadened its soil ecosystem ideas to include other aspects of ecosystems. Our understanding of issues such as biodiversity or carbon cycling would be increased by closer cooperation. He makes a plea for closer integration in the examples he cites of ecosystem studies. Addiscott's discussion in Chapter 17 of nitrogen cycling in agroecosystems complements Binkley's discussion of undisturbed ecosystems.

Berthelin and colleagues tell the story of soil biology, how the different organisms living in the soil determine functions of soils in ecosystems. Understanding in microbiology developed at the end of the 20th century, stimulated by scientists such as Pasteur. Studies of the role of soil fauna, the meso and macro soil inhabitants came even later. An exception would be early studies by Darwin of earthworms and earthworm ecology. Soil biology has developed from the study of individual species to studies of soil ecology. This chapter traces the history of some of these changes, and the details the contributions made by many individuals.

Chemistry was the first laboratory science applied to soils questions. Under the name agricultural chemistry, it included chemical, physical and biological soil properties. The earliest soil studies, about 200 years ago, dealt with the sources of nutrient supply to plants. Liebig, mentioned in several chapters, made a major contribution to the development of agricultural chemistry, both organic and inorganic. His laboratory was active in identifying the inorganic elements required by plants. His strength was in spreading the knowledge of his and other people's work. His book on plant nutrition published around 1850 and widely distributed had a large influence in many countries, including the USA where "worn out" soils had become a concern. The strength of his arguments with his peers, and the force of his personality also stimulated studies that answered many questions in crop production. The chapter by Addiscott in Section IV has more details. Sparks discusses the history of two major topics in soil chemistry, ion exchange and soil pH. Ion exchange developed from an observation by a practitioner, Thompson, to studies of surface properties of clays by 21st century methods. Soil acidity has a large influence on how soils function in ion exchange, supply of nutrients to plants, on decomposition and recycling. The story of how acidity develops in soils needed to be elucidated.

Application of methods from physics and mechanics to soils have come to be known as soil physics in soil science and as soil mechanics in engineering. While celestial mechanics found early application in describing the solar system, little application of mechanics was made to understanding soils until the last 200 years. Methods of physics were used to define static properties of soil such as volume weight, porosity or grain size distribution. From these measurements, information was inferred about soil properties more difficult to measure e.g. rate of drainage of water, aeration, root room for roots, or resistance to compaction. A major achievement in soil physics was the description of water retention and water movement in partly saturated soils where both air and water share the pore space. Hasegawa reviews the main steps that have led during the last 100 years to our understanding of water flow in unsaturated soils, where the water is under negative pressure, or suction. This work began about 100 years ago. The concept of effective stress, subtracting pore water pressure from total stress, was developed by Terzaghi for use in soil mechanics. But this did not work for unsaturated soils (partly saturated) where the negative pressure of water could not simply be added to the total stress e.g. in compaction. The description of these conditions has been worked out in the last 20 years.

A second major concern of soil physics studies has been to bring the surface soil to a physical condition that provided the best environment for seedling germination and root growth. This is the elusive soil tilth. Practitioners have had to deal with this since cultivation began. Stirring the soil and making finer soil aggregates or crumbs was achieved by ploughing and breaking up clods. Tilth results from the combination of many physical factors, and there is still no adequate measurement for evaluating the effect of different tillage operations. The negative effects of excessive stirring in decreasing biological activity have recently been recognized, leading currently to less tillage. Warkentin tells this story of soil structure as tilth from early emphasis on the plough to the present concept of structure as habitat for soil organisms.

10
Soils In Ecology and Ecology In Soils

Dan Binkley[1]

"Many ecologists glibly designate soil as the abiotic environment of plants, a phrase that gives me the creeps." Hans Jenny, 1984.

Robert P. McIntosh's 1985 book on the history of ecology did not include the word "soils" in the index; soils are almost absent from his view of the development of ecology. McIntosh did mention the soil scientist Hans Jenny and his factors of soil formation, but only in the context of Jenny using the ecosystem concept without naming it, rather than in the context of Jenny's ideas advancing ecosystem ecology. Frank B. Golley's 1993 history of ecosystem ecology contains just three references to soil: soils are mentioned on one page as being parts of terrestrial ecosystems; another page notes that two ideas of the Russian soil scientist Dokuchaev (from the 19th century) were relevant to landscape geochemistry; and the last reference notes that a French project in Africa happened to include soil animals as a very small part of a very large research project. Leslie A. Real and James H. Brown (1991) compiled 40 classic papers in ecology, and only one paper dealt with soils as a key component. Two other papers in their collection dealt with soils as repositories of pollen grains (which revealed past patterns of vegetation distribution), and the final classic paper examined effects of forest cutting and herbicide application on streamwater, without any explicit examination of the soil that was actual the protagonist of the story. Has the interaction of soil science and ecology been as sparse as these histories of ecology suggest?

Terrestrial ecosystems are intimate associations of plants that fix carbon and supply energy to animals and micro-organisms that in turn recycle nutrients from plant detritus into forms that plants can use for fixing carbon. These biological and chemical interactions are joined by physical processes; if plants could obtain water only during periods when rain was actually falling, their production would be greatly reduced compared to situations were the water-holding ability of soil sustained supplies of water over periods of days and weeks after rainfall events.

---

[1] Department of Forest, Rangeland and Watershed Stewardship, Colorado State University, Ft. Collins, CO 80523 USA.

The science of ecology deals with living components, chemical and physical features of the environment, and how these interact to shape communities and ecosystems. Almost all features of terrestrial ecology are shaped directly or indirectly by processes and organisms that are found in the soil. The response of grasses to droughts depends in part on the storage of water in the soil. The ability of an antelope to outrun a wolf depends on the overall health of the antelope, which depends in large part on the quantity and quality of food consumed by the antelope, which in turn is a function of the plant community's interactions with soil.

Dependent interactions between soils, plants, and animals were recognized and appreciated by scientists before we had names for the scientific disciplines of ecology and soils. The subsequent development of these sciences has been a story of occasional interaction, frequent segregation, and profitable integration. This chapter explores some parts of the history of interaction, and non-interaction, between ecology and soil science.

Integration Before the Disciplines Emerged: Darwin and Pasteur

Charles Darwin read a paper before the Geologic Society of London in 1837, titled "On the Formation of Mould" (Darwin 1883). A friend had pointed out that a layer of rock fragments placed on the surface of a meadow had become buried after a few years, and surmised that the activity of earthworms had deposited fine materials on top of the rocks:

> I was thus led to conclude that all the vegetable mould over the whole country has passed many times through, and will again pass many times through, the intestinal canals of worms. Hence the term 'animal mould' would be in some respects more appropriate than that commonly used of "vegetable mould."

Darwin used the term "mould" to refer to the layer of material, dark in color, that rests on various types of subsoils. Darwin's last book (1883) provides fascinating reading, and a short passage illustrates the interweaving of soils, animals and plants in his ecological view:

> Beneath large trees few castings can be found during certain seasons of the year, and this is apparently due to the moisture having been sucked out of the ground by the innumerable roots of the trees; for such places may be seen covered with

castings after the heavy autumnal rains. Although most coppices and woods support many worms, yet in a forest of tall and ancient beech-trees in Knole Park, where the ground beneath was bare of all vegetation, not a single casting could be found over wide spaces, even during the autumn. Nevertheless, castings were abundant on some grass-covered glades and indentations which penetrated this forest.

Louis Pasteur's integrative thinking about micro-organisms and human health also led to an interest in earthworms, and the key role worms played in the transmission of anthrax. In 1888, Pasteur's laboratory work verified that the bacillus identified by Robert Koch was responsible for anthrax in sheep (Debré 1994). But how were anthrax spores transmitted to sheep? Farmers told Pasteur that sheep would develop anthrax after grazing in some pastures, but not in others. Pasteur toured some of the pastures, and noticed earthworm casts were prominent in the pastures where sheep developed anthrax; the worms were most active right above the carcasses of the anthrax-killed sheep (or cattle) the farmers had buried. Pasteur thought the worms might be bringing spores up from carcasses, and the sheep contracted anthrax by eating the spore-tainted grasses. However, sheep did not develop anthrax when fed spore-laden feed; Pasteur found that the spores were effective only when they entered the sheep's bloodstream through wounds in the mouth (which might happen when eating thistles). This classic story of a disease has all the components of an ecological story (animals, plants, microbes, and humans), with a key role played by soils and soil animals.

Early Themes in the Development of Ecology and Soil Science

The early development of ecology and soil science required systematic evaluation of patterns across landscapes or regions, and explanations for the causes of those patterns. Taxonomists dealing with plants and animal species described species and higher levels of integration by relying (unknowingly) on the shared patterns of ancestry that provided common structural features to use in drawing up phylogenic classification. What patterns could be used to describe and classify ecological communities and soils?

Ecology needed to develop and define units of study that went beyond individual organisms, leading to ideas of plant assemblages, communities, and even superorganism climax groups. Soil science needed to go beyond the topics of geology and plant nutrition, where soils were viewed as either surficial noise or passive vessels, to see soils as interacting natural bodies with complex structure and processes. Both sciences settled on a central role of climate in shaping plant assemblages and soils, and at times both strove to develop idealized pictures of how Nature ought to be, rather stress observations and validation/falsification of theories.

Alexander von Humboldt played a key role in spurring scientists to consider patterns, or associations, in the growth forms of plants in relation to latitude and elevation (Humboldt and Bonpland 1807, McIntonsh 1985). Within a few decades, some botanists (such as the Austrian Anton Kerner von Marilaun) imbued assemblages of plants with the integrity of species:

> The horizontal and vertical assorting of large plant communities is by no means accidental in spite of its apparent lack of order . . . In every zone the plants are gathered into definite groups which appear either as developing or as finished communities, but never transgress the orderly and correct composition of their kind. Science has given to such groups the name Plant Formations. (Kerner von Marilaun, quoted in Conard 1951, McIntosh 1985)

The enthusiasm for seeing plant assemblages as integral units progressed to the point in Europe of providing Latinized names for plant communities, with "-etum" as a suffix on the generic name of the dominant plant spieces (such as Picetum for a community dominated by Norway spruce, *Picea abies*).

In the United States, the enthusiasm for seeing plant assemblages as units went beyond the analogy of communities and species to embrace the idea of communities as super organisms. Frederic E. Clements was an ecologist in the early 20th century who promoted the idea of succession from "pioneer" to "climax" communities as being the same type of process as the maturation of juvenile to mature organism:

> The unit of vegetation, the climax formation, is an organic entity. As an entity, the formation arises, grows, matures, and dies . . . Succession is the process of reproduction of a formation, and this reproductive process can no more fail to terminate in the adult form than it can in the case of the individual. (Clements 1916)

The climax community concept was a dynamic view of vegetation, where regular changes in species composition over time represented a process of maturation (called succession) that would lead to the adult stage (called climax) that would 1) persist indefinitely unless disturbed, and 2) be in equilibrium with the regional climate (McIntosh 1985). Clements felt the description and classification of vegetation units required a new lexicon, including new words such as lamiation, station, socies, locies, consociule, eoclimax and panclimax.

Clements thought that the "unity" of climax ecosystems was a reflection of the influence of climate, and that these climatic influences might be so subtle that only plants as bioindicators could be sensitive enough to measure the climatic influence ("neither physical nor human measures of a climate are adequately satisfactory" Clements 1936). This assertion seems to remove Clements' ideas about climatic control of vegetation development from the realm of scientific testability and into the domain of tautology: if only plants can discern how climate drives plant communities, then the only evidence of climate influences on plants can be the community patterns themselves.

Clements' ideas (and proliferation of terminology) dominated much of the ecological thinking of his colleagues, even though dissenting opinions were common. For example, Henry C. Cowles (quoted in Cooper 1935) advocated a dynamic view of plant communities, but did not embrace climaxes: "The condition of equilibrium is never reached . . . we have a variable approaching a variable rather than a constant." Clements defined areas with unusual conditions of water, temperature, light, and soil as abnormal; vegetation that did not progress through the expected series of species replacements to reach a final, climatically balanced climax community were defined as proclimaxes, serclimaxes, disclimaxes, polyclimaxes, and even postclimaxes. Few if any ecologists would now describe themselves as "Clementsian" (McIntosh 1985), yet the allure of climax ecology remains prevalent in the public's view of ecology, and even in many introductory ecology texts.

Some of the themes that shaped the early development of ecology as a science were also evident in soil science. Humans had been relying on the fertility of soils to feed societies for millennia, but the systematic study of soils and their properties developed only in the 18th and 19th centuries, facilitated by the fundamental gains in chemistry and microbiology. The scientific investigation of soils blossomed in the late 19th century, in the United States with insights of Eugene W. Hilgard (including the soil as a natural body worthy of investigation, and the relationships between climate and soils), and in Russia under the charismatic leadership of Vasily V. Dokuchaev. Earlier work with soils tended to focus on geology, or on plant nutrition, and features of soils were examined individually, such as physical properties or the supply of nutrients. Viewing soils as holistic units to be studied separately from geology and botany, led to great excitement:

> The still young discipline of these relations is of an exceptional inspiring scientific interest and meaning. Each year it makes greater and greater strides and conquests; gains daily more and more of active and energetic followers, eager to devote themselves to its study with the passionate love and enthusiasm of adepts (Dokuchaev 1898, quoted by Jenny 1961a).

One of the key ideas of Dokuchaev and colleagues (especially N.M. Sibirtsev) was the idea of the "zonal" soil. Soils developed under the influence primarily of climate; with enough time, a soil would come to reflect the dominate features of the climate (and the vegetation that also responded primarily to climate). For example, the vast grasslands of Russia grew on Chernozem soils, with organic-rich topsoil and subsoil with high concentrations of calcium carbonate. The humus-rich topsoil resulted from the effects of climate as mediated by the lifeform and growth of grasses, and the carbonate reflected the moderate level of precipitation that transported salts to depth, but not completely out of the soil profile. Many soils did not match the zonal concept, and these were considered deviants that were called azonal (or intrazonal); azonal soils resulted from factors such as insufficient time for climate to shape the soil, or from abnormal topography (such as slopes). The idea of a zonal soil became essentially trans-scientific; observations that did not support the ideal of the correct zonal characteristics of a given climate and vegetation simply indicated that the soil was azonal. A map of soils of the United States in the 1920s indicated that about half the soils in the country were abnormal! The development of soil science through the 20th century included vacillation between the allure of the zonal concept (and many of its assumed features of soil development processes), and a more empirical approach of simply describing the observable features of a soil with little or no assumption about how the characteristics came to be. The zonal concept has given way in the U.S. to a largely empirical approach to soil classification, but a quick search of the world wide web reveals that many (most?) introductory soils courses still feel that the zonal concept is important.

These grand climax and zonal ideas of ecology and soil science represented what Darwin might have referred to as "argued from inner consciousness and not from observation." Nonetheless, the zonal and climax ideas provided an organized world-view that spurred tremendous research into the details of how soils and ecological systems work. The idea of zonal soils included major questions of how the influence of climate would be expressed in fully developed soils, and how deviations from normal conditions would appear. The idea that vegetation communities change over time, in predictable ways, fueled many studies of vegetation change through time under the influence of varying climate and soils.

Soils and Ecosystem Ecology

Terrestrial ecosystem ecology and soil science overlap in dealing with terrestrial landscapes, with organisms and the abiotic influences of the environment. The units of study of both disciplines are somewhat vague; boundaries are often fuzzy and changes over time make it difficult to identify when one is dealing with the same system or a new system. Both disciplines use the tools of chemistry, physics, plant physiology, microbiology and information science to advance understanding.

Given these common subjects, interests, and methods, one might expect that the two disciplines would have integrated smoothly over the past century. The rest of this chapter explores that expectation, using three case studies that illustrate key conceptual syntheses.

Jenny's State Factor Equation

Many soil scientists were fascinated by the patterns of soils across landscapes, and how biotic and abiotic processes led to the diversity of soils over time. An impressive list of 19th century scientists from Russia (Dokuchaev, Sibirtsev, K.D. Glinka), and the US (Hilgard) were trying to discern these patterns, and infer the processes responsible. The exposure of the broader audience of soil students to the ideas of these researchers was spotty in the U.S.

The first two editions of the classic textbook from Cornell University, *Soils: Their Properties and Management* (Lyon et al. 1909, 1915) dealt with processes that affected soil development, such as water and frost, but did not fully develop ideas about the overall effect of climate (beyond the geochemical effects). Arnold (1980) felt these books viewed soils as "mainly . . . a geologic product, indicating that soil comes from rocks and returns to rock." Lyon et al. (1915) cited the work of Hilgard on 17 of the 740 pages, but the work of Dokuchaev and colleagues was omitted.

The first three editions of the subsequent Cornell text, *The Nature and Properties of Soils* (Lyon and Buckman 1922, 1929, 1938) developed the themes of soil formation a bit more fully, concentrating on more details of processes (such as illuviation, the movement of compounds or particles from the upper soil to the lower soil), and on classification of soils into types. Hilgard was now mentioned on only one page of the third edition, and the Russians not at all. By the fourth edition (Lyon and Buckman 1947), a coherent view of soil formation was hinted at (but not developed), in a footnote on page 258: "For a thought-provoking treatise . . . see Jenny 1941." Interestingly, in the 12th edition of this series, the Russian pioneers in soil science are referred to as "brilliant" (on one page); Hilgard's contributions are not mentioned, and Jenny's books were still highlighted only in a footnote (p. 37; Brady and Weil 1999). This recent edition does not list "ecosystem" in its glossary, and the uses of "ecosystem" in the text are usually narrowed to the "soil ecosystem" (rather than how soils fit into a more complete ecosystem).

In contrast to the Cornell series of textbooks, Wilbert W. Weir (1936; a forest ecologist with the US Department of Agriculture, with experience at the Universities of Wisconsin, and California at Berkeley) wrote a textbook that highlighted the factors of soil formation. The factors described (pp. 101–108) included climate (rainfall and temperature), biota (grasses, animals and microbes), "edaphic" (defined as parent material characteristics), and topographic. Weir explicitly considered the "conjoint" action of these soil forming agents, and then devoted 6 pages to general processes of soil formation including podzolization and alkalization. Time was not explicitly included in the discussion of soil forming factors (and time was not listed in the index), though the discussions of soil formation clearly included ideas about changes over time.

Were ecologists integrating insights from soils into their descriptions of ecosystem changes over time? In the first few decades of the 20th century, American plant ecologists seem to have paid scant attention to soils or to ideas about soils. Clements largely overlooked soils, with the exception of noting that vegetation could fail to achieve climax because of unsuitable edaphic factors (Clements 1936). Henry C. Cowles' (1899) description of vegetation succession on sand dunes refers to "soil" simply as quartz sand, and not as a developing component of the landscape. Cooper's research on plant succession following recession of glaciers in Alaska also omitted soils, although one paper (Cooper 1923) described in detail a buried soil horizon that predated the most recent glacial advance. By the late 1920s, Cowles began thinking about soils as driving major patterns in vegetation. Citing soil concepts from Curtis F. Marbut and K.D. Glinka, Cowles (1928) proposed that the boundary between prairies and forests might have resulted from positive feedback effects where one type of vegetation altered soils in ways that fostered its own continuance. Shifts of the prairie/forest boundary would relate to climate changes that favored grasses or trees.

The second issue of *Ecology* in 1930 included an article on *Potent Factors of Soil Formation* by Charles F. Shaw, a professor of soil technology at the University of California at Berkeley. Shaw summarized these factors in the form of an equation:

$$S = M(C + V)^T + D$$

were S = soil, M = parent material, C = climate, V = vegetation, T = time and D = deposition or erosion. He thought parent material played a stronger role in soil formation than some other scientists at the time. Shaw called this equation an "expression" of the prime factors that control the trend of soil formation and development, but it is not clear that he intended it to be an equation that could be parameterized and solved.

Shaw was not aware of a similar equation advocated by Dokuchaev in 1898 (Dokuchaev 1951):

$$S = f(cl, o, p)t^o$$

where S = soil, cl = climate, o = organisms, p = geologic substrate, and t^o is a measure of relative age (this version of the equation was related by Jenny 1961b).

The science of soil formation took a major leap forward in 1941 with the publication of Hans Jenny's *Factors of Soil Formation: A system of quantitative pedology*. The book was a major advance not because it advanced new concepts about soil formation, but because it revised Dokuchaev's equation and suggested that the factors driving soil formation could actually be quantified and explored. Jenny summarized the factors that account for differences among soils, including climate, organisms, topography, parent material and time:

$$S = f(cl, o, r, p, t, \ldots)$$

where S = soil, cl = climate, o = organisms, r = topography, p = parent material, t = time, and "$\ldots$" is a place holder for other factors that might later be determined to be important. Jenny referred to these parameters as independent variables, and as state factors. At present, we might emphasize that organisms are not independent of climate (organisms influence soils, and are influenced by soils), and that time may be thought of more usefully as a dimension over which the other factors operate rather than as a factor or independent variable (Billings 1952, Lewis 1969).

Two major advances with Jenny's book were 1) greatly expanding interest and thought into soil formation by both ecologists and soil scientists, and 2) his emphasis on the quantifiability of the contribution of each factor. He suggested that the contributions of each factor could in fact be quantified by finding soils where all the factors were similar except one, such as glacial moraines in the same vicinity that differ in age, or old soils that developed on adjacent substrates of different parent material. Jenny freely cited a range of ideas from his predecessors in soil science, and from major figures in plant ecology such as Henry Gleason, Clements, Christen Raunkiaer, and Josias Braun-Blanquet. His book included dozens of graphs that synthesized the variation of soils in relation to factors such as elevation, rainfall, topography, and vegetation.

Jenny's book has probably had more influence on ecologists than any other soils book; if an ecology book cites only one soils book, it is usually one of Jenny's. He contributed to the thinking of such ecologists as W.D. Billings (1952) and Eugene Odum (1953); these (and other) ecologists in turn shaped the world views of generations of ecologists. Jenny also influenced the way ecologists considered long-term development of ecosystems. Before his 1941 book, the succession studies (following deglaciation) in Alaska looked just at the vegetation. After his book, these sorts of studies included soil development as a key component (Crocker and Dickson 1957, Ugolini 1968, Chapin et al. 1994, Bormann and Sidle 1990). Jenny's state factor approach has also been used to try and understand the drivers of patterns in ecosystems across gradients in parent material, topography, and climate (cf. Van Cleve et al.

1991, Vitousek et al. 1983, 1994). Despite this widespread influence on ecology and soil science in North America, Jenny's books are not routinely cited worldwide; for example, Jenny is not cited in Australian soil science texts (cf. Uren and Leeper 1993, Paton et al. 1995).

Vernadsky's Biosphere

Ideas about plantetary-scale modification of the Earth's surface by Life go back at least to James Hutton, who wrote in 1789 "I consider the Earth to be a super-organism and that its proper study should be by physiology" (cited by Lovelock 1988). In 1844, Jean-Baptiste-André Dumas and Jean Baptiste Joseph Dieudonné Boussingault (1841) described the global-scale effect of life on the chemistry of the earth and its atmosphere:

> "By the agency of light, carbonic acid yields up its carbon, water its hydro-gen, nitrate of ammonia its nitrogen. These elements combine, organic mat-ters are formed, and the earth is clothed with verdure."

Thirty years later Eduard Suess coined the term "biosphere" to denote the zone at the Earth's surface where biotic processes connect the lithosphere and the atmos-phere (Smil 2002). Fifty years after that, Vladimir I. Vernadsky published his detailed description of the biosphere as a global system where biotic processes shape the chemistry of the Earth's surface (Vernadsky 1998, cited in Smil 2002):

> Activated by radiation, the matter of the biosphere collects and redistributes solar energy, and converts it ultimately into free energy capable of doing work on Earth … This biosphere plays an extraordinary planetary role … The gases of the biosphere are generatively linked with living matter which, in turn, determines the essential composition of the atmosphere.

Vernadsky continued the development of ideas from Dumas, Boussingault, and Seuss to view the biosphere a driver of the development of the Earth's surface regions rather than simply a zone occupying the surface of the planet. He described major biogeochemical functions of the biosphere, particularly the formation of gases in the atmosphere (especially oxygen), the variety of oxidative and reductive reactions with compounds of sulfur, hydrogen, oxygen, and carbon, and the over-all concentration of many elements to levels far above the background in the abi-otic environment. Vernadsky's work was part of the mainstream scientific establishment of Russia and parts of Europe (Smil 2002), but it received little if any notice in North America before the ecologist and limnologist G. Evelyn

Hutchinson introduced Vernadsky's work (1945; with translations by Vernadsky's son George, a history professor in the U.S.).

What impact did Vernadsky's ideas have on the development of ecology and soil science in the second half of the 20th century? The impact on science in general was variable; Lovelock (1988) pointed out that the monumental *Earth's Earliest Biosphere* (Schopf 1983) did not mention either Hutton or Vernadsky. Odum's (1953) ecology text referred to Vernadsky's work; McIntosh's (1985) history of ecology mentioned him on one page; and Golley's history of ecosystem ecology devoted almost one page to Vernadsky. The largest impact on the ecological community probably resulted from a special issue of *Scientific American* edited by Hutchinson in 1970; this issue synthesized work that had been developing over decades into a coherent, quantitative picture of planetary-scale biogeochemistry. This synthesis spurred research, education, and enthusiasm for global-scale issues much like Jenny's 1941 synthesis book on soil formation spurred quantitative studies on soil development. One of the most intriguing extensions of Vernadsky's biosphere was the Gaia hypothesis of James Lovelock and Lynn Margulis (1974), where the overall development of the atmosphere was characterized as not only shaped by life on earth, but shaped *for* life on earth. Margulis saw Vernadsky's contributions as fundamental to the development of the Gaia Hypothesis, but Lovelock thought Vernadsky did not seem to have a "feeling for . . . the tight-coupled feedback between life and its environment" (Lovelock 2000).

The impact of Vernadsky's biosphere on soil science appears to be minimal or non-existent. Soil textbooks and reference volumes on soil formation do not mention Vernadsky, the biosphere, or global biogeochemistry. Even the integrative views of Hans Jenny did not appear to have any connection to Vernadsky's writings. One of the most integrative, global discussions of soils was a series of papers in Geoderma in 1993 that examined soils in light of the Gaia hypothesis (van Breemen 1993), but none of these papers mentioned Vernadsky.

Likens and Bormann's Small Watershed

In the 21st century in North America, the term "ecosystem" is firmly in use to describe units of landscapes, with a focus on the biotic components, the abiotic features, and their interactions. At the beginning of the 20th century, ecologists did not have a clearly agreed upon term for referring to these units. The term "ecosystem" was proposed by Tansley in 1935 to integrate the biotic and abiotic features of landscapes; he expressly included soils as part of the ecosystem. This contrasts sharply with the earlier view of some plant ecologists, who focused on vegetation and considered soil more as an environmental feature rather than a component of an integrated system. This distinction between vegetation and soils was less pronounced in Russia, where several of the classic scientists working in forestry and plant ecology (such as G.F. Morozov and Vladimir N. Sukachev) were influenced by the strong tradition of soil science (Remezov and Pogrebnyak 1969). In Russia, the term biogeocoenose (Sukachev and Dylis 1968) was used for essentially the same purpose as ecosystem in western Europe and the United States.

The development of ecosystem ecology in the United States has threads leading from many sources, including Raymond Lindeman's paper on energy flow in a lake (Lindeman 1942), a classic textbook by Eugene Odum (1953), the development of analog energy diagrams for ecosystems (H.T. Odum 1960), a series of nutrient cycling papers by J.D. Ovington (1962), and the combination of computers and large, coordinated research in the International Biological Programme of the 1960s (Golley 1993). Soils began to be a routine part of many ecosystem studies, though underrepresented in others; soils were an explicit focus of some of the IBP biome studies in the U.S. (such as the grassland and Arctic biomes), but less so in some others (such as the eastern deciduous forest biome).

A key part of the development of ecosystem ecology was the development of biogeochemical studies at the scale of forest stands (such as Ovington 1962, Cole et al. 1968), and then small watersheds (Likens et al. 1970). The small-watershed approach to studying ecosystem biogeochemistry began with the Hubbard Brook Ecosystem Study in New Hampshire. The view of soils in the early phase of the Hubbard Brook studies was similar to the views that were widespread in the mid-19th century: soils were simply the location where rocks weathered and organic matter decomposed. The nascent ideas about small watershed biogeochemistry were captured in a letter from F. Herbert Bormann (a forest ecology professor at Dartmouth College) to a colleague in the U.S. Forest Service:

> The other day while discussing the problem of mineral cycling through ecosystems, the thought came to me that your installation at Hubbard Brook represents a veritable research gold mine in regard to fundamental studies on mineral cycling.

> One of your small watersheds with a weir at the outlet represents a perfect area for controlled research. If one were able to select one or several minerals, such as K^+, it would be possible, by taking weekly water samples and analyzing them, to determine quantitatively the amount of K^+ leaving the system ... Some minerals may be added by rain or snowfall, therefore both rain and snow would have to be analyzed for the mineral(s) in question.

> By subtracting the total amount added from the total amount lost, it would be possible to estimate the steady-state losses from the system ... (F.H. Bormann, 1960, cited in Golley 1993).

Bormann independently discovered an idea developed at the Rothamsted Experimental Station in Britain almost a century before. John Benet Lawes and John Henry Gilbert realized in 1866 that each of their long-term plots was situated over a separate tile drain, and that they could analyze drainage waters to determine retention of applied fertilizers, the importance of soil cation exchange capacity, and other processes (described by Johnstone 1994). Like Rothamsted, the Hubbard Brook Ecosystem Study developed into a large, integrated project that shaped the world view of many ecologists. The study was developed and managed by a team, led by Bormann and a limnologist, Gene Likens, and including other ecologists and a geochemist. A devegetation (cutting and repeated herbicide application) experiment demonstrated the importance of vegetation on the ecosystem's ability to retain N (Likens et al. 1970), and monitoring of the chemistry of precipitation led to the first realization of acid precipitation in North America (Likens 1976).

Early work in the Hubbard Brook watershed focused on quantifying the cycle of nitrogen, including deposition, accumulation in vegetation, and losses. Bormann et al. (1977) provided a thorough budget on the N cycle, including inputs, outputs, and internal fluxes. The first major point of their paper was a full accounting of the cycle. When scientists report only parts of a full budget, their overall accounting is not constrained by the need to balance all sources and sinks for an element. A full accounting provides a quality assurance check on the believability of the overall budget.

Another key feature of the Bormann et al. (1977) budget was an attempt to estimate the rate of change in some pools, such as the organic horizon at the top of the soil, using the chronosequence approach first advocated by Hans Jenny. Covington (1981) had sampled the O horizon of 17 forests between the age of 1 year (just after clearcutting of the prior forest) and 200 years, carefully choosing the sites to be as similar as possible except for age. The average trend across the 17-forest chronosequence indicated that the O horizon of the 55-year-old Hubbard Brook forest should be increasing by about 500 kg of organic matter annually, which equaled about 15% to 20% of the annual input from litterfall (Covington and Aber 1980, Yanai et al. 2003).

How did their budget add up? Bormann et al. (1977) estimated that deposition from the atmosphere added 6.5 kg N/ha annually to the forest, and that streamwater losses removed 4.0 kg N/ha annually. The vegetation accumulated 9.0 kg N/ha annually, and they estimated that the O horizon of the soil was increasing its pool of N by 7.7 kg N/ha annually. The complete mass balance of the budget indicated that 14.2 kg N/ha annually must come from a "missing input" to account for the N accumulation in the biomass and O horizon.

An earlier estimate of complete N budget for a deciduous forest in Tennessee (Henderson and Harris 1975) had balanced the budget by assuming the net flow of N into vegetation was balanced by a net removal from the soil (O horizon +

mineral horizons; this was also the first N budget to consider the requirement for annual root production). In contrast, Bormann et al. (1977) assumed that the N content of the mineral horizons would be in a "steady state," acting as neither a net source or sink for N. This assumption led to the conclusion that the source of the missing input of 14.2 kg N/ha annually was likely non-symbiotic fixation of nitrogen by microorganisms, perhaps in decaying wood. The estimates they had available for the likely rates of N fixation in decaying wood were about an order of magnitude too low to account for this missing source of N, and one author noted that the budget would balance (without any need for N fixation) if just 0.4% of the mineral soil N pool (3600 kg N/ha in the 0–45 cm depth) was transferred to the vegetation and O horizon annually (J. Melillo, personal communication 1978).

Nonetheless, Bormann thought that an unexplained source of N was more likely than a net transfer of N from the mineral soils, and he set out to construct a more precise "watershed" budget by creating mesocosm ("sandbox") pine forests 7.5 × 7.5 m × 1.5 m volumes of sand within a rubber-lined hole in the ground. Bernard Bormann and colleagues (1993) then estimated the net change in the N contained in the mesocosms, and estimated that the actual rate of accumulation of unexplained N was about 3–4 times *greater* than the missing N from the large watershed's budgets. Though the precision of the estimates has been debated (Bormann et al. 2002, Binkley 2002), the value of experimenting with mass balance of ecosystems (both soil and vegetation) is great.

In later phases of the Hubbard Brook Ecosystem Study, including its years as one of the National Science Foundation's Long-term Ecological Research sites, the soils have received more explicit inclusion in the ecosystem concept. The mineral soils have been sampled repeatedly over time in hopes of detecting net changes in nutrient pools (cf. Johnson et al. 1991). Despite these gains at Hubbard Brook, soils often remain an under-investigated black box of ecosystem ecology; for example, McIntosh's (1985) historical review provided moderate depth in discussing ideas about changes in nutrient cycling as part of ecosystem succession, but he didn't mention soils.

Views on Ecology and Soil Science in the 21st Century

In current soils textbooks, soils are referred to as soil ecosystems, but not always connected very well with the complete ecosystem (cf. Brady and Weil 1999) in a way that would please Hans Jenny or Eugene Odum. On the other hand, chapters on soils are now found in books that focus on plant ecology (cf. Barbour et al. 1998), and a volume on the terrestrial vegetation of North America mentioned soils on an astounding 101 of its 688 pages (Barbour and Billings 2000).

In the United States, the teaching of soil science has been concentrated in departments of agriculture and natural resources, largely at universities with a land-grant tradition. A substantial portion of the teaching of ecology is found in biology departments, often within colleges that have little if any depth in soil science. Across the country, students majoring in natural resources or agriculture are required to take coursework in biology and often ecology. Students majoring in biology, however, are not always expected to take classes in natural resources or agriculture. These differences have been evident for decades, and will probably remain important in coming decades, preventing our undergraduate liberal education system from reaching its potential.

Several unifying themes have the potential to foster closer interaction between ecologists and soil scientists. Global-scale issues of biogeochemistry include the effects of increasing atmospheric CO_2 on both climate and organisms, and these are central issues for globally focused ecologists. Soils are of course a key player in the net balance in the C cycle, and soils are dramatically responsive to CO_2-driven changes in climate and organisms.

Biodiversity is another zone of major overlap; ecologists are very interested in any connections between biodiversity and ecosystem function, and the vast majority of ecosystem diversity is found in soils. A recent overview of soil biodiversity (Hooper et al. 2000) emphasized the importance of soil animals, but even the soil animal community appears simple indeed in light of the almost unimaginable diversity in the microbial community indicated by RNA and DNA methods (see Wardle 2002 for the best overview). Microbial biodiversity may include 1000 to 10000 species per g of soil (Torsvik et al. 1990, Borneman et al. 1996). The biodiversity outside the confines of the soil is far smaller than the uncertainty estimates of the biodiversity within the soil.

Perhaps the most important area of integration that can be made between ecology and soil science is improving our understanding of human modifications of ecosystems, from individual forests and farms up to the globe, over long periods of time. Richter and Markewitz (2001) noted:

> Ironically, enormous amounts of biogeochemical data have been collected that describe soil and ecosystem processes as they operate over one to a few years . . . but nearly all of these studies are limited in their ability to predict changes in soils and ecosystems over timescales of decades.

Coordinated, long-term studies are needed to understand ecology and soils in non-agronomic portions of the globe (Richter and Markewitz 2001, Wardle 2002). Over 100 million dollars have been spent on the Long-term Ecological Research program (http://lternet.edu/) of the National Science Foundation in the United States, but few if any of these sites would be able to tell if major changes developed in their soils over time. Ecological monitoring needs to incorporate precision monitoring of the soil foundation of the ecosystems.

The integration of ecology and soils will also need collaboration from the social science, economics, and the humanities (Turner et al. 1990). Scientists and students need to read broadly, and strive to reduce their ignorance of fields outside their specialties.

Acknowledgements

The information and ideas presented here arose from conversations over the years with many colleagues, especially Indy Burke, Dale Johnson, Gene Kelly, William Reiners, and Dan Richter.

References

Arnold, R.W. 1980. Concepts of soils and pedology. pp. 1–21. *In* Pedogenesis and soil taxonomy I. Concepts and interactions (L.P. Wilding, N.E. Smeck, and G.F. Hall, eds.) Elsevier, Amsterdam.

Barbour, M.G. and W.D. Billings. 2000. North American terrestrial vegetation, 2nd edition. Cambridge University Press, Cambridge.

Barbour, M.G., W.D. Pitts, J.H. Burk, F. Gilliam, and M.W. Schwartz. 1998. Terrestrial plant ecology, 3rd edition. Benjamin/Cummings, Menlo Park.

Billings, W.D. 1952. The environmental complex in relation to plant growth and distribution. Quarterly Review of Biology 27:251–264.

Binkley, D. 2002. Response to: "Lessons from the sandbox: is unexplained nitrogen real?" Ecosystems 5:734–735.

Bormann, B.T. and R.C. Sidle. 1990. Changes in productivity and distribution of nutrients in a chronosequence at Glacier Bay National Park, Alaska. Journal of Ecology, 78:561–578.

Bormann B.T., C.K. Keller, D. Wang, and F.H. Bormann. 2002. Lessons from the sandbox: is unexplained nitrogen real? Ecosystems 5:727–733.

Bormann, F.H., G.E. Likens, and J.M. Melillo. 1977. Nitrogen Budget for an Aggrading Northern Hardwood Forest Ecosystem. Science 196:981–983.

Borneman, J., P.W. Skroch, K.J. O'Sullivan, J.A. Paulu, N.G. Rumjanek, J.L. Jansen, J. Nienhuis, and E.W. Triplett. 1996. Molecular Microbial Diversity of an Agricultural Soil in Wisconsin. Applied and Environmental Microbiology 62:1935–1943.

Brady, N.C. and R.R. Weil. 1999. The nature and properties of soils, 12th edition. Prentice-Hall, Upper Saddle River.

Buol, S.W., R.J. Southard, R.J. McCracken, and F.D. Hole. 1997. Soil genesis and classification. Iowa State University Press, Ames.

Chapin, F.S., L.R. Walker, C.L. Fastie and L.C. Sharman. 1994. Mechanisms of primary succession following deglaciation at Glacier Bay, Alaska. Ecological Monographs, 64:149–175.

Clements, F.E. 1916. Plant succession: an analysis of the development of vegetation. Carnegie Institution of Washington, Washington, D.C.

Clements, F.E. 1936. Nature and structure of the climax. Journal of Ecology 24:252–284.

Cole, D.W., S.P. Gessel, and S.F. Dice. 1968. Distribution and cycling of nitrogen, phosphorus, potassium, and calcium in a second-growth Douglas-fir forest. Primary production and mineral cycling in natural ecosystems. pp. 197–213. *In* (H.E. Young, ed.) University of Maine Press, Orono, Maine.

Conard, H.S. 1951. The background of plant ecology. Iowa State University Press, Ames.

Cooper, W.S. 1923. The recent ecological history of Glacier Bay, Alaska: the inter-glacial forests of Glacier Bay. Ecology 4:93–128.

Cooper, W.S. 1935. Heny Chandler Cowles. Ecology 16:281–283.

Covington, W.W. 1981. Changes in the forest floor organic matter and nutrient content following clear cutting in northern hardwoods. Ecology 62:41–48.

Covington, W.W. and J.D. Aber. 1980. Leaf production during secondary succession in Northern Hardwoods. Ecology 61:200–204.

Cowles, H.C. 1928. Persistence of prairies. Ecology 9:380–382.

Cowles, H.C. 1899. The ecological relations of the vegetation on the sand dunes of Lake Michigan. The Botanical Gazette 27:95–117, 167–202, 281–308, 361–291.

Crocker, R.L. and B.A. Dickson. 1957. Soil Development on the Recessional Moraines of the Herbert and Mendenhall Glaciers, South-Eastern Alaska. Journal of Ecology 45:169–185.

Darwin, C.W. 1883. The formation of vegetable mould, through the action of worms with observations on their habit. John Murray, London. Available at: http://pages.britishlibrary.net/charles.darwin/texts/vegetable_mould/mould.html.

Debré, P. 1994. Louis Pasteur. Johns Hopkins, Baltimore.

Dokuchaev, V.V. 1951. Writing. Akademia Nauk, Moscow, 6:381.

Dumas, J.B.A. and M.J.B. Boussingault. 1841. Lecon sur la statique chimique des estres organizes. Philosophical Magazine 19:337–347, 456–469.

Golley, F.B. 1993. A history of the ecosystem concept in ecology: more than the sum of the parts. Yale University Press, New Haven.

Henderson, G.S. and W.F. Harris. 1975. An ecosystem approach to characterization of the nitrogen cycle in a deciduous forest watershed. *In* Forest Soils and Forest Land Management. (Bernier B and Winget CH, eds.) Laval University Press, Ste Foy, Quebec.

Hooper, D.U., D.E. Bignell, V.K. Brown, L. Brussaard, J.M. Dangerfield, D.H. Wall, D.A. Wardle, D.C. Coleman, K.E. Giller, P. Lavelle, W.H. van der Putten, P.C. De Ruiter, J. Rusek, W.L. Silver, J.M. Tiedje, and V. Wolters. 2000. Aboveground and belowground biodiversity in terrestrial ecosystems: patterns, mechanisms, and feedbacks. BioScience 50:1049–1061.

Humboldt, A. von and A. Bonpland. 1807. Essai sur la Geographie des Plantes. Librarie Lebrault Schoell, Paris.

Hutchinson, G.E. (ed.). 1970. The biosphere. Scientific American 223, issue 3.

Jenny, H. 1941. Factors of soil formation: a system of quantitative pedology. McGraw-Hill, New York.

Jenny, H. 1961a. E.W. Hilgard and the birth of modern soil science. Collana Della Revista "Agrochemica," Pisa.

Jenny, H. 1961b. Derivation of state factor equations of soils and ecosystems. Soil Science Society of America Proceedings 25:385–388.

Jenny, H. 1984. My friend, the soil. A conversation with Hans Jenny, interview by K. Stuart. Journal of Soil and Water Conservation 39:158–161.

Johnson, C.E., A.H. Johnson, and T.G. Siccama. 1991. Whole-tree clear-cutting effects on exchangeable cations and soil acidity. Soil Science Society of America Journal 55:502–508.

Johnstone, A.E. 1994. The Rothamsted classical experiments. pp. 9–37. *In* Long-term experiments in agricultural and ecological sciences (R.A. Leigh and A.E. Johnstone, eds.) CAB International, Wallingford.

Lewis, J.K. 1969. Range management viewed in the ecosystem framework. pp. 97–187. *In* The ecosystem concept in natural resource management (G.M. Van Dyne, ed.) Academic Press, New York.

Lindeman, R.L. 1942. The tropic-dynamic aspect of ecology. Ecology 23:399–418.

Likens, G.E. 1976. Acid precipitation. Chemical Engineering News 54:29–44.

Likens, G.E., F.H. Bormann, N.M. Johnson, D. Fisher, and R.S. Pierce. 1970. Effects of forest cutting and herbicide treatement on nutrient budgets in the Hubbard Brook Watershed-Ecosystem. Ecological Monographs 40:23–47.

Lovelock, J.E. 1988. The ages of Gaia: a biography of our living Earth. Norton, New York.

Lovelock, J.E. 2000. Homage to Gaia: the life of an independent scientist. Oxford University Press, Oxford.

Lovelock, J.E. and L. Margulis. 1974. Atmospheric homeostasis by and for the biosphere: the Gaia hypothesis. Tellus 22:2–9.

Lyon, T.L. and H.O. Buckman. 1922, 1929, 1938, 1947. The nature and properties of soils. MacMillan, New York.

Lyon, T.L., and E.O. Fippin, and H.O. Buckman. 1909, 1915. Soils, their properties and management. MacMillan, New York.

McIntosh, R.P. 1985. The background of ecology: concept and theory. Cambridge University Press, Cambridge.

Odum, E.P. 1953. Fundamentals of ecology. W.B. Saunders, Philadelphia.

Odum, H.T. 1960. Ecological potential and analogue circuits for the ecosystem. American Scientist 48:1–8.

Ovington, J.D. 1962. Quantitative ecology and the woodland ecosystem concept. Advances in Ecological Research 1:103–192.

Paton, T.R., G.S. Humphreys and P.B. Mitchell. 1995. Soils, A new global view. Yale University Press, New Haven.

Real, L.A. and J.H. Brown. 1991. Foundations of ecology: classic papers with commentaries. University of Chicago Press, Chicago.

Remezov, N.P. and P.S. Pogrebnyak. Forest soil science. U.S. Department of Commerce, Springfield.

Richter, D.D., Jr., and D. Markewitz. 2001. Understanding soil change: soil sustainability over millennia, centuries, and decades. Cambridge University Press, Cambridge.

Schopf, J.W. (ed.). 1983. Earth's Earliest Biosphere. Princeton University Press, Princeton.

Shaw, C.F. 1930. Potent factors in soil formation. Ecology 11:239–245.

Smil, V. 2002. The Earth's biosphere. MIT Press, Cambridge.

Stevens, P.R. and T.W. Walker. 1970. The chronosequence concept and soil formation. Quarterly Review of Biology 45:333–350.

Sukachev, V.N. and N. Dylis. 1968. Fundamentals of forest biogeocoenology. Oliver and Boyd, London.

Torsvik, V., J. Goksoyr, and F.L. Daae. 1990. High diversity in DNA of soil bacteria. Applied and Environmental Microbiology 56:782–787.

Turner, B.L. II, W.C. Clark, R.W. Kates, J.F. Richards, J.T. Mathews, and W.B. Meyer. 1990. The Earth as transformed by human action. Cambridge University Press, Cambridge.

Ugolini, F. 1968. Soil development and alder invasion in a recently deglaciated area of Glacier Bay, Alaska. pp. 115–140. In Biology of Alder (J.M. Trappe, J.F. Franklin, R.F. Tarrant, and G.M Hansen, eds.) Pacific Northwest Forest and Range Experiment Station, Portland.

Uren, N.C., and G.W. Leeper. 1993. Soil science: an introduction. Melbourne University Press, Melbourne.

van Breemen, N. 1993. Soils as biotic constructs favoring net primary productivity. Geoderma 57:183–211.

Van Cleve, K., F.S. Chapin III, C.T. Dyrness, and L.A. Viereck. 1991. Element cycling in taiga forests: state-factor control. BioScience, 41:78–88.

Vernadsky, V.I. 1945. The biosphere and the noosphere. American Scientist 33:1–12.

Vernadsky, V.I. 1998. The biosphere, translated by D.B. Langmuir. Copernicus, New York.

Vitousek, P.M., D.R. Turner, W.J. Parton, and R.L. Sanford. 1994. Litter decomposition on the Mauna Loa environmental matrix, Hawaii: patterns, mechanisms, and models. Ecology 75:418–429.

Vitousek, P.M., K. Van Cleve, N. Balakrishnan, and D. Mueller-Dombois. 1983. Soil development and nitrogen turnover in montane rainforest soils on Hawaii. Biotropic 15:268–274.

Wardle, D.A. 2002. Communities and ecosystems: linking the aboveground and belowground components. Princeton University Press, Princeton.

Weir, W.W. 1936. Soil science: its principles and practice. J.B. Lippincott, Chicago.

Yanai, R., W.S. Currie, and C.L. Goodale. 2003. Soil carbon dynamics after forest harvest: an ecosystem paradigm reconsidered. Ecosystems 6:197–212.

11
History of Soil Biology

Jacques Berthelin[1], Ulrich Babel[2], François Toutain[1]

The Dawn and Lag Phase of Soil Biology

In prehistoric, probably from the Magdalenian culture (25,000–30,000 years ago), and ancient civilizations (Sumarian, Egyptian 2,000–5,000 years ago), different animals of the soil fauna, such as beetles, were represented and appeared in different man-made things, as reported by Kevan (1985).

At this period in the history of humanity, such representations were much associated with veneration or superstition or belief. It is now known that in ancient Egyptian civilization, scarabaeids, in particular those commonly called "dung beetles," were considered as sacred (or venerable) animals, because "it pushed or brought the sun between its legs" when it rolled and then buried balls of vertebrate fecal material in which the female lays her eggs. It symbolized the day and night sun cycle, and it was also the symbol of the egg of the world or of the earth from which originated life. The scarabaeid which buries itself in soil symbolized Osiris (the past) and the new scarabaeid which emerged symbolized Horus (the future) (Cambefort 1994). The Greek, Aristotle (384–322 BCE), called earthworms "the intestine of the earth" by observing their behavior, and the Egyptian Queen Cleopatra (69–30 BCE) instituted a law to forbid the export of earthworms. So, it was then not really soil biology—it was much more soil mythology!

Soil at the dawn of humanity was also associated with the origin of man (through the clay material) and was considered as "the spring of life" in the Bible. Also prehistoric man discovered, around 8500 years ago, that soil can receive seeds to produce wheat grains for their subsistence, but it was very far from modern agronomy!

---

[1] Laboratoire des Interactions Microorganismes-Minéraux-Matiéres Organiques dans les Sols (LIMOS) UMR 7137 C.N.R.S., Université H. Poincaré-Nancy I, Faculté des Sciences B.P.239, 54506 Vandoeuvre-les-Nancy Cedex, France.
[2] c/o Institute of Soil Science, University Hohenheim D 70593 Stuttgart 70, Germany.

It was only recently, at the end of the 19th century, that a real interest was shown and that progress was made in soil biology, for both soil fauna and soil microorganisms. Before the end of the 19th century, the "story" of soil biology, in association with the story of soil science, emerged indirectly with the history of plant and vegetation and was focused mainly on plant growth. Some early thinkers, e.g., Olivier de Serres (1539–1619) had written "Le theatre d'Agriculture et mesnage des champs" (i.e., The agricultural scene and field management) which can be considered as a key intermediate between Latin (Roman) agronomy and the establishment of a modern objective way of thinking about agronomy (Henin 2001). In this first historical period, before approximately 1750, different questions arose about how plant growth? What are the origins of their growth? What substances are involved? These fundamental scientific questions were at the origin of researches on what was called the "vegetation principle" or the "vegetation natural law." According some authors such as Bacon (1627) mentioned by Boulaine (1989) or Woodward (1699), this vegetation principle was attributed to compounds present in soil water or in "particular juice" provided by soil mineral or soil organic substances. Some ideas suggested that these compounds could be specific to a group of plants or to "terrestrial elements" or "terrestrial matter" which enter plant composition. Such ideas were established on the results of simple experimental studies performed on plant growth in different conditions. The flemish Van Helmont (1577–1644) as reported by Boulaine (1989) planted a willow of 5 pound weight in a pot containing 200 pounds of soil. He collected five years later a tree of 179 pounds and the soil weight has not changed significantly. He concluded that water was responsible for the willow growth. The conclusion erred because he had not measured gas and has neglected soil composition and weight variation, but the principle of the experiment was good! In an other type of experiment, in England, Woodward (1699) cultivated mint in different types of water (rain water, water from the river Thames, water containing humus) and observed that plant growth was larger when waters contain impurities. He concluded that the plant was formed of water and different substances present in water. Again the observation was good but as previously, the explanations were incorrect by the inability to make measurements of the main parameters involved.

From approximately 1750 to 1860, which can be considered as a second historical period, soil biological science was focused on plant nutrition and plant growth, i.e., on agronomic problems, and was developed using field experiments, pot experiments, and chemical analysis of plants, air, and soils. Agronomists such as Home (1757), introduced the pot experimental method. The work during this period was reported by Boulaine (1989), Pochon and de Barjac (1958), and Denis (2001), who also emphasized that it was mainly de Saussure, Boussingault and Liebig who, later laid the foundation of biological soil science. The first concepts of plant nutrition for sources of mineral carbon and nitrogen presented by de Saussure (1804a,b) were not well accepted, for instance by Berzelius (1838), Von Thaer (1809, 1812) (reported by Pochon and de Barjac, 1958), and Davy (1814) until Boussingault (1843–1844) published the results of his field experiments in France on plant nutrition and fertilization. Similar studies were also developed independently by Lawes and Gilbert at Rothamstead (U.K.) (1843) (mentioned by Pochon and de Barjac 1958, Boulaine 1989). During this period Liebig, in Germany, developed, presented, and popularized his theories on mineral nutrition of plants (Liebig 1842, 1846) and justified the concepts of de Saussure.

Around 1855, the fundamentals of what can be considered as the first or "lag phase" of soil biology were established, essentially on principles of plant physiology and agricultural practices to maintain soil fertility using both fallow land practices and fertilizers. These first studies were published from the end of the 18th to the middle of the 19th century by chemists (see Boulaine 1989, Robin and Blondel-Megrelis 2001, Denis 2001). During this period, and mainly in Europe, experiments on plant nutrition done by agricultural chemists, introduced the first preliminary form of biological soil science, as they raised some new questions, such as plant nitrogen nutrition of leguminous plants, that further encouraged the "birth" of soil bacteriology. The questions concerning plant nutrition of nitrogen, proposed since 1775, remained controversial for an entire century until the origin of soil microbiology at the end of the 19th century.

During this same historical period and not only in urban areas (Barles 1999) questions arose about the relation between "earth or soil properties" and human health and human constitution. As an example, studies tried to obtain or define "a topographical and medical map of France where temperament, constitution, and troubles or illnesses of the inhabitants of every district and province would be considered in relation to the nature and exposure of soil" (e.g., Brieude 1782–1783, Menuret de Chambaud 1786, Barles 1999). Similar studies were done in Germany, United Kingdom, and the United States of America (e.g., Jordanova and Porter 1979, Barles 1999). However, this "ecological" approach to human health and constitution in relation to soil properties was essentially halted by the emergence of scientific medicine and urban and hygienic engineering. Nevertheless, such questions are today again of great interest and importance but with new concepts and tools to define the relation between environmental parameters and human health.

During this period of the 18th century, soil fauna were not really considered as "soil organisms" or included in soil biology. Although Linné, in his 10th edition of <u>Systema naturae</u> in 1758, had classified all earthworms as *Lumbricus terrestris*, during this period there remained a controversy over how to classify these animals. Some "agronomists" considered earthworms as harmful animals because they thought that they ate the roots of plants in gardens. Some, however, emphasized the beneficial role that earthworms played in the improvement of soil fertility. As reported by Buch (1991),

White wrote in 1789, "that without earthworms, the earth would become cold, hard and would not be enough ploughed and, as a consequence, would be sterile."

The True Beginning of Soil Biology

This early lag phase of soil biology was followed, in the second part of the 19th century, by major findings in different scientific domains and fields of biology concerning soil fauna and soil microorganisms, and by the development of different and more specific approaches using, adapting, or developing specific methods and new technologies (e.g., microscopy and micromorphology). It is also important to emphasize that all these findings mainly originated from questions arising in agronomy or forestry and from the interests in natural sciences and their development.

Discovery and Evidence of the Involvement of Microorganisms in the Functioning of the Soil-Plant System

After 1865—and Liebig's theory (1842, 1865) that ammonia was the sole unique nitrogen plant nutrient—other agricultural and plant chemists, as well as agronomists (e.g., Boussingault 1843–1844, 1860–1874) established the importance of nitrate for plant growth, and as a consequence, for global soil fertility. Such an assessment, however, raised a new problem: what was the origin of the nitrate formation in soil? The experiments of Schloesing and Muntz (1877–1879) provided an answer, which was verified by Warrington (1878–1888), and was cited and discussed by Winogradsky (1890, 1949), and Pochon and de Barjac (1958). They confirmed Pasteur's earlier hypothesis (who had suggested in 1862 to verify the possible role of a specific microbial oxidation in nitrification process) by demonstrating the occurrence of bacteria in the nitrification process. It was Winogradsky (1890), however, who provided the conclusive proof for this hypothesis when he isolated two species of nitrifying bacteria.

During the same period, the question of the nitrogen nutrition of leguminous plants was elucidated by the works of Berthelot (1885), and Helbriegel and Wilfarth (1886–1888), as reported by Pochon and de Barjac (1958), and subsequentially by the isolation of the first rhizobium, *Bacillus radicicola*, by Beijerinck (1888). In addition, other discoveries observed the association between fungi and plant roots (Pfeffer 1877), and the term mycorrhizae (of greek origin, which means fungi and root), was proposed by Frank (1885), who later distinguished ecto- from endo-mycorrhizae.

Thus, by 1890, major and fundamental progress in microbiology, and in particular in soil microbiology, had affirmed that bacteria were indispensable and essential for life and the productivity of soils. Bacteria had come to be considered the agents involved in and responsible for decomposition—the transformation of organic and nitrogeneous matter. Also, scientists had come to recognize associations between plant roots and microorganisms, even if the function of these associations were not completely or at all understood.

Discovery of the Implication of Soil Fauna in Soil Biology

The progress in knowledge concerning the role and impact in soil functions of soil fauna was slow because they were not considered as important as the role of microorganisms. It was Darwin (1881), who first emphasized the role of earthworms in soil mould formation. He wrote that, like a "gardener," earthworms prepare the soil in a remarkable way for plant growth because they mixed intimately the fine earth. Darwin emphasized also that earthworms participate significantly in the increase of the content of soil organic matter because they bury decomposed leaves from the soil surface to two to three inches depth in their tunnels. Drummond (1887) suggested that in tropical soils termites play the same or a similar role; that is, they break down and blend surface soil horizons.

Sergei Winogradsky and Martinus Beijerinck, founders of general and environmental microbiology

The Russian microbiologist, Sergei Winogradsky (1856–1953), took an early interest in soil microbiology when microbiology was concerned primarily with pathology (e.g., Louis Pasteur (1822–1895) and Robert Koch (1843–1910)). Winogradsky began his research in 1880 with Andrei Famintsyn at the University of St. Petersburg, and continued it in two Western European apprenticeships; first at Anton DeBary's botanical laboratory at the University of Strasbourg (1886–1888), and second, at the Hygeine Institute in Zurich (1888–1890). His research on sulfur-, iron-, and nitrogen bacteria led him to discover autotrophism. Between 1891–1912, he worked at the Institute of Experimental Medicine in St. Petersburg, Russia where he investigated anaerobic nitrogen fixing bacteria, cellulose transformation, bacterial morphology, and aerobic nitrogen fixation. After the October Revolution and Russian Civil War, he emigrated to France, where in 1922 he became the director of the Pasteur Institute Laboratory of Agricultural Microbiology near Paris. For thirty years, he continued to study not only the areas that had made him a founder of soil microbiology, but also developed novel methods for microbial ecology. In 1949, he republished his life's work in this new ecological perspective in the French monograph *Soil Microbiology: Problems and Methods, Fifty Years of Research*, which was translated into Russian, Polish, and Chinese. Ackert's thesis (2004) provides new light on Winogradsky work.

The Dutch microbiologist, Martinus Beijerinck (1851–1931), was professor at Delft, where he was among the first to study microorganisms associated with plants. His investigation of root nodules led him to discover symbiotic nitrogen-fixing bacteria, which he called *Bacillus radicicola* (1888). For this work, drawing on Winogradsky's elective culture method, he developed a selective culture (now called enrichment culture) technique. He made fundamental discoveries on aerobic nitrogen-fixing bacteria, sulfate-reducing bacteria, sulfur oxidizing bacteria, and *lactobacilli* ... He also showed that the infectious agent of tobacco disease was not bacterial but an "organism or something" incorporated in the plant host, thus providing one basis for the development of virology. Based on their fundamental discoveries and their novel methods, Winogradsky and Beijerinck set the foundation of general microbiology and environmental microbiology.

The 19th Century Had Initiated Major Concepts of Soil Biology

By the end of the 19th century and the beginning of the 20th, soil microbiology had not only benefited from the discoveries and advances in microbiology in general, but it also developed its own methods and made major discoveries concerning nitrification, nitrogen fixation, organic matter decomposition, and plant-microorganism associations (e.g., rhizobium, mycorrhizae). Soil microbiology was established and recognized. New concepts had been clearly presented: e.g., the enrichment culture and the use of specific selective culture media by Beijerinck (1895–1901), the chemolithotrophy by Winogradsky (1889–1890) (see box). As emphasized by Brock (1979) and his successors (e.g., Madigan *et al.*, 2000) in the book *Biology of Microorganisms*, "Beijerinck and Winogradsky can be considered not only as the founders of soil and environmental microbiology but also of the field of general microbiology."

Development and Expansion of Soil Microbiology

From the beginning to the end of the 20th century, the development of new techniques, methods, and concepts, and the adaptation of discoveries in other sciences (e.g., physics, chemistry, biochemistry, and biology) have led to the emergence and development of soil microbial ecology, soil biology, and soil ecology in many ways to answer different questions, to develop multidisciplinary, interdisciplinary, and multiscale approaches.

From Soil Microbiology to Soil Microbial Ecology

During the first part of 20th century, soil microbiology still was mainly focused on what can be called agronomic microbiology. Other fields, however, were also developed for application in animal and human health, looking at pathogenic bacteria (e.g., *Clostridium tetani and Bacillus anthracis*) present as resting cells in soil, for taxonomic studies and antibiotic production by isolating microorganisms, in the transformation of mineral elements and in geochemical cycles (weathering of minerals, leaching, extraction, accumulation and deposits of metals), and involvement in pedogenetic processes (e.g., soil formation, soil processes, soil degradation).

During this period, the main objectives of soil microbiology were to study the impact of soil microorganisms (bacteria, actinomycetes, fungi, algae, protozoa) in all soil processes. Three main concepts and approaches were developed: (1) to understand as completely as possible the biological "status" at the soil surface and in the different soil horizons using quantitative studies of the number and activity of the "total" microbiota and of the main physiological groups responsible for defined functions (e.g., nitrogen fixation and cellulose fermentation); (2) to study the variation (dynamic) of these physiological groups with changes in different parameters (e.g., seasons, soil evolution, soil sequence, agricultural practices, and soil physico-chemical properties); and (3) to compare soil fertility or soil dysfunctioning in relation to microbial status.

Such studies involved primarily isolated pure strains of microorganisms. However, as stated by Kluyver (1956), "Many of the pure cultures are worthy of the name of physiological artefacts." Such pure culture research did provide valuable information, and the results must be considered carefully (Pochon and de Barjac 1958). Studies of soil microbiology have to be based, as emphasized by Winogradsky (1949), "not on the behavior of isolated species outside their natural environment, but on reactions and activities of the microbial communities in their environments."

The book Winogradsky published in 1949 is a compendium of his life's work since 1888, providing for a new generation of microbiologists his fundamental research and concepts: on the first autotrophic bacteria (sulfur and iron bacteria), the morphology and pleomorphism of bacteria, nitrification, anaerobic and aerobic nonsymbiotic nitrogen fixation, symbiotic nitrogen fixation, the degradation of cellulose, methods in soil microbiology, and, last but not least, a chapter entitled

"The principles of microbial ecology." In this chapter he stated, "I think that I have demonstrated that the application of the classical method (i.e., works on isolated microorganisms) to the microbial ecology of natural environments, in particular the soil, would produce (or will produce) inaccurate results or at least doubtful results. It is time to distinguish physiological and chemical studies from the ecological research on the microbial dynamics in natural environments. This new method does not want to take the place of the pure culture method, but will be a development, devoted to specific studies of ecological problems, that the classical method is not able to handle and treat."

At this period, the main methods and techniques were the measurement of CO_2 evolved during the biodegradation of organic matter, the enumeration of total microbiota, the measurements of activity, and the enumeration of physiological groups using specific media and reactions (e.g., nitrification, ammonification, aerobic or anaerobic cellulolytic activity). The significance of the results was based on comparisons between countings and measurements of activities, and on the observation and comparison of kinetics.

As indicated by Paul and Clark (1989), "despite considerable effort in making census counts of soil organisms and in attempting to use counts as indices of soil fertility, the concept failed because the number of propagules capable of forming viable colonies on agar plates represent only a small percentage of the total microbial population." However, much progress in knowledge has been made by developing and comparing countings and measurements of the activity of physiological groups and of global activity (e.g., carbon and nitrogen mineralization), both *in situ* and under laboratory conditions.

At the beginning of 20th century, Hiltner (1904) coined the term rhizosphere to describe the interactions between bacteria and roots of legumes. Later, the concept of rhizosphere was widened to consider all the soil and living organisms under the influence of root systems. As a consequence, new fields of research were opened first to symbiotic associations, then to nonsymbiotic ones, and also to beneficial, harmful and/or neutral or variable effects. "Rhizosphere effects" and rhizosphere microbiology were developed in the period just before the First World War, and they have had larger growth after 1960 (see box).

These various studies marked the origin of the development of soil microbial ecology i.e., the study of the relation between soil microorganisms and all the biotic and abiotic parameters of the soil systems (e.g., Pochon and de Barjac 1958, Dommergues and Mangenot 1970, Alexander 1971) and soil biochemistry i.e., the chemical activity of soil organisms (e.g., McLaren and Peterson 1967). During the period from before the First World War to after the Second World War, different scientists contributed to the development of soil microbiology (Winogradsky, Beijerinck, Kluyver, as already mentioned). Russel (1923), Waksman (1927), and

First book and journal in soil microbiology

The first major textbook in soil microbiology, entitled *Handbuch der Landwirtschafliche Bakteriologie*, publisher Borntraeger, Berlin, was published in German by Lohnis (1910). It underwent a total of 19 editions and was translated into four languages (Paul and Clark 1989). Waksman (1927) considered this publication a monumental work. The first journal with a specific subtitle of soil microbiology was also edited in Germany: *Centralblatt für Bakteriologie und Parasitenkunde Infektionskrankheiten und Hygiene* entitled its Abteilung 2, *Allgemeine Landwirtschaftliche und Technische Bakteriologie, Gärungsphysiologie und Pflanzenpathologie*.

Waksman and Starkey (1931), emphasized the existence not only of bacteria but also of other microorganisms in the soil (e.g., fungi, algae, protozoa) and of complex soil populations and communities that were continuously interacting. The observation of such interactions was responsible, in part, for the discovery of antibiotics: actinomycetin by Gratia and Dath in 1924, thyrotricine by Dubos in 1925 who later isolated from soil *Bacillus brevis* which produces gramicidine, penicillin by Fleming in 1929, and streptomycin by Waksman and Schatz in 1944.

In the later part of the 20th century, soil microbiology and soil biochemistry expanded to many topics and approaches using different techniques and tools. Also, major new questions arose from environmental concerns (e.g., plant production, plant pathology, quality of water, air pollution, greenhouse effect and global change).

Soil Microbiology in Specific Applications

To discuss all the main fields of soil microbiological and biochemical research would require too many pages, but the major and new topics must be emphasized. Carbon and nitrogen cycles were and are always major topics. They were considered first with respect to the decomposition and transformation of soil organic matter, and major results have been obtained on the processes of lignin and cellulose transformation, humus formation, the production of CO_2 and CH_4, nitrogen mineralization (NH_4^+, NO_3^- production), nitrogen fixation, denitrification, and nitrogen immobilization. Applications concern primarily agronomy and forestry, as well as water quality (e.g., nitrate release to waters), and soil structure. Different scales of atmospheric release of CO_2, CH_4, and N_2O are being considered for the problem of greenhouse gas production. Modeling of these processes has been proposed, and some connections have appeared between soil microbiology and soil physics, although not yet enough.

The rhizosphere

The rhizosphere, defined first by Hiltner (1904) as the soil under the influence of plant roots and associated with bacteria, is, in fact, soil sites under the influence of interactions between plant roots, soil constituents, and soil microorganisms. Plants roots, by their growth, modify the physical, chemical, and physicochemical properties of soil. The roots have mechanical effects, take up nutrients, and release organic and mineral compounds, including CO_2. The soil provides water, nutrients, and organized solid support. Dead roots and root exudates provide favorable energetic and nutritional conditions for microorganisms and some microfauna. In such environments, the soil microbiota have interactions among themselves and with roots and soil constituents.

The rhizosphere has provided a fantastic landscape for physical, chemical, and biological studies concerning soil structure, soil aeration, soil mineral weathering, soil humification, water and nutrient dynamics, and plant growth and quality (Dommergues and Mangenot 1970, Lynch 1990, Berthelin *et al.* 1994, Gobat *et al.* 2003). All the interactions and processes in the rhizosphere are under the control of microbial behavior and activities, i.e., symbiosis, antibiosis, nitrogen fixation, plant growth promotion, soil stabilization, water uptake, nutrient flux, enzyme release, allelopathy, competition, phytotoxicity, and infection. The rhizosphere has fostered research in different fields of biology e.g., plant physiology, plant pathology, microbial ecology, microbial genetic, and general ecology. An example of less well- known interactions and processes, and of the development of interdisciplinary studies, is provided by the plant growth promoting rhizobacteria (PGPR), that are involved in plant growth by their influence on plant nutrition e.g., inorganic phosphorus dissolution, production of siderophores, nitrogen fixation, and promotion of nutrient uptake or their involvement in protection from diseases e.g., competition against pathogens, improvement or induction of resistance, and production and release of inhibitors of parasites.The rhizosphere is, along with upper soil horizons, a major soil site involved in the transfer of matter and energy in humification and weathering processes. However, there is a need for more knowledge (Lynch 1990, Berthelin *et al.* 1994, Gobat *et al.* 2003) and the rhizosphere is the focus of world-wide active research networks.

In relation to studies about carbon and nitrogen, studies of the degradation and behavior of organic pollutants, first pesticides then polyaromatic and chlorinated compounds, e.g., diffuse or chronic and localized pollution, were developed. Studies were expanded to include urban or anthropic as well as agricultural or forest soils. Problems of management of sewage sludges and wastes of different origins were also new and "sensitive." What to do with domestic, agricultural, and industrial wastes that are not treated or managed by specific physical, chemical, physicochemical, or biological engineering processes?

Soil enzymes have been studied in a biochemical approach. Major developments have been made on their relationship to microbial activity, on their binding and persistence on clays and humic substances (Burns 1978, Stotzky 1986).

Cycling and transformation of major and trace elements (e.g., P, S, Fe, Mn, Al, K, Hg, Cd, As, and Se) have been studied to understand weathering and soil functioning processes, as well as processes involved in metal extraction and pollution control (Dommergues and Mangenot 1970, Alexander 1977, Berthelin and Toutain 1982, Berthelin *et al.* 1994).

Microorganisms and their activities have appeared as possible tools in ecotoxicology, in remediating degradation and pollution of soil (Taradellas *et al.* 1997).

Interactions among microbes, microbes and fauna, and microbes and plants, were studied for the improvement of plant production and health (e.g., nitrogen fixation, mycorrhizal association, plant pathology). The problematic development of plant growth promoting rhizobacteria (PGPR) emerged. The relations and dynamics between predators and prey were studied, for instance, between protozoa and bacteria. Protozoa were also studied as symbiotic organisms in the guts of earthworms and termites (e.g., Darbyshire 1994). Interactions among microorganisms-minerals-organic matter were also developed (Hattori 1973, Huang and Schnitzer 1986, Huang *et al.* 1995, Berthelin *et al.* 1999). Numerous studies concerning interactions between microorganisms and clay minerals, microorganisms and water dynamics, and microorganisms and soil structure enabled the definition of major interactions that have a strong impact on the behavior and activities of microorganisms and on the activities of enzymes in soil. (e.g., Stotzky 1986).

All these fields opened and developed with the help of new techniques and methods. The use and availability of radioactive isotopes (e.g., 14C, 109Cd, 42P, and 35S) and of stable isotopes (e.g., 13C and 15N) allowed obtaining both qualitative and quantitative information on metabolic transformations in complex systems. A better knowledge of the chemical and physical properties of soil organic matter (Flaig *et al.* 1975, Haider and Martin 1975, Schnitzer and Khan 1978) were of great interest and importance for soil biology. Improvements in microscopic methods, in particular, electron microscopy and, more recently, confocal scanning laser microscopy, enabled obtaining clear and accurate observations and pictures associated with specific stainings or labelings. The recent use of molecular biology and molecular biochemistry techniques has provided new information and promises to be very productive not only in taxonomy but in population dynamic and activity.

Soil Fauna (Meso and Macrobiota) in the Development and Study of Soil Biology and Soil Ecology

Darwin (1881), Drummond (1887), and Müller (1887) were the first to observe, propose and emphasize, the role of soil fauna (i.e., earthworms, termites, and enchytraeidae) in the transformation of organic matter originating from plant roots and litter, and in humus formation in soil. Until after the middle of the 20th century, different groups of the fauna in soil e.g., nematoda, gastropoda, lumbricidae, arthropoda, oligocheta, crustacea, myriapoda, and insecta (among them collembola, termites, diptera and coleoptera) were studied mainly by taxonomists, more from a taxonomic and biologic than from an ecological and/or functional approach (Bachelier 1978, Dindal 1990). Selective methods for the extraction of soils animals were developed (as reported by Berthelin and Toutain 1982, Berthelin et al. 1994). The Berlese device in 1905 enabled the recovery of a large proportion of these animals. The technique was simplified and improved by Tüllgren in 1918.

In 1990, Dindal edited a book, entitled *Soil Biology Guide*, that was a multi-author general overview on soil fauna with one group of organisms or one taxon per chapter. In the introductory chapter, Dindal discussed the ecological and taxonomical approaches and emphasized their specific and complementary interests. In the second part of the 20th century, integrated concepts and approaches to the interactions between different organisms, meso- macro- and microbiota, and with the soil constituents emerged and developed. In the same period, studies at different scales (e.g., cm in field observation, mm in micromorphologic observation, µm in ultra-structural and submicroscopic observation) have provided results on the specific activities of representative groups of the soil fauna. Their implication in different processes have been recognized and, for some (e.g., earthworms, termites, oribates, isopods) their organization in structural and functional communities and involvement in major pedogenetic processes have been proposed, as summarized below.

The variable and diverse feedings (saprophytic, phytophagic, bacteriophagic, and carnivory) of nematodes have been well defined (Dunger 1983, Saur and Arpin 1989). Similarly, the enchytraeids and lumbricids have been studied. The phyto-saprophitic enchytraeids (oligochaeta) have been recognized to be largely involved in the dynamic (by feeding and spreading) of fungal and bacterial populations (Toutain *et al.* 1982, Wolters 1988). Their droppings can form a major part of the OH layer of humus of moder type as defined here after (Babel 1969, Toutain 1987a). A more complete definition of all the types of humus is presented by Babel (1969), Toutain (1987a), and in Berthelin *et al.* (1994) or in Brêthes *et al.* (1994). The importance of lumbricid (earthworms) was recognized long ago, but their mechanical, physico-chemical, and ecological impacts have been relatively well documented only recently. In their gut they mix mineral and organic constituents and microorganisms; they contribute significantly to soil fabric (soil structure) (Boumaza *et al.* 2002); they increase soil porosity with their tunnels (Figure 1A) and they develop a specific habitat, the drilosphere as defined by Bouché (1975), Lavelle (1984). They are one of the major animal groups in soil (Lavelle and Spain 2001, Gobat *et al.* 2003).

Different classes of the arthropod group are also involved in various processes such as humus formation and dispersion of microorganisms. Isopods (wood lice), which are numerous in many soils (Bornebusch 1980, Molfetas 1982), mainly fragment and ingest leaf litter (Neuhauser and Hartenstein 1978). The myriapods and, in particular, the diplopods are also involved in the transformation of plant material (Geoffroy *et al.* 1987, Tajosky *et al.* 1991) and in humus formation as indicated by Kubiena (1953). The oribates (arachnida) (Krantz 1978) are involved in fragmentation and mixing of plant residues (Figure 1D). They feed on plant residues, fungal mycelia, and spores, and are involved in the population dynamic of fungi (Hartenstein 1962, Ponge 1988).

Among the insects, the apterygotes (insects without wings), especially collembola, are present in most soils and litters (Rusek 1975, Benckiser 1997) where their droppings (30–100 μm) (Figure 1D) indicate that they are feeding mainly on fungal mycelia, leaves, and root residues (Mac Milan 1975, Ponge and Prat 1982). The pterygotes (winged insects), represented primarily by termites, play a major role in the functioning of tropical soils (Lee and Wood 1971, Grassé 1982), and they form specific habitats designated as termitospheres (Lavelle 1984, Lavelle and Spain 2001). They are strongly involved in the consumption, transformation, and humification of plant residues and in the dynamics of carbon and nitrogen by their enzymatic activities (Garnier-Sillam *et al.* 1989). They have associations and symbiosis with microorganisms (especially protozoa) (Grassé 1978, Lavelle and Spain 2001), and they are involved in the formation and organization of organo-mineral associations (Eschenbrenner 1986, Garnier-Sillam *et al.* 1988). Other pterygotes, such as the numerous species of coleoptera that live in soils (Scarabids, Carabids, Staphylinds, Silphids), are predators, and their larvae are rhizophagic, saprophagic, or coprophagic (Crowson 1981).

The hymenoptera, especially ants, are essential soil organisms that are present in temperate and intertropical soils and are carnivorous, herbivorous, or omnivorous (Levieux 1973, Lavelle and Spain 2001, Gobat *et al.* 2003). Some ants such as the

Atta in South America farm fungi to produce proteases that they use for their own digestion (e.g., Cherrett *et al.* 1989).

Some other groups e.g., diptera (Figure 1B) and gasteropods appear to have less important functions in soil. The roles of the soil fauna (meso and macrobiota) alone or in association with microorganisms are now better known. Their mechanical (physical), chemical, and biochemical impacts on soil organization (structure) and functions emphasize their importance in different specific soil habitats or subsystems [e.g., the drilosphere (earthworms) and the termitosphere (termites)]. Their involvement in the litter transformation and evolution and in the functioning of the rhizosphere is now well known (Lavelle 1984, Lavelle and Spain 2001). Lavelle suggested also some major fundamental functions of the soil fauna, such as ecosystem "engineers," litter "transformers," and micro predators. However, the association and interaction of the soil microfauna with microorganisms, soil constituents, etc. in soil are not well defined and need attention.

An interesting emergent aspect is the use of soil animals (e.g., earthworms and snails) in the evaluation of "soil quality," where they are of interest to determine parameters of ecotoxicity.

Contribution of Micromorphology and Microscopy in the Study of Soil Biology

The first microscope, probably built by a dutch optician Jansen was described by Galileo in 1609. Almost a century later, in 1684, van Leeuwenhoek presented drawings of "animalcules" probably bacteria he observed with his microscope. However the "golden age" of light microscopy began two centuries later with the making of the modern microscopes from 1880. Somewhat later, soils were subjected to microscopic studies. Müller (1887), a forest scientist, using methods from work with petrographs, developed direct microscopic observation of soils. In his studies on the "natural humus forms and their impact on vegetation and soil" he observed fungal mycelia and other organic soil components under the microscope. He was also the first to establish a relation between the presence of earthworms, enchytraeidae, and the type of forest humus. From historical point of view it is interesting to note that Müller's interest was related to the publication of Darwin's book on earthworms (1881), as indicated by Müller himself in the foreword of his german translation. However, after these first studies, the direct observation of living organisms and of indicators of their presence or activity in natural habitats declined, and further studies then emerged relatively slowly.

Kubiena (1933, 1938) is considered the founder of the study of soil micromorphology, because he began to work with undisturbed samples of soils "i.e., in natural arrangement," using impregnation of friable soil material with hardening agents (resins) and observations of thin sections cut from these samples (Figures 1A, 1B, 1C, 1D). This method, which originated in petrography, was developed by Kubiena 1933, who later told in his book (1938) that after he had developed this method he came to know that Ross had already developed a similar one (cited in Kubiena, 1938). Nevertheless, Kubiena was the discoverer of the microfabric (microscopic fabric) of soils and its relation to processes and soil types. Considerable improvements in the methods used to study soil micromorphology were developed later (e.g., Altemüller 1962 a and b, Murphy 1982). A specific terminology of microscopic "soil fabric" was instituted by Kubiena and more recently developed by Brewer (1964) and Bullock *et al.* (1985), who proposed a guideline for a complete description of any thin section of soil. An international working group on soil micromorphology was founded in 1970.

Kubiena (1967) also adapted the estimation of "volume percentages" of fabric components (e.g., voids) by applying the so-called point counting method, which was used in mineralogy and petrology. Progress in quantification became possible by further development of stereology, a set of mathematical methods for obtaining 3D-answers from counting and measurements on 2D-preparations (Weibel 1979). Despite the considerable value of the development of quantification methods, it must be emphasized that qualitative approaches prevail in soil micromorphology.

Study of Biological Habitats

Around 1930, Kubiena was in contact with the zoologist, Kühnelt, who had written a book, *Bodenbiologie*, in 1961. They found that, in many cases, the excrements of soil animals are so specific in size, shape, and composition that they are good indicators of the activity of a given animal group (Figures 1A to 1C) and Zachariae (1965) has specified such observations.

Thin sections of soil showed the spaces available for colonization by soil organisms, e.g., Figure 1A. Animals in soil are seldom observed in thin sections, but the results of their activities (e.g., tunnels, specific consumption of plant material, and droppings) are detectable (Figures 1A, 1B, 1C). From recognition of certain excrements, conclusions may be drawn on the effects of the activities of the related animal groups, e.g., earthworms and enchytraeids (Figures 1A-1B) cause and intensify the mixture of organic and mineral matter, oribates do not mix these two soil materials (Figure 1C), they feed exclusively on organic matter. Similarly, bacteria in soil are usually not recognized in normally prepared thin sections. They may, however, become visible by fluorescent staining (Altemüller and Van Vlietlanoe, 1990). Bacteria are most frequently present on surfaces of pores and on organic particles, in or on aggregates and around roots (Figures 1E to 1G). Fungal mycelia are observed primarily by their primary fluorescence.

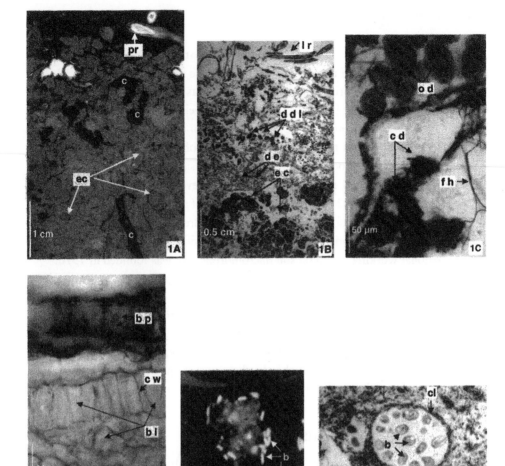

Figure 1. (A) Vertical section of undisturbed soil showing the result of earthworm activity with formation of channels (c) and production of casts (e c) that constitute the major part of the humus form Mull layer of a meadow, in a deep loamy soil with plant residue (p r) at the surface. Photo corresponds to a upper part of A-horizon, (photography U. Babel). (B) Soil thin section showing superposition of droppings of diptera larvae (d d l), of enchytraeids (d e) and earthworm casts (e c) in a humus (name Tangel in german) with leaf (needle) residue (l r) at the surface. The lower part is the A horizon of a spruce forest soil. (photography U. Babel). (C) Soil thin section of the OF layer in the upper part of a humus form Tangel showing on the top oribate droppings (o d) lying inside a spruce needle residue. From center to bottom and right, dark cell residue from droppings of collembola (c d) and dark fungal hyphae (f h). (photography U. Babel). (D) Soil thin section observed in transmitted light. In the upper part one beech leave residue with brown pigments (b p) and in the lower part another leaf bleached by white-rot fungi which have degraded the brown pigments. Cell wall structures are still preserved (c w). (humus form Moder) (photography U. Babel). (E) Bacteria (b) on an aggregate of acid brown forest soil A horizon after fluorescent staining. (photography J. Berthelin in light microscopy) (bacteria size 0.5 to 1.0 μm). (F) Bacterial (b) aggregate surrounding by clay particles (cl) formed during the gut transfer of earthworm. A horizon of a brown forest soil. (photography F. Toutain in TEM)

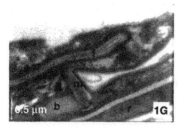 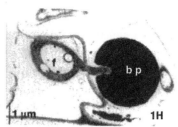

Figure 1. (*Cont'd*) (G) Bacteria and fungi in rhizosphere of pine. f (fungi), m (mineral), b (bacteria), r (root). (photography C. Leyval in TEM) (bacteria size 0.5 to 1 μm). (H) Colonization and biodegradation by hyphea of white rot fungi (f) of a senescent beech leaf containing brown pigment (b p). (photography F. Toutain in TEM). Color version (see the color plate section at the end of the book).

Soil Aggregates and Structures Resulting from Faunal Activities

Activities of soil animals often result in changes in fabric that are visible to the naked eye, other changes occur at lower levels. The effects of earthworm activities have been investigated in detail with the microscope. Best known and most important are the conclusions made from the occurrence of specific aggregates composed of excrements, "droppings," or casts around the actual or former presence of a given soil animal group. However, wrong conclusions can be drawn; e.g., small aggregates are not always attributable to a specific animal group; presence of many droppings does not always mean high activity (e.g., presence of fossil droppings or droppings degraded at lower rate than they are produced) (Zachariae 1965). The nature of droppings may indicate the ability of the animals to mix mineral and organic materials, as with earthworms, or to ingest only organic materials, as with oribates. Channels of soil animals are often less frequent than would be expected, because many animals use existing ones and don't dig their own channels, e.g., collembola or predatory chilopoda.

Soil micromorphology has established useful connections and relations at original scales around 1 mm to 10 μm between the activities of soil organisms, microscopic observations, and macroscopic findings in the knowledge of forest microbial ecology, as illustrated in Figures 1A to 1H. Moreover, it can provide suggestions for studies in other fields and other scales from small sizes at ultramicroscopic or molecular scale to large sizes.

Diversity in Forest Humus Ecology as the Result of Different Microbial and Faunal Activities

Thin sections show a high diversity of organic particles (Babel 1975): plant residues, organ residues, tissue residues, cell residues, amorphous particles, and the so called "punctuations" (particles around 1 μm) (Figures 1B, 1C, 1D, 1E). Soil organic matter will often appear as pigments of the mineral matter. Different abiotic (physical and chemical) and biological (microbial, faunal) processes occur which can be observed. Tissue residues in thin sections of soil exhibit several types of alteration, originating at least partially from defined microbial activities: Browning, which corresponds to the formation of tannin-protein complexes after the death of the cell (Handley 1954) (Figure 1D); blackening is typical for wet conditions, and occurs in a type of humification whose chemistry is unknown. Bleaching, which is an effect of the disappearance of the protein-tannin complexes of brown leaves and needles, is a result of the activity of some basidiomycete fungi (e.g., white rot fungi) (Figure 1D). Liquefaction, corresponds to the activity of brown rot fungi, with decomposition of cellulose and the loss of the cellular structure of the tissue residues.

These microbial alterations of organic matter are well recognized in the formation of the different types of forest humus profiles. In the humus form Moder, the OL layer, consists essentially of brown or bleached complete leaves; in the OF layer the animal comminution is important (enchytraeids, collembola, diptera larvae, mites ...) (Figures 1B, 1C); in the OH layer the plant residues are comminuted with the production of darkened pellets and organic residues; the A organic horizon is made mainly of pellets of enchytraeids and collembola and fragments of plant residues. In the humus form Mull, only OL and a possible tiny OF layer are present due to earthworm activity (Figure 1A). The A horizon is made by blending of organic substances and mineral particles that are mixed in the earthworm gut (Figure 1A).

The total of horizons OL, OF, OH, and A characterize the "humus forms," a term of great importance for ecological conditions in forests and which goes back to Müller (1887) and is the foundation of several humus classification in different countries (e.g., Canada, France, Germany).

Micromorphology has also been of great help in measuring root length and density and in investigations of the root system and surrounding soil. Using fluorescent staining, it has enabled the observation of living or dead roots, as well as such parameters as root size and distribution, soil utilization by roots, relation with oxygen and water supplies, root faunal consumption, and microbial decomposition of roots.

Illustration of the Contributions of Microscopic and Submicroscopic Studies in Soil Biology

Studies of biological habitats in soil and of the activity of soil organisms have taken advantage of the development of microscopic and ultramicroscopic methods associated or not with spectroscopic methods. Some photographs in light microcopy (Figures 1A to 1D), provide illustrations of micromorphology at different scales. Four other Figures (1E to 1H) present illustrations at micrometric and submicrometric scales. Figure 1E shows examples of light fluorescence microscopy and the presence of bacteria (small fluorescent rods or cocci 0.5 to 1.0 μm) on a soil aggregate.

From a historical point of view, the first ultramicroscopic images obtained using a transmission electron microscope (TEM) were shown in 1932 by Ruska (Nobel Prize 1986) and Knoll. The ultrathin slide technique was developed by Pease and Baker (1948) (with thickness of about 0.3 μm) and by Sjöstrand, who in 1952 made, an ultramicrotome that enabled obtaining "slides" of 0.02 μm thickness. The techniques of transmission electron microscopy (TEM) and scanning electron microscopy (SEM), discovered in 1938 by Van Ardenne, were used to observe for the first time in 1934 biological materials and later to study the evolution of plant material during the humification of organic matter, the weathering of soil minerals, the presence and localization of microorganisms. For example, Reisinger and Kilbertus (1973) observed the production by bacteria of melanin particles originating from degradation of fungal cell walls, and which could be a source of humic substances (Kang and Felberck 1965). Figures 1F to 1H are transmission electronic microscopic views. Figure 1F presents the organization of bacterial aggregate surrounding by clays; bacteria are located inside an aggregate made in the earthworm gut. Figure 1G shows a bacterium, a fungal hypha, a part of root, and a mineral particle in a plant rhizosphere, suggesting strong interaction in such microhabitats. Figure 1H presents a white-rot fungus degrading brown pigment in leaf litter. As in light microscopy, specific coloration or fixation (e.g., polysaccharide staining as in Figure 1G) improve the quality of the observation.

The development of spectroscopic methods, as the microanalysis of X-Ray emission using for instance the microprobe made by Castaing in 1951, or more recently, the new techniques of EELS (Electron Energy Loss Spectroscopy) of Egerton in 1986, have contributed to significant progress in the understanding of different phenomena in soil (e.g., biodegradation of organic matter, humification, organo-mineral associations, biological and microbial weathering of soil minerals e.g., Berthelin et al. 1994).

In the last two decades, methods have emerged to obtain three-dimensional imaging. Atomic Force Microscopy (AFM) and Confocal Scanning Laser Microscopy provide three-dimensional information on microorganisms and small animals themselves but also on their environment (see box) and processes related to their activity.

"Modern" Soil Biology

Modern soil biology started essentially at the beginning of the second half of the 20th century but new concepts, approaches, and tools emerged after 1980–1990 and are more or less developed now. Many microbial processes in soil have been observed and were already known, but there has been little information on communities and populations involved in these processes and on the interactions and parameters controlling them. Discoveries in genetics, and in antibody reactions, etc. (see Van Elsas *et al.* 1997, Nalin *et al.* 1998, Madigan *et al.* 2000) have been applied to environmental questions, including soils, to obtain, for example, finger prints of microorganisms. For example, the Enzyme-Linked Immunosorbent Assay (ELISA) or the Enzyme-Linked Immuno-Filtration Assay (ELIFA) have been developed in soil microbial ecology to study the dynamics of bacterial populations in the rhizosphere, or the attachment on mineral particles (e.g., Dziurla *et al.* 1998). Another specific approach to identify and quantify microorganisms in nature is the use of nucleic acid probes made to fluorescence by an attached colouring. More recently, especially after the discovery in 1988 by K. Mullis of Polymerase Chain Reaction (PCR) methodology and its development for application in natural environments, new sensitive methods for the detection of soil microbes have been developed (see box).

These molecular approaches have been applied to microorganisms, but they can also be applied to meso and macrobiota (e.g., Van Elsas *et al.* 1997, Prosser 2002, Gobat *et al.* 2003, Legay and Barbault 1995, Redecker 2000, Kirk J.L. *et al.* 2004). These molecular approaches in soil microbial ecology and soil ecology do not

New promising tools for soil biology: molecular biology and confocal laser microscopy

Traditional laboratory cultivation of microorganisms on nutrient media was unable to isolate and cultivate the major portion of soil microbial communities. Recent methods, developed around 1990, to obtain molecular profiles of microbial populations and functional genes from DNA extracted and purified from soil, followed by amplification of the DNA sequences using primers and DNA polymerase, have improved knowledge of noncultivable bacteria in soil. The amplification of ribosomal RNA and functional genes provided new information on both the dynamics and activities of *in situ* bacterial populations.

The different microscopic and submicroscopic methods introduced various artifacts due to sample preparation, limited focus area, image analysis, and data treatment. The scanning confocal laser microscopy (SCLM) improves observations by using focused laser beams and pinholes that eliminate unfocused parts and enable the reconstruction of an object by a series of three dimensional optical sections. The combination of SCLM with fluorescent labeled molecular probes, allows *in situ* observations of specific bacteria in their 3D environment.

constitute a "new science," but they are new tools that enable the opening of new "horizons" and new "windows" in the knowledge of the relations between processes, parameters, and population dynamics and activities. As Gobat *et al.* (2003) wrote, the "windows" already opened by former scientists, such as Winogradsky, remain open and are still valuable and used e.g., specific nutrient medium for isolation and cultivation of microorganisms; measurements of biochemical activities of communities and functional groups.

The use of molecular techniques with specific molecular probes has already provided progress in the knowledge of different mechanisms, such as relations between populations and activities involved in nitrogen fixation, nitrification, denitrification, and biodegradation of organic pollutants. However, culture methods, field and laboratory experimental devices, etc. should not be abandoned. They need to be developed in interdisciplinary research. Previous studies, as reported by Dommergues and Mangenot (1970), Hattori (1973), Alexander (1977), Burns (1978) and Stotzky (1986) have provided much information on the impacts of physical or physico-chemical parameters (e.g., clay type and contents) on microbial communities and activities. Attempts have been made (e.g., Huang and Schnitzer 1986, Huang *et al.* 1995, Berthelin *et al.* 1999) to enhance knowledge of interactions between minerals, organics, and microbes in soil processes by promoting pluridisciplinary approaches. However, explanations, quantification, threshold values, and interactions still need to be defined.

Involvement of the soil fauna in ecology and interactions has also progressed in concepts and knowledge concerning their habitats and impacts on soil functions (e.g., Lavelle and Spain, 2001), but connections with other disciplines have to be improved such as larger interactions with microbial ecology or soil physics.

Recent development in analytical methods, such as in microscopy and spectroscopy (e.g., Nuclear Magnetic Resonance, NMR; Fourier – Transform Infrared, FTIR; Extended X-ray Absorption Fine Structure, EXAFS), especially in soil chemistry and mineralogy, have been most helpful in studies of the location and activities of microorganisms in soil, both in the laboratory and in the field. Experimental designs are progressing to mimic as closely as possible natural conditions. Finally, complementing new methods, interdisciplinary and integrated approaches are enabling the emergence of new concepts and knowledge in soil biology and biochemistry.

The history of soil biology shows first a long period of observation of plant growth and of plant production and their possible relations with the soil properties. Then in a recent historical period, since the 18th–19th centuries, it was the questions and the works of isolated and specialized scientists, chemists and agronomists then microbiologists and later zoologists which were at the origin of progress and new concepts. Very recently in the second part of the 20th century, the development of interdisciplinary and pluridisciplinary approaches that emerged with the help of new techniques allowed development of the integration of biological and non biological parameters, and to consider their interdependency in the functioning of terrestrial ecosystems at different scales of time and space.

References

Ackert, L. 2004. From the thermodynamics of Life to Ecological Microbiology: Sergeî Winogradsky and The Cycle of Life, 1850–1950, Ph.D. Thesis, The Johns Hopkins University, 2004.

Alexander, M. 1971. Microbial Ecology, Wiley, New York.

Alexander, M. 1977. Introduction to soil microbiology, 2nd edition; John Wiley and Sons Inc., New York.

Altemüller, H.J. 1962a. Contribution to the micromorphological differentiation of leached para-braunerde, podzol-braunerde, and humus podzol. Z. PflErnähr. Düng. 98:247–258.

Altemüller, H.J. 1962b. Improvement of the technique of embedding and grinding in the preparation of thin sections of soil by means of vestopal. Z. PflErnähr.Düng. 99:164–177.

Altemüller H.J. et Van Vlietlanoe. 1990. Soil thin section fluorescence microscopy. pp. 565–579. In: Douglas, L.A.: Soil micromorphology. Elsevier, Amsterdam.

Babel, U. 1969. Enchytraeen-Losungsgefüge in Loess. Geoderma 2:57–63.

Babel, U. 1975. Micromorphology of soil organic matter. pp. 369–473. In: Gieseking, J.E. (ed.) Soil components Vol.1, Organic components. Springer-Verlag, New-York, Heidelberg, Berlin.

Bachelier, G. 1978. La faune des sols, son écologie et son action. Documentations techniques 38, ORSTOM, (IRD), Paris.

Barles, S. 1999. La ville délétère. Médecins et ingénieurs dans 1'espace urbain. XVIIIeXIXe siecles. Editions Champ Fallon, Seyssel.

Beijerinck, M.W. 1888. Die Bakterien der Papillonaceen-Knollchen. Bot. Ztg. 46:724–735.

Benckiser, G. 1997. Fauna in soil ecosystems. Marcel Dekker Inc., New York.

Berlese, A. 1905. Apparecchio per raccogliere presto, ed in gran numero piccolo arthropodi. Redia 2:85–89

Berthelin, J. and F. Toutain. 1982. Soil Biology, 140–183. In: Bonneau, M. and Souchier, B. (eds.) Constituents and properties of soils, Academic Press, London.

Berthelin, J., P.M. Huang, J.M. Bollag, and F. Andreux (eds.). 1999. Effect of Mineral-Organic-Microorganism Interactions on Soil and Freshwater Environments. Kluwer/Plenum Publ., New York, London.

Berthelin, J., C. Leyval, and F. Toutain. 1994. Biologie des sols. Role des organismes dans l'altération et l'humification. Ch. 7, 143–237. In: Bonneau M. et Souchier B. Pédologie 2. Constituants et propriétés du sol. Masson, Paris.

Bornebusch, C.H. 1980. The fauna of the forest soil. Forstl. Forsoksv. Danm. 2:1–125.

Bouché, M.B. 1975. Action de la faune sur les états de la matiére organique dans les écosystèmes. pp. 157–168. In: Kilbertus, G., Reisinger, O., Mourey, A. (eds.) Biodegradation et humification. Pierron Publi. Sarreguemines.

Boulaine, J. 1989. Histoire des pédologues et de la science des sols. Institut National de la Recherche Agronomique, Paris.

Boumaza, O., F. Toutain, J.-C. Pargney. 2002. Apports of soil's thin section in study of *Tuber Mesentericum* and its edaphic environment, Bull. Acad. Lorr. Sc. 41(3–4):105–114.

Boussingault, J.B. 1843–1844. Economie rurale considérée dans ses rapports avec la chimie, la physique et la météorologie, 2 vol. Bechet jeune, Paris.

Boussingault, J.B. 1860–1874. Agronomie, Chimie agricole et Physiologie. 5 Vol. Mallet-Bachelier, Paris.

Brêthes, A., J.J. Brun, B. Jabiol, J.F. Ponge, and F. Toutain. 1994. Classification of humus forms: a french proposal. Ann. Sc. For. 52:535–546.

Brewer, R. 1964. Fabric and Mineral Analysis of Soils. Willey and Sons, London, New York.

Brieude, de. 1782–1783. "Topographie médicale de la Haute-Auvergne". Mémoire de la Société Royale de Médecine. 257–340.

Brock, T.D. 1979. Biology of microorganisms, Prentice-Hall, Inc. Upper Saddle River, New Jersey.

Buch, W. 1991. Le ver de terre au jardin. Arts Graphiques Européens, Le Plessis Robinson.

Bullock, P., N. Fedoroff, A. Jongerius, G. Stoops, and T. Tursina. 1985. Handbook for soil thin section description. Waine Research Publications, Wolverhampton.

Burns, R.G. 1978. Soil enzymes, Academic Press, London.

Cambefort, Y. 1994. Le scarabée et les Dieux. Editions Boubée. Paris.

Cherrett, J.M., R.J. Powell, and D.J. Stradling. 1989. The mutualism between leaf-cutting ants and their fungus. pp. 92–120. In: Wilding, N. *et al.* (eds) Insect fungus Interactions. Wilding, N. *et al.* (eds) 14th Symp. Roy. Entomol. Soc., London.

Crowson, R.A. 1981. The biology of the Coleoptera. *Academic Press*, London.

Darbyshire, J.F. (ed.). 1994. Soil protozoa. CAB International, Wallinford.

Darwin, C.R. 1881. The formation of vegetable mould through the action of worms, with observations on their habit. John Murray and Co., London.

Davy, H. 1814. Elements of agricultural chemistry. Longman, London.

Denis, G. 2001. Du physicien agriculteur du dix-huitième à l'agronome des dixneuvième et vingtième siècles: Mise en place d'un champ de recherche et d'enseignement. C.R. Acad. Agric. Fr. 87(4):81–103.

Dindal, D.L. (ed.). 1990. Soil Biology Guide, John Wiley and sons, New-York.

Dommergues, Y. and F. Mangenot. 1970. Ecologie microbienne du sol. Masson, Paris.

Drummond, H. 1887. On the termite of the tropical analogue of earthworm. Proc. R. Soc. Edinb. 13(1884–85):137–146.

Dunger, W. 1983. Tiere im Boden. Neue Brehm. Bücherei, A. Ziemsen Verlag, Wittemberg-Lutherstadt.

Dziurla, M.A., W. Achouak, B.T. Lam, Th. Heulin, and J. Berthelin. 1998. Enzyme-linked Immunofiltration assay to estimate attachment of Thiobacilli to pyrite. Appl. Environm. Microb. 64:2937–2942.

Egerton, R.F. 1986. Electron Energy Loss Spectrometry in the Electron microscope, Plenum, New York.

Eschenbrenner, V. 1986. Contribution des termites à la microagrégation des sols tropicaux. Cah. ORSTOM, sér. Pédologie 22:397–408.

Flaig, W., H. Beutelspacher, and E. Rietz. 1975. Chemical composition and physical properties of humic substances. pp. 1–211, In: Gieseking, J.E. (ed.) "Soil Components" Vol. 1. Springer-Verlag, Berlin and New York.

Frank, A.B. 1885. Ueber die auf Wurzelsymbiose beruhende Ernährung gewisser Bäume durch unterirdische Pilze. Ber. deut. bot. Ges. 3:123–145.

Garnier-Sillam, E., F. Toutain, and J. Renoux. 1988. Comparaison de l'influence de deux termitières (humivore et champignonniste) sur la stabilité structurale de sols forestiers tropicaux. Pedobiologia 32:89–97.

Garnier-Sillam, E., J. Renoux, and F. Toutain. 1989. Les composés humiques des termitières de Thoracotermes macrothorax (humivore) et Macrotermes mülleri (champignonniste), Soil Biol. Biochem. 21:499–505.

Geoffroy, J.J., M.L. Celerier, I. Garay, S. Rherissi, and P. Blandin. 1987. Approche quantitative des fonctions de transformation de la matière organique par des macroarthropodes saprophages (isopodes et diplopodes) dans un sol forestier à moder. Protocoles expérimentaux et premiers résultats. Rev. Ecol. Biol. Sol 24:573–590.

Gobat, J.M., M. Aragno, and W. Matthey. 2003. Le sol vivant. *Presses polytechniques et universitaires romandes*. Lausanne. 2ème édition.

Grassé, P.P. 1978. Sur la véritable nature et le rôle des meules à champignons construites par les termites. C.R. Acad. Sci., Paris, série D 287:1223–1226.

Grassé, P.P. 1982. Termitologia. Tomes I à III. Masson, Paris.

Haider, K. and J.P. Martin. 1975. Decomposition of specifically carbon-14 labelled benzoic and cinnamic acid derivatives in soil. Soil Sci. Soc. Am. Proc. 39:657–662.

Handley, W.R.C. 1954. Mull and mor formation in relation to forest soils. Forestry Comm., Bull. 23, London.

Hartenstein, R. 1962. Soil Oribatei. VII. Decomposition of conifer needles and deciduous leaf petioles by *Staganacarus diaphanum* (*Acarina: Phtiracaridae*). Ann. Ent. Soc. Am. 55:713–716.

Hattori, T. 1973. Microbial life in the soil. An introduction. Marcel Dekker Inc., New York.

Henin, S. 2001. Olivier de Serres "Le théâtre d'Agriculture et mesnage des champs". A key stage between the latin agronomists and the setting of the modern objective thought (in french). C.R. Acad. Agric. F. 87:23–29.

Hiltner, L. 1904. Uber neuere Erfahrungen and problem auf dem Gebeit der Bodenbakteriologie and unter besonderer Berucksichtigung der Grundungung and Brache. Arb. Dtsch. Landwirt. Ges. 98:59–78.

Home, F. 1757. The principles of agriculture and vegetation. G. Hamiton and J. Balfour, Edinburgh.

Huang, P.M. and M. Schnitzer (eds.). 1986. Interactions of soil minerals with natural organics and microbes. SSSA Special Publication N° 17. Madison.

Huang, P.M., J. Berthelin, J.M. Bollag, W.B. McGill, and A.L. Page. 1995. Environmental impact of soil component interactions. Vol. 1. Natural and Anthropogenic Organics, Vol. II. Metals, Other inorganics, and microbial Activities, CRC, Lewis Publ. Boca Raton.

Jordanova, L.J. and R.S. Porter (eds.). 1979. Images of earth: Essays in the history of the Environmental Sciences. Bucks: *British Society for the History of Science, coll.* "BSHS monographs", London.

Kang, K.S. and G.T. Felberck. 1965. A comparison of the alkaline extract of tissues of *Aspergillus niger* with acids from three soils. Soil Sci. 99:175–181.

Kevan, D.K. McE. 1985. Soil zoology, then and now-mostly then. Quaest. Entomol. 12(4):371.7–472.

Kirk, J.L., L.A. Beaudette, M. Hart, P. Moutoglis, J.N. Kliromos, H. Lee, and J.T. Trevors. 2004. Methods of studying soil microbial diversity. J. Microbiological Methods 58:169–188.

Kluyver, A.J. 1956. The microbe's contribution to biology. Harvard University lecture. Cambridge.

Krantz, G.W. 1978. A manual of Acarology. Oregon State University Book Store, Corvallis.

Kubiena, W. 1933. Mikropedologische untersuchungen über kristallneubildungen in Bodenhohlräumen. Z. PflErn. Düng., Bodenk. 31(4/6):255–278.

Kubiena, W.L. 1938. Micropedology. Collegiate Press, Ames.

Kubiena, W.L. 1953. The soils of Europe. Thomas Murby & Co. London.

Kubiena, W. 1967. Die mikromorphometrische Bodenanalyse. Enke, Stuttgart.

Kühnelt, W. 1961. Soil Biology. Faber, London.

Lavelle, P. 1984. The soil system in the humid tropics. Biology International 9:2–15.

Lavelle, P. and A.V. Spain. 2001. Soil ecology. Kluwer Academic Publishers, Dordrecht.

Lee, K.E. and T.G. Wood. 1971. Termites and Soil. Academic Press, London and New York.

Legay, J.M. and R. Barbault. 1995. La revolution technologique en ecologie. Masson, Paris.

Levieux, J. 1973. Etude du peuplement en fourmis terricoles d'une savane préforestiére de Côte d'Ivoire. Rev. Ecol. Biol. Sol 10:379–428.

Liebig, J. Von. 1842. Bemerkungen zu vorstehenden Versuchen de Saussure's. Annalen der Chemie 42:291–297.

Liebig, J. Von. 1846. Des engrais artificiels. Vve Bouchard-Huzard, Paris.

Liebig, J. Von. 1865. Die chemie in ihrer Anwendung auf Agrikultur and Physiologie, Braunschweig.

Lynch, J.M. (ed.). 1990. The rhizosphere. John Willey and sons, Chichester.

Mac Milan, J.H. 1975. Interspecific and seasonal analyses of the gut contents of three Collembola. Rev. Ecol. Biol. Sol 2:295–300.

Madigan, M.T., J.M. Martinko, and J. Parker. 2000. Brock Biology of microorganisms. Prentice Hall International, U.K. Limited, London.

Mc Laren, A.D. and G.H. Peterson (eds.). 1967. Soil Biochemistry (Vol. 1). Marcel Dekker Inc. New York.

Menuret de Chambaud, J.J. 1786. Essais sur l'histoire médico-topographique de Paris, ou lettres à M. D'Aumont, Professeur de Médecine à Valence, sur le climat de Paris, sur l'état de la médecine, sur le caractére & le traitement des maladies, & particuliérement sur la petite vérole & l'inoculation. Paris.

Molfetas, S. 1982. Etude d'un écosysteme forestier mixte. VIII. Les Isopodes. Rev. Ecol. Biol. Sol 19:427–438.

Müller, P.E. 1887. Recherches sur les formes matérielles de 1'humus. Annales de la Science Agronomique. 85:423.

Murphy, C.P. 1982. Thin section preparation of soils and sediments. AB Academic Publishers, Berkhamsted.

Nalin, R., L. Ranjard, S. Nazaret, and P.W. Simonet. 1998. La biologie moléculaire en écologie microbienne du sol: application à l'analyse de la structure des communautés bactériennes. Bull. Soc. Fr. Microbiol. 13(1):21–36.

Neuhauser, E.F. and R. Hartenstein. 1978. Phenolic content and palatability of leaves and wood to soil isopods and diplopods. Pedobiologia 18:99–109.

Paul, E.A. and F.E. Clark. 1989. Soil microbiology and biochemistry. Academic Press. San Diego.

Paul, E.A. and J.N. Ladd (eds.). 1981. Soil biochemistry. Vol. 5, Dekker, New-York.

Pfeffer, W. 1877. Ueber fleischfressende Pflanzen. Landw. Jahrb. 6:969–998.

Pochon, J. and H. de Barjac. 1958. Traité de microbiologie des sols. Applications agronomiques. Dunod, Paris.

Ponge, J.F. 1988. Etude écologique d'un humus forestier par l'observation d'un petit volume. III. La couche Fl d'un moder sous *Pinus sylvestris*. Pedobiologia 31:1–64.

Ponge, J.F. and B. Prat. 1982. Les collemboles, indicateurs du mode d'humification dans les peuplements résineux, feuillus et mélangés. Résultats obtenus en forêt d'Orléans. Rev. Ecol. Biol. Sol 19:237–250.

Prosser, J.I. 2002. Molecular and functional diversity in soil microorganisms. Plant and Soil 244:9–17.

Redecker, D. 2000. Specific PCR primers to identify arbuscular mycorrhizal fungi (*Glomales*) within colonized roots. Mycorrhiza 10:73–80.

Reisinger, O. and G. Kilbertus. 1973. Biodégradation et humification – III. Libération des granules. Modèle experimental en présence de bactéries: conclusions générales. Soil Biol. Biochem. 5:187–192.

Robin, P. and M. Blondel-Megrelis. 2001. 1800 et 1840, Physiologie végétale et chimie agricole. 1. Saussure, une publication à ressusciter. C.R. Acad. Agric. Fr. 87(4):31–59.

Rusek, J. 1975. Die Bodenbildende Funktion von Collembolen and Acarina. Pedobiologia 15:299–308.

Russel, E.J. 1923. The microorganisms of the soil. Longmans, Green. London and New-York.

Saur, E. and P. Arpin. 1989. Ultrastructural analysis of the intestinal contents of *Clarkus papillatus (Nemata: Mononchina)* ecological interest of the survey. Rev. Nematol. 12:413–422.

Saussure, T. de. 1804a. Recherches chimiques sur la végétation. Veuve Nyon, Paris.

Saussure, T. de. 1804b. Recherches chimiques sur la végétation. Journal de Physique, de Chimie et d'Histoire naturelle et des Arts 58:393–405.

Schnitzer, M. and S.U. Khan. 1978. "Soil Organic Matter". Elsevier, Amsterdam.

Stotzky, G. 1986. Influence of soil mineral colloids on metabolic processes, growth, adhesion and ecology of microbes and viruses. pp. 305–428. In: Huang, P.M. and Schnitzer, M. Interactions of soil minerals with natural organics and microbes. Soil Sc. Soc. Am. Special Publication n°17, Madisson, USA.

Tajosky, K., G. Villemin, and F. Toutain. 1991. Microstructural and ultrastructural changes of the oak leaf litter consumed by millipede *Glomeris hexasticha (Diplopoda)*. Rev. Ecol. Biol. Sol 28(3):287–302.

Tarradellas, J., G. Bitton, and D. Rossel (eds.). 1997. Soil ecotoxicology CRC Lewis Publishers, Boca Raton.

Toutain, F. 1987a. Biological activity of soils, modalities and mineral dependance. Biol. Fertil. Soils 3:31–38.

Toutain, F. 1987b. Les litières: siège de systèmes interactifs et moteur de ces interactions. Rev. Ecol. Biol. Sol 24:231–242.

Toutain, F., G. Villemin, A. Albrecht, and O. Reisinger. 1982. Etude ultrastructurale des processus de biodégradation. II. Modèle Enchytraeides-litières de feuillus. Pedobiologia 23:145–156.

Tüllgren, A. 1918. Ein sehr einfacher Auslese apparat für. terricole tierformen. Zeit. Ange. W.Ent. 4:149.

Van Elsas, J.D., J.T. Trevors, and E.M.H. Wellington (eds.). 1997. Modern Soil Microbiology. Marcel Dekker Inc., New York.

Waksman, S.A. 1927. Principles of soil microbiology. Williams and Wilkins, Baltimore.

Waksman, S.A. and R.L. Starkey. 1931. The soil and the microbe. Wiley, New York.

Weibel, E.R. 1979. Stereological Methods, Volume 1, Pratical methods for biological morphometry. Academic Press, London.

Winogradsky, S. 1890. Recherches sur les organismes de la nitrification. Ann. Inst. Pasteur 4:213–231, 257–275, 760–771.

Winogradsky, S. 1949. Microbiologie du sol. Problèmes et mèthodes. Masson, Paris.

Wolters, V. 1988. Effects of *Mesenchytraeus glandulosus* (oligochaeta, enchytraeidae) on decomposition processes. Pedobiologia 32:287–298.

Woodward, J. 1699. Some thoughts and experiments concerning vegetation. Philosophical Transactions of the Royal Society of London 21:193–227.

Zachariae, G. 1965. Spuren tierischer Tätigkeit im Boden des Buchen-waldes. Beihefte zum Forestwiss. Cbl., H.20, 68 p., Hambourg, Parey.

12
Historical Aspects of Soil Chemistry

Donald L. Sparks[1]

The Origins of Agricultural Chemistry: The Forerunner of Soil Chemistry

In ancient times Aristotle proposed that plants derive nourishment from their roots that obtain preformed organic matter. This view was held for more than 2000 years, and ultimately resulted in the humus theory of plant nutrition that was discounted by Sprengel and later by Liebig (Browne, 1944; van der Ploeg et al., 1999). Agricultural chemistry in the beginning focused on practical observations and experience, and did not have a real scientific basis until over 1000 years after the fall of the Roman Empire (Browne, 1944). Sir Humphrey Davy (1778–1829) used the title "Elements of Agricultural Chemistry" for a book containing lectures he presented before the British Board of Agriculture in 1802–1812. He noted that "Agricultural chemistry has for its objects all those changes in the arrangements of matter connected with the growth and nourishment of plants-the comparative values of their produce is food; the constitution of soils; and the manner in which lands are enriched by manure, or rendered fertile by the different processes of cultivation." In his treatise, Davy listed 47 chemical elements with details on their properties. Elements that were named included Na, K, Ca, Si, Ba, and Mg. He also noted the importance of rocks and parent material in soil formation (Browne, 1944).

The Phlogistic period was the time when the humus theory was first proposed by J.G. Wallerius (1709–1785), a Swedish professor of chemistry at the University of Uppsala. This theory stated that plants obtain nutrients from humus-derived extracts (in German Extraktivstoff) that contain water soluble compounds of C, H, O, and N from which the plants could rebuild more complex plant tissue. From these four elements they could also form, using an internal vital source (in German Lebenskraft and in Latin, vis vitalis), other vital elements such as Si and K (van der Ploeg et al., 1999). The theory further assumed that while salts and lime

---

[1] Department of Plant and Soil Sciences, University of Delaware, Newark, Delaware, 19717-1303, USA.

were important to plants, they primarily served the role of promoting the decomposition of humus and the dissolving or dissolution of organic matter into the soil solution (van der Ploeg et al., 1999).

Contrary to prevalent thinking, the debunking of the humus theory, the formulation of mineral nutrition of plants, and the Law of the Minimum, were first proposed by a German scientist Phillip Carl Sprengel (1787–1859), not the celebrated chemist, Justus von Liebig (Browne, 1944; Wendt, 1950; Böhm, 1987; van der Ploeg et al., 1999). Sprengel was born near Hannover, Germany and earned his Ph.D. degree for the University of Gottingen. He became an instructor in agriculture and agricultural chemistry at Gottingen and gave courses in soil analyses, fertilizers, and crops (Browne, 1944). Sprengel's early experiments focused on the humus theory and the sole role of soil organic matter (SOM) as the source of plant nutrients. He determined the water-soluble components in humus extracts of soils and found salts of alkali nitrates, sulfates, chlorides, and phosphates. He concluded that the soluble salts were plant nutrients, and thus disproved the humus theory. However, the humus theory persisted until 1840 when the great influence and stature of Liebig resulted in its demise. Sprengel also investigated inorganic compounds found in the root zones of soils (Sprengel, 1828). He considered 15 elements as important. These included O, C, H, N, S, P, Cl, K, Na, Ca, Mg, Al, Si, Fe, and Mn. Based on these findings he developed the Law of the Minimum (Sprengel, 1838) which stated that if only one of the elements needed for plant growth is lacking, the plant will suffer, despite the fact that all the other essential elements are present in sufficient quantity. Sprengel also noted that elements such as As, Pb, and Se could be detrimental to plants. He stated: "When a plant needs 12 substances to develop, it will not grow if any one of these is missing."

Liebig in his many writings and experiments made these ideas generally accepted. Browne (1944), in his book on the history of agricultural chemistry, stated that "in Liebig's books he was more a promulgator and defender of truths that had already been announced than a discoverer of new knowledge." Despite these landmark discoveries by Sprengel, Justus von Liebig (1803–1873) was a world class organic chemist and is considered by most to be the patriarch of agricultural chemistry as well as the most important scholar of the Modern Period of soil science (Fig. 1). Liebig was born in Darmstadt, Germany and first studied chemistry at the University of Bonn, later transferring to Erlanger where he received his doctoral degree. He then went to Paris where he studied with Gay-Lussac and Thenard. On the recommendation of Alexander von Humboldt he was appointed as Professor of Chemistry at the University of Giessen in Germany in 1824 (Browne, 1944). He rapidly became a star in organic chemistry and students and scholars from around the world were attracted to his laboratories. His early research resulted in the development of techniques for determining the

Figure 1. Justus von Liebig (1803–1873).

elementary composition of organic materials such as malic, aspartic, and uric acids and plant alkaloids.

Liebig published the book that rapidly became famous, "Organic Chemistry In Its Applications to Agriculture and Physiology." In this book Liebig mentions three chemical requirements of crops: 1) substances containing C and N that can provide these elements to the plants; 2) water and its elements; and 3) soil that can provide the inorganic matters vital to plant growth. He de-emphasized the role of humus in soil fertility and plant growth and the humus theory by stating: "Humus does not nourish plants, by being taken up and assimilated in its unaltered state, but by presenting a slow and lasting source of carbonic acid, which is being absorbed by the roots and is the principal nutrient of young plants at a time when being destitute of leaves, they are unable to extract food from the atmosphere."

It can arguably be stated that Liebig established the field of agricultural chemistry. One piece of evidence that supports his stature and influence in the field of agricultural chemistry was that by 1880, there were nine professorial chairs in agricultural chemistry in Germany that were primarily held by Liebig's students (Yaalon, 1997). An example of the influence of Liebig's ideas in the work initiated by Lawes and Gilbert at Rothamsted, England. This is discussed in the chapter by Addiscott in this volume.

Major Themes in Soil Chemistry

As soil became the focus of agricultural chemistry, the term soil chemistry became a more accepted term. There is a vast amount of literature about various historical developments in soil chemistry. I have chosen to focus on a few topics that have profoundly affected the research directions and leitmotifs in soil chemistry, and at the same time have been characterized by lively debates and scientific excitement. I have also chosen to discuss what I believe are some of the frontiers in the field over the next decade. I apologize ahead of time to those soil chemists whom I do not cite or topics that are not included. There is certainly no intention to deliberately slight any piece of work.

Ion Exchange

Soil chemistry, as a subdiscipline of soil science, had its roots in the early 1800s in the observations of two Italian scientists, Giuseppe Gazzeri (1828) and Lambruschini (1830) (see Boulaine, 1994), as well as the German pharmacist Johann Bronner (1837), who noted the importance of the ability of soils to exchange elements. As early as 1819 Gazzeri showed that clay decolored and deodorized liquid manure and retained soluble substances that were available to plants. In 1845, a British farmer, H.S. Thompson (Thomas, 1977), began among the first detailed studies documenting the ability of soils to exchange ions. Thompson, knowing that soils might absorb gaseous ammonia conducted a series of experiments to ascertain: 1) if ammonia, absorbed to soil, could be leached out by rainfall, 2) if soils differed in their ability to absorb ammonia, and 3) if soils can absorb ammonia in the form of NH_4SO_4, which was being used at the time as a N fertilizer. Some of his major findings were that Ca and SO_4 appeared in the leachates examined after rainfall, suggesting that NH_4 was exchanged by Ca and SO_4, and was not retained. Thompson told a consulting chemist to the Royal Agricultural Society, J. Thomas Way, about his discoveries but did not publish his work until 1850, at which time Way had already conducted more in-depth studies (Thomas, 1977).

Way conducted what is now recognized as pioneering research on ion exchange, which has for the most part stood the test of time. He can be considered a patriarch of soil chemistry. Way conducted a remarkable group of experiments and discovered that soils can exchange Ca and Mg and to a lesser extent NH_4 and K. The term base exchange was used to describe this process. At the time it was not known that H, in most cases resulting from hydrolysis of Al, could also be exchangeable. Even after exchangeable H was recognized, the term base exchange was used for many years. It also should be noted that the base exchange process was referred to as absorption until the studies of Wiegner began in 1912. Based on his comprehensive studies, Way drew the following conclusions (Way, 1850, 1852; Kelley, 1948): 1) The cation of a neutral salt solution is absorbed on the soil with an equivalent displacement of Ca from the soil, but the anion of the salt is not absorbed, remaining in solution, unless an insoluble Ca salt is formed; 2) absorption was not due to sand, organic matter, $CaCO_3$ or free alumina, but by the clay fraction of the soil: 3) the ability of the soil to absorb ions was reduced by preheating the sample, but not completely eliminated except at high heating temperatures; 4) absorption was rapid, almost instantaneous; 5) $(NH_4)OH$ and $(NH_4)_2CO_3$ were absorbed by the soil in their entirety, with no exchange of the cation; 6) the ability of soils to exchange bases includes NH_4, K, Na, and Mg but if any bases are applied in the OH or CO_3 forms, the base is absorbed without exchange; 7) Ca, added as sulfate, chloride, or nitrate, passes through the soil without alteration, but $Ca(OH)_2$ and $Ca(HCO_3)_2$ are absorbed completely, like alkaline compounds of other cations; 8) the ability of soils to absorb ammonium increases with the concentration of the ammonium solution and as the ratio of solution to soil increases; and 9) base absorption is irreversible.

The majority of these conclusions have proven to be correct. However, the conclusions that ion exchange is irreversible and that SOM is not important in ion exchange are clearly incorrect. Additionally, although the soils that Way studied indicated that anions were not absorbed, this was probably due to their relative lack of metal oxides. Of course, we now know that soils contain variable-charge minerals such as metal oxides and clay minerals such as kaolinite. As Sumner (1998) has noted, Way's conclusion that soils did not adsorb anions caused most scientists to dismiss the importance of anion exchange capacity until the work of Gedroiz (1925), and particularly Mattson (1927). In 1852, Way published a second paper in which he determined which soil components were important in cation exchange. He proposed that soil silicates were important exchangers and was among the first to propose an order of replacing power of one cation for another. Assuming Ca was the absorbed cation and the exchanger was reacted with a 0.01 N solution of Cl, he proposed the following replacing power order: Na<K<Ca<Mg<NH_4. Liebig, being the ever critical scientist, said about Way's findings: "Agriculture cannot be advanced by agricultural experiments of this kind."

Following Way's remarkable experiments, a number of studies were carried out that confirmed his results. Henneberg and Stohmann (1858) found that the amount of NH_4 adsorbed by a soil depended on the concentration of the NH_4 and the ratio of solution to soil employed. Peters (1860) showed that K could be replaced by Na, Ca, or Mg ions and most easily by NH_4. He found the order of Mg, NH_4 and Na to replace Ca was: Mg>NH_4> Na. He also confirmed Way's finding that cation exchange appeared to be a rapid process. Rautenberg (1862) concluded that organic matter was important in cation exchange. Knop (1872, 1874) hypothesized that certain soil sorbents, such as Al and Fe silicates were important for absorption of NH_4, that the process was chemical in nature, and that the degree of NH_4 absorption was roughly proportional to the quantity of fine-grained soil materials present. Although he did not know it at the time, Knop had correctly pointed out the role of clay minerals and metal oxides in cation exchange and that soil components could have different abilities to retain cations. Pillitz (1875) discovered that soils differed in their abilities to absorb NH_4, but that under constant pressure and temperature, every soil has a maximum "absorptive power." He was thus the first scientist to demonstrate that under a given set of conditions, the cation exchange capacity (CEC) of each soil is a definite quantity. van Bemmelen (1878) concluded that the ion exchange process is a true chemical reaction and he and Eichhorn concluded that zeolites or zeolitic materials absorbed cations and were the primary soil component responsible for cation exchange. For the next 50–70 years, most scientists believed that zeolites were the primary soil exchanger material.

Ion Exchange Studies in the 1900s

For almost 20 years after van Bemmelen's 1888 paper, no important work appeared on ion exchange. In the early 1900s some seminal studies were carried out by the German chemist, G. Wiegner, the Russian scientist, G. Gedroiz, and the Dutch scientist, Hissink. Weigner and his students studied cation exchange from a colloidal chemistry viewpoint. They concluded that permutite in soils and certain adsorption compounds were important sorbents and that exchangeable ions were held on particle surfaces and in capillaries by an adsorption process and not by chemical attraction. Wiegner also made the perceptive conclusion that the anions form a swarm around the particle and in aqueous medium are more or less hydrated and proposed an order in which ions would be preferred on soil exchangers of Li<Na<K<Rb<Cs and Mg<Ca< Sr<Ba, which is the Hofmeister lyotrophic series. Along with his student, Hans Jenny, they emphasized the importance of ion size and hydration in ion exchange selectivity, concluding that the smaller the ion the greater its selectivity and replacing power.

Beginning in 1912, Gedroiz (1872–1933), a Russian, published an amazing group of papers over the next 13 years (1925) that further advanced our understanding of ion exchange. The following studies were conducted: 1) the influence of a number of exchangeable cations on the physical, chemical, and plant nutritional properties of soils; 2) the kinetics of the exchange reaction and the replacing power of different cations; 3) methods for determining exchangeable cations and determining the CEC of soils; 4) ascertaining the relation of exchangeable

Na to alkali and saline soils; 5) the role of exchangeable cations in Chernozem and Podzol soils; 6) base unsaturation and degradation of soils; and 7) the nature of the exchange materials in soils. He, like Wiegner, felt that ion exchange was not a chemical reaction but rather a physicochemical reaction. Gedroiz also reported the following replacing powers of cations: $Li<Na<K<Mg<Rb<NH_4<Co<Al$.

Gedroiz, A. De Dominicis in Italy and J.A. Prescott in Africa (Kelley, 1948) independently proposed that the base exchange principle proposed by Way played a major role in alkali soils. Eugene Woldemar Hilgard (1833–1916), the great soil scientist at Berkeley, who had trained with R.W. Bunsen at the University of Heidelberg in Germany, had identified soils in California as white alkali or salty soils and black alkali soils, and pioneered processes for reclaiming the soils including leaching and adding gypsum. Kelley and coworkers confirmed the role that base exchange played in alkali soils by their finding that Na adsorbed on clay minerals caused the black alkali soils. In Budapest, De' Sigmind (1873–1939) contributed to research on exchange phenomena.

During the 1930s Walter P. Kelley (1878–1965, Fig. 2), a soil chemist trained under Eugene Hilgard at the University of California at Berkeley, and later a faculty member at UC-Riverside and Berkeley, along with his students and colleagues carried out many studies on ion exchange, the cation exchange capacities of soils and seminal research on the chemistry of alkali and saline soils (Kelley and

Figure 2. Walter P. Kelley (1878–1965).

Brown, 1924; Kelley, 1937; Kelley et al., 1939a,b). Much of Kelley's research on cation exchange is summarized in his landmark books, "Cation Exchange in Soils" in 1948, and "Alkaline Soils: Their Formation, Properties and Reclamation," in 1951.

Beginning in 1929, a Swedish soil chemist, Sante Mattson (1886–1980), published a series of papers that at the time were not well accepted, in the journal "Soil Science" with a lead title of "Laws of Soil Colloidal Behavior." In these landmark papers, Mattson (Fig. 3) explained cation exchange on the basis of ionization. He also correctly proposed the idea that soil components have an isoelectric point (IEP) in which equivalent quantities of cations and anions are dissociated, while below the isoelectric point, the dissociation of OH anions exceeds that of cations and in this pH range, anion adsorption can take place. At pHs above the isoelectric point, cation dissociation predominates and cation adsorption predominates. Mattson had pioneered the idea of variable or pH-dependent charge and correctly noted that CEC could vary in soils with pH changes and that pH should be stated when the CEC is reported (Mattson, 1927, 1931, 1932). In fact, he beautifully illustrated that soils from the Southeastern USA contained variable charge colloids that carried a positive charge at acid pHs. Important studies conducted by Schofield and Samson (1953) and Mehlich (1952) also appeared on pH-dependent or variable charge soils.

Figure 3. S.E. Mattson (1886–1980).

Unfortunately, the perceptive observations of Mattson were largely ignored until the late 1950s and 1960s when Coleman and Rich and coworkers rediscovered the importance of Mattson's seminal studies (Coleman et al., 1959; Rich, 1968). However, because much of the soil chemistry research was being conducted in areas where constant (permanent) charge soil minerals dominated, the chemistry of variable charge soils did not receive the attention it deserved (Sumner, 1998). This resulted in practices such as liming and soil analyses methodologies being transferred to parts of the world such as Africa, Asia, and South America where they were not appropriate. For example, the idea of measuring the CEC at pH 7 by extracting with 1 M ammonium acetate solution (a standard and widely used procedure), while appropriate for soils dominated by constant charge minerals such as montmorillonite and vermiculite, is not proper for CEC measurements on highly weathered soils high in kaolinite and Al- and Fe-oxides.

In the first few decades of the 20th century many attempts were also made to derive an equilibrium constant for a binary exchange reaction and numerous ion exchange equations were developed to explain and predict binary reactions (reactions involving two ions) on clay minerals and soils. These were named after the scientists who developed them and included the Kerr, Vanselow, Gapon, Schofield, Krishnamoorthy and Overstreet, Donnan, and Gaines and Thomas equations. However, from a thermodynamic standpoint, it was Albert Vanselow (Vanselow, 1932) who correctly established that a true exchange equilibrium constant should express the activity on the solid phase as the product of the mole fraction and the adsorbed phase activity coefficient. In the early 1950s two seminal papers appeared (Argersinger et al., 1950; Gaines and Thomas, 1953) that provided a theoretical framework for determining a thermodynamic exchange constant, K_{ex}, for a binary process. All of the aforementioned landmark research laid the groundwork for numerous studies throughout the world on ion exchange processes in soils (Sparks, 1995, 2002) and truly established ion exchange as one of the hallmarks of soil chemistry. The reader is further referred to a number of reviews and books dealing with ion exchange history and developments in soil chemistry (Kelley, 1948; Thomas, 1977; Sposito, 1981a,b, 2000; Sparks, 1995, 2002).

Clay Mineralogy

One of the major landmarks in soil chemistry and the identity of soil mineral components was the discovery by Hendricks and Fry (1930) and Kelley et al. (1931) that the major portion of the inorganic fraction of soils was crystalline (Cady and Flach, 1997). In the paper of Hendricks and Fry (1930) they presented results of total chemical analyses, optical studies, and X-ray diffraction on geological specimens of montmorillonite, bentonite, halloysite, kaolinite, and dickite and related them to data on 23 samples of colloids from soils. They indicated that the diffraction patterns showed "the presence of only one crystalline component in each soil" and that "none of the primary soil minerals such as micas, feldspars, and quartz are shown as crystalline components of the colloid fraction." Over the next 12 years, Hendricks and his associates published many papers on clay mineral structures and their physical properties (e.g., Hendricks and Alexander, 1939; Ross and Hendricks, 1945). Kelley et al. (1931), Hoffman et al. (1933), Marshall (1935), Kelley and co-workers (1939 a,b) and M.L. Jackson and his students and colleagues (Coleman and Jackson, 1945) also made significant contributions in delineating the crystal structures of clay minerals and the types of clay minerals found in USA soils M.L. Jackson and his coworkers also developed many standard procedures for soil particle fractionation and separation and phyllosilicate identification (Jackson, 1956). Jacob et al. (1935) analyzed the clay minerals found in soils around the world and Nagelschmidt et al. (1940) determined the clay minerals in soils of India. From these investigations, it was found that in most soils more than two clay minerals were found with kaolinite, montmorillonite, and mica being common. These researchers and others, based on clay mineralogical analyses, concluded that weathered soils were dominated by kaolinite while grassland soils were dominated by montmorillonite.

Clay mineralogy of soils was later developed with more detailed methods of identification of the minerals and their significance in soil chemistry.

Soil Acidity

Another significant leitmotif in soil chemistry in the 20th century was the merry-go-round on soil acidity (Jenny, 1961). Perhaps the first person to lime soils for the correct reason was Edmund Ruffin (1794–1865). His paper "An Essay on Calcareous Manures" greatly influenced soil science in the USA. Ruffin (Fig. 4), as Grant Thomas in his outstanding paper "Historical Developments in Soil Chemistry: Ion Exchange" (1977) so eloquently and eruditely noted, Ruffin was a gentleman farmer, an amateur chemist, politician, philosopher, and rebel who was trying to making a living farming near Petersburg, Virginia, USA. Ruffin fired the first Confederate shot at Fort Sumter, South Carolina, which began the U.S. Civil War, and committed suicide after Appomattox because he did not wish to live under the "perfidious Yankee race." Ruffin added oyster shells to his soils to correct the acidity problem and also accurately described zinc deficiencies in his journals (Thomas, 1977). He concluded that all calcareous soils are naturally fertile and durable and all soils that are naturally poor are devoid of calcareous materials. He speculated that soils on the Eastern US coast were low in $CaCO_3$, which was correct (Jenny, 1961). Ruffin can be considered the father of soil chemistry in the USA.

Figure 4. Edmund Ruffin (1794–1865).

It was 70 years after Ruffin's work before research on soil acidity was initiated again. Questions about whether acidity was primarily attributed to hydrogen or aluminum were the basis for many of the studies. Fierce arguments ensued in the early 1900s and continued for over five decades. As Jenny (1961, Fig. 5) astutely

Figure 5. Hans Jenny (1899–1992).

noted the debates were like a merry-go-around, and as Thomas (1977) has so wonderfully and entertainingly noted, some of the giants of soil chemistry were caught up in the often pointed arguments. F.P. Veitch (1902) found that titration of soils that had been equilibrated with $Ca(OH)_2$ to a pink endpoint with phenolphthalein was a good test for predicting whether lime (e.g., $CaCO_3$) was needed to neutralize acidity, that would be detrimental to crop growth (Thomas, 1977). Hopkins et al. (1903) developed a lime requirement test based on the titration of a soil equilibrated with 1 N NaCl. Veitch (1904) showed that a 1 N NaCl extract, while not replacing all the soil's acidity, was a good lime requirement test. A very important finding by Veitch (1904), that was not recognized at the time, was that the acidity replaced by 1 N NaCl was $AlCl_3$, not HCl.

After Veitch's work a number of soil chemists started to study soil acidity and to debate whether acidity was caused by Al or H. Bradfield (1923, 1925) titrated clays and observed that their pK_a values were similar to those found for weak acids. Kelley and Brown (1926) and Page (1926) hypothesized that "exchangeable Al" was dissolved by exchangeable H^+ during the extraction with salt. Paver and Marshall (1934) believed that the exchangeable H^+ dissolved the clay structure, releasing Al which in turn became a counterion on the exchange complex. This was indeed an important discovery that was not definitively proved and accepted until the 1950s and early 1960s, as we shall see.

Chernov (1947) had shown that electrodialyzed clays and naturally acid clays were primarily Al-saturated. Shortly thereafter, Coleman and Harward (1953) found that H resin-treated clays or clays leached rapidly with 1 M HCl had quite different properties from clays that were slowly leached, leached with dilute acid solutions, or electrodialyzed. They concluded, based on their studies, that hydrogen clays were strongly acid. Low (1955), employing potentiometric and conductometric titration analyses, proved that an electrodialyzed clay was Al-saturated. Coleman and Craig (1961) confirmed the earlier finding of Coleman and Harward (1953) that H clays are unstable and rapidly convert to Al clay, with temperature having a dramatic effect on the transformation rate. The research on H vs Al clays was very important in that it showed that Al is more important in soil acidity than H.

Also in the 1950s and 1960s there were some important discoveries made about the types of Al found in soils. Rich (Fig. 6) and Obenshain (1955) showed that in some Virginia soils, formed from mica schist, there was not only exchangeable Al^{3+}, but also nonexchangeable Al, with the latter blocking exchange sites and thus lowering the cation exchange capacity (CEC) of the soils. The nonexchangeable Al also kept vermiculite from collapsing (Rich, 1964; Rich and Black, 1964) and was referred to as interlayer hydroxy-Al. It was thus Coleman and coworkers (Coleman and Thomas, 1967) and Rich and co-workers (Rich and Obenshain, 1955; Hsu and Rich, 1960) who, based on landmark studies, concluded that

Figure 6. C.I. Rich (1918–1975).

aluminum, including trivalent, monomeric, and polymeric hydroxy Al, was the primary culprit in soil acidity.

During the period of the 1950s and 1960s H. Jenny, P.F. Low, N.T. Coleman, C.E. Marshall, M. Peech, and others also vociferously argued over the cause of the suspension effect (the often observed phenomenon that the pH of a soil suspension was lower than the pH of the overlying solution) and in the end, most agreed that a junction potential, due to differences in ion mobility on soil particle surfaces, was the primary reason for the suspension effect.

Studies on the chemistry of Al and soil acidity reached their zenith in the 1950s and 1960s. In the 1970s and 1980s, as concerns about acid rain escalated, fundamental experiments employing NMR spectroscopy and other analytical techniques and methods were conducted to explore the speciation of Al in solution and soils (Barnhisel and Bertsch, 1982).

Sorption Processes

As mentioned earlier, most of the early research on ion sorption focused on cation sorption and exchange on permanent-charge minerals. This can be ascribed to research being conducted on relatively young soils that were dominated by permanent charge minerals that were negatively charged over a wide pH range. During the 1960s and 1970s, increasing attention was given to chemistry of variable charge systems and anion sorption on variable charge minerals and soils (Sumner, 1963a,b; van Raij and Peech, 1972; Gillman, 1974, 1979; Fey and le Roux, 1976; Thomas, 1977). These studies contributed to our understanding of the effects of pH and ionic strength on the chemistry of variable charge soils and the importance of anion adsorption. The first textbook devoted to the chemistry and physics of variable charge soils appeared in 1981 (Uehara and Gillman, 1981).

In the last 25 years, major advances have been made in modeling sorption reactions at the mineral/water interface and elucidating the kinetics and mechanisms of sorption/desorption phenomena in soils. In the past, soil chemists had employed classical adsorption models such as the Freundlich and Langmuir equations to describe an array of reactions on soils and soil components. While these models often describe experimental data quite well and can be useful in making comparisons about sorption behavior among soils, they provide no mechanistic information. Many investigators, based on parameters calculated from these equations, made conclusions about binding sites and strengths and types of sorption mechanisms (e.g., adsorption, precipitation).

However, in the 1970s and early 1980s, a number of investigators began to question whether the original assumptions of the Langmuir equation were valid for heterogeneous soil materials and the wisdom in making conclusions about sorption mechanisms (Harter and Baker, 1977; Veith and Sposito, 1977; Sposito, 1981b; Harter and Smith, 1981). In a landmark paper, Veith and Sposito (1977) showed that the Langmuir equation could equally well describe both adsorption (accumulation of a substance or material at an interface between the solid surface and the bathing solution) and precipitation (formation of three-dimensional products). Thus, one cannot differentiate between adsorption and other sorption processes, such as surface precipitation, using an "adsorption" isotherm, and clearly mechanistic information cannot be derived from macroscopic models (Sposito, 1986; Scheidegger and Sparks, 1996).

To provide some molecular description to sorption processes at the soil mineral/water interface, a series of surface complexation models (SCM) were developed in the 1970s by Stumm and Schindler and their coworkers, and others, and included the constant capacitance (Schindler and Gamsjager, 1972; Hohl and Stumm, 1976; Schindler et al., 1976; Stumm et al., 1976, 1980), and triple-layer (Davis and Leckie, 1978, 1980) models. Especially detailed reviews on the applications and theoretical aspects of other surface complexation models such as the Stern variable surface charge-variable surface potential (Bowden et al., 1977, 1980; Barrow et al., 1980, 1981), general two-layer model (Dzombak and Morel, 1990) and the one-pK and MUSIC models (Bolt and van Riemsdijk, 1982;

Hiemstra et al., 1987) can be found in Davis and Kent (1990) and Goldberg (1992). Goldberg and Sposito (1984a,b) were among the first soil chemists to apply a SCM (i.e., constant capacitance model) to study the sorption of phosphate on goethite. Since that time, SCM have been widely used by soil chemists and other scientists to describe an array of chemical reactions including proton dissociation, metal cation and anion adsorption reactions on oxides, clay minerals, and soils, organic ligand adsorption on oxides, and competitive adsorption reactions on oxides. Many of these studies are described and alluded to by Goldberg (1992).

While surface complexation models have proved useful in soil chemical investigations, the limitations of the models must be appreciated. One of the major problems with many of them is that multiple adjustable parameters are often used to fit the experimental data. Consequently, it may not be surprising that equilibrium data will fit a number of the models equally well as was beautifully illustrated by Westall and Hohl (1980). Additionally, it has been shown that surface precipitation of metals can occur on soil minerals in addition to purely adsorption phenomena. Most of the SCM do not consider surface precipitation as a possible sorption mechanism, although some attempts have been made to include precipitation processes (Farley et al., 1985; Katz and Hayes, 1995a,b).

Over the past 15 years, major advances have occurred in elucidating sorption mechanisms at the mineral/water interface. Based on pioneering studies by Stumm and Schindler and their coworkers it was assumed that major adsorption complexes were outer-sphere (a water molecule is present between the surface functional group and the bound ion or molecule) and inner-sphere (no water molecule is present between the ion or molecule and the surface functional group to which it is bound). However, until molecular scale studies were performed, conclusions about sorption mechanisms were purely speculative. With the advent of in-situ molecular scale techniques such as Fourier-transform infrared (FTIR), nuclear magnetic resonance (NMR), and x-ray absorption (XAS) spectroscopies and scanning probe microscopy (SPM), definitive information on sorption products and mechanisms have been gleaned (Sparks, 2002).

Kinetics of Soil Chemical Processes

During the late 1970s, and particularly in the 1980s, soil chemists increasingly realized that while the study of equilibrium processes and reactions in soils were important, most field reactions were seldom at equilibrium. It became apparent that one needed to understand the rates of soil chemical processes to accurately predict the fate and transport of ions and organic compounds in soil and water environments, including the development of predictive models.

Perhaps there are two reasons why equilibrium approaches dominated soil chemistry for so long. J. Thomas Way had conducted some kinetic studies during his classic ion exchange research and concluded that the rate of exchange was almost instantaneous. This was later verified by Hissink and Gedroiz. However, Kelley (1948), in his beautiful book, *Cation Exchange in Soils*, had clearly raised the point that perhaps ion exchange and sorption processes would not be so rapid on soil minerals like mica and vermiculite where ion diffusion could occur, resulting in slow reaction rates. However, few soil chemists appeared to have paid attention to Kelley's perceptive observations.

It was left to the classic study of Boyd et al. (1947) and later the seminal studies of Helfferich (1962) and coworkers who demonstrated that diffusion was the rate-limiting process for ion exchange. Later, Scott and coworkers (Scott and Reed, 1962; Scott and Smith, 1966, 1967) and Mortland and coworkers (Mortland, 1958; Mortland and Ellis, 1959) clearly showed that K release from soil minerals was diffusion-controlled. I recall quite vividly that during my Ph.D. research, a number of scientists questioned my findings that K sorption on soils was quite slow, which I attributed to interparticle diffusion into micropores of soil clay minerals. With time, my results were accepted, and many other soil chemists began to study the kinetics of soil chemical reactions.

Another major impediment in promotion of kinetics as a major area in soil chemistry was the complexity of the topic, particularly as kinetics is applied to heterogeneous soil systems, and the lack of good methods to measure both rapid and slow soil chemical reactions. Over the past two decades, soil chemists have developed and employed an array of batch and flow methods to measure reactions over time scales of minutes and longer, and have used chemical relaxation methods, such as pressure-jump and concentration-jump, to study reaction rates on millisecond time scales. These methods are discussed in a number of books and reviews (Sparks, 1989; Amacher, 1991; Sparks and Zhang, 1991; Sparks et al., 1996). Without question, in the last two decades, kinetics has become a major theme in soil chemistry. There is not space to discuss all the significant contributions that soil chemists have made in the area of kinetics, however, the reader can consult a number of references authored by soil chemists (Sparks, 1989; Sparks and Suarez, 1991; Sposito, 1994; Sparks, 1999, 2000, 2002).

The Road in Soil Chemistry, Summary

It is fair to say that the primary impetus in soil chemistry over the first 120 years was to better understand the reactions and processes of plant nutrients in soils with a major emphasis on ion exchange, clay mineralogy, and soil acidity. In the 1970s, major societal concerns centered around the environment, including air, soil, and water quality as impacted by greenhouse gases, trace metals, radionuclides, pesticides and other organic chemicals, and nutrients such as nitrogen. As a result, the emphasis in soil chemistry has shifted to studies on environmental soil chemistry/geochemistry. Over the past 25 years, major studies have been conducted on: acid rain effects on soils and waters; effects of trace metals/metalloids in land applied sludge materials and from other sources on metal/metalloid retention/release in soils, plant uptake and bioavailability; speciation of contaminants in soils using chemical extractions and molecular scale techniques; nitrate and phosphate, organic chemical, and radionuclide fate/-transport in surface and groundwaters; facilitated colloid transport of metals and organic chemicals; elucidation of humic substances structure and C sequestration; residence time effects on contaminant sequestration; redox transformations and mechanisms of inorganic and organic contaminants in soils; development of surface complexation models, based on classical double layer theory, to describe metal and ligand sorption on soils and soil components; soil remediation using chemical and phytoremediation approaches and microbially-mediated transformations of contaminants.

In the past 25 years, advanced analytical techniques became available for the study of reactions at the soil mineral/water interface. Of particular importance was the advent of molecular scale in-situ techniques in which soil chemical reactions can be monitored in aqueous environments. In fact, a new multidisciplinary field of soil and environmental sciences, which often employs the aforementioned techniques, has been created—molecular environmental science. Molecular environmental science can be defined as the study of the physical and chemical forms and distribution of contaminants in soils, sediments, waste materials, natural waters, and the atmosphere at the molecular level. Without question, the employment of in-situ molecular scale techniques, along with macroscopic, equilibrium and kinetic studies has created a revolution in soil and environmental chemistry research. Along with rapid developments in analytical tools, soil chemistry research is becoming highly multi- and inter-disciplinary. Soil chemists are collaborating with chemists, physicists, biologists, geochemists, engineers, material scientists and marine scientists like never before. These interactions and the breathtaking advances and developments in analytical techniques, information technology, and biotechnology will forever change the field of soil chemistry.

References

Aharoni, C., and D.L. Sparks. 1991. Kinetics of soil chemical reactions: A theoretical treatment, pp. 1–18, *In* D.L. Sparks and D.L. Suarez, eds. Rates of Soil Chemical Processes, SSSA Spec. Publ. No. 27. Soil Sci. Soc. Am., Madison, WI.

Ainsworth, C.C., J.L. Pilou, P.L. Gassman, and W.G. Van Der Sluys. 1994. Cobalt, cadmium, and lead sorption to hydrous iron oxide: Residence time effect. Soil Sci. Soc. Am. J. 58:1615–1623.

Amacher, M.C. 1991. Methods of obtaining and analyzing kinetic data, pp. 19–59, *In* D.L. Sparks and D.L. Suarez, eds. Rates of Soil Chemical Processes, SSSA Spec. Publ. No. 27. Soil Sci. Soc. Am., Madison, WI.

Amacher, M.C., and D.E. Baker. 1982. Redox Reactions Involving Chromium, Plutonium, and Manganese in Soils DOE/DP/04515-1. Penn. State University, University Park, PA.

Argersinger, W.J., A.W. Davidson, and O.D. Bonner. 1950. Thermodynamics and ion exchange phenomena. Trans. Kans. Acad. Sci. 53:404–410.

Backes, C.A., R.G. McLaren, A.W. Rate, and R.S. Swift. 1995. Kinetics of cadmium and cobalt desorption from iron and manganese oxides. Soil Sci. Soc. Am. J. 59:778–785.

Ball, W.P., and P.V. Roberts. 1991. Long-term sorption of halogenated organic chemicals by aquifer material: 1. Equilibrium. Environ. Sci. Technol. 25:1223–1236.

Bargar, J.R., G.E. Brown, and G.A. Parks. 1995. XAFS study of Pb(II) sorption at the α-Al_2O_3-water interface. Physica B 209:455–456.

Barnhisel, R., and P.M. Bertsch. 1982. Aluminum, pp. 275–300, *In* A.L. Page, ed. Methods of Soil Analysis: Chemical Methods, Agron. Monogr. 9, Part 2. Am. Soc. Agron., Madison, WI.

Barrow, N.J. 1986. Testing a mechanistic model: II. The effects of time and temperature on the reaction of zinc with a soil. J. Soil Sci. 37:277–286.

Barrow, N.J. 1998. Effects of time and temperature on the sorption of cadmium, zinc, cobalt, and nickel by a soil. Aust. J. Soil Res. 36:941–950.

Barrow, N.J., J.W. Bowden, A.M. Posner, and J.P. Quirk. 1980. An objective method for fitting models of ion adsorption on variable charge surfaces. Aust. J. Soil Res. 18:37–47.

Barrow, N.J., J.W. Bowden, A.M. Posner, and J.P. Quirk. 1981. Describing the adsorption of copper, zinc and lead on a variable charge mineral surface. Aust. J. Soil Res. 19:309–321.

Barrow, N.J., J. Gerth, and G.W. Brummer. 1989. Reaction kinetics of the adsorption and desorption of nickel, zinc and cadmium by goethite: II. Modeling the extent and rate of reaction. J. Soil Sci. 40:437–450.

Bartlett, R.J. 1988. Manganese redox reactions and organic interactions in soils, pp. 59–77, *In* R.D. Graham et al., eds. Manganese in Soils and Plants. Proc. Int. Symp. Kluwer Academic, Dordrecht.

Bartlett, R.J., and B.R. James. 1979. Behavior of chromium in soils. III. Oxidation. J. Environ. Qual. 8:31–35.

Bartlett, R.J., and B.R. James. 1993. Redox chemistry of soils. Adv. Agron. 50:151–208.

Bloom, P.R., and E.A. Nater. 1991. Kinetics of dissolution of oxide and primary silicate minerals, pp. 151–190, *In* D.L. Sparks and D.L. Suarez, eds. Rates of Soil Chemical Processes, Soil Sci. Soc. Am. Spec. Publ. 27. Soil Sci. Soc. Am., Madison, WI.

Bochatay, L., P. Persson, and S. Sjoberg. 2000. Metal ion coordination at the water-manganite (γ-MnOOH) interface I. An EXAFS study of cadmium(II). J. Colloid Interf. Sci. 229:584–592.

Böhm, W. 1987. Der Thaer-Schüler Carl Sprengel (1787–1859) als Begründer der neuzeitlichen Pflanzenernahrung [The Thaer pupil Carl Sprengel (1787–1859) as founder of the modern science of plant nutrition]. Jahresheft der Albrecht-Thaer-Gesellschaft 23:43–59.

Bolt, G.H., and W.H. Van Riemsdijk. 1982. Ion adsorption on inorganic variable charge constituents, pp. 459–503, *In* G.H. Bolt, ed. Soil Chemistry. Part B. Physico-Chemical Methods. Elsevier, Amsterdam.

Boulaine, J. 1994. Early soil science and trends in the early literature, pp. 20–42, *In* P. McDonald, ed. The Literature of Soil Science. Cornell University Press, Ithaca, NY.

Bowden, J.W., A.M. Posner, and J.P. Quirk. 1977. Ionic adsorption on variable charge mineral surfaces. Theoretical charge development and titration curves. Aust. J. Soil Res. 15:121–136.

Bowden, J.W., S. Nagarajah, N.J. Barrow, A.M. Posner, and J.P. Quirk. 1980. Describing the adsorption of phosphate, citrate and selenite on a variable charge mineral surface. Aust. J. Soil Res. 18:49–60.

Boyd, G.E., A.W. Adamson, and J.L.S. Meyers. 1947. The exchange adsorption of ions from aqueous solutions by organic zeolites. II. Kinetics. J. Am. Chem. Soc. 69:2836–2848.

Bradfield, R. 1923. The nature of the acidity of the colloidal clay of acid soils. J. Am. Chem. Soc. 45:2669–2678.

Bradfield, R. 1925. The chemical nature of colloidal clay. J. Am. Soc. Agron. 17:253–370.

Bronner, J.P. 1837. Der Weinbau im Konigreich Wurtenberg. C.F. Winter, Heidelberg.

Browne, C.A. 1944. A Source Book of Agricultural Chemistry. The Chronica Botanica Co., Waltham. MA.

Bruemmer, G.W., J. Gerth, and K.G. Tiller. 1988. Reaction kinetics of the adsorption and desorption of nickel, zinc and cadmium by goethite: I. Adsorption and diffusion of metals. J. Soil Sci. 39:37–52.

Cady, J.G., and K.W. Flach. 1997. History of soil mineralogy in the United States Department of Agriculture. Advances in GeoEcology 29:211–240.

Charlet, L., and A. Manceau. 1992. X-ray absorption spectroscopic study of the sorption of Cr(III) at the oxide-water interface. II: Adsorption, coprecipitation and surface precipitation on ferric hydrous oxides. J. Colloid Interface Sci. 148:443–458.

Charlet, L., and A. Manceau. 1993. Structure, formation, and reactivity of hydrous oxide particles: Insights from x-ray absorption spectroscopy, pp. 117–164, In J. Buffle and H.P. v. Leeuwen, eds. Environmental Particles. Lewis Publishers, Boca Raton, FL.

Chernov, V.A. 1947. The Nature of Soil Acidity (English translation furnished by Hans Jenny – translator unknown). Press of Academy of Sciences, Moscow. Soil Sci. Soc. Am. (1964).

Chisholm-Brause, C.J., P.A. O'Day, G.E. Brown Jr., and G.A. Parks. 1990a. Evidence for multinuclear metal-ion complexes at solid/water interfaces from X-ray absorption spectroscopy. Nature 348:528–530.

Chisholm-Brause, C.J., A.L. Roe, K.F. Hayes, G. E. Brown Jr., G.A. Parks, and J.O. Leckie. 1990b. Spectroscopic investigation of Pb(II) complexes at the γ-Al$_2$O$_3$/water interface. Geochim. Cosmochim. Acta 54:1897–1909.

Coleman, N.T., and D. Craig. 1961. The spontaneous alteration of hydrogen clay. Soil Sci. 91:14–18.

Coleman, N.T., and G.W. Thomas. 1967. The basic chemistry of soil acidity, pp. 1–41, In R.W. Pearson and F. Adams, eds. Soil Acidity and Liming, Agron. Monogr.12. Am. Soc. Agron., Madison, WI.

Coleman, R., and M.L. Jackson. 1945. Mineral composition of the clay fraction of several Coastal Plain soils of Southeastern United States. Soil Sci. Soc. Am. Proc. 10:381–391.

Coleman, N.T., and M.E. Harward. 1953. The heats of neutralization of acid clays and cation exchange resins. J. Am. Chem. Soc. 75:6045–6046.

Coleman, N.T., E.J. Kamprath, and S.B. Weed. 1959. Liming. Adv. Agron. 10:475–522.

Comans, R.N.J., and D.E. Hockley. 1992. Kinetics of cesium sorption on illite. Geochim. Cosmochim. Acta 56:1157–1164.

Davis, J.A., and D.B. Kent. 1990. Surface complexation modeling in aqueous geochemistry, pp. 177–260, In M.F. Hochella and A.F. White, eds. Mineral-Water Interface Geochemistry, Vol. 23. Mineralogical Society of America, Washington, DC.

Davis, J.A., and J.O. Leckie. 1978. Surface ionization and complexation at the oxide/water interface. II. Surface properties of amorphous iron oxyhydroxide and adsorption of metal ions. J. Colloid Interface Sci. 67:90–107.

Davis, J.A., and J.O. Leckie. 1980. Surface ionization and complexation at the oxide/water interface. III. Adsorption of anions. J. Colloid Interface Sci. 74:32–43.

Davy, H., G. Sinclair, and J.R. Bedford. 1813. Elements of Agricultural Chemistry, in a Course of Lectures for the Board of Agriculture. Longman, Hurst, Rees, Orme, and Brown, London.

Dzombak, D.A., and F.M.M. Morel. 1990. Surface Complexation Modeling. Hydrous Ferric Oxide. Wiley, New York.

Elzinga, E.J., and D.L. Sparks. 1999. Nickel sorption mechanisms in a pyrophyllite-montmorillonite mixture. J. Colloid Interf. Sci. 213:506–512.

Farley, K.J., D.A. Dzombak, and F.M.M. Morel. 1985. A surface precipitation model for the sorption of cations on metal oxides. J. Colloid Interface Sci. 106:226–242.

Fendorf, S.E. 1992. Oxidation and sorption mechanisms of hydrolyzable metal ions on oxide surfaces. Ph.D. Dissertation, University of Delaware, Newark.

Fendorf, S.E., M.J. Eick, P.R. Grossl, and D.L. Sparks. 1997. Arsenate and chromate retention mechanisms on goethite. 1. Surface structure. Environ. Sci. Technol. 31:315–320.

Fendorf, S.E., M. Fendorf, D.L. Sparks, and R. Gronsky. 1992. Inhibitory mechanisms of Cr (III) oxidation by γ-MnO_2. J. Colloid Interface Sci. 153:37–54.

Fendorf, S.E., G.M. Lamble, M.G. Stapleton, M.J. Kelley, and D.L. Sparks. 1994a. Mechanisms of chromium (III) sorption on silica: 1. Cr(III) surface structure derived by extended x-ray absorption fine structure spectroscopy. Environ. Sci. Technol. 28:284–289.

Fendorf, S.E., and D.L. Sparks. 1994. Mechanisms of chromium (III) sorption on silica: 2. Effect of reaction conditions. Environ. Sci. Technol. 28:290–297.

Fendorf, S.E., D.L. Sparks, G.M. Lamble, and M.J. Kelley. 1994b. Applications of X-ray absorption fine structure spectroscopy to soils. Soil Sci. Soc. Am. 58:1583–1595.

Fendorf, S.E., and R.J. Zasoski. 1992. Chromium (III) oxidation by γ-MnO_2: I. Characterization. Environ. Sci. Technol. 26:79–85.

Fey, M.V., and J.L. Roux. 1976. Electric charges on sesquioxidic soil clays. Soil Sci. Soc. Am. J. 40:359–364.

Fitts, J.P., G.E. Brown, and G.A. Parks. 2000. Structural evolution of Cr(III) polymeric species at the γ-Al_2O_3/water interface. Environ. Sci. Technol. 34:5122–5128.

Ford, R.G., A.C. Scheinost, K.G. Scheckel, and D.L. Sparks. 1999. The link between clay mineral weathering and the stabilization of Ni surface precipitates. Environ. Sci. Technol. 33:3140–3144.

Ford, R.G., A.C. Scheinost, and D.L. Sparks. 2001. Frontiers in metal sorption/precipitation mechanisms on soil mineral surfaces. Adv. Agron. 74:41–62.

Ford, R.G., and D.L. Sparks. 2000. The nature of Zn precipitates formed in the presence of pyrophyllite. Environ. Sci. Technol. 34:2479–2483.

Fuller, C.C., J.A. Davis, and G.A. Waychunas. 1993. Surface chemistry of ferrihydride: Part 2. Kinetics of arsenate adsorption and coprecipitation. Geochim. Cosmochim. Acta 57:2271–2282.

Gaines, G.L., and H.C. Thomas. 1953. Adsorption studies on clay minerals. II. A formation of the thermodynamics of exchange adsorption. J. Chem. Phys. 21:714–718.

Gazzeri, G. 1828. Compendio d'un trattato elementare di chimica. 3rd ed. Stamperia Piatti, Firenze, Italy.

Gedroiz, K.K. 1925. Papers on Soil Reactions. USDA, Washington, D.C.

Ghosh, K., and M. Schnitzer. 1980. Macromolecular structure of humic substances. Soil Sci. 129:266–276.

Gillman, G.P. 1974. The influence of net charge on water dispersible clay and sorbed sulphate. Aust. J. Soil Res. 12:173–176.

Gillman, G.P. 1979. Proposed method for the measurement of exchange properties of highly weathered soils. Aust. J. Soil Res. 17:129–139.

Goldberg, S. 1992. Use of surface complexation models in soil chemical systems. Adv. Agron. 47:233–329.

Goldberg, S., and G. Sposito. 1984a. A chemical model of phosphate adsorption by soils. I. Reference oxide minerals. Soil Sci. Soc. Am. J. 48:772–778.

Goldberg, S., and G. Sposito. 1984b. A chemical model of phosphate adsorption by soils. II. Noncalcareous soils. Soil Sci. Soc. Am. J. 48:779–783.

Grossl, P.R., M.J. Eick, D.L. Sparks, S. Goldberg, and C.C. Ainsworth. 1997. Arsenate and chromate retention mechanisms on goethite. 2. Kinetic evaluation using a pressure-jump relaxation technique. Environ. Sci. Technol. 31:321–326.

Harter, R.D., and D.E. Baker. 1977. Application and misapplications of the Langmuir equation to soil adsorption phenomena. Soil Sci. Soc. Am. J. 41:1077–1088.

Harter, R.D., and G. Smith. 1981. Langmuir equation and alternate methods of studying "adsorption" reactions in soils, pp. 167–182, In R.H. Dowdy et al., eds. Chemistry in the Soil Environment. Am. Soc. Agron./Soil Sci. Soc. Am., Madison, WI.

Hayes, K.F., A.L. Roe, J.G.E. Brown, K.O. Hodgson, J.O. Leckie, and G.A. Parks. 1987. In situ x-ray absorption study of surface complexes: Selenium oxyanions on α-FeOOH. Science 238:783–786.

Helfferich, F. 1962. Ion exchange kinetics. III. Experimental test of the theory of particle-diffusion controlled ion exchange. J. Phys. Chem. 66:39–44.

Hendricks, S.B., and L.T. Alexander. 1939. Minerals present in soil colloids: I. Descriptions and methods for identification. Soil Sci. 48:257–268.

Hendricks, S.B., and W.H. Fry. 1930. The results of X-ray and mineralogical examination of soil colloids. Soil Sci. 29:457–476.

Henneberg, W., and F. Stohmann. 1858. ueber das Verhalten der Ackerkrum gegren Ammoniak und Ammoniaksalze. Ann. Chem. Pharm. 107:152–174.

Hiemstra, T., W.H. Van Riemsdijk, and M.G.M. Bruggenwert. 1987. Proton adsorption mechanism at the gibbsite and aluminum oxide solid/solution interface. Neth. J. Agric. Sci. 35:281–293.

Hingston, F.J., A.M. Posner, and J.P. Quirk. 1972. Anion adsorption by goethite and gibbsite. I. The role of the proton in determining adsorption envelopes. J. Soil Sci. 23:177–192.

Hoffman, U., K. Endell, and D. Wilm. 1933. Kristalstruktur und quellung von Montmorillonit. Z. Krist. 86:340–348.

Hohl, H., and W. Stumm. 1976. Interaction of Pb^{2+} with hydrous γ-Al_2O_3. J. Colloid Interface Sci. 55:281–288.

Hopkins, C.G., W.H. Knox, and J.H. Pettit. 1903. A quantitative method for determining the acidity of soils. USDA Bur. Chem. Bull. 73:114–119.

Hsu, P.H., and C.I. Rich. 1960. Aluminum fixation in a synthetic cation exchanger. Soil Sci. Soc. Am. Proc. 24:21–25.

Huang, P.M. 1991. Kinetics of redox reactions on manganese oxides and its impact on environmental quality, pp. 191–230, In D.L. Sparks and D.L. Suarez, eds. Rates of Soil Chemical Processes. Soil Sci. Soc. Am., Madison, WI.

Hug, S.J. 1997. In situ Fourier transform infrared measurements of sulfate adsorption on hematite in aqueous solutions. J. Colloid Interf. Sci. 188:415–422.

Hunter, D.B., and P.M. Bertsch. 1994. In situ measurements of tetraphenylboron degradation kinetics on clay mineral surfaces by FTIR. Environ. Sci. Technol. 28:686–691.

Jackson, M.G. 1956. Soil Chemical Analysis – Advance Course. Published by author.

Jacob, A., U. Hofmann, H. Loofmann, and E. Maegdefrau. 1935. Chemische und röntgenugraphische untersuchungen über die mineralische Sorptions-substanz in Boden. Z. Ver. deut. Chem. Beiheffe 21:11–19.

James, B.R., and R.J. Bartlett. 2000. Redox phenomena, pp. B-169–B-194, In M.E. Sumner, ed. Handbook of Soil Science. CRC Press, Boca Raton.

Jenny, H. 1961. E. W. Hilgard and the Birth of Modern Soil Science. Collana Della Rivista "Agrochimica", Pisa, Italy.

Karickhoff, S.W., D.S. Brown, and T.A. Scott. 1979. Sorption of hydrophobic pollutants on natural sediments. Water Res. 13:241–248.

Katz, L.E., and K.F. Hayes. 1995a. Surface complexation modeling I. Strategy for modeling monomer complex formation at moderate surface coverage. J. Colloid Interf. Sci. 170:477–490.

Katz, L.E., and K.F. Hayes. 1995b. Surface complexation modeling II. Strategy for modeling polymer and precipitation reactions at high surface coverage. J. Colloid Interf. Sci. 170:491–501.

Kelley, W.P. 1937. The reclamation of alkali soils. California Agr. Exp. Sta. Bull. 617:3–40.

Kelley, W.P., A.O. Woodford, W.H. Dore, S.M. Brown. 1939a. Comparative study of the colloids of a Cecil and a Susquehanna soil profile. Soil Sci. 47:175–193.

Kelley, W.P., W.H. Dore, A.O. Woodford, S.M. Brown. 1939b. The colloidal constituents of California soils. Soil Sci. 48:201–255.

Kelley, W.P. 1948. Cation Exchange in Soils. Reinhold Pub. Corp., New York.

Kelley, W.P. 1951. Alkaline Soils. Their Formation, Properties and Reclamation. Reinhold, New York.

Kelley, W.P., and S.M. Brown. 1924. Replaceable bases in soils. California Agr. Exp. Sta. Tech. Paper 15:1–39.

Kelley, W.P., and S.M. Brown. 1926. Ion exchange in relation to soil acidity. Soil Sci. 21:289–302.

Kelley, W.P., W.H. Dore, and S.M. Brown. 1931. The nature of the base exchange material of bentonite, soils and zeolites, as revealed by chemical investigation and X-ray analysis. Soil Sci. 31:25–55.

Knop, W. 1872. Die Bonitirung der Ackererde. 2nd ed., Leipzig.

Knop, W. 1874. Methode der chemischen Analyse der Ackererden. Landw. Vers. Stat. 17:70–87.

Liebig, J.v., and L.P. Playfair. 1840. Organic Chemistry in its Applications to Agriculture and Physiology. Taylor and Walton, London.

Low, P.F. 1955. The role of aluminum in the titration of bentonite. Soil Sci. Soc. Am. Proc. 19:135–139.

Manceau, A., and L. Charlet. 1994. The mechanism of selenate adsorption on goethite and hydrous ferric oxide. J. Colloid Interface Sci. 164:87–93.

Manceau, A., B. Lanson, M.L. Schlegel, J.C. Harge, M. Musso, L. Eybert-Berard, J.L. Hazemann, D. Chateigner, and G.M. Lamble. 2000. Quantitative Zn speciation in smelter-contaminated soils by EXAFS spectroscopy. Am. J. Sci. 300:289–343.

Marshall, C.E. 1935. Layer lattices and the base-exchange clays. Z. Krist. 91:433–449.

Mattson, S. 1927. Anionic and cationic adsorption by soil colloidal materials of varying $SiO_2/(Al_2O_3 + Fe_2O_3)$ ratio. Trans. 1st. Int. Cong. Soil Sci. pp. 199–211.

Mattson, S. 1931. The laws of soil colloidal behavior: VI. Amphoteric behavior. Soil Sci. 32:343–365.

Mattson, S. 1932. The laws of soil colloidal behavior: VII. Proteins and proteinated complexes. Soil Sci. 33:41–72.

McBride, M.B. 1994. Environmental Chemistry of Soils. Oxford University Press, New York.

McKenzie, R.M. 1967. The sorption of cobalt by manganese minerals in soils. Aust. J. Soil Res. 5:235–246.

McKenzie, R.M. 1980. The adsorption of lead and other heavy metals on oxides of manganese and iron. Aust. J. Soil Res. 18:61–73.

McLaren, R.G., C.A. Backes, A.W. Rate, and R.S. Swift. 1998. Cadmium and cobalt desorption kinetics from soil clays: Effect of sorption period. Soil Sci. Soc. Am. J. 62:332–337.

Mehlich, A. 1952. Effect of iron and aluminum oxides on the release of calcium and on the cation-anion exchange properties of soils. Soil Sci. 73:361–374.

Mortland, M.M. 1958. Kinetics of potassium release from biotite. Soil Sci. Soc. Am. Proc. 22:503–508.

Mortland, M.M., and B.G. Ellis. 1959. Release of fixed potassium as a diffusion-controlled process. Soil Sci. Soc. Am. Proc. 23:363–364.

Myneni, S.C.B., J.T. Brown, G.A. Martinez, and W. Meyer-Ilse. 1999. Imaging of humic substance macromolecular structures in water. Science 286:1335–1337.

Nagleschmidt, G., A.D. Desai, and A. Muir. 1940. The minerals in the clay fractions of a black cotton soil and a red earth from Hyderabad, Deccan State, India. J. Agr. Sci. 30:639–653.

O'Day, P.A., G.A. Parks, and G.E. Brown, Jr. 1994a. Molecular structure and binding sites of cobalt(II) surface complexes on kaolinite from X-ray absorption spectroscopy. Clays Clay Miner. 42:337–355.

O'Day, P.A., G.E. Brown, Jr., and G.A. Parks. 1994b. X-ray absorption spectroscopy of cobalt (II) multinuclear surface complexes and surface precipitates on kaolinite. J. Colloid Interface Sci. 165:269–289.

O'Day, P.A., C.J. Chisholm-Brause, S.N. Towle, G.A. Parks, and G.E. Brown, Jr. 1996. X-ray absorption spectroscopy of Co(II) sorption complexes on quartz (α-SiO_2) and rutile (TiO_2). Geochim. Cosmochim. Acta. 60:2515–2532.

Oscarson, D.W., P.M. Huang, and W.K. Liaw. 1980. The oxidation of arsenite by aquatic sediments. J. Environ. Qual. 9:700–703.

Page, H.J. 1926. The nature of soil acidity. Int. Soc. Soil Sci. Trans. Comm. 2 A:232–244.

Papelis, C., and K.F. Hayes. 1996. Distinguishing between interlayer and external sorption sites of clay minerals using X-ray absorption spectroscopy. Coll. Surfaces 107:89.

Paver, H., and C.E. Marshall. 1934. The role of aluminum in the reactions of the clays. Chem. Ind. (London):750–760.

Peak, D., R.G. Ford, and D.L. Sparks. 1999. An in-situ ATR-FTIR investigation of sulfate bonding mechanisms on goethite. J. Colloid Interf. Sci. 218:289–299.

Peak, D., G.W. Luther, and D.L. Sparks. 2003. ATR-FTIR spectroscopic studies of boric acid adsorption on hydrous ferric oxide. Geochim. Cosmochim. Acta 67:2551–2560.

Peters, C. 1860. Ueber die Absorption von Kali durch Ackerede. Lardwirtsch Vers. Stn. 2:131–151.

Pignatello, J.J. 2000. The measurement and interpretation of sorption and desorption rates for organic compounds in soil media. Adv. Agron. 69:1–73.

Pignatello, J.J., and L.Q. Huang. 1991. Sorptive reversibility of atrazine and metolachlor residues in field soil samples. J. Environ. Qual. 20:222–228.

Pillitz, W. 1875. Studien über die Bodenabsorption. Z. Anal. Chem. 14:55–71; 283–297.

Rautenberg, F. 1862. Über die Absorptionsfahigkeit verschiedener Bodenarten und das geognostische Vorkommen derselben. J. Landw 7:49.

Rich, C.I. 1964. Effect of cation size and pH on potassium exchange in Nason soil. Soil Sci. 98:100–106.

Rich, C.I. 1968. Mineralogy of soil potassium, pp. 79–96, In V.J. Kilmer et al., eds. The Role of Potassium in Agriculture. Am. Soc. Agron., Madison, WI.

Rich, C.I., and W.R. Black. 1964. Potassium exchange as affected by cation size, pH, and mineral structure. Soil Sci. 97:384–390.

Rich, C.I., and S.S. Obenshain. 1955. Chemical and clay mineral properties of a red-yellow podzolic soil derived from muscovite schist. Soil Sci. Soc. Am. Proc. 19:334–339.

Roberts, D.R., A.M. Scheidegger, and D.L. Sparks. 1999. Kinetics of mixed Ni-Al precipitate formation on a soil clay fraction. Environ. Sci. Technol. 33:3749–3754.

Roberts, D.R., R.G. Ford, and D.L. Sparks. 2003. Kinetics and mechanisms of Zn complexation on metal oxides using EXAFS spectroscopy. J. Colloid Interf. Sci. 263:364–376.

Roberts, D.R., A.C. Scheinost, and D.L. Sparks. 2002. Zinc speciation in a smelter-contaminated soil profile using bulk and microspectroscopic techniques. Environ. Sci. Technol. 36:1742–1750.

Roe, A.L., K.F. Hayes, C.J. Chisholm-Brause, G.E. Brown Jr., G.A. Parks, K.O. Hodgson, and J.O. Leckie. 1991. In situ X-ray absorption study of lead ion surface complexes at the goethite/water interface. Langmuir 7:367–373.

Ross, C.S., and S.B. Hendricks. 1945. Minerals of the montmorillonite group, their origin and relation to soils and clays. U. S. Geol. Surv., Tech. Pap. 205B.

Ruffin, E. 1832. An Essay on Calcareous Manures. J.W. Campbell, Petersburg, Va.

Scheckel, K.G., A.C. Scheinost, R.G. Ford, and D.L. Sparks. 2000. Stability of layered Ni hydroxide surface precipitates – A dissolution kinetics study. Geochim. Cosmochim. Acta 64:2727–2735.

Scheckel, K.G., and D.L. Sparks. 2001. Dissolution kinetics of Ni suface precipitates on clay mineral and oxide surfaces. Soil Sci. Soc. Am. J. 65:719–728.

Scheidegger, A.M., and D.L. Sparks. 1996. A critical assessment of sorption-desorption mechanisms at the soil mineral/water interface. Soil Sci. 161:813–831.

Scheidegger, A.M., G.M. Lamble, and D.L. Sparks. 1997. Spectroscopic evidence for the formation of mixed-cation, hydroxide phases upon metal sorption on clays and aluminum oxides. J. Colloid Interf. Sci. 186:118–128.

Scheidegger, A.M., D.G. Strawn, G.M. Lamble, and D.L. Sparks. 1998. The kinetics of mixed Ni-Al hydroxide formation on clay and aluminum oxide minerals: A time-resolved XAFS study. Geochim. Cosmochim. Acta 62:2233–2245.

Scheinost, A.C., R.G. Ford, and D.L. Sparks. 1999. The role of Al in the formation of secondary Ni precipitates on pyrophyllite, gibbsite, talc, and amorphous silica: A DRS Study. Geochim. Cosmochim. Acta 63:3193–3203.

Schindler, P.W. 1981. Surface complexes at oxide-water interfaces, pp. 1–49, *In* M.A. Anderson and A.J. Rubin, eds. Adsorption of Inorganics at Solid-Liquid Interfaces. Ann Arbor Sci., Ann Arbor, MI.

Schindler, P.W., and H. Gamsjager. 1972. Acid-base reactions of the TiO_2 (anatase)-water interface and the point of zero charge of TiO_2 suspensions. Kolloid-Z. Z. Polym. 250:759–763.

Schindler, P.W., B. Fürst, B. Dick, and P.U. Wolf. 1976. Ligand properties of surface silanol groups. I. Surface complex formation with Fe^{3+}, Cu^{2+}, Cd^{2+}, and Pb^{2+}. J. Colloid Interface Sci. 55:469–475.

Schnitzer, M. 2000. A lifetime perspective on the chemistry of soil organic matter. Adv. Agron. 68:1–58.

Schnitzer, M., and H.-R. Schulten. 1995. Analysis of organic matter in soil extracts and whole soils by pyrolysis-mass spectrometry. Adv. Agron. 55:168–218.

Schofield, R.K., and H.R. Samson. 1953. The deflocculation of kaolinite suspensions and the accompanying change-over from positive to negative chloride adsorption. Clay Miner. Bull. 2:45–51.

Schulten, H.-R., and M. Schnitzer. 1993. A state of the art structural concept for humic substances. Naturwissenschaften 80:29–30.

Schulze, D.G., and P.M. Bertsch. 1995. Synchrotron x-ray techniques in soil, plant, and environmental research. Adv. Agron. 55:1–66.

Scott, A.D., and M.G. Reed. 1962. Chemical extraction of potassium from soils and micaceous minerals with solution containing sodium tetraphenylboron. II. Biotite. Soil Sci. Soc. Am. Proc. 26:41–45.

Scott, A.D., and S.J. Smith. 1966. Susceptibility of interlayer potassium in micas to exchange with sodium. Clays Clay Miner. 14:69–81.

Scott, A.D., and S.J. Smith. 1967. Visible changes in macro mica particles that occur with potassium depletion. Clays Clay Miner. 15:357–373.

Scribner, S.L., T.R. Benzing, S. Sun, and S.A. Boyd. 1992. Desorption and bioavailability of aged simazine residues in soil from a continuous corn field. J. Environ. Qual. 21:115–120.

Sparks, D.L. 1989. Kinetics of Soil Chemical Processes. Academic Press, San Diego.

Sparks, D.L. 1995. Environmental Soil Chemistry. Academic Press, San Diego.

Sparks, D.L. 1998. Kinetics of sorption/release reactions on natural particles, pp. 413–448, *In* P.M. Huang et al., eds. Structure and Surface Reactions of Soil Particles, Vol. 4. John Wiley and Sons, New York.

Sparks, D.L. 1999. Kinetics of soil chemical phenomena: Future directions, pp. 81–102, *In* P.M. Huang et al., eds. Future Prospects for Soil Chemistry. Soil Sci. Soc. Am., Madison, WI.

Sparks, D.L. 2000. Kinetics and mechanisms of soil chemical reactions, pp. B-123–B-168, *In* M.E. Sumner, ed. Handbook of Soil Science. CRC Press, Boca Raton.

Sparks, D.L. 2002. Environmental Soil Chemistry. 2nd ed. Academic Press, San Diego.

Sparks, D.L., S.E. Fendorf, I.C.V. Toner, and T.H. Carski. 1996. Kinetic methods and measurements, pp. 1275–1307, *In* D.L. Sparks, ed. Methods of Soil Analysis: Chemical Methods. Soil Sci. Soc. Am., Madison, WI.

Sparks, D.L., and D.L. Suarez (eds.) 1991. Rates of Soil Chemical Processes, SSSA Spec. Publ. No. 27. Soil Sci. Soc. Am., Madison, WI.

Sparks, D.L., and P.C. Zhang. 1991. Relaxation methods for studying kinetics of soil chemical phenomena, pp. 61–94, *In* D.L. Sparks and D.L. Suarez, eds. Rates of Soil Chemical Processes, Soil Sci. Soc. Am. Spec. Publ. 27. Soil Sci. Soc. Am., Madison, WI.

Sposito, G. 1981. Cation exchange in soils: An historical and theoretical perspective, pp. 13–30, *In* R.H. Dowdy et al., eds. Chemistry in the Soil Environment. Am. Soc. Agron. Soil Sci. Soc. Am., Madison, WI.

Sposito, G. 1981. The Thermodynamics of the Soil Solution. Oxford Univ. Press (Clarendon), Oxford.

Sposito, G. 1986. Distinguishing adsorption from surface precipitation, pp. 217–229, *In* J.A. Davis and K.F. Hayes, eds. Geochemical Processes at Mineral Surfaces, Vol. 323. Am. Chem. Soc., Washington, DC.

Sposito, G. 1994. Chemical Equilibria and Kinetics in Soils. Wiley, New York.

Sposito, G. 2000. Ion Exchange Phenomena, pp. B-241–B-264, *In* M.E. Sumner, ed. Handbook of Soil Science. CRC Press, Boca Raton.

Sprengel, C. 1828. Von den Substanzen der Ackerkrume und des Untergrudes (About the substances in the plow layer). Journal für Technische und Ökonomische Chemie 2.

Sprengel, C. 1838. Die Lehre von den Urbarmachungen und Grundverbesserungen (The science of cultivation and soil ameiloration). Immanuel Müller Publ. Co., Leipzig, Germany.

Steinberg, S.M., J.J. Pignatello, and B.L. Sawhney. 1987. Persistence of 1,2 dibromo-ethane in soils: Entrapment in intra particle micropores. Environ. Sci. Technol. 21:1201–1208.

Strawn, D.G., A.M. Scheidegger, and D.L. Sparks. 1998. Kinetics and mechanisms of Pb(II) sorption and desorption at the aluminum oxide-water interface. Environ. Sci. Technol. 32:2596–2601.

Strawn, D.G., and D.L. Sparks. 1999. The use of XAFS to distinguish between inner- and outer-sphere lead adsorption complexes on montmorillonite. J. Colloid Interf. Sci. 216:257–269.

Strawn, D.G., and D.L. Sparks. 2000. Effects of soil organic matter on the kinetics and mechanisms of Pb(II) sorption and desorption in soil. Soil Sci. Soc. Am. J. 64:144–156.

Stumm, W., H. Hohl, and F. Dalang. 1976. Interaction of metal ions with hydrous oxide surfaces. Croat. Chem. Acta 48:491–504.

Stumm, W., K. Kummert, and L. Sigg. 1980. A ligand exchange model for the adsorption of inorganic and organic ligands at hydrous oxide interfaces. Croat. Chem. Acta 53:291–312.

Suarez, D.L., S. Goldberg, and C. Su. 1998. Evaluation of oxyanion adsorption mechanisms on oxides using FTIR spectroscopy and electrophoretic mobility, pp. 136–178, In D.L. Sparks and T.J. Grundl, eds. Mineral-Water Interfacial Reactions: Kinetics and Mechanisms. Am. Chem. Soc., Washington, D.C.

Sumner, M.E. 1963a. Effect of alcohol washing and pH value of leaching solution on positive and negative charges in ferruginous soils. Nature (London) 198: 1018–1019.

Sumner, M.E. 1963b. Effect of iron oxides on positive and negative charge in clays and soils. Clay Min. Bull. 5:218–226.

Sumner, M.E. 1998. Soil chemistry: Past, present and future, pp. 1–38, In P.M. Huang et al., eds. Future Prospects for Soil Chemistry. Soil Sci. Soc. Am., Madison, Wis.

Taylor, R.M. 1984. The rapid formation of crystalline double hydroxy salts and other compounds by controlled hydrolysis. Clay Minerals 19:591–603.

Thomas, G.W. 1977. Historical developments in soil chemistry: Ion exchange. Soil Sci. Soc. Am. J. 41:230–238.

Thompson, H.A., G.A. Parks, and G.E. Brown Jr. 1999. Dynamic interactions of dissolution, surface adsorption, and precipitation in an aging cobalt(II)-clay-water system. Geochim. Cosmochim. Acta 63:1767–1779.

Thompson, H.A., G.A. Parks, and G.E. Brown, Jr. 1996. Structure and composition of U(VI) sorption complexes at the kaolinite-water interface. Abstracts of Papers of the American Chemical Society 211:99-Geoc.

Thompson, H.A., G.A. Parks, and G.E. Brown Jr. 1999. Ambient-temperature synthesis, evolution, and characterization of cobalt-aluminum hydrotalcite-like solids. Clays and Clay Min 47:425–438.

Tonner, B., and S.J. Traina. 1998. Microorganisms, organic contaminants, and plant-metal interactions. Workshop on Scientific Directions at the Advanced Light Source, pp. 170–172.

Towle, S.N., J.R. Bargar, G.E. Brown, Jr., and G.A. Parks. 1997. Surface precipitation of Co(II) (aq) on Al_2O_3. J. Colloid Interf. Sci. 187:62–82.

Uehara, G., and G. Gillman. 1981. The Mineralogy, Chemistry, and Physics of Tropical Soils with Variable Charge Clays. Westview Press, Boulder, CO.

van Bemmelen, J.M. 1878. Das Adsorptionvermögen der Ackererde. Landw. Vers. Stat. 21.

van Bemmelen, J.M. 1879. Das Adsorptionvermögen der Ackererde. Landw. Vers. Stat. 23.

van Bemmelen, J.M. 1888. Die Absorptionsverbindungen und das Absorptionsvermögen der Ackererde. Landw. Vers. Stat. 35:69–136.

van der Ploeg, R.R., W. Böhm, and M.B. Kirkham. 1999. On the origin of the theory of mineral nutrition of plants and the law of the minimum. Soil Sci. Soc. Am. J. 63:1055–1062.

van Raij, B., and M. Peech. 1972. Electrochemical properties of some Oxisols and Alfisols of the tropics. Soil Sci. Soc. Am. Proc. 36:587–593.

Vanselow, A.P. 1932. Equilibria of the base exchange reactions of bentonites, permutites, soil colloids, and zeolites. Soil Sci. 33:95–113.

Veitch, F.P. 1902. The estimation of soil acidity and the lime requirement of soils. J. Am. Chem. Soc. 24:1120–1128.

Veitch, F.P. 1904. Comparison of methods for the estimation of soil acidity. J. Am. Chem. Soc. 26:637–662.

Veith, J.A., and G. Sposito. 1977. On the use of the Langmuir equation in the interpretation of "adsorption" phenomena. Soil Sci. Soc. Am. J. 41:497–502.

Way, J.T. 1850. On the power of soils to absorb manure. J. Royal Agric. Soc. Engl. 11:313–379.

Way, J.T. 1852. On the power of soils to absorb manure. J. Royal Agric. Soc. Engl. 13:123–143.

Weesner, F.J., and W.F. Bleam. 1997. X-ray absorption and EPR spectroscopic characterization of adsorbed copper(II) complexes at the boehmite (AlOOH) surface. J. Colloid Interf. Sci. 196:79–86.

Wendt, G. 1950. Carl Sprengel und die von ihm geschaffene Mineraltheorie als Fundament der neuen Pflanzenernährungslehre (Carl Sprengel and his mineral theory as foundation of the modern science of plant nutrition). Ernst Fisher Publ. Co., Wolfenbüttel, Germany.

Westall, J.C., and H. Hohl. 1980. A comparison of electrostatic models for the oxide/solution interface. Adv. Colloid Interface Sci. 12:265–294.

Xia, K., W. Bleam, and P.A. Helmke. 1997. Studies of the nature of binding sites of first row transition elements bound to aquatic and soil humic substances using X-ray absorption spectroscopy. Geochim. Cosmochim. Acta 61:2223–2235.

Yaalon, D.H. 1997. Introducing soils as an object of study, In D.H. Yaalon and S. Berkowicz, eds. History of Soil Science. Catena Verlag GMBH, Reiskirchen, Germany.

Yamaguchi, N.U., A.C. Scheinost, and D.L. Sparks. 2001. Surface-induced nickel hydroxide precipitation in the presence of citrate and salicylate. Soil Sci. Soc. Am. J. 65:729–736.

Zhang, P.C., and D.L. Sparks. 1990. Kinetics and mechanisms of sulfate adsorption/desorption on goethite using pressure-jump relaxation. Soil Sci. Soc. Am. J. 54:1266–1273.

13

The Changing Understanding of Physical Properties of Soils: Water Flow, Soil Architecture

Shuichi Hasegawa[1] and Benno P. Warkentin[2]

Knowledge in the discipline we now call soil physics accumulated over millennia from the observations and reasoning of both tillers of the soil and of philosophers thinking about the natural world. Physical descriptions of the universe and of the earth were available many centuries before chemical or biological studies began. The accumulated knowledge, for example of soil water, soil temperature or tillage of soils, was available in written records, in books, and in lectures presented to philosophical and agricultural societies. This situation prevailed until the 1800s, with interest from the philosophers decreasing with time and that from the agriculturalists increasing. Agricultural chemistry, which included studies of physical properties of soils, developed as a scientific discipline beginning about 200 years ago. Soil physics then separated in the 1900s into three main streams—soil water, soil structure and soil physical properties related to plant growth. Physics became less central as a background science. Hydrology, mechanics, plant physiology, and ecology began to contribute more of the background.

The striking changes in physical and mechanical properties with changing water content set soils apart from other natural materials put to human use. The changes in plasticity with changing water content make soil suitable for fashioning containers, and pottery was an early use. Similarly, earth for building shelters depends on malleability when moist but hardness and strength when dry, as in bricks. Seeds are planted in moist soil, with the water slowly released for growth of the plant.

Interacting with water content in determining these properties is the arrangement of the soil grains and hence of the spaces between, the voids or pores. This arrangement

[1] Laboratory of Soil Amelioration, Graduate School of Agriculture, Hokkaido University, Kita 9 Nita 9, Sapporo, Hokkaido 060-8589 Japan.

[2] Department of Crop and Soil Sciences, Oregon State University, Corvallis, Oregon 97331.

we call soil structure or soil architecture. In the interaction of soils with plants, and as habitat for the diverse soil biota, the sizes and arrangement of voids are the important parts of the soil. Much of human manipulation of soil has as its purpose the rearrangement of these spaces and the change in their sizes. Improved plant growth has been the purpose since humans began to nurture plants rather than harvesting "wild" plants. The story is told that early cultivators would have noticed improved plant growth in areas where pigs had rooted. Let us imagine that an early gatherer made this observation; she realized its importance and tillage for seed bed preparation has been with us since. It's a good story, let's accept it.

For this chapter we have chosen two aspects of this long and many faceted story of our changing understanding of physical properties of soils. These are how our understanding of the transmission of water, specifically water movement in partly saturated soil, has changed, and how our concepts of soil architecture in the surface or cultivated layers of the soil have changed from emphasis on the solid grains to emphasis on the voids. The voids are habitat for biota, as well as the spaces for the many processes of flow, sorption and biochemical changes that occur.

The material is taken largely from the literature in the English language. The water flow studies have developed as quantitative descriptions with a 100-year history; the structure studies largely as qualitative and descriptive with a several thousand year history. The changed style of presentation from one section to the next reflects this difference.

PHYSICS OF WATER MOVEMENT IN PARTLY SATURATED SOILS

Introduction

Retention and movement of water in soils is basic to the practice of irrigation, an ancient technology vital to the beginning of civilization. Managing water in soils continues to be vital for global food production, both in irrigated and dryland agriculture. Knowledge about water in soils accumulated from the experiences of practitioners, of the irrigators who had to store and convey water, and then worry about how different soils reacted to the water. This technology was well developed when physics was applied by Darcy in 1856 to water movement in sands saturated with water, i.e. to saturated flow as illustrated in Fig. 1. Water pressure in the sand column is always positive (higher than the atmospheric pressure) in saturated flow. Darcy experimentally demonstrated the following relation.

$$q = k\,\frac{(h_A + z_A) - (h_B + z_B)}{L} = k\,\frac{\Delta H}{L} \tag{13-1}$$

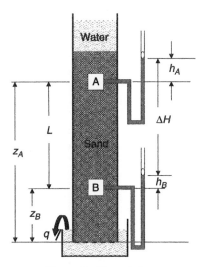

Figure 1. Darcy's Law.

where q is the flow rate from a unit area of sand column, k is the saturated hydraulic conductivity, h is the pressure head, z is the elevation head from the bottom of the column, L is the length between A and B, and ΔH is the difference of total head (pressure + elevation) between A and B. This equation, now called Darcy's law, relates quantity of water flow to the driving force or head. Drainage theory to accompany practice was developed on the basis of Darcy's equation.

But saturation is the unusual condition for surface soils, which are almost always partly saturated, i.e. the pores contain both air and water. Water additions and removals continually occur. The terms partly saturated and unsaturated are used interchangeably. Studies of the physics of water retention and movement under partly saturated conditions were initiated only 100 years ago.

This section describes the important experimental and theoretical studies that over the past century led to our present understanding of this important function of soils—to store and release water.

Capillary Potential and Extension of the Darcy Equation

When the bottom of a cylinder containing dry soil is placed in water, the soil sorbs water and conducts it upward. The rate of sorption and the consequent rise of water decreases with time and finally becomes zero at equilibrium. The soil is saturated with water near the water surface, but only partly saturated further up (Fig. 2(a)). As no water moves, the total water potential at the water surface and at some distance from the water surface, e.g. A in Fig. 2(b) must be the same. Taking the water surface as reference of elevation and the water potential of free water

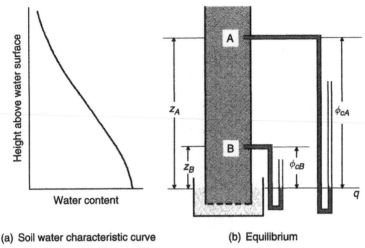

(a) Soil water characteristic curve   (b) Equilibrium

Figure 2. Water content above water surface and equilibrium.

being zero, total water potential at A is also zero because of the equilibrium condition. The point A however has a gravitational potential z_A. The units of potential are expressed as energy per unit weight of water or head. This means that the point A has a potential ϕ_{cA} whose magnitude is z_A but of negative value.

Buckingham (1907) following Briggs (1897) introduced this idea and named this negative potential as capillary potential. Later Gardner et al. (1922) showed that negative soil water pressure relative to atmospheric pressure was equal to capillary potential.

The capillary potential has been known by several names, including soil water potential and matric suction. Matric indicates that the potential arises from the soil matrix, but not necessarily as a capillary pressure. The soil water potential can be varied experimentally. In addition to varying the gravitational potential by changing the height in Fig. 2(b), the air pressure on the soil could be increased above atmospheric pressure. The relationship between the water content of the soil and the soil water potential, shown in Fig. 2(a), is known as the soil water characteristic curve. This relationship shows hysteresis, the water content at equilibrium on drying a soil is higher than on wetting the soil (Haines, 1930).

Buckingham extended the idea to water flow due to capillarity based on an analogy with heat and electricity flows. The difference in capillary potential between two points in the soil was the driving force for flow. His original flow equation is in differential form but can be written simply as,

$$Q = \lambda \frac{\Delta \phi_c}{\Delta x} = \lambda \frac{\Delta \phi_c}{\Delta \theta} \frac{\Delta \theta}{\Delta x} \tag{13-2}$$

where Q is capillary current density (amount of water which passes in one unit of time through a unit area of an imaginary plane surface perpendicular to the direction of flow), λ is capillary conductivity and $\Delta\phi_c$ is the difference in capillary potential across a small distance of Δx. The extension of equation 13-2 shows that soil water can also be considered to move from a wetter side to a dryer side by the gradient of soil water content ($\Delta\theta/\Delta x$), where θ is volumetric soil water content, the volume of water per volume of soil. Buckingham further pointed out that mathematical results obtained from Fourier's law of heat flow and Ohm's law of electricity could not be applied directly to capillary flow of water because λ and $\Delta\phi_c/\Delta\theta$ in the above equation are not constant, but vary enormously with water content. Solution of this equation, and hence prediction of water flow in partly saturated soil, was not straight forward.

Richards (1931) reasoned that if air spaces in a partly saturated porous medium such as soil were instead filled with solid, the condition of flow would be unchanged and the proportionality between the flow and the water-moving force (difference in potential) would still hold because Darcy's law is independent of the size of particles or the state of packing. As the experimental law by Darcy was analogous to Fourier's and Ohm's laws, Richards introduced the following mathematical relation. He explicitly showed the driving force of water for vertical flow is the sum of capillary potential and gravitational potential, z. Flow can be upward or downward, depending on the capillary potential. His equation can be written,

$$q = -k\left(\frac{\Delta\Phi}{\Delta z}\right) = -k\left(\frac{\Delta\phi_c}{\Delta z} + 1\right) \tag{13-3}$$

where q is the volume of water crossing a unit area perpendicular to the flow in unit time, k is a proportionality factor commonly called unsaturated hydraulic conductivity and Φ is the sum of ϕ_c and z. The minus sign in this equation means that upward flow is positive. Partly water-saturated soils are commonly referred to by the shorter term, unsaturated.

For soil water potentials at points A and B in the soil column in Fig. 3 $\Phi_A = \phi_{cA}' + z_A$ and $\Phi_B = \phi_{cB}' + z_B$ respectively, Equation 13-3 becomes,

$$q = -k\left(\frac{\Phi_A - \Phi_B}{z_A - z_B}\right) = -k\left(\frac{\phi_{cA'} - \phi_{cB'}}{\Delta z} + 1\right) \tag{13-4}$$

As the value of $\phi_{cA}' - \phi_{cB}'$ is smaller than Δz, the difference in heights, water flow is negative indicating vertical downward flow.

Soil water is held not only by capillarity, so the term matric potential (ϕ) resulting from interaction between the soil matrix and water is now used in place

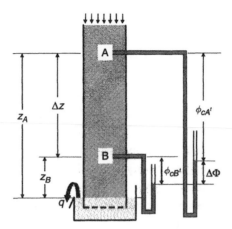

Figure 3. Driving force of unsaturated soil.

of capillary potential (ϕ_c). As Buckingham first proposed that water flow in unsaturated soil was written in the same form as Darcy's law, Equation 13-3 is now often called the Buckingham-Darcy equation.

Water Movement as a Diffusion Phenomenon

When similar soil blocks with different water content are in contact, water in a wetter block moves toward the dryer block. This phenomenon suggests that soil water content can also be considered a driving force for water movement in unsaturated soils. Gas diffuses along a gas concentration gradient, and its rate is expressed as a product of gradient and gas diffusivity described by Fick's law. Childs (1936) considered that water movement could be described by the water concentration gradient and proposed using soil water diffusivity. There is no suggestion that water moves through the pore space by molecular diffusion. Rather, this approach provides for easier solution of the flow equation. Horizontal water movement by Darcy's law can be expressed by a diffusion-type equation as,

$$q = -k \frac{\Delta \phi}{\Delta x} = -D \frac{\Delta \theta}{\Delta x} \qquad (13\text{-}5)$$

Where D is soil water diffusivity, expressed as, $D = k\Delta\phi/\Delta\theta$, and the slope $\Delta\phi/\Delta\theta$ can be obtained from the soil water characteristic curve (Fig. 2(a)).

From the mass conservation or continuity equation, the change of water volume in a volume of soil during Δt is equal to the volume of water entering the soil during Δt minus the water leaving the soil volume during Δt. This statement, written in differential form for vertical water movement is,

$$\frac{\partial \theta}{\partial z} + \frac{\partial q}{\partial z} = 0 \tag{13-6}$$

Substituting the Buckingham-Darcy equation yields,

$$\frac{\partial \theta}{\partial t} = \frac{\partial}{\partial z}\left(k(\phi)\frac{\partial \phi}{\partial z} + 1\right) \tag{13-7}$$

This equation was first derived by Richards (1931), and is called the Richards' potential equation. Equation 13-7 could not be solved directly because it contained two unknowns, θ and ϕ.

Klute (1952), substituting Darcy's law with diffusion (Equation 13-5) into Equation 13-6, introduced the following equation,

$$\frac{\partial \theta}{\partial t} = \frac{\partial}{\partial z}\left(D(\theta)\frac{\partial \theta}{\partial z}\right) + \frac{\partial k(\theta)}{\partial z} \tag{13-8}$$

This is a second order nonlinear partial differential equation that can generally be solved by a numerical method. Unlike the analogous thermal diffusivity, soil water diffusivity is highly dependent on soil water content. Development of numerical methods using high-speed computers in the latter half of 20th century enabled the solution of unsaturated water movement under various initial and boundary conditions.

Prediction of Unsaturated Flow Based on Pore Size Distribution

A number of theoretical and empirical studies have sought to predict the hydraulic factor k in Darcy's law. Soils are composed of various sizes and shapes of solid particles and, therefore, pore size distribution and pore continuity are very complex. Most early studies modeled soils as homogeneous porous materials consisting of grains such as sand or as bundles of capillary tubes.

Kozeny (1927) considered the balance of forces between hydraulic resistance exerted on a unit volume of water and the driving force to move water under saturated conditions. In his model, saturated hydraulic conductivity (k_s) is expressed as a function of the 3rd power of porosity (n), the pore space per unit volume of soil.

$$k_s = \alpha \, n^3 \tag{13-9}$$

where α is related to the shape of the granular materials, and is constant for a given material. Kozeny considered soils formed from single-size particles like sand. Later his equation was also introduced assuming soils consisting of cylindrical channels of uniform size. The Kozeny and similar models could explain saturated hydraulic conductivities of granular materials whose range of pore sizes was relatively narrow, but was not useful for fine-grained soils.

Hydraulic conductivity shows a maximum value at saturation, k_s and decreases with decrease of soil water content. Kozeny's type of equation was expanded to apply to unsaturated hydraulic conductivity, where effective saturation was employed instead of porosity. Effective saturation (S_e) is the ratio of the mobile water content to the total amount of mobile water, based on the idea that water in fine pores is immobile. Unsaturated hydraulic conductivity (k) was expressed as the nth power of effective saturation.

$$k = k_s S_e^n \qquad (13\text{-}10)$$

where k/k_s, the relative hydraulic conductivity, has the advantage of eliminating the effect of water viscosity. The values of n were examined either theoretically or empirically, and found to be lower than 3 for granular soils and higher for fine-grained soils.

The models discussed above are macroscopic approaches, they use pore volume either as porosity or degree of saturation for estimating hydraulic conductivity. Following the macroscopic approaches, there were studies considering pore size and pore continuity in water flow. These require measurement of pore sizes. Flow rate in a tube is proportional to the 4th power of its radius, from Poiseuille's equation. This means that the larger pores have a disproportionally large influence on flow rate. Pore size distribution is more important than total porosity in evaluating hydraulic conductivity.

Pore size distribution can be calculated from the soil water characteristic curve, the relationship between water content and soil water potential. The assumption in the calculation is that the pores are bundles of capillaries of varying sizes. Soil water content at a soil water potential h_1 is the amount of water held by pores with radii smaller than $2\sigma/h_1$, where σ is surface tension. Consider a bundle of capillary tubes of varying diameter placed with ends in water (Fig. 4), the relation between the amount of water in the tubes and height above the water surface is similar to the soil water characteristic curve. This is the capillary model of soils. The diameter of the largest water-filled tube decreases with increase of soil water suction or decrease of matric potential.

Pore size distribution obtained by this method gives a number that can be used for calculation of hydraulic conductivity and water flow. The difficulty lies in relating this number to the morphological pore distribution seen in microscopic examination of a thin section of the soil.

In 1950, Childs and Collis-George developed a model considering pore size distribution and pore water continuity at a cross section perpendicular to water flow

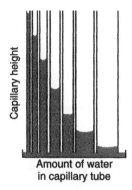

Figure 4. Capillary model.

within a soil. When a vertical soil column is divided horizontally into two parts, the new exposed surfaces have similar pore size distributions. As water is drained from larger pores to finer ones, a water filled pore group on the upper surface can only make water contact with water-filled pore groups at the lower surface whose diameters are smaller than those of the upper surface. Using this idea and Poiseuille's equation, they calculated saturated and unsaturated hydraulic conductivities of soils. The Childs and Collis-George model was a microscopic approach, because it took account of pore radius in addition to pore volume. Since then their model has been improved by many researchers. Mualem (1978) using pore size distribution and effective saturation, was able to calculate the power term in Equations 13-9 and 13-10. A later model by van Genuchten (1980) presents relative hydraulic conductivity as the product of effective saturation and a pore continuity term. The macroscopic approach of Mualem uses pore size distribution, which is microscopic, for evaluating the power terms. The microscopic approach of van Genuchten uses effective saturation, which is macroscopic. So the macroscopic approach now uses the microscopic concept, and the microscopic approach uses the macroscopic concept for predicting hydraulic conductivity.

Infiltration from Green and Ampt to Philip

Infiltration is the phenomenon where water enters partly saturated soils either from rain, irrigation, or ponded water. Infiltration rate, the volume of water infiltrated per unit surface area in unit time, is large at the beginning of infiltration and converges to a lower and nearly constant value with time (Fig. 5). Infiltration has been an attractive subject not only for basic research but also for applied sciences such as irrigation engineering, and has been studied theoretically as well as experimentally. When water infiltrates a dry soil we can observe a

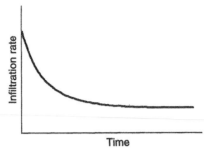

Figure 5. Vertical infiltration rate decrease with time and converge to a certain value.

distinct difference in color between the wetted part and the dry part; this boundary is named the wetting front. There are two dominant approaches to describe and predict infiltration.

A pioneering study by Green and Ampt in 1911 assumed that capillarity exerted at the wetting front, along with gravity, were the driving forces of infiltration, and that the soil between the water source and the wetting front was saturated with water. This model assumed infiltration occurs due to the difference of water potential and the fundamental equation expressing infiltration is the same as Darcy's law. When ponded water infiltrates vertically into a dry soil having uniform soil water content (θ_i) throughout depth, infiltration increases soil water content to θ_f to the distance of the wetting front (l) as shown in Fig. 6. Infiltration rate (i), downward positive, is thus expressed as,

$$i = k_s \frac{h_o + l + h_c}{l} = k_s \left(\frac{h_o + h_c}{l} + 1 \right) \tag{13-11}$$

where k_s is saturated hydraulic conductivity of the wetted zone, h_o is depth of water ponded at the surface and h_c is capillarity (a positive value in this equation) at the wetting front. Equation 13-11 describes the infiltration phenomenon. Infiltration rate during the early stage is large because l is small, and the contribution of capillarity and ponding water depth is dominant. As l becomes long with time the first term in brackets becomes small and finally i converges to k_s. This means that gravity is the dominant driving force when time is large.

For horizontal infiltration, the driving force for infiltration is only capillarity, and infiltration rate is proportional to $t^{-1/2}$. Numerous experimental studies have shown the theory of Green and Ampt is sufficiently accurate to predict infiltration for practical use.

Infiltration can also be described as water entry caused by the difference in soil water content instead of soil water potential. This is similar to heat diffusion in a

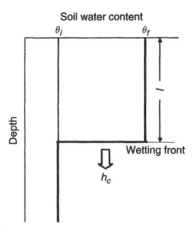

Figure 6. Schematic diagram of vertical infiltration by Green and Ampt (1911).

solid. Philip (1957) applied the diffusion equation shown as Equation 13-8 to infiltration problems and solved them analytically in a series of papers in Soil Science from 1957 to 1958. He described infiltration into a homogeneous semi-infinite medium with constant initial water content and water available in excess at the surface. For horizontal infiltration, he derived the infiltration rate as,

$$i = St^{-1/2} \tag{13-12}$$

where S is a coefficient named sorptivity which is determined by initial soil water content and soil water content in the wetted zone. Equation 13-12 shows that infiltration rate is proportional to $t^{-1/2}$, leading to the same result obtained by Green and Ampt.

For vertical infiltration Philip developed the solution of cumulative infiltration (I) in a power series in $t^{1/2}$.

$$I = \int_{\theta_i}^{\theta_o} z \, d\theta = St^{\frac{1}{2}} + At + A_3 t^{\frac{3}{2}} + A_4 t^{\frac{4}{2}} + \cdots \tag{13-13}$$

where θ_o is soil water content at the soil surface during infiltration, θ_i is initial soil water content of the soil and coefficients A, A_3, A_4 are calculated from hydraulic conductivity and soil water diffusivity. Infiltration rate is obtained by differentiating Equation 13-13 with respect to t. It is generally sufficient for practical use to use the first two terms.

$$i = \frac{1}{2} St^{-1/2} + k_s \tag{13-14}$$

This simple expression obtained by solving the diffusion equation may be considered as the sum of horizontal infiltration by capillarity and infiltration by gravity. During short times, the first term on the right hand side is dominant, which means vertical infiltration is expressed in the same form as horizontal infiltration, i.e. infiltration in the early stages is governed by capillarity. However at longer times or longer depth of the wetting zone, gravity rather than capillarity becomes dominant. At $t \to \infty$ in Equation 13-14 the infiltration rate approaches k_s.

Philip also showed that the theory of Green and Ampt was a special case. They had assumed that the wetted part of the soil is saturated, and water content changes abruptly at the wetting front to the initial water content. According to Philip their assumption indicates that soil water diffusivity is zero except at saturation where it goes to infinity and hydraulic conductivity is also zero except at saturation where it is k_s. Studies of Green and Ampt, and Philip taught us the physical process of water movement into a partially saturated soil and the physical properties of soils affecting water movement.

Concluding Remarks

This chapter has drawn attention to significant studies that developed our ideas on water movement in partly saturated soils. Our understanding developed from the concept of a capillary potential of soil water, which was used to solve the equations for water flow.

A more detailed discussion of unsaturated or partly saturated flow of water in soils is available in soil physics textbooks, e.g. Hillel, and in published papers in journals of soil science, geotechnical engineering and hydrology. This literature also contains the large number of studies that applied unsaturated water flow to problems of irrigation, drainage, and movement of solutes in soils.

References for Physics of Water Movement in Partly Saturated Soils

Briggs, L.J. 1897. The mechanics of soil moisture. U.S. Dept. Agr., Bur. of Soils, Bull. 10.

Buckingham, E. 1907. Studies on the movement of soil moisture. U.S. Dept. Agr. Bur. Bull. 38.

Childs, E.C. 1936. The transport of water through heavy clay soils: I. J. Agr. Sci. 26:114–127.

Childs, E.C. and N. Collis-George. 1950. The permeability of porous materials. Proc. Royal Soc. Series A. 201:392–405.

Gardner, W., O.W. Israelsen, N.E. Edlefsen and H. Conrad. 1922. The capillary potential function and its relation to irrigation practice. Phys. Rev. 2, 20:196.

Green, W.H. and G.A. Ampt. 1911. Studies on soil physics. I. The flow of air and water through soils. J. Agr. Sci. 4:1–24.

Haines, W.B. 1930. Studies in the physical properties of soil. V. The hysteresis effect in capillary properties and the modes of moisture distribution associated therewith. J. Agric. Sci., Camb. 20:97–116.

Hillel, D. 1998. Environmental Soil Physics. Academic Press, San Diego.

Klute, A. 1952. A numerical method for solving the flow equation for water in unsaturated materials. Soil Sci. 73:105–116.

Kozeny, J. 1927. Über kapillare Leitung des Wassers im Boden. Sitzungsbericht der Wissenschaften in Wien. Mathem. Naturw. Klasse. Abt. IIa. Band 5 u. 6.

Mualem, Y. 1978. Hydraulic conductivity of unsaturated porous media: generalized macroscopic approach. Water Resour. Res. 14:325–334.

Philip, J.R. 1957. The theory of infiltration: 1. The infiltration equation and its solution. Soil Sci. 83:345–357.

Richards, L.A. 1931. Capillary conduction of liquids through porous mediums. Physics. 1:318–333.

Van Genuchten, R. 1980. A closed-form equation for predicting the hydraulic conductivity of unsaturated soils. Soil Sci. Soc. Am. J. 44:892–898.

SOIL ARCHITECTURE: FROM TILTH TO HABITAT

Introduction

Soil structure for most of recorded agricultural history meant achieving a finely-divided soil by fragmenting clods using cultivation tools. It was tilth of surface soil for a desirable seedbed, and plowing to achieve it. Tilth is the condition of a seedbed desirable for plant growth. The present concept of soil structure or soil architecture is the agglomeration of soil particles to build an arrangement of solids and pores at all scales from 1 μm to 5 mm. This is achieved at aggregate and sizes less than 100 μm, by natural forces of bonding, fostered through wetting and drying. Particles are pushed into specific arrangements and held by different bonding mechanisms. The arrangement of larger sizes of aggregates provides the voids suitable for a seed bed, for the habitat for biota, and for all the soil functions in unmodified (natural) and in agro ecosystems. This arrangement can be modified by tillage. The development of this concept of structure

Table 1. Periods for study of surface soil structure

Period	Dominant concepts
Pre-Renaissance	Plowing, fragmenting clods, crumbling of soil
1450–1850	Tilth, fineness of soil, plowing
1850–1930	Grain size distribution, organic matter, clay bonding
1930–1970	Aggregate size, stability, tilth related to physical properties
After 1970	Habitat for soil functions, heirarchial size arrangement of aggregates and voids

can be followed through the increased understanding in the last century of the significance of compound particles and how they are formed. The terms soil structure and soil architecture have the same meaning and will be used interchangeably.

This section examines the approaches to soil structure, based on writings in Europe for five periods of different thinking in the last two millennia (Table 1). The Renaissance marked a change in Europe in people's interest in the soil resource. The date 1850 is taken as the beginning of scientists' interest in and contribution to soils knowledge. Before that soils knowledge had been accumulated by practitioners. After 1930 studies of soil structure as we now understand the term multiplied rapidly, with measurements of properties of aggregates. After 1970 the present concept was developed of the hierarchial build up of structure to form voids allowing soil processes to function.

Thinking About Soil Architecture in Different Periods

Soil Structure in Pre-Renaissance Writing

The concept of soil structure in the surface soil was soil tilth, stirring the soil to prepare a fine seed bed. The plow was key from prehistoric times. Plowing had often to be followed by breaking up clods with a mallet. Early Roman writers such as Cato (234-149 BPE) in his *De agricultura* wrote, "What is it to till the land well? It is to plow well. What next? To plow. What is third? To manure" (from Olson, 1944b). The beneficial effects on tilth of grass crops in the rotation were known and stressed by writers. Vergil (70-19 BPE) in his poems *Georgics* wrote that grasslands have a natural crumb-like structure, "such as plows make by art" (from Olson, 1944a). Columella (1st century AD) accumulated and analyzed the writings then available in his *De re rustica*. The contributions of the early Roman writers are discussed in a series of papers by Olson (1943, 1944a,b) on the early literature. The translations quoted above are from her papers.

These ideas on soils and soil management were repeated in writings on "husbandry" in the next thousand years, generally including the range of activities about which a "husbandman" needed to have knowledge—soils, plants, and animals. Few new ideas or observations were apparently added. Many of the writings between the Roman period and the Renaissance have been lost, but some authors are quoted in an Arab book of the 12th century. Ibn Al Awam (see Clemènt-Mullet, 1864) in Seville, wrote a book on agriculture, probably about 1150. It contains the experience of Moorish writers on agriculture during the time when that culture flourished. It is important because most of the Moorish literature in Spain was destroyed during the "Reconquest," completed in the late 15th century. Ibn Al Awam describes compact soils having low permeability, where water runs off the surface. There is an emphasis on structure related to water, expected in the dry climate of Spain. He incorporates the writings and experience of many earlier authors. Olson and Eddy (1943) discuss the times in which he lived and some of his contributions.

In England, the first writing was probably Walter of Henley's manuscript on husbandry, written around 1250 in an anglicized French (see Lamond, 1890). He recommends shallow plowing, and recommends against removing stubble: "If you take away the least, you will lose much." He must have experienced deterioration of structure of the bare surface soil. This writing, and indeed books written until the end of the 17th century are difficult for us to read because of the many interspersed details. This arrangement, with lengthy comments on side issues, apparently dates back to Herodotus and his "Histories."

The concept of structure, then, was the finely divided soil created by the plow—tilth for a seedbed. Certain crops in the rotation made it easier to get this better tilth after plowing.

From 1450–1850

By the middle of the 17th century, need for food production in Europe increased as populations recovered after the plagues of the 13th and 14th centuries and the Black Death. Landowners began to take an interest in cultivation. Preparation of a good seedbed was the objective, and fineness of the surface soil was seen as a prerequisite. Gentlemen farmers solved practical problems in their fields, with fewer concerns by philosophers about the nature of soil. Different ploughs were developed for effective tilling of different soils. Breaking up clods to achieve a finely divided soil was still the goal, and still difficult to achieve. Fitzherbert (1534) has 14 out of 180 pages on descriptions of different ploughs, how to use them, and

how to plough for different crops. He made observations on the efficacy of different cultivation methods. "Than is the ploughe the moste necessaryest instrumente
that an husbande can occupy."

Most books were reviews of what previous authors had written plus some personal experiences of the author. Mortimer (1707) states of his book ". . . being a
full collection of what has been writ, either by ancient or modern authors, with
many additions . . ." He remarks on the great difference in plows used in different
regions of England. Part of this would be due to differences in soils and crops, but
part, he claimed was because ". . . every place almost being wedded to their
particular fashion, without any regard to the goodness, convenience, or usefulness
of the sort they use." It was this situation that experiments such as those Tull, and
the import of improved plows from Flanders, sought to correct.

Tull (1731) is remembered for his practice of repeated plowing, horse-hoeing.
He emphasized the importance of fineness of soil, because plant nutrients were at
the surfaces of very fine particles. These attempts to find rational reasons for small
soil particles were in line with the concepts that tillage was necessary to produce a
finely-divided soil. His system, involving tilling several times during the season,
gave him increased yields and allowed for continuous annual crops of wheat. Tull
also developed a practical seed drill that allowed sowing in rows rather than
broadcasting seeds. The spaces between rows, or groups of rows, could be tilled
for weed control. The system had adherents, and many detractors, well into the
19th century. Warkentin (2000) has reviewed some of this work.

In the early 18th century, when Tull was farming and writing, prices for mainstream agricultural crops, the cereals, were depressed. His drill and his ploughing
were a response to the situation he found, an attempt to remain solvent by achieving higher yields. After 1750 when prices improved with the beginning of an industrial revolution, his drills and cultivation ideas were put to use in intensive
cultivation to produce more grain. This would have had a deleterious effect on soil
structure. At the end of the 20th century, in another period of depressed prices for
mainstream crops, we look to the opposite of intensive cultivation, to minimum
tillage or no-till.

Many other landowners in the 18th century were experimenting with soil management. Agricultural societies, where gentlemen farmers met and contributed
papers on their findings, were formed in European countries as well as in North
America. Many of these reports were on designs for improved plows, and evaluation of their effectiveness. Major contributions to husbandry came from writings
in Italy in the 18th and 19th centuries, carrying on the Italian tradition of detailed
prescriptions for various farming operations. The Accademia dei Georgofili
(friends of the land) was founded in Florence in 1753 to improve agriculture
through studies and well-designed experiments.

The general outlines of maintaining good tilth (structure) were known. The effect of grasses in the rotation was appreciated, although a disadvantage was the difficulty of achieving a good seedbed (fineness of soil) through plowing of the grass sward for a subsequent crop. The effect of manures (compost) on tilth was recognized. But arguments about "worn out" soils centered on chemical treatments. Soil fertility was seen as the key to crop production; deterioration of soil structure was not recognized.

Soil structure was still synonymous with tilth. Some scientific studies of soil were carried out toward the end of this period, e.g. Davy (1821), but it was the middle of the 19th century before the "scientists" took over soil structure.

From 1850–1930, the Beginnings of Soil Physics

About the middle of the 19th century in Western countries scientists took over the study of physical properties of soils from the practitioners (Warkentin, 1999). Schübler (1838) attempted to bring science to the aid of practice. He reported measurements of physical properties that affect structure, such as volume weight, adhesion and shrinkage, but no measurements of aggregates. He justifies the study of physical properties by pointing out that they can decrease crop productivity even where chemical conditions are optimum—the importance of structure. Chemical studies of soils had begun slightly earlier, based on advances in chemistry (e.g. Nye, 1996). Steam power began to be used for cultivation, a boon for plowing clay (heavy) soils where cereal crops responded to deeper cultivation. But soil structure did not become a major topic in soils books well into the 20th century, until the end of this period.

Grain size distribution, which could be easily and reproducibly measured, became a dominant physical measurement after 1900 (Table 2). Many methods were developed to measure the proportion of different size classes of grains in soil. Keen (1931) devoted 51 pages out of a total of 354 to methods for mechanical

Table 2. Emphasis on structure in selected soil physics books

	Grain size distribution	Pages discussing		
		Clay properties	Soil structure	Cultivation
King, 1897	6	5	—	40
Keen, 1931	51	55	—	61
Baver, 1948	38	44	68	35

analyses. Temporal variations in grain size were minor, so results were repro-
ducible over time. Keen had no separate chapter on soil structure, but had a discussion of soil factors bearing on cultivation. The contents of sand, silt and
clay-size grains were only broadly related to different tilth. While a good index
property for draft requirement or water permeability, grain size does not relate
well to structure.

Toward the end of this period interest developed in the factors determining soil
structure. The content of clay size particles was seen to be important for the cohesion needed to attain a desirable structure. It was realized that agglomeration was a
necessary process, and that clay particles had large cohesion. Studies on clay properties increased after 1900, on the basis that the clay grains needed to be flocculated
to get good tilth (structure). Measurements were made of flocculation of clay suspensions, plastic behavior of pastes (rheology), and effects of exchangeable cations
on clay behavior. Keen (1931) had 55 pages on soil and clay pastes and suspensions.
Atterberg (1911) classified soils on the basis of plastic properties, developing what
are now called the Atterberg limits—plastic and liquid limit. They are not useful
index measurements for predicting soil structure because they are made on remolded
soil, where important aspects of structure are destroyed. They are, however, widely
used in soil engineering as an index property for engineering behavior of clay soils.

Changing Concepts of Soil Structure

The changing concepts of structure from the late 1800s to 1930 and beyond can
be followed in the several editions of E.J. Russell's book, "Soil Conditions and
Plant Growth." The first edition in 1912 has no entries specifically for structure or
porosity. Mechanical analysis for grain sizes, and the effects of grain size on plant
growth are discussed. He does state that compound particles make a quantitative
interpretation of a mechanical analysis for tilth impossible. The properties of soil
are modified by this union, but we do not know how. These ideas are expanded in
the 3rd edition of 1917. ". . . the components of soil do not form a mere casual
mixture," but, again, little is known about the compound particles. A new chapter
on colloidal properties was inserted, indicating the importance of the clay fraction
in soils. Mechanical analysis is still the measurement of choice, but the interpretation for soil management is being questioned.

The 5th edition in 1927 has a page on total pore space, but still no index list-
ing for structure. Under texture he includes tilth, e.g. crumbly vs. sticky and
lumpy, as well as grain size to make up the composition of the soil. The properties of soil depend on the state of the colloids as well as the quantity. There is,
however, no extension yet of this idea to arrangement of the clay particles.
While cultivation is a large concern in soils books of this time, Russell states,

"the science of cultivation hardly exists." This concern will come up again in the next section of this chapter.

The 7th edition in 1932 was the last one written by E. John Russell. By this time, E. Walter Russell, who revised the later editions, had begun his studies of soil structure. The 1932 edition has a subheading on soil structure, with the idea of particle arrangement. He states, ". . . pore spaces determined by sizes, shapes, and modes of arrangement of soil particles," but this idea was not carried to soil structure. Cultivation leaves the soil in condition for weather to make crumbs. This introduced the idea of natural conditions being responsible for desirable structure. The 8th edition in 1950 has a 23-page chapter on "Soil Structure and Soil Tilth," with 27 index items under structure.

Warington (1900) was early in his discussions of how soil structure was formed. He pointed out that tilth was not always the result of tillage. For example, tilling clay soils does not produce good tilth. Tilth is due to formation of compound particles, which can arise spontaneously under certain natural conditions such as frost, volume change or root action, without the aid of tillage. Russell (1957) concedes this point, but adds that the build up of good tilth, crumbs from 0.5 to 5 mm, by natural processes under grass and clover is too slow for most cultivators and the crumbs have to be formed by cultivation. Warington's book was based on a series of lectures he gave in 1896. Warington's own research dealt with chemical properties of soils; it is often insights from a distance that provide new thoughts. He stated that while scientists recommended fertilizers, farmers knew that good tilth was vital.

The Early Textbooks in the USA

Most soils textbooks published in the USA around 1900, the first flurry of soils books, accepted the need for tillage and continued the earlier tradition of discussing tilth in chapters on cultivation. This was a period of depressed prices for agricultural products, but the practice of plowing with its high energy, time and labor requirements was not questioned. Plowing was still the key to cereal crop production.

King (1897) used the term texture as synonymous with structure or tilth, e.g. p. 279 ". . . stirring the soil to improve texture . . ." He quotes studies showing that lime improves clay soils, presumably by flocculating the colloids. He speculates on the structure of clay soils, p. 75, ". . . open types must have some sort of structure." Later King (1907) defined texture as size of grains and the way they are grouped into composite clusters, kernels or crumbs. His discussion of pore space is restricted to the model of packing of spheres developed by Slichter. Emerson (1930) has two pages on soil structure as part of soil morphology. It is remarkable that a book on soil technology had essentially nothing on soil structure.

Burkett (1907) in his book entitled "Soils" writes extensively on tillage and tillage tools with emphasis on the plow and on making the soil finer. In essence he expounds Tull's ideas; Burkett's ideas were based on his experience with crop production in Kansas. Aeration of the soil is an important benefit of tillage. After recounting the old story of a father tricking his sons into thorough tillage of a plot by telling them there is a buried pot of gold, he writes, "Deep breaking of the soil, frequent and intelligent tillage—those are the foundations of soil restoration" (p. 285). The plow was still of primary importance.

The dominant theme during this period was tilth achieved by tillage: the measurements were of grain size distribution, rather than of aggregates or pores. It was accepted that tillage was beneficial for crop production. The occasional voice questioned whether tillage as practiced was excessive and detrimental to soil productivity. For example Lee (1849) concluded that plowing in the southeast of the USA exhausted the soils due to increased oxidation of organic matter and leaching of soluble nutrients. The best known of the voices chiding use of the plow came later, e.g. Faulkner (1943).

From 1930–1970, Aggregates

By the 1930s the broad outline of how soil architecture is formed and the factors influencing it were known. Aggregates of soil particles were necessary, arrangement was the key. Russell (1938) summarized the understanding of soil structure at that time. Good tilth required an air and water regime suitable for a growing plant, and one that would be stable against disruptive forces. This necessary distribution of pore sizes is controlled by the grain size distribution (texture) and the distribution and position of these particles in the soil (architecture). Several methods were in use for specification of soil structure: permeability as an index, pore space at different water contents, and size distribution of aggregates—first by dry sieving and later by sieving in water to measure stable aggregates. Numerous methods were developed between 1920 and 1940 for aggregate analysis, largely by Russian and USA scientists, most of them variations on the method of wet sieving proposed by Tiulin (1928).

Methods to measure pore size distribution by direct observation were described (Kubiena, 1938). Apparent pore size distribution could also be measured from water content vs. energy measurements, the soil water characteristic curve (see section on water movement). It had always been difficult to measure size, shape, arrangement, and continuity of pores; this difficulty led to concentration on measurements of properties depending upon pores, e.g. hydraulic or air conductivity and diffusion. Relating the results to tilth or soil structure, which was the objective, became no easier.

Organic matter and iron oxides were the known cements holding soil particles together. The mechanisms for action of organic matter were not known; it was considered that decomposition produced sticky substances that acted as glue. Clay was another important component required to achieve aggregation. Adherence of clay particles was a key, and depended upon whether the clay was flocculated and the influence of drying. Cultivation of loam soils at an intermediate water content, but not in wet or dry soils, could bring smaller units together to form aggregates.

Structure was defined by Baver (1940) as the arrangement of soil particles, both grains and aggregates. Micro structure depends on the arrangement of particles of a size not visible to the naked eye. He devoted nearly 20% of his book to a chapter on soil structure. Russell (1938) drew attention to the role of microaggregates as units from which larger aggregates are built up. This concept of a hierarchal buildup of soil structure was fleshed out fifty years later (e.g. Oades, 1987). Warington (1900) had discussed the formation of compound particles that arise spontaneously under certain natural conditions, e.g. volume change or frost. Russell (1933) advanced the hypothesis that clay particles were held together by oriented molecules of water around cations through dipole-cation-dipole links, a modern concept. The water held the cation away from the clay surface. Crumbs were formed when the water was removed on soil drying. Hénin (1935) in France, in a series of studies on effects of clay properties on soil structure, discussed orientation of clay particles affecting secondary particle formation.

The considerable body of field studies carried out by Russian soil scientists in the early 1930s on formation of aggregates and the effects of different crops on soil structure was evaluated by Russell (1938) and summarized by Baver (1940). The Russian work stemmed from the interests and leadership of Williams (1935) who taught that structure was the key to soil productivity. He was a strong proponent of the desirable effects of a grassland system of agriculture on soil structure.

The interest in soil structure, to judge from the number of research papers published, peaked about 1950 (Table 3). While the importance of structure was appreciated, it was not a productive area for research. There seemed to be few new ideas. Aggregate size distribution and its stability in water had been measured on many soils, and correlated with the known aggregating agents—clay, organic matter, iron, and metal oxides. The wet sieving method correlated with field observations of good structure in soils where organic matter was the main aggregating agent, and under climatic conditions were the cultivated soils had originally been under forest. It was not a useful method for arid area soils or for soils with clay content above about 25%.

A blip in interest occurred in the 1950s with the introduction of synthetic polyacrylic molecules for aggregate stabilization. It was known that the aggregating action of organic matter resided in the decomposable fractions, e.g. polysaccharides. But these materials decomposed rapidly in soils. The idea then was to produce a

Table 3. Published papers on soil structure

	Structure	Porosity
Soil Science		
1916–1928	7	0
1941–1953	32	2
1954–1972	1	10
Soil Science Society of America Journal		
1957–1966	17	10
1972–1976	2	4
1977–1981	4	5

molecule with similar action, but stable in soils (Quastel, 1952). In 1951 such a material, Krilium© was announced by the Monsanto Co. Numerous studies showed it stabilized aggregates, although it did not form aggregates from dispersed soils (e.g. Hedrick and Mowry, 1952). However, it also became evident that nature produced aggregating materials at a small fraction of the price charged by the chemists. These polyacrylic and other materials were used in specialized applications, but were too expensive for use on field crops. The chain of history leads from manure to synthetics and back to manure.

Relating aggregation to tilth continued to be difficult. All the methods in common use, e.g. size distribution of stable aggregates, were indirect measurements of soil structure, which then had to be related to soil tilth or to the effects of cultivation. This continues to elude us today. A Joint Committee on Tilth in the U.S.A. reported in 1943 that empirical experimentation would not tell us which type of tillage was superior, because we did not know what physical state is desired for a grain cropping system (Shaw, 1952). They recommended a systematic approach to measurements of physical properties. The book "Soil Physical Conditions and Plant Growth" edited by B.T. Shaw (1952) was a review of the state of knowledge, based on this recommendation.

Comprehensive studies include the 15-year study by De Leenheer (1977) and colleagues on structure changes by compaction on high-silt soils in Belgium. Significant differences between soil types and treatments could not generally be established because the variability was too high. Part of this was due to year-to-year variability. Similarly, a study in Germany (Hartge and Stewart, 1995) measuring various properties in different research laboratories again concluded that we do not have a sufficiently complete understanding of aggregate stability or of the processes of aggregation to use as a guide in evaluating tillage.

The alternative approach has involved modeling the processes leading to soil structure. The complex nature of structure and lack of understanding of many of the processes involved hinders the usefulness of modeling. It still has to be admitted that we have not achieved the aim of relating aggregation to soil tilth.

The questions asked about tillage changed in the last decades of the 1900s from tillage required to create tilth (structure) to a concern that tillage destroys structure, that the changes in soil architecture are undesirable. Warkentin (2001) has argued that this change in the efficacy of tillage began when machine power became available, and led to excessive tillage. With human or animal power, cultivation probably improved the seedbed. The interest in decreased tillage (no-till or direct drilling) in cropping systems has kept alive the attempts to predict changes in soil structure from different tillage practices. The desire to understand what happens to soils in the change from intensive cultivation to no-till has added to the interest in soil structure. Success is still eluding us. "It is not possible to predict the resulting soil condition from any given tillage operation" (Dexter, 1988). "None of the models for predicting the changes in soil structure of the tilled layer at present available can be used to forecast the effects of a cropping system on these changes over space and time, especially when the soil is ploughed" (Roger-Estrada et al., 2000).

The renewed impetus for work on soil architecture was to come after 1970 with considerations of soil as a habitat for biological activity.

Architecture for Soil Functions

During the last third of the 20th century soil scientists recognized that soils performed necessary and unique functions in ecosystems. Ecologists had developed these ideas earlier for unmodified ecosystems, but now they were also being applied to agro ecosystems. The change in research funding in Europe and America away from agricultural production and towards environmental quality was an important driver for soil scientists. The critical role of soil functions in quality (and quantity) of water is one example. Concepts of soils in an ecological context did appear sporadically in the earlier literature. King (1897) in the introductory chapter to "The Soil" begins with ecological concepts of cycles, growth, life in the soil, energy, movement, water, and sunshine. Further discussion in this book is based on processes, e.g. soil ventilation rather than pore space. Pedology in Switzerland developed with an emphasis on soil functions in ecosystems (Schticher, 2001). But these are minority examples. The overwhelming impetus for considering soils in an ecological context came in the last thirty years of the 20th century.

Table 4. Hierarchy of structure units

Compound particles	Particle size	Voids
Domains	1 μm	Micropores
Microaggregates	10 μm	Mesopores
Aggregates	1 mm	Macropores
Clods	10 cm	

The focus on soil structure changed from aggregates, their size and stability, to pores or voids. The "empty" spaces were determined by the hierarchial arrangement of solid structure units, from clay grains through microaggregates to aggregates visible in the field (e.g. Oades, 1987; Dexter, 1988; Edwards and Bremner, 1967; Warkentin, 1982). Arrangement of the smaller units formed the next larger ones (Table 4). The size of a biological organism determined where in the void size distribution it would exist. The size and continuity of voids would determine flux of gases, and hence whether the habitat was aerated or anoxic.

The term "soil architecture," used first by E.J. Russell, seemed apt to describe different void sizes, connected by doorways of varying sizes, with walls available for attachment of water, chemicals and microorganisms. From grand ballroom spaces to closets under the stairs, they provided habitat for different organisms and different functions. And all of this depended upon a stable structure or arrangement of solids. The emphasis in soils, as in architecture of buildings, changed to spaces. "The Reality of the Building does not consist in roof and Walls but in the space within to be lived in" (Lao Tse, 6th century, BPE). This could be the motto for soil scientists.

A stable soil structure, resulting in a diversity of habitats, was necessary to allow important soil functions to proceed. The "ecological approach" to soil science has led us to think in terms of the functions that soil performs, both in unmodified (natural) ecosystems and in agro ecosystems. These functions—biochemical and geochemical, recycling of carbon and nutrients, partitioning of water at the soil surface, physical and chemical buffering, storage and release of water and nutrients, energy partitioning at the surface, and a physical base for plants, animals, and man-made structures, all depend upon a stable structure or soil architecture within which they can proceed. The different sizes, shapes, and connections of voids allow these functions. Dexter (1988) defines structure as "the spatial heterogeneity of the different components or properties of soil." This definition accommodates the different aspects of the many size scales. ". . . spatial heterogeneity = spatial variability = structure."

Structure as habitat has re-vitalized the study of soil structure, led largely by soil biologists and landscape ecologists looking at structure as habitat for biota. Six et al. (2004) have a detailed review of the work leading up to this concept. The focus is on natural processes leading to association (combination) of soil particles rather than fragmentation processes resulting from tillage. This story of the next phase in soil structure, which promises to be both fascinating and productive in increasing our knowledge of soils, is waiting to be played out.

Summary

Soil structure was, for millennia, tilth of the seedbed and plowing to achieve it. This was the concept until about 1850 during the time practitioners wrote about soils; then the soil scientists defined it as aggregation. With the advent of soil science laboratories the dominant concern became measurement of static properties relating to tilth and stability of structure: porosity, grain-size distribution, aggregate-size distribution, stability of aggregates, shape and size of aggregates. The importance of organic matter and clay in soil structure was generally recognized. All these properties were related to cultivation and to the effects of cultivation on tilth and yield. Cultivation was assumed to be necessary for crop production. The appreciation in the late 1900s of the role of soil in ecosystem functions led to the concept of soil structure determining habitat for soil functions. Structure came to be seen as a built up in hierarchal fashion from smaller units arranged to form larger units, from clay crystals to aggregates (peds). The voids associated with the different size units were related to functions of storage or transmission and provided the diverse habitat for microbial and other biota. This approach revitalized the study of soil structure in the latter third of the 20th century. Forces associated with different sizes could be hypothesized. This was a concept of soil architecture with solid particles forming walls, with spaces of different sizes, and different halls and doorways connecting them.

References for Soil Architecture

Atterberg, A. 1911. Die Plastizität der Tone. Int. Mitt. für Bodenkunde 1:7–9.
Baver, L.D. 1940. Soil Physics. Wiley, New York. [The 2nd (1948) and 3rd (1956) editions contained only a few additional paragraphs on soil structure.]
Brewer, R. 1964. Fabric and Mineral Analysis of Soils. Wiley, NY.
Burkett, H. 1907. Soils. Orange Judd, NY.

Clemènt-Mullet, J.-J. 1864. Le Livre de l'Agriculture. Translation into French of Ibn-Al-Awam manuscript. Libraire A. Franck, Paris, 2 volumes.

Davy, H. 1813. Elements of Agricultural Chemistry. Longman et al., London.

DeLeenher, L. 1977. Structure Fertilite de Sols Limoneux Sur Fermes Mecanisees. Faculteit der Lanndbouwwwetenschappen, Rijksuniversiteit, Gent.

Dexter, A.R. 1988. Advances in characterization of soil structure. Soil Till. Res., 11:199–238.

Edwards, A.P. and J.M. Bremner. 1967. Microaggregates in soils. J. Soil Sci. 18:64–73.

Emerson, P. 1930. Soil Technology. Macmillan, New York.

Faulkner, E.H. 1943. Plowman's Folly. U. of Oklahoma Press.

Fitzherbert. 1534. The Boke of Husbandry. Thomas Berthelet, London, 90 sheets. First published in 1523.

Hall, A.D. 1931. The Soil. Dutton, NY (the 1st edition was published in 1912).

Hartge, K.H. and B.A. Stewart (eds.) 1995. Soil Structure: Its Development and Function. Adv. in Soil Sci., CRC Press.

Hedrick, R.M. and D.T. Mowry. 1952. Effect of synthetic polyelectrolytes on aggregation, aeration, and water relationships of soil. Soil Sci. 73:427–441.

Hénin, S. 1935. Soil Structure. Ann. Agron. 5:44–50 (seen in Russell, 1938).

Kay, B.D. 1990. Rates of change of soil structure under different cropping systems. Adv. Soil Sci. 12:1–52.

Keen, B.A. 1931. The Physical Properties of the Soil, Longmans Green, London.

King, F.H. 1897. The Soil. Macmillan, New York.

King, F.H. 1907. A Textbook on the Physics of Agriculture, 4th Edition, published by the author.

[King published several editions of this book, but they were printings without changes rather than editions in the modern sense.]

Kubiena, W.L. 1938. Micropedology. Collegiate Press, Ames, IA.

Lamond, E. 1890. Walter of Henley's Husbandry. Translation of "Le Dite de Hosebondrie." Longmans Green, London.

Laotse (Li, Erh, 6th century BPE). Seen on a plaque at Taliesin West, Phoenix, AZ.

Lee, D. 1849. The philosophy of tillage. Trans. N.Y. Agric. Soc. 8:342–358 (seen in Sewell, 1919).

Mortimer, J. 1707. The Whole Art of Husbandry. Mortlock at the Phoenix, London. (J.M. Esq. FRS, is listed as the author on the title page.)

Nye, M.J. 1996. Before Big Science. Twayne, New York.

Oades, J.M. 1987. Association of colloidal materials in soils. Trans. 13th ISSS Congress, Hamburg (1986), vol. 6:660–674.

Olson, L. 1943. Columella and soil science. Agric. History 17:65–72.

Olson, L. 1944a. Vergil and conservation. Agric. History 18:153–155.

Olson, L. 1944b. Cato's views on the farmer's obligation to the land. Agric. History 19:129–132.

Olson, L. and H.L. Eddy. 1943. Ibn-Al-Awam—a soil scientist of Moorish Spain. Geographical Review 33:100–110.

Quastel, J.H. 1952. Influence of organic matter on aeration and structure of soil. Soil Sci. 73:419–426.

Roger-Estrada, J., G. Richard, H. Boizard, J. Boiffin, J. Caneill, and H. Manichon. 2000.

Modeling structural changes in tilled topsoil over time as a function of cropping systems. Eur. J. Soil Sci. 51:455–474.

Russell, E.J. 1912. Soil Conditions and Plant Growth. Longmans Green, London. Seven editions to 1932.

Russell, E.W. 1933. The binding forces between clay particles in a soil crumb. Trans. 3rd Int. Congress Soil Sci. 1:26–29.

Russell, E.W. 1938. Soil Structure. Imp. Bur. Soil Sci., Tech. Comm. No. 37, 40 pp.

Russell, Sir E.J. 1957. The World of the Soil. Fontana Library, London.

Schticher, H. 2001. Bodenkunde und Bodencundler in der Schweiz. 1855–1962. Bodenkundliche Gesellschaft der Schweiz, Juris Druck u Verlag.

Schübler, G. 1838. Grundsätze der Agrikultur Chemie. 2nd Edition, Baumgärtner, Leipzig.

Shaw, B.T. (ed.) 1952. Soil Physical Conditions and Plant Growth. Agron. Monograph No. 2, Academic Press, New York.

Six, J., H. Bossuyt, S. Degryze, and K. Denef. 2004. A history of research on the link between (micro) aggregates, soil biota, and soil organic matter dynamics. Soil & Tillage Research 79:7–31.

Tiulin, A.F. 1928. Seen in Baver, 1948. Soil Physics, 2nd Edition, Wiley, p. 173.

Tull, J. 1731. The New Horse-Houghing Husbandry. Published by the Author. London.

Tull, J. 1951. Horse-hoeing Husbandry, 3rd Edition, Miller, London.

Warington, R. 1900. Lectures on some of the physical properties of soil. Oxford, Clarendon.

Warkentin, B.P. 1982. Clay soil structure related to soil management. Tropical Agric. 59:82–91.

Warkentin, B.P. 1999. The return of the other soil scientists. Can. J. Soil Sci. 79:1–4.

Warkentin, B.P. 2000. Tillage for soil fertility before fertilizers. Can. J. Soil Sci. 80:391–393.

Warkentin, B.P. 2001. The tillage effect in sustaining soil functions. J. Plant Nutr. Soil Sci. 164:1–6.

Williams, V.R. 1935. (seen in Baver, 1948).

Section IV
Soil Utilization and Conservation

Soils, like water and air, have many attributes that make them usable and useful to humans. Soils functions include storage and gradual release of water and nutrients to plants, and decomposition of a vast range of materials from organics to synthetics. Their ready availability makes them almost unnoticed—soils are always under foot. Growing plants for food and fodder has been the dominant soil use for millennia. In the process of using soils, humans have changed them in small and large ways—adding nutrients that increased crop yields but also leaving soils exposed to large erosion events. Major changes have resulted from water control measures. Irrigation, and especially drainage on small and large scales, have altered soils in major ways. Construction of terraces, an ancient practice, involves major soil displacement. In the last half century, the addition of chemicals that leach through the soil and make ground water unsuitable for many uses has become a major source of soil alteration and of soil degradation. Soil erosion and soil conservation are topics of wide public interest. When a soil scientist in a crowd of non-scientists replies "soil science" to the question of "What do you do?" the likely rejoinder will be "Oh, you do soil conservation."

Showers takes a global view of soil erosion, and the practices that humans have used to prevent movement of soil, i.e. to achieve soil conservation. Soil erosion processes and human responses are central to this chapter. The large role of anthropology in understanding responses to erosion is an important part of her story. Why are different responses found in different regions? She reviews different land use systems, and gives examples of how soil conservation methods were transferred, both the successes and failures. Soil conservation as a profession developed largely in the USA, beginning in the early 20th century. The knowledge acquired was then widely used in other national soil conservation programs, and influenced soil conservation efforts around the world.

Historically, and in the foreseeable future, soil has been the source of food and fodder. This is the main use that humans make of soils. In the process, they have modified soil properties to increase the production of food. Finck tells the story of the gradual increased use of inputs of plant nutrients. The early gatherers only gathered, but with the advent of horticulture and agriculture, various amendments were added to soils to increase productivity. The literature of Roman agriculture, discussed by Winiwarter in Chapter 1, had gathered previous and current experience as guidelines for the use of a range of amendments. With the advent of agricultural chemistry, experiments on nutrition of plants gradually showed how nutrients could be used, and led to the development and finally extensive use of chemical fertilizers. Practical experience and experimentation in the 20th century developed the guidelines for use of chemical fertilizers as they gradually replaced organic byproducts such as composts as convenient and cost-effective sources of nutrients. In the last half of the century undesirable effects on soil ecosystem functions became evident. Fertilizers in amounts beyond the plant's needs became soil pollutants. But that is another story discussed by Addiscott in Chapter 17.

Gregorich and co-authors describe some responses designed to prevent soil degradation. The "land care" movement is a response that involves all members of a community in identifying land degradation problems, planning for their correction, and mobilizing the various resources available for making changes. Concepts of soil quality and soil health, although difficult to define precisely, have an intuitive appeal to the public that is concerned with soil degradation and its cost to future generations. Keeping soils healthy, i.e. able to perform their functions in ecosystems, is the road to the sustainable use of soils. The concepts of how we want the earth to be, and of humans as part of ecosystems, are important components of thinking about sustainable use of soils.

Adding to the story told by Finck, Addiscott discusses the environmental issues of amendment additions to soils in crop production. The effects were seen for many years before the full implications were realized. Soil pollution was considered by the mid 20th century as toxic heavy metal accumulation, e.g. cadmium or lead. Later the concern was water quality, where nutrients such as nitrogen or phosphorus leached into ground water or surface water. Use of fertilizers had become one of the main ways humans interacted with soils. Addiscott describes the experiments, beginning 150 years ago, on the nitrogen cycle in agro ecosystems. The increases in soil knowledge from these experiments are the footprints that led to our increased understanding. The value of long term experiments, following effects of crop production practices on the same plots of land for over 100 years, are strikingly illustrated.

Terracing to increase the area of cultivable land, to control water, and to check soil erosion was a major and ancient practice of human modification of soils to aid soil uses. Sandor discusses different forms of terraces and how they have altered soil properties. Some terraces have been successful in meeting their purposes; others have not. Terraces can increase soil erosion by channeling water, or, soils that become more susceptible to erosion after disturbance can lead to failure if used in terraces. Studies of older terraces on different soils display indigenous knowledge in matching type of terrace to soil properties.

Benno P. Warkentin

14

Soil Erosion and Conservation: An International History and a Cautionary Tale

Kate B. Showers[1]

Introduction

Constructing a history of soil conservation is complex because the term has meant different things to different people, and because it has not been common for the notion of soil conservation to exist apart from ideas about land use management. In the late 20th century, soil conservation included both 'promoting the optimum use of land in accordance with its capability' and 'halting the decline of soil properties and restoring productivity' (Dudal in Morgan 1981, p. 8). Plant nutrient supplies, water holding capacity, soil particles, and soil organic matter are the aspects of soil most usually recognized as needing conservation. At different times and in different places, the need to maintain or improve one or more of these soil properties has been identified and addressed through plant, soil and/or water management techniques. The distinction between the loss of soil fertility over time and the actual loss of soil particles has not always been made clearly. In the 20th century many soil conservation professionals were most concerned about preventing or stopping the movement of soil particles, or soil erosion. For this reason, consideration of soil erosion processes will be central to this chapter.

The practice of soil conservation was imbedded in many land use traditions. This is not to say that traditional land use methods are always soil conserving, or that in the past soils were universally respected and protected. Even a superficial survey of world history from a soil perspective reveals that societies have had different awarenesses of the need for, and ability to, conserve soil. Yet so polarized were the late 20th century soil debates that assertions of the inherently conserving nature of many traditional practices, or recognition of the existence of ancient awareness of soil erosion processes and conservation methods, were labeled

---

[1] Senior Research Associate, Centre for World Environmental History, University of Sussex.

'romantic'. This has been particularly true in discussions of the African environment (e.g. Adams and Hulme, undated). If soil is to be elevated in status and given the protection it deserves, then there must be agreement that there are many ways to accomplish this task, and that different societies at different points in time have had both good and bad ideas of how to proceed.

Soil erosion control can be imagined from two quite different perspectives: vegetation and structures, or biology and physics. Biological approaches work with natural and introduced vegetation to reduce the transportation capacities of wind and water, as well as to protect the soil surface from raindrop impact and overland flow. Foresters advocate afforestation, permaculture agriculture, and the planting of shelter belts. Livestock and range management specialists are concerned with grazing patterns that least disrupt 'natural cover' and plant species best suited for pastures. Agronomists consider crop canopy, various mulching techniques, and arrangement of plants to provided maximum soil protection. Production techniques are analyzed for their effects on soil disturbance. Since organic matter is known to bind soil particles and enhance soil structure, great emphasis is placed on its maintenance or increase.

Conservation conceptualized as a physics problem emphasizes the need to reduce the velocity (energy of transport) of moving wind and water. This can be accomplished by diverting or dissipating flow. Fences and other structures can deflect wind, and ditches and drainage systems can collect and remove water from the soil's surface. The erosive capabilities of water moving down a slope can be reduced by decreasing a slope's length. Terraces or ridges effectively create a series of short slopes. While vegetational approaches to erosion control were aspects of the general practices of forestry, range management and agriculture, structural control belonged to the new field of agricultural engineering. At times, tensions have existed between advocates of each approach to conservation; in some places emphasis was placed primarily on structural approaches, and in others distinctions between biological and structural techniques have been blurred as both were implemented in an integrated fashion.

To understand fully human societies' interactions with soil, answers to basic questions are required. What are the elements of soil conservation? What has been tried with success, and what has failed? More fundamentally, what social activities cause or promote soil erosion, and how can they be curtailed? The first set of questions are essentially technical, the second relate to social processes and institutions. Both can be answered using historical analysis. We want to know, basically, who did what, when, why and how – and what the consequences were. In this chapter, the inherently soil conserving aspects of traditional land use systems will be discussed, a popular world history of soil erosion and conservation experiences will be suggested, and the rise of the 20th century profession of soil conservation will be traced. Most of the world's experience with soil erosion and soil protection occurred before the profession existed, and so it will be discussed first. The very specific experience of soil conservation after it had been professionalized will be examined separately, with special foci its rise in the United States of America (USA) and implementation in British Colonial Africa. This will lead to some conclusions about what usefully can be done in future to preserve and protect the earth's quite fragile outer surface. Because of the extensive literature on these subjects, references in the following paragraphs will be suggestive rather than exhaustive.

Conceptualizing Soil Erosion

If there were no human beings on the earth, soil erosion and deposition would still occur in events such as glaciation, avalanches, shoreline erosion and severe storms. Blowing wind and flowing water pick up exposed sediments. They are transported until the motion slows, dissipating the requisite energy for transport. This mechanism results in the formation of loess (wind blown silt from glacial floodplains) soils and sand dunes, the loss of soil on hillslopes and from riverbanks, and the creation or enrichment of soils when streams expand into their flood plains. All are examples of natural (which some conservationists have referred to as 'normal') soil erosion, sedimentation and deposition. When human interventions in landscapes increase rates of erosion or stimulate its initiation, the result is commonly referred to as 'accelerated erosion'.

It is not always easy to distinguish natural (normal) from accelerated erosion, particularly when someone from one ecosystem attempts to judge the appearance and processes of a very different one. What is expected in one environment is abnormal in another (Showers, 1989, 2005). This basic confusion has been a constant component of writings about landscapes and land use around the world. Whatever the intentions of the observer, bias often creeps in. A semi-arid landscape can look deficient or barren to someone from a humid climate, just as trees without leaves in a northern hemisphere winter can appear dead to someone from the tropics. Perceptions of 'foreign' landscapes are often based on comparisons with 'home' – wherever that might be.

Unless it occurs in a catastrophic event like a flood or avalanche, or takes the dramatic form of a gully, the movement of soil in landscapes is usually slow. Losses from a particular place are often so gradual as to go unnoticed. For some cultivators, particularly those using a plow that mixes surface soil, loss is evident only when visually or texturally different subsoil is exposed.

An individual soil's susceptibility to erosion (erosivity) depends upon its position in the landscape as well as on its physical, chemical and structural properties. Any soil on a slope or exposed to wind is more likely to erode than one on a flat surface or sheltered from wind. Organic matter binds soil mineral particles, making them less susceptible to erosion, but an organic soil, when dry, is more susceptible to blowing wind than a mineral soil. Highly weathered soils with iron oxide structures can be more physically stable than less weathered, more agriculturally productive soils. Water movement within soils can be erosive, especially when there is a flow across a dense layer (a hardpan or thick layer of semi-permeable clay). When an outlet is provided, this internally mobile water can cause tunneling and piping – often revealed only when the structures cave in. Dramatic gullies can, thus, apparently form overnight.

Given the great variety of soil properties, landscape formations, and climate, it should come as no surprise that generalizing about 'good soil management practices' is difficult. One observation appears to be universal: soil movement is reduced when soil is covered, and increases when soil is exposed. Protective cover can be living or dead – vegetation, stones, organic matter – and is most effective when closest to the ground. Research has shown that it is tree litter, rather than the canopy, that protects forest soils.

Valuing Soil

The European industrial revolution and the ancient Chinese and Japanese feudal states were predicated upon changing fundamental attitudes about, and relations with, the natural world (for Japan: Oyadomari, 1989; England: Maltby, 1991; China: Ronan and Needham, 1995). Instead of understanding human beings as part of a larger whole, human beings were seen to be separate from, and superior to, the natural world. Ecosystem components became 'natural resources' that only had value if they could be used for human benefit – and were finite (Lemons and Saboski, 1994; Showers, 2000). Soil, water and air were largely considered to be limitless, and thus without value, so they could be used without concern for loss or degradation. Those who placed non-utilitarian value on the environment and did not seek to control it were considered to be 'backwards', 'lazy' and 'superstitious' – certainly in need of improvement and education. When these attitudes were combined with Biblical beliefs in racial superiority or with elements of social Darwinism, a virulent form of racism resulted (Grove, 1989, 1997; Showers, 2005). Educated people from industrial societies who believed in non-utilitarian value systems were dismissed as 'romantics' or ridiculed as being 'anti-modern'.

The early European settlers of North America, Africa and Australia imagined they had discovered vast continents of limitless natural resources waiting to be turned into wealth. They harvested, produced and sold, on small scale and large. When productivity dropped, or political and economic pressures encouraged them, they moved inward from the coasts, cutting trees, making fields, grazing stock – often leaving eroded land behind them (e.g. for Australia: McTainsh and Boughton, 1993; USA: Mc Donald, 1966; South Africa: Drought Investigation Commission, 1923). Soil was valued only for its ability to support crops and pasture.

Concern about soil grew with declining yields, obvious erosion and, most of all, drought and deforestation (e.g. Saberwal, 1999). The 19th and 20th century concept of soil conservation arose when individuals from industrial northern hemisphere societies recognized that soil was threatened by conventional land use practices, and needed protection. Some soil conservationists tried to teach both urban dwellers and practitioners of industrial agriculture that soil has inherent and intrinsic value, and should be protected for these reasons. But often they were reduced to making utilitarian, economic arguments. Lost soil was so much lost money, and protected or restored soil represented money saved or gained (e.g. Bennett, 1929). Periods of drought and associated blowing soil in the USA, Australia, Kenya and South Africa in the early 1930s helped create an impression among settler societies that something urgent had to be done. The need to alert the public to the value of soil and the catastrophe of its loss led to exaggerations so as to 'shock people into awareness' (e.g. Morgan, 1965; Rasmussen, 1982; for discussion of crisis narratives see Roe, 1991; 1995; Grove, 2000). Exaggerated claims came to characterize programs for soil conservation (e.g. El-Swaify in Morgan 1981; Evans, 1995; Boardman, 1998). Agricultural scientists (particularly soil chemists and agricultural engineers) who had been trying to alert the farming public and their governments to the need for attention to soil erosion suddenly were heard. Initially these claims galvanized public support and government funding, but ultimately they devalued the cause and decreased the credibility of soil conservationists. Both funding and political debates soon engulfed scientific concerns about soil movement and the emerging profession of soil conservation.

Traditional Land Use Systems

Traditional land use systems evolve over time as societies gain specialist knowledge of plants and animals in particular soil, topographic and climatic conditions. They are often combinations of observations made in a particular place and ideas introduced from afar that have been modified for local conditions (for discussion: Reij et al., 1996). When people or a person move(s) to a new and different location, they become what Beach refers to as 'environmental pioneers', encountering a 'new, misunderstood environment' (Beach, 1998, p. 400). Throughout the earth's human history there are examples of environmental pioneers wreaking havoc by applying their traditional methods of land use to new ecosystems. In some instances, over time, land use practices were modified and ecosystems recovered, while in others there was no learning, recovery was not possible, or economic and political forces made conserving practices impossible (for Australia: McTainsh and Boughton, 1993; Guatemala: Beach, 1998; China: Perdue, 1987).

The response to environmental change – including soil erosion – is often determined by the rate of change. It must be rapid enough to be noticed, and there must be some expectation of its recurrence, before people will respond (Bell in Bell and Boardman, 1992). The same processes are at work when people are confronted with changes in basic environmental parameters, such as the onset of drought or wetter weather, the amount of land available, or access to different types of land. Only when a recognizable pattern sets in is there likely to be a consistent response. Whether people move by choice or by force, and whether factors beyond their control cause fundamental change in their environment, the need to adapt traditional methods to new ecosystem parameters is the same, although the amount of time and resources available for adaptation may differ. Much of environmental history consists of recording learning processes of environmental pioneers, be they 19th North Americans moving from forests to prairies, ancient Mayans adjusting their agriculture to a tropical rainforest, the waves of North China refugees and immigrants moving south to Hunan Province 1000–2000 years ago, or colonial powers establishing themselves in distant lands.

Some traditional systems had/have inadvertently soil conserving components, some had/have conscious soil management aspects – and others were/are not conserving at all. Just as there are examples of indigenous soil conservation, there are also examples of indigenous soil exploitation throughout the world and throughout time. When assessing land use systems, it is useful to keep in mind some common soil conserving features.

Inherently Soil Conserving Practices

Many traditional farming systems increased soil fertility by encouraging microbial synthesis of plant available nutrient forms through cultivation of legumes and plants with mycorrhizal fungal associations. Nutrient cycling was ensured by leaving plant material on the soil surface or incorporating it into the soil, transporting manure to fields, grazing animals on stubble, burning vegetation, and/or carrying ash to fields. Not only do these practices conserve plant nutrients, but many also protect the soil surface as well as improve water holding and infiltration capacity.

Water conservation resulted from many aspects of traditional management systems. Water from precipitation, snow melt, or runoff enters the soil through surface openings called pores. To enter a soil pore, water must move slowly across the soil surface, or be held in place on the surface. Seed bed preparation that produces a field with clean, loose soil encourages soil erosion and nutrient loss through increased runoff. In contrast, traditional systems that leave crop residues on the soil surface, allow weeds to grow until crop plants have been established, or use a hoe or plow to create a rough soil surface rather than a smooth one, reduce overland flow and increase infiltration of water, thus minimizing soil erosion. That these practices had a soil conserving function may not have been known by practitioners.

Crop/Fallow Rotation Systems

One of the most widespread agricultural systems involved the rotation of land between agricultural crops and natural vegetation (sometimes called fallow or bush fallow). Typically, a plot was cleared from forest or treed savannah, cropped for 3–5 years, and then allowed to revert to natural vegetation for 15–20 years. A new plot was cleared when the old one was no longer in use. Nineteenth and twentieth century European observers called these systems 'swidden agriculture', 'shifting cultivation', or 'slash and burn agriculture', and usually characterized them as being wasteful, primitive or destructive (for southern Africa: Grove, 1990; South Asia: Pouchepadass in Arnold and Guha, 1995; Indonesia: Boomgaard, 1997). However, researchers at Zambia's (then Northern Rhodesia) Abercorn Agricultural Station found the local Chitemene, or tree lopping, system a useful way to control weeds (N. Rhodesia Dept. Ag. An. Rept., 1929/30). Nye and Greenland's 1960 study in West Africa, *The Soil Under Shifting Cultivation*, found that 'the traditional system of shifting cultivation in the forest admirably protects the soil from erosion in spite of steep slope and heavy rainfall' (Nye and Greenland, 1960, p. 91). In the humid tropics, pedologist Wright (1962) argued, the soil fertility of 'stable' (non-weathering, non-mobile) soils is maintained through organic matter. Shifting cultivation provided a mechanism for organic matter management that permitted long term use of an area's soils. Systematic study since then has shown such systems to be 'technologically complex and ecologically sound and flexible', as well as 'enormously diverse .... or .. specialized' (Sutton, 1989, p. 100). A.T. Grove (2000) sites a number of technical studies in the colonial era that demonstrated the viability of traditional African agricultural systems.

As long as these systems were not stressed by population increase or land shortage, and the farmers remained in the location where the system had been elaborated, rotations between crop and fallow sustained human populations, wildlife, plant diversity and ecosystems. In many parts of Asia, Africa and Latin America rotations between crops and natural vegetation supported human populations for generations or centuries without ecological or social collapse. When population pressures, changes in agricultural technology or market stimulation caused the length of time allocated for cropping and fallow to change, conserving aspects such as nutrient cycling, protective crop cover, tree stump preservation, non-crop field border plants, and tree preservation were diminished or eliminated. Many of the negative accounts of these rotating systems were written when population and land pressures were affecting them, when settlers sought to acquire the land being restored in fallow, or when state forestry interests were threatened by the practice.

Livestock Management

In many ecosystems, livestock provides both efficient use of local vegetation and stability in the face of marginal agricultural conditions. While some traditional systems involved corralling stock at night, many relied on herding animals to water and pastures for dispersed grazing daily or seasonally. Specific areas were allocated for grazing at particular times of the year. Where seasonal movement of stock to another region (transhumance) was involved, a predictable 'calendar' of events was determined by climate, plant growth and water availability. In order to sustain grazing by numerous herds, regulation was achieved through water and grazing rights and responsibilities. In regions of transhumance, there are often reciprocal relations with distant cultivators and land users. Such rhythms and relationships have been documented in Africa and South Asia (e.g. for West Africa: Powell et al., 1996; South Asia: Arnold and Guha, 1995 various chapters). Although a substantial literature exists asserting that traditional grazing systems caused 'degradation' and erosion around the world, late 20th century research questioned its accuracy. Long term research has shown that a degraded rangeland can recover just as quickly under moderate grazing as when not grazed. Dispersed herding and transhumance can minimize, if not prevent, excessive trampling and overgrazing. The existence of local protective customs, rules, and regulations – although well understood by the users – have not always been appreciated by foreign observers.

Agricultural Terraces

Soil will accumulate behind any barrier placed across a slope, resulting in a relatively flat surface on which crops can be grown. This is the essence of agricultural terracing (Spencer and Hale, 1961). There is no evidence for a single origin of the concept; multiple independent origin sites are most likely (Spencer and Hale, 1961; Wright, 1962). Dating of terrace structures shows them to have been used in some form in China and on the island of Cyprus from at least 3000 BP; in Lebanon from 2500 BP; Peru, Mexico and Guatemala from approximately 2000 BP; the southwestern USA from around 1000 BP; and in Tanzania and Ethiopia from at least 500–300 BP.

The development of terracing traditions is distinct from that of hydraulic engineering. Terraces were not constructed in the great river valleys where flood control engineering developed. According to pedologist A.C.S. Wright (1962), in the western hemisphere and on Pacific Islands, terraces were commonly built on alluvial and colluvial soils of steeply inclined valley floors and on the skeletal soils, lithosols and steep land soil of valley slopes. All of these soils can be described as 'unstable', since they continuously gain or lose mineral soil particles. Terracing was most advantageous in drier tropical regions, where permanent settlement required the creation of agricultural soils of adequate depth near fresh water supplies. In the humid tropics, terracing offered fewer advantages, and little or none in non-tropical humid regions.

Terraces were built for many reasons, but erosion control was not their primary purpose in most places. Construction was usually stimulated by the desire to create agricultural fields, raise fields above standing water or wet ground, and/or to facilitate irrigation (Spencer and Hale, 1961). Terraces functioned, often inadvertently, as anti-erosion devices. Recent studies have shown that the ability of terraces to reduce erosion has been over-estimated; well-maintained terraces were found to reduce soil loss by little more than 50% (Foster and Highfill cited in Beach and Dunning, 1995). The effectiveness of terraces in erosion control depends upon their design, maintenance, water control/drainage systems, and other conservation practices used in conjunction with them. If not well maintained, terraces can exacerbate, rather than reduce, soil erosion as previously dispersed water is collected and concentrated before flowing downslope (e.g. Sandor et al., 1986; Showers, 1989; Zurayk, 1994).

Twentieth century terraces built for soil conservation were constructed differently than traditional terraces. In many instances modern terraces have proven to be less sturdy or functional than those built using traditional methods (e.g. for Morocco: Hamza in Reij et al., 1996; Indonesia: Huszar, 1997; Mexico: Mountjoy and Gliessman, 1988). However, even if superior in construction, ancient terracing has been implicated in long-term gully erosion and the impoverishment of soils in some locations (Sandor et al., 1986).

Techniques Affecting Soil Movement Before the Era of Soil Conservation Professionals

There is a long history of human responses to soil movement in the forms of removal (erosion) or deposition (sedimentation). Some societies attempted to minimize soil erosion, while others sought to maximize sedimentation. Still others ignored the processes altogether, only slowly (if ever) connecting soil loss or stream sediment loads to land use practices. Because of the variation of environmental, social and economic conditions, generalizations are very difficult – and dangerous – to make. Regional or even continental summaries of a history of interactions between societies and soil are beyond the scope of a book chapter. However, a history can be suggested by examining examples of responses to sedimentation from particular places at specific times of responses to sedimentation, the relationships among changes in vegetation, agriculture and soil movement, and structure building in fields.

Sedimentation

When the human species left Africa for Asia, they also left a relatively benevolent set of climate, topographic, flora and fauna conditions for harsher ones (Oliver, 1991). Intricate systems of hydrology and river control were devised to provide irrigated land and reduce inundation of inhabited flood plains. That this knowledge system developed separately from soil conservation is illustrated by the increasing need for sediment management. Modern Iraq and China have at least 6,000 years of agricultural experience. Along Iraq's Tigris and Euphrates Rivers ('the Fertile Crescent'), irrigation systems evolved from simple cuts in levees for flooding fields to sophisticated engineering works during the Sassanian Dynasty (2152–1776 BP) (Jacobsen and Adams, 1958). The weakness of this system was sedimentation generated by woodland clearance, overgrazing and cultivation. The expanding system of canals moderated the rivers' natural flood surges and reduced its ability to transport sediment, which accumulated rapidly. Any soil conserving practices during this period were ineffective, because when the social structure that organized maintenance collapsed approximately 1900 years ago, the central canal was choked with sediment, and could no longer supply irrigation water (Jacobsen and Adams, 1958).

A clearer relationship between agriculture, sedimentation, and river control can be seen in China, where pollen and river sediment analysis established dates of land-use changes (Zhu and Ren, 2000). Thirteen thousand years ago the vegetation of China was predominantly grasses, with some forest. The Yellow River's sediment load was one fifth of that in the late 20th century. The advent of cultivation in China dates back between 7000 and 5000 years – 6000 years on the Yellow River's Loess Plateau. Until about 4,700 years ago grazing predominated, the vegetation (largely grasses and shrubs) was not destroyed, and the Yellow River's sediment load was relatively low.

The promotion of agriculture began in the Zhou (Chow) Dynasty (4124–2258 BP); silt deposition was recognized as a fertilizing process (Ronan and Needham, 1995). The Yellow River remained 'clean' until the end of the Confucian Era (2500 BP) (Kleine, 1997). Four hundred years later, after a period of 'exceptional development' of irrigation and transport projects, two books (*Meng Zi* and *Huai Nan Zi*) described deforestation and landscape degradation (Ronan and Needham, 1995).

The Chinese expressed their specialist knowledge of soil movement in terms of hydrological engineering, not soil conservation. The first quantitative estimate of a river's sediment load was made by hydrologist Zhang Rong 2001 years ago (Ronan and Needham, 1995). Reports on the dangers of silt were published in the following 22 years. The importance of geographical information to the government was signaled 1610 years ago when a Regius professor of geographical communications was created; the relationship between denudation, erosion, and flooding was well recognized within thirty years (Ronan and Needham, 1995). Despite the fact that the literate elite and government officials understood both the causes of soil erosion and its serious consequences, the subsequent history of China is one of relentless environmental exploitation. Government policies of land taxation, agricultural intensification; and civil strife resulted in the cutting of forests, plowing of grassland and cultivation of hillsides – to say nothing of the wetlands and floodplains essential for river function. As population grew and land was degraded, people moved to new frontiers and repeated the processes of short term gain and long term degradation.

The development of agricultural terraces must be understood in this context. Terracing began in some areas of the Loess Plateau 3000 years ago to control water for crop production (Kleine, 1997). The mastery of wet rice cultivation and terracing in central and southern China 900–1300 years ago made settlement possible in what had been frontier regions (Stross, 1986). Terracing was a technology that facilitated further exploitation of the landscape, and was not developed to control soil erosion. In south and southeast Asia water control for crop production, rather than soil erosion, was the primary reason for the spread of this technology (Spencer and Hale, 1961).

In the Americas, the use of sedimentation and terracing to regenerate fields or create new agricultural land was widespread, and is discussed in Chapter 17 in detail. It should be noted that not only was sediment laden water on alluvial fans or steep slopes manipulated but, in some locations, erosion was deliberately induced to rejuvenate or create a downslope agricultural field (e.g. Bocco, 1991).

Agricultural Practices

Incidents of increased erosion in Europe have been linked to major shifts in climate in the last 10,000 years, and to specific changes in agricultural technologies and land use practices in the last 2000 (VanVliet-Lanoë et al. in Bell and Boardman, 1992). The earliest forms of agriculture were variations of shifting cultivation. When plows were introduced, larger and more permanent fields could be cultivated, changing vegetation patterns. Strip lynchets (terrace like structures) developed, but there is no evidence that these were conscious anti-erosion devices (Spencer and Hale, 1961; Bell, 1992).

When viewed from the last ice age to present, the single most erosive land use activities were found to be those associated with agricultural mechanization (VanVliet-Lanoë et al. in Bell and Boardman, 1992). The Woburn Experiment Farm in England was established in 1875 to determine payment due to displaced tenant farmers for the soil improvements they had made (Catt in Bell and Boardman, 1992). The focus of the experiments was, therefore, soil quality under various management systems. It wasn't until 1950 that soil erosion was noted in the farm records; by the late 1960s it was a fairly frequent problem. These erosion increases were thought to be due to the replacement of pasture and winter cereals with potatoes; the use of heavy soil cultivating machinery (soil compaction); and herbicides that eliminated weeds that had helped stabilize the soil (Catt in Bell and Boardman, 1992).

Russian agricultural expansion also caused erosion. Three periods of accelerated erosion have been identified: from the 12th–15th centuries near fortified cities, after the 1718 reforms of Peter the Great, when taxation policy encouraged ploughing up of land, and after 1861 when serfdom was abolished (Sobolev, 1947). Settlers arriving in Central Russia during the 17th and 18th centuries cut forests and burned the steppe; dust storms and deflation of sandy soils were observed along the Don River in the 18th century (Stebelsky, 1974). The population of the region doubled from six to twelve million people, and two-thirds to three quarters of the land was cultivated in the 19th century (Stebelsky, 1974). Traditional Russian farming techniques of shallow ploughing, peasant fields in scattered narrow strips, and the loss of pasture, aggravated soil erosion. Many of these 'strips' were not on the contour, and gully erosion became a problem (Stebelsky, 1974).

Russian scientists such as P.A. Kostychev (1845–1895), A.A. Izmailsky (1851–1914) and V.V. Dokuchaev (1846–1903) urged action to regulate the flooding of small rivers, control gully erosion, and develop water supplies for times of drought (Jackson, 1980). Tree planting to conserve soil moisture and protect the steppe from blowing winds were also recommended (Jackson, 1980). In 1898 Tsarist government scientists made a planimetric survey of sandy wastes and began a grass planting program to hold the lower Volga River basin's soil (Jackson, 1980). Mapping soil erosion began with the first soil map of European Russia (Jackson, 1980). Published in 1900, it showed the extent of wind erosion. By 1913 a soil conservation program had been established (Jackson, 1980).

North American grasslands similarly suffered with the arrival of European agricultural production. The mid-19th century Canadians who pioneered western Canada settled land that should have been left in prairie, and discovered that agricultural methods suitable in humid eastern Canada needed adjustment. These environmental pioneers were provided assistance by neighbors from the USA with experience from working their own grasslands (McConkey, 1952). By the 1830s crop failure, soil erosion, and low prices had brought both poverty and the realization that farming methods had to change and erosion be controlled in Canada and the USA (McConkey, 1952; Hopkins, 1937). In a very different ecosystem, similar patterns of land clearance for agriculture followed by erosion was associated with Japanese migrants to Indonesia (Boomgaard, 1997).

Elsewhere, some established farmers—distinct from the environmental pioneers just described—developed techniques for integrating perennial plants with annual crops to minimize soil loss. In Africa's West Cameroon Highlands, the traditional distribution of cultivated land, fallow land, trees and pasture promoted stability (Temato and Olson, 1997). If rills began on cultivated fields, the land was covered with brush or returned to pasture. Trees in the fallow area broke the slope length and inhibited erosion. When this vegetative management system changed under late 20th century pressures, rapid erosion resulted. Farmers in the Mexican highlands of Tlaxcala also had a farming system that incorporated native vegetation, fallowing, crop rotation and intercropping to stabilize their fields (Mountjoy and Gliessman, 1988).

Structures in Fields

In African locations where topography and rainfall created a need for water control or the prevention of soil movement, locally evolved, specialist techniques developed over the centuries that included constructing mounds, ridges, pits and terraces (Grove and Sutton, 1989). These interventions almost always were temporary, so that water was not concentrated in the same place from year to year. Mounds and ridges, usually prepared by hoe, were components of agricultural systems in regions of medium rainfall, especially on lighter soils. These techniques emerged from crop production needs, but one function was the prevention of soil erosion in savanna soils. Chapters in Reij et al. (1996) provide detailed examples. A form of mounding is pit construction on steep slopes in the Matengo Hills, Tanzania (Temu and Bisanda in Reij et al., 1996). A series of pits dug across and down the slope collect water for slow infiltration, preventing erosive overland flow (photograph included with text). This pitting system was a variant on shifting cultivation; late 20th century pitted fields were cultivated for 4–6 years, with a fallow of 2–10 years. In the Ethiopian Highlands, water control and erosion prevention were achieved by ditching (Alemayehu in Reij et al., 1996). Farmers varied the position and depth of ditches in their fields each year to avoid gradual widening and deepening over time. For generations stone bunds or lines were constructed in Ethiopia's Harerge Highlands on the contour, their spacing determined by the slope (Asrat et al. in Reij et al., 1996). In addition, soil bunds were used to control erosion at planting time. They were removed when the crop canopy protected the soil and weeding was essential. Most of these structures were temporary, but some persisted for as long as 20 years.

Terraces

Terraces fascinated 19th and 20th century northern hemisphere observers. Those of Asia and the Americas are well known examples of excellent engineering skills. Perhaps the most controversial have been those of the Mediterranean region. For some, the ancient terrace structures represented a perfect response to soil erosion; to others, the terraces were a response to difficult crop production conditions (Lowdermilk, undated; Grove and Rackham, 2001). Agricultural terraces in the Vasilikos Valley of Cyprus have been estimated to be 3600–4000 years old (Wagstaff in Bell and Boardman, 1992). They appear to have been built up the valley wall from the valley floor in a piecemeal fashion, as check dams across ephemeral streams to trap sediment and control water. Lebanon is known for terraces that are more than 2500 years old, most dating back to the Phoenician era (Zurayk, 1994). Phoenician terraces were constructed after forests had been cleared from mountain slopes. Whether constructed to create agricultural land or to reduce soil erosion, their abandonment in the 20th century has been associated with the initiation of a cyclical erosion process resulting in enormous soil loss. In this cycle, the untilled soils crust, increasing runoff into unmaintained waterways, which increases erosivity. Reduced discharge from the waterways results in over-saturated terraces, which drain through – and destabilize – the stone walls. Plant roots add to the pressure on the walls, which ultimately collapse, leaving the entire slope vulnerable to rill and sheet erosion (Zurayk, 1994). On the Mediterranean's southern coast, in Morocco's Rif Mountain, shifting cultivation of grain together with arboriculture of raisins, figs and plums, olives or almonds (depending upon soil type) on terraced hillsides, goat raising and charcoal production sustained a population for centuries (Heusch, 1981). When the system was disrupted by 20th century land re-allocation and forest protection, serious erosion began.

Further south on the African continent, in Konso, Ethiopia, cooperatively constructed and managed stone terraces dating from at least the 16th century (and in use at the end of the 20th) were built for erosion control, land creation and water management (Amborn, 1989). In the Mandara Mountains of northern Cameroon, stone terraces have been constructed for generations as a component of water control management, along with ridging and drainage canals (Hiol Hiol et al. in Reij et al., 1996). However, they were not built as fixed, permanent structures, but rather were continually modified by adjustment of height of wall or even location on the slope. Stone-walled terraces have been used on steep slopes in Maku, Nigeria to create agricultural fields (Igbokwe in Reij et al., 1996). Oral tradition states that construction began as fortification of villages in the era of slave raids. Soil that accumulated behind the walls was used as small fields, and did not form a continuous terrace system on the hillsides. Evidence of terracing constructed with stone walls also exists in Nigeria, Cameroon, Darfur and Kordofan (Sudan), Tanzania, Kenya, and Zimbabwe (Grove and Sutton, 1989).

The 18th and 19th centuries provide examples of terraces used specifically as anti-erosion measures after land use practices had triggered massive soil loss. The environmental pioneers in the British colonies forming the core of what became the USA had serious erosion problems by the 18th century. Deforested New England lost topsoil, and travelers reported gullies in the southern plantation region (Hall, 1937; McDonald, 1966). The old English practice of contour ploughing had been introduced to Virginia, and was popularized by environmentally concerned plantation owner (and former president) Thomas Jefferson in the early 1800s. However, by the 1830s deep horizontal ploughing of any kind was not considered to be sufficient for erosion control in the south. Farmers experimented with hillside ditching and terrace construction as a means of draining erosive water from fields (Hall, 1937). From 1860–1920, hillside ditching and terracing became almost universal in the hill sections of the southeast, as was concern that their construction could be associated with increased, rather than decreased, soil erosion (Hall, 1940).

In late 19th century Japan, narrow based terraces were selected for use in the restoration of denuded and severely eroded slopes. For centuries, Japan's steep mountain slopes had been protected by forests. However, in the feudal period that developed 800 years ago, power struggles and battles resulted in forests being cut and burned in many locations, leaving slopes without protective vegetation (Encylopaedia Britannica, 1929). Mountain streams eroded their banks and flooded lowlands, while hillsides eroded. When the Meiji era began in 1867, attention was paid to the serious problems of hillside erosion (Lowdermilk, undated). Japanese engineers began to restore vegetative cover in order to control flooding. In 1871 Dutch engineer Johann Dorehk was asked to collaborate with Japanese engineers in the establishment of a research program to identify techniques for reducing both soil loss from slopes and the transportation capacity of streams. They found that before afforestation could take place, slope stabilization was required. Check dams covered with vines and narrow terraces cut into slopes and covered with grass sod proved highly effective. The 1897 Forestry Act of Japan specifically set aside protective forests for erosion prevention. By the turn of the century, controlling erosion was considered to be important for all of society, to be carried out whatever the costs. Japanese engineers engaged in soil conservation – building dikes along rivers to protect alluvial areas from sand deposition, and restoring denuded and eroded hillsides with narrow terraces and vegetation (Lowdermilk, undated).

Evolution of the Soil Conservation Profession

The preceding section indicates that over hundreds – if not thousands – of years, cultures in many parts of the world arrived at understandings of soil-water relations that, when applied, resulted in soil conserving practices. In some cultures, specialist soil knowledge evolved that was not known by everyone. There were experts who could calculate sediment transport in streams or design terrace systems in Chinese, Japanese and Incan empires, 18th century English and European agricultural writers realized some of the problems of runoff and erosion, and in the 19th century Russian scientists began mapping soils and erosion (www.soilmuseum.narod.ru/first_en.htm for Russia). Although they were concerned with soil-water interactions and soil movement, these people could not be considered to be soil conservation professionals. Institutionalized and professionalized soil conservation only emerged when aspects of agriculture itself became professionalized, when soil loss was significant enough to be noticed on a large scale, and when soil itself was valued.

In the USA the profession of soil conservation evolved over a period of about 150 years. Within a hundred years of independence, soil erosion was an obvious problem. Farmers in the southeastern states in particular began to experiment with different erosion control techniques, and the study and promotion of agriculture were being institutionalized by the government. In 1862 the United States Department of Agriculture (USDA) was created, and federal funds were provided for the establishment of land-grant universities to educate the 'sons and daughters of farmers and mechanics'. Twenty-five years later federal legislation mandated agricultural research and extension activities at these public universities, including publication and mailing of free bulletins and circulars to all who requested them. This resulted in a system of state agricultural experiment stations that addressed local agricultural problems. Initially the focus was on increased crop production, often neglecting its effects on soil and water conservation. In 1889 the USDA began publishing and distributing *Experiment Station Record*, which abstracted bulletins and reports from state experiment stations and the USDA, as well as non-governmental books and journals. The Division of Agricultural Soils, established in the Weather Bureau in 1894, became an independent division in the USDA in 1895, the Division of Soil in 1897, and the Bureau of Soils in 1901. When a Division of Soil Erosion was created in 1908, increased emphasis was placed on 'the use of drainage, dams and proper tillage methods' (Harding, 1947).

Soil erosion rose to prominence in the United States through its observation. In 1899 Division of Agricultural Soils scientists working with their land grant university counterparts began making soil maps and writing survey reports in which rampant soil destruction was noted (Trimble, 1985). Hugh Hammond Bennett, who started mapping soils for the USDA in 1903, became Inspector of Soil Surveys in 1909. That same year, the National Conservation Commission surveyed 30,000 farmers, documenting the abandonment of 1659 square miles of land and erosion of 6,076 square miles. Bennett observed the consequences of contemporary farm practices (including terracing) during twenty years of soil mapping across the country. Evidence from these diverse climate and soil conditions persuaded him that soil erosion caused by bad farming practices was a 'menace' threatening both the landscape and the very future of farming itself (Bennett and Chapline, 1928), and that soil conservation measures – especially terracing – required systematic study.

The national soil mapping effort had shown no large geographical region of the country free from the problems of soil and water losses, and there was no knowledge of either the cause of erosion, or the best methods for its control. The first comprehensive American effort to quantify soil loss was made by M.F. Miller when he established soil erosion plots at the University of Missouri in 1914. Bennett was a personal friend of Miller's and knew of his research (Browning, 1977). None of the soil conservation techniques in use in the early 20th century had been systematically studied. The only information available came from individual experience and observation. There was no certainty about transferability – or even utility – of any particular technique (Meyer and Moldenhauer, 1985; Harding, 1947). Unresolved was the growing debate about which was more effective: biological or physical approaches to erosion control (Morgan, 1965).

In the decade before the photogenic dust clouds of the 1930s 'Dust Bowl' attracted national and international attention to the problem of soil erosion, Bennett argued for more research to explore the mechanisms of erosion and its prevention. The result of his campaigning was the 1929 Congressional allocation of $160,000 to establish a national network of erosion research stations across the country under a wide range of soil and climate conditions. By 1933 eight had been established, one inaugurated and a tenth one was being planned (Lipman, 1933). For the first time the various technologies of soil conservation were to be evaluated systematically at a national level. Soil conservation as a research-based profession had been born.

Despite its growing recognition in the United States, soil erosion and soil conservation were not central to the growing international soils knowledge base through the 1920s. Although Iceland had been the first to establish a national soil conservation service in 1907 (see http://www.gm-unccd.org/FIELD/Bilaterals/Iceland/scs.htm), Icelandic observations of soil loss from excessive tree cutting and sheep grazing and measures devised for its control were not widely known. When the First International Congress of Soil Science was held in Washington, DC in June 1927, soil classification and mapping, physical and chemical analysis, and aspects of soil water dominated the 10 days of sessions. Only three papers

concerned soil erosion – all from the United States. Those of H.H. Bennett and
W.R. Chapline, subsequently published together by the USDA as the influential
1928 Circular *Soil Erosion a National Menace*, argued for the importance of
proper management of vegetation and soil cover.

The British government convened an Imperial Agricultural Research Conference
in 1927 to assess the needs of its empire. The recommendation for establishing
information clearing houses was the origin of the Imperial Bureau of Soil Science.
Its 1929 review bulletin *Soil Erosion*, marked the beginning of Colonial Office –
rather than individual territorial government – information about soil erosion
(Imperial Bureau of Soil Science, 1929). So dominant were U.S. publications that
of *Soil Erosion*'s 202 citations, 117, or 59.3% were American. The Colonial
Office's first Agricultural Adviser to the Secretary of State for Colonies, Frank
Stockdale, had been concerned with soil erosion as Director of Agriculture in
Ceylon (modern Sri Lanka) during the 1920s and was aware of the soil conservation campaign in the United States (e.g. Stockdale, 1937). He made great efforts
to publicize the problem of soil erosion and build support for its control from
his appointment in 1929 through the 1930s. However, the Colonial Office did not
formulate programs or implement projects; this was the responsibility of each
colonial government.

Despite the lack of a uniform British Colonial Empire soil conservation policy
or program, publications helped create consensus about erosion. The influential –
and sensational – 1938 literature review by G.V. Jacks and R.O. Whyte, first
published as the technical bulletin 'Erosion and Soil Conservation', was widely
circulated in popular editions with photographs under the titles *The Rape of the
Earth* (Britain) and *Vanishing Lands* (USA) (Jacks and Whyte, 1939). A more tech-
nically based – but less widely circulated – publication was Lord Hailey's 1938
African Survey (Hailey, 1938). Meant to be the first ever continental assessment,
the report mentioned the rapidity with which soil erosion had become a problem
in the colonies, discussed the practicality and functionality of traditional African
land use practices, and described the erosive nature of forest clearing and plantation agriculture. While French policies to limit tree cutting and grass burning in its
territories and serious erosion problems in Belgian Ruanda-Urundi (modern
Rwanda and Burundi) were noted, Hailey found that the British territories with
significant settler populations faced the greatest crisis. Eleven years later the former
Colonial Office Agricultural Adviser Harold A. Tempany reviewed colonial
responses to soil erosion. 'The Practice of Soil Conservation in the British Colonial
Empire' was published in 1949 (Tempany, 1949). It described programs in different regions, ranging from earth structures (ridges, terraces, trenches, pits and
drains) to cultural practices (contour ploughing, tied or boxed ridges, mulching,
strip cropping and rotations, protective covers, wind breaks and afforestation) and

grazing management schemes. The paper noted that the most extensive use of terraces for soil conservation was in the USA and colonies in South and East Africa; elsewhere they were still understood to be experimental or demonstration techniques. It was narrow based terraces (called 'ridges' or 'contour banks') that were most usually constructed in Africa, despite the fact that Americans had abandoned them for the more easily managed broad based terraces.

The various African colonial governments convened collaborative meetings to consider the problems of soil (Ross, 1963). In 1948 the first Inter-African Conference on Soil Conservation and Land Utilization was held at Goma in the former Belgian Congo (modern Democratic Republic of Congo). Representatives from virtually all of the territories south of the Sahara attended. A second conference was held in Leopoldville (modern Kinshasa, Democratic Republic of Congo) in 1954, and a third in Dalaba, French West Africa (modern Republic of Guinea) in 1959. After the first Conference, permanent organizations were established to promote inter-territorial cooperation and collaboration: Bureau Interafricain des Sols et de l'Economie Rurale – Inter-African Bureau of Soil and Rural Economy (BIS), Service Pédologique Interafricain – Inter-African Pedological Service (S.P.I.), and Reginal Committees for the Conservation and Utilisation of the Soil. The BIS was an Information Bureau on Soil Conservation, Land Utilization and Rural Economy based in Paris. It published a monthly Bibliographical Bulletin and the quarterly scientific journal *African Soils/Sols Africains*. The S.P.I., based in Brussels, worked to achieve closer coordination and uniformity among soil scientists working in areas of analysis, survey, nomenclature, classification and mapping of soils. S.P.I. also compiled data for a soil map of Africa south of the Sahara.

The permanent regional soil conservation organization for southern Africa was called SARCCUS – Southern African Regional Committee for the Conservation and Utilization of Soil (Ross, 1963). When formed in 1950, it was the largest of four designated regions in Africa. Covering 2.5 million square miles, it included Katanga and Kasai Provinces, Belgian Congo (modern Democratic Republic of Congo); the Portuguese territories of Angola and Mozambique; the Federation of Rhodesia and Nyasaland (modern Zimbabwe, Zambia and Malawi); the High Commission Territories of Basutoland, Bechuanaland and Swaziland (modern Lesotho, Botswana and Swaziland), and the Republic of South Africa, including the United Nations Trust Territory of South West Africa (modern Namibia). The South African government played a leading role in this organization throughout the 20th century. The inaugural meeting was held in Pretoria (1950); subsequent meetings were in Pretoria (1952), Luanda (1953), Salisbury (modern Harare) (1954), Mbabane (1955), Windhoek (1957), Lourenço Marques (modern Maputo) (1958) and Salima (1960) (Ross, 1963).

National Soil Conservation Programs: 20th C USA and Southern Africa

Since the United States of America had a long history of soil conservation as a response to agriculturally induced erosion, implemented national research and action programs in response, and published prolifically on the topic, it was extremely influential internationally. Not only were American publications read and used for guidance, but officials and scientists from around the world visited the United States on study tours. American philanthropic foundations such as the Carnegie Corporation (which had funded Hailey's *African Survey*) sponsored many of these trips, contributing to—or paying in full—travel and expenses (e.g. Hailey, 1938; Jeffries, 1938). Carnegie study tours typically lasted three months, so that visitors could tour all (or most) of the research centers and demonstration projects. Because of its influence, the origins of the American program will be discussed in some detail in the following paragraphs. This will be followed by a review of the implementation of soil conservation programs in southern Africa, where conservation structures built on African land using American designs had severe unintended consequences. This negative legacy today clouds efforts to confront what continues to be a serious problem.

The United States of America

The major force in the United States for a national focus on soil erosion and the creation of a soil conservation bureaucracy was Hugh Hammond Bennett, a soil chemist who grew up on an eroded farm and spent his early professional life mapping soil. His observations led him to believe that bad farming techniques were ruining soil, and he was one of the first to recognize that sheet erosion contributed to declining crop yields (Bennett and Chapline, 1928; Harding, 1947). He believed the remedy lay in restoring vegetation on unused land, increasing organic matter in cultivated soils, and changing the basic techniques of farming. Bennett had seen enough gully erosion associated with terracing to be leery of its widespread application. Most of all, he believed that different techniques were required for different soil and topographic conditions. However, there was no popular support for a soil conservation program. Farmers still abandoned their fields when productivity failed, and moved west – or believed that this was what they could do – and did not want the government telling them how to farm (Rasmussen, 1982).

The first publicly funded soil conservation projects in the USA were, therefore, part of the emergency public works programs established as unemployment relief during the 1930s depression. The National Industrial Recovery Act of June 16, 1933 authorized soil erosion control work in national forests and national parks.

Bennett, working with A.J. Pieters of the Bureau of Plant Industry, urged the use of a combination of vegetative, engineering and other methods. At the same time, another soil conservation program was to be administered by the Bureau of Agricultural Engineering, funded by a $5 million allocation to the Public Works Administration. So concerned was Bennett that this would result in a major program limited to terracing, that he successfully lobbied government officials to have the funds transferred from the Public Works Administration to the U.S. Department of the Interior for use by the Civilian Conservation Corps (CCC) (Morgan, 1965). This program became the Soil Erosion Service (SES) on Sept. 19, 1933. Hugh Hammond Bennett was named its Chief.

Bennett had a long-standing desire to implement soil conservation on a watershed basis, but there was no public support. However, on Federal land, only the consent of the responsible government agency was required. American Indian Reservations were Federal land, and the new Commissioner for Indians, John Collier, wanted to display his new policy and approach by implementing a CCC project. The Navajo Tribal Council agreed to allow a demonstration project at Mexican Springs, Arizona. Bennett persuaded Collier to add a soil conservation component to reduce the siltation of the Colorado River. The Project involved dams, terraces, water spreading devices, reseeding, arroyo (gully) stabilization, and rodent control. It was the largest SES project, consuming 1/5 of the budget, and showcased Bennett's ideas about conservation on a watershed basis (Kelly, 1985).

However, Bennett wanted to create a permanent national soil conservation program. To do this he had to build support among the general public so that they would participate, and influence the legislators who could create the necessary institutions and release operating funds. He decided to mount a three part campaign: 1) shock the nation into believing that soil erosion was a serious threat in need of immediate national action; 2) accelerate research into proper methods of control; and 3) gain support of the influential leaders of every community to work with soil conservation idealists. To 'shock the nation', emphasis was to be laid on the amount of soil that had been lost, how much more would be lost, and how it affected agriculture. 'Erosion was often equated with sin – something everyone was against with no question . . . . . This was the crusading era – and no crusader can stop to weigh opposing evidence and arguments for fear of diluting or losing his crusade' (Held and Clawson, 1965, p. 63, 64). Primary emphasis was placed on terraces as a device for erosion control, even though Bennett and many of his closest colleagues knew that a broader approach was required. The Bureau of Agricultural Engineering believed that terraces were the only solution. Terracing became emblematic of the 'fight' or 'battle' against soil erosion. The argument for an increased research effort was supported by publication, in 1934, of a National Erosion Survey in both map and statistical forms (Held and Clawson, 1965).

Bennett's plan to 'shock the nation' into awareness was aided by the weather. In 1934 the first of a series of major dust storms occurred in Texas, Oklahoma, Colorado and Nebraska. That year he submitted a 'seven point plan of action' for a voluntary program of erosion control to the President's Natural Resources Board. The following year Bennett was scheduled to testify before the U.S. Senate's Public Lands Committee (Morgan, 1965). Knowing that a major western dust storm's massive cloud of dust was blowing towards the East Coast, he agreed to testify on April 2, the day the dust cloud was calculated to reach Washington, D.C. As Senators questioned him, the sky darkened with dust. Three and a half weeks later, on April 27, 1935, the U.S. Congress passed the Soil Conservation Act, establishing a permanent soil conservation program to be led by a new agency within the Department of Agriculture, the Soil Conservation Service (SCS). Most of the USDA's soil related activities were consolidated in the new SCS, which absorbed the old SES.

Not surprisingly, the SCS was headed by Hugh Hammond Bennett. As Assistant Chief, he chose Walter Clay Lowdermilk, the internationally recognized authority on forest hydrology and soil erosion research. Lowdermilk had been the Assistant Chief of the Soil Erosion Service, responsible for the research program. In the new SCS, his mandate was more to promote conservation. He was a great advocate of terraces, having seen their successful use in the stabilization of catastrophically eroded hillsides in China and Japan. Over time, these two men became extremely influential internationally. As concern about soil erosion and interest in soil conservation methods spread in the 1940s and 1950s, national and colonial governments sought Bennett's and Lowdermilk's approval and advice for projects and programs throughout the world.

The new Soil Conservation Service shifted emphasis from research and demonstration on a watershed basis (the Soil Erosion Service mandate) to direct assistance to farmers (Rasmussen, 1982). This was prompted both by the high cost of demonstration projects and the need for increased farmer participation. Because a federal government agency could not order such a program to exist at a state level, model legislation establishing soil Conservation Districts as the basis of soil conservation work was drafted. The Districts were to be established only if a majority of 'land occupiers' agreed, and were to be governed by five supervisors – three of whom would be elected. The model law was sent to all states in 1937; four years later, forty-one states had passed their own versions, and by 1947 all states had enacted such laws.

The Conservation Districts were supposed to make cooperative agreements with local farmers. An SCS specialist would work with a farmer to make a conservation plan for the entire farm based on the land's requirements and the farmer's resources, and helped with implementation. The Conservation District then channeled help

Hugh Hammond Bennett

Hugh Hammond Bennett was born in the southeastern, humid temperate state of North Carolina in 1881 on a farm that had been impoverished by continual cotton production and tree clearing. After earning a degree in soil chemistry from the University of North Carolina in 1903, he was hired as a lab assistant at the U.S. Bureau of Soils. Bennett was promoted to soil scientist, assigned to soil survey work, and began a twenty year career in soil mapping. In 1905 he was deeply impressed when surveying the badly eroded section of the Virginia Piedmont in Louisa County. His 1910 Report on Lauderdale County, Mississippi included an extended discussion of soil erosion, with cross hatchings on the map to indicate eroded locations. This was one of the first American attempts to map erosion. Soil conservation was a religious matter to Bennett and the small farmers of the southeast. He was often asked to preach in rural churches. He had an almost religious intolerance of non-believers of the importance of conservation; the cause of soil conservation became a lifetime crusade. To promote his cause, Bennett 'combined science with showmanship', according to his contemporary Santford Martin (Held, 1999).

In 1909 Bennett was made Inspector of Soil Surveys. His expertise was recognized when he was asked to serve on the committee to study agriculture in the Panama Canal Zone in 1909, head an expedition to explore agricultural regions of Alaska in 1914, participate in the 1919 Guatemala-Honduras Boundary Commission and, in 1925–26, work on the Reconnaissance Survey of Cuba. He traveled internationally to promote conservation, visiting Venezuela, South Africa, Mexico and Canada among other places. By the time he retired in 1951, more than 1,100 technicians from 88 countries had visited the United States to study his methods. Bennett received several honorary degrees, was president of the Association of American Geographers in 1943; was awarded the Frances K. Hutchinson Award by the Garden Club of America in 1944, the Distinguished Service Medal by the USDA and the Audubon Medal by the National Audubon Society in 1947, and the American Geographical Society's Cullum Geographical Medal in 1948. At his death in 1960, Hugh Hammond Bennett was called 'The Father of Soil Conservation'.

Hugh Hammond Bennett Papers Summary can be found at http://www.lib.unc.edu/mcs/inv/b/Bennett,_Hugh_H.html.

Walter Clay Lowdermilk

Walter Clay Lowdermilk was born in humid North Carolina, grew up in Missouri, and graduated from semi-arid Arizona's University of Arizona in 1912. Named a Rhodes Scholar to Oxford, he earned a Bachelor's degree in 1914 and a Masters degree in 1915. During the summers he studied forestry in Germany. Upon returning to the US in 1915, Lowdermilk joined the Forest Service as a research officer in the Northern Rocky Mountain region.

In 1922 Lowdermilk married an old friend, Inez Marks, who had just returned from five years overseas service with the Methodist Church Board for Overseas Service in China. She suggested he combine his professional interests with work in China. His application for a teaching and research grant at Nanking's American Union University to use his experience in forestry in the interest of famine prevention was accepted. The Lowdermilks lived and worked in China from 1922–1927.

Lowdermilk concentrated on soil erosion in the Yellow River watershed. He and his Chinese colleagues compared runoff and erosion from bare hillsides with that from hillsides covered by ancient, protected temple forests, publishing their results in scientific journals. Upon leaving China in 1927, Lowdermilk had an international reputation for erosion research. He rejoined the Forest Service as Project Leader for California Forest Experiment Station's erosion and streamflow studies, and completed a PhD in Forestry at the University of California, Berkeley. He designed the San Dimas Experiment Station in California to study forest hydrology. In 1933 he was named Assistant Chief of the new Soil Erosion Service (responsible for research), and continued in the Soil Conservation Service (SCS) at its 1935 formation as an 'ambassador' for conservation.

From August 1938 to November 1939 the SCS sent Lowdermilk on a tour of western Europe, North Africa, and the Middle East to seek soil erosion and conservation practices that might be appropriate for American erosion problems. His lectures about the trip, illustrated with lantern slides, became the basis for *Conquest of the land through seven thousand years*". An earlier book, *Palestine: Land of Promise*, argued that with proper use of the Jordan River, a greater population could be supported. From 1942–1943 Lowdermilk and Chinese colleagues traveled through the Yellow River watershed looking for indigenous soil conserving farming methods and establishing demonstrations of American practices.

After retiring in 1947, Lowdermilk became a conservation consultant. He advised French and British African Colonial governments about soil conservation, recommending terracing in North Africa. Most of the 1950s was spent in Israel working for the Food and Agriculture Organization (FAO) or Israeli universities. Haifa's Technion University named its school of agricultural engineering after him. In 1957 he worked on the UN's river basin development plan for (then) Yugoslavia's Cetina River, and from 1960–1969 was a consultant for the Save-the Redwoods League in California.

Lowdermilk served as President of the American Geophysical Union and the Soil Conservation Society of America, was a Fellow of the Society of American Foresters, and was appointed Associate Editor, Journal of Forestry, in 1929. He served for eleven years. Walter Clay Lowdermilk died in 1974. He has been called the founder of the scientific basis of soil conservation.

such as technical aid, seed and, in some cases, machinery. The recommended measures varied from farm to farm, depending upon soils, topography and the type of farm operation. There was no mandate for any particular conservation technique (Bennett, 1946).

This was the US program that visitors from around the world toured: experiment stations, demonstration sites, and farmer-controlled, voluntary programs of vegetation management and structures. What the visitors chose to take home with them along with pamphlets and memories was beyond the control of the Americans. For many southern African government officials, it was the experimental terrace technology, and not its adaptive experimental research, that was noticed.

Southern Africa

When Europeans arrived in Africa, they were environmental pioneers. They knew nothing about the soils, vegetation or climates, and had little idea of ecosystem interactions. Their traditional agricultural, livestock and forest management systems had been developed in humid, temperate, glacially rejuvenated Europe. Yet the Europeans were certain of these systems' suitability to tropical and sub-tropical Africa. For some, the African landscapes appeared to be deficient and to most, the indigenous land use systems seemed inefficient or unproductive. Whether they arrived with settlers, missionaries or officials, European agricultural and land management practices changed the vegetative cover and reduced both organic matter and soil structure. Monocultures replaced intercrop, relay crop, and fallow rotations; plows replaced hoes, and rows replaced broadcast planting. Clean weeding from planting time was advocated. Livestock was confined rather than disbursed. Settlers, plantations, and forest and game reserves caused intensification of land use by removing substantial amounts of land from African control. Finally, transport systems (road and rail) concentrated people, livestock and water. All of these activities contributed to the acceleration of soil erosion processes. The nature and extent of accelerated erosion varied with the type of intervention, soils and climate, and responses varied from place to place, and year to year. It was not until the 1930s that national soil conservation programs began to be implemented.

As in North America, southern African awareness of soil erosion resulted from the experience of eroding domestic spaces (Showers, 1989, 1994). Noticed as a problem by a few at the end of the 19th century and early in the 20th, concern became more widespread in the 1920s and 1930s. When local government officials looked for information about techniques that might be useful, they invariably found American government literature. Later, as local programs developed and the American national soil conservation program became established, southern African officials made study tours to the United States, and Hugh Hammond Bennett toured South Africa and Basutoland (modern Lesotho) to review progress and make recommendations to the South African government. Despite this strong influence, there was not direction from the United States in southern Africa (Showers, *forthcoming*).

Soil erosion in southern Africa cannot be discussed in isolation from European settler's land claims. Indigenous land use systems were extensive, and thus required large amounts of land – land the Europeans wanted. As the number of settler farmers and ranchers increased, the land available to Africans decreased. Treaties signed in the 19th century transferred the rich farmland of the Basotho (residents of Lesotho) to the Afrikaner farmers of the Orange Free State (modern free state province, South Africa), and designated two thirds of the land used by Swazi people as Title Deed Land owned by the British Crown, leaving one third of the land as Swazi Nation Land under control of the King (Germond, 1967; Swaziland. *Col. Ann. Report*, 1956). Large blocks of the best watered land in the desert nation of Bechuanaland (modern Botswana) were declared Crown Lands available for European settlement (Tlou and Campbell, 1984). South African legislation passed in 1913 and 1936 crowded the African population onto 13% of the territory, giving the tiny European population control of the remaining 87% of the land, including the most favorable agricultural regions of the country (Platzky and Walker, 1985). In Southern Rhodesia (modern Zimbabwe), the boundaries of Native Reserves for Africans were formalized in 1913 on marginal land with infertile soils and low rainfall, again leaving the more favored land in the hands of European settlers (Mpofu, 1987).

The result was the disruption of indigenous land use systems. The details and dates vary from nation to nation, but the overall form is of stable and sustainable systems for the provision of food, shelter and surplus for trade having reduced land bases. First there was no surplus, and then not enough for self-sufficiency. This process resulted in increasing – and inevitably unbearable – pressures on the landscape; pressures resulting in the cultivation of marginal land, degradation of pastures, deforestation and soil erosion. In many areas accelerated soil erosion had been an unknown and unnamed concept, in others traditional agricultural systems included soil conservation practices. 'Overcrowding' or 'over population' of people and their livestock was identified as a cause of erosion by most southern Africa colonial governments by the late 1920s. A land problem created by the arrival of Europeans pre-dated the erosion problem.

Just as land had not been equitably divided, soil conservation programs were not the same for European settlers and Africans. Initially programs for settlers were voluntary, involving choices of techniques or components. These conservation programs were all based on information, with major education campaigns to explain the various options. Officials wrote popular and technical articles and bulletins, gave lectures, and made demonstrations to encourage farmers to adopt soil conserving technologies. The importance of agricultural terraces on plantations and large farms (tea and tobacco in particular) was increasingly promoted. The *Rhodesia Agricultural Journal* carried a series of articles on erosion and its control in the 1920s. Agricultural engineers responsible for Southern Rhodesia (Zimbabwe)'s earliest soil conservation efforts advised farmers on the construction of terraces. The primacy of this approach was confirmed in 1929 when a Conservation Unit was created within the Irrigation Department.

Despite the education campaigns and availability of technical advice, European land users were slow to adopt anti-erosion practices. In South Africa, subsidies were deemed essential when conservation schemes were introduced in 1933. By 1941, less than 10 percent of the European farmers had participated, and of these, only one third had built erosion control structures. It was not until the 1946 Soil Conservation Act, No. 45 that soil conservation programs were mandatory on European farms (Ross, 1963). Further north, in spite of advice and visits from Agricultural Engineers, only two of Southern Rhodesia's districts had more than 40% of cultivated land protected by contour ridges in 1938. The Southern Rhodesian 1939 'Report of the Commission to Enquire into the Preservation of the Natural Resources of the Colony' stated that soil erosion was still a serious problem on European farmland (McIlwaine, 1939). Despite the creation of a Natural Resources Board in Swaziland, and meetings and advice to employ conservation techniques, the government had to resort to fines to enforce reclamation and erosion control measures on European farms as late as the mid-1950s.

African farmers were equally uninterested in soil conservation programs, but choice was not an option. Soil conservation measures were imposed, often without explanation. Each government devised its own program. Most included some kind of engineering structures (diversion ditches and narrow-based terraces – referred to as contour banks, ridges or bunds), grass strips and mechanisms for the regulation of grazing. Compulsion and punishment for non-compliance were central to the soil protection programs on African land (Showers, 1994).

The first coordinated soil conservation programs in the region were implemented on African land in the early 1930s by South Africa's Native Affairs Department. In 1933 the Director, R.W. Thornton, used funding for a famine relief program to create a soil conservation public works program with workers paid in food (Showers, 2005). This was, perhaps, the first food-for-work soil conservation program, a mechanism still used in the early 21st century.

Once the program had been approved, the Director and his Chief Engineer, L.H. Collett sent away for booklets from around the world to learn about conservation techniques. They were particularly impressed by the amount of information available from the U.S.A. The program they designed included terracing fields and pastures; reallocating fields; fencing stressed areas; digging dams; and creating small-scale irrigation systems. Local white shopkeepers were hired as foremen and trained to supervise African laborers. The contour banks (terraces) collected and concentrated water, increasing its erosive potential. They broke repeatedly during heavy rainstorms, releasing water onto the areas they were supposed to protect. Maintenance became a regular activity, and new banks were dug to protect existing ones. Dams made in gullies and at low spots in the landscape similarly collapsed. Terrace outlets were problematic in construction and function. Erosion increased, rather than decreased, with this conservation work. By 1940 over 1,200 miles of contour banks had been constructed. In 1952, the Chief Engineer recollected that leaving the engineering design to untrained persons had been 'a tragedy' (Showers, 1994).

The residents of the Herschel District of the Transkei, where this program was first implemented, did not cooperate. They deliberately plowed out or otherwise destroyed the contour banks in their fields. By 1941, 800 land users had been charged, but attempts to fine the resisters were not successful. A survey in 1940 showed that while the contour banks near the main roads had been left in place, those further away had completely disappeared (Showers, 1994).

In the mid-1930s, South Africa's Thornton was appointed Director of Agriculture in adjacent Basutoland (Lesotho); his Chief Engineer, L.H. Collett was also hired. Together they designed the Basutoland national soil conservation programme to reclaim and protect the ten percent of the land affected by erosion, in order to prevent erosion in the unaffected 90% of the country. Their proposal called for a structural approach, similar to what they had supervised in South Africa. Collett was sent on a Carnegie sponsored study tour of the USA. What began as a demonstration project in 1936 expanded to the entire country beginning in 1937 (Showers, 1989, 2004).

The Agriculture and Livestock Officer of each District was given training in the layout and construction of contour banks, and was then responsible for all the work in his District. Conservation officials would arrive at a location, notify the chief, assemble a work crew, and begin construction. There was usually no advance warning or explanation, and there certainly was no choice about having contour banks in one's field (Showers, 2005). As in Herschel, European shopkeepers were trained to use a level, and hired to oversee contour construction by African labourers. While the shopkeepers were paid in cash, the Basotho received only food for their efforts.

Also as in Herschel, the poorly designed contour banks were overtopped and broke, they were destroyed by hailstorms and rainstorms, and gullies formed in their outlets (Showers and Malahleha, 1992; Showers, 1989, 2005). Attempting to reduce damage, the design of the banks was changed each year – but the existing ones were not replaced or redesigned. Collapse due to design failure was blamed on the Basotho's failure to engage in 'proper maintenance'. What began as a limited ten year reclamation project became a decades long program of installing conservation structures with constantly changing designs. By 1964 a total of 519,681 acres had been terraced with 26,717 miles of terraces (contour banks) and, 1,597 miles of diversion furrows had been excavated. In addition, 737,346 acres of buffer strips and 1,409 miles of meadow strips had been planted. The country was judged to be protected. Nevertheless, at the end of the 20th century, soil scientist Rooyani wrote 'Gullies are the most spectacular feature of the landscape in lowland zone of Lesotho' (Rooyani, 1982).

Basutoland's was the first national soil conservation program in British Africa, and became a showcase for the new concept. Visitors from around Africa and around the world came on tour – including American Hugh Hammond Bennett. Not all visitors were impressed, however. In 1938 visiting Kenyan colonial officials H.L.G. Gurney, D.C. Edwards and R.O. Barnes noted that "the campaign against soil erosion was launched on a scale which was large in comparison with the extent of knowledge of the best methods of attack" (Gurney et al., 1938, p. 4).

After observing the effects of the contour banks on their fields, and attempting to mitigate them, many Basotho shared the Kenyan official's assessment. However, there was broad consensus about the need for discretion (Showers, 2005). The Basotho had asked for a protective alliance with the British government in the late 19th century to block land claims of the settlers in adjacent South Africa, and feared that obvious rejection of the government's conservation program would be misunderstood as a rejection of the British government. So, as in neighboring Herschel, contour banks out of view were removed or altered, and those most visible were left in place. When Chiefs were ordered to fine those who destroyed conservation works, no money was collected. The Basotho's consistent and persistent passive resistance of, and lack of enthusiasm for, soil conservation programs has been interpreted repeatedly as evidence of their ignorance of the existence of erosion, disinterest in protection of the soil, and, finally, as an example of bad farmers who willfully destroy their land (Showers, 1989, 2005).

In contrast to South Africa and Basutoland officials, Swaziland's Agriculture officials chose to use vegetation rather than structures in the national conservation programme (Showers, 1994). As in the rest of southern Africa, participation was optional on European land, but King Sobhuza II commanded in 1949 that all Swazi Nation Land be treated with grass strips. Unlike the rest of southern Africa, however, by the end of the 20th century, Swazi farmers had accepted grass strips as important, and incorporated them voluntarily into their farming systems – although not as originally designed (Osunade and Reij in Reij et al., 1996).

Soil conservation in Southern Rhodesia became so notorious that it helped fuel the bloody war for independent Zimbabwe (Showers, 1994). The Natural Resources Board in 1941 mapped out Intensive Conservation Areas (ICAs) on European land, and charged local farm groups with responsibility for management. Controlled by the farmers, the ICAs proved to be both popular and effective. Soil conservation work only began in earnest in 1947, but within four years, European farm areas had been fully protected by contour ridges (narrow based terraces). Soil protection programs began on the land allocated to Africans (African Reserves) in the 1930s, but intensified in the 1940s. Contour ridges, storm drains, and grass buffer strips were put in fields, and ditches, contour drains, check dams, tree planting and fencing were used on grazing lands. By the end of 1949 approximately 22% of the agricultural land had been treated. To increase compliance, the 1951 Native Land Husbandry Act was passed. Enforced between 1956–1962, it included rules about proper farming techniques, livestock numbers, land tenure, and the requirement for all land users to construct conservation works in their fields. As in Herschel and Basutoland, African farmers in Southern Rhodesia who observed the function of contour ridges criticized their effects on the landscape (Showers, 1994). In the dry south of the country, farmers reported that contour banks had contributed to the silting of streams. Once again, poorly trained supervisors ensured improperly aligned ridges, making erosion even worse (Wilson, 1995). Most of all, rural people resented being forced to do the hard manual labor of construction without explanation. People were imprisoned for refusing to dig contour banks (Staunton, 1990). Independence movement leaders promised an end to soil conservation in the new Zimbabwe.

Summary and Conclusions

Wherever agricultural societies have existed, soil movement has been a companion. Over long periods of time, some societies developed techniques for minimizing agriculturally induced erosion.

Some found ways to incorporate normal geomorphic erosion processes into their farming systems. This learning process resulted in both broad cultural understandings of appropriate ways to intervene in landscapes and specialist knowledge about soil and water processes. Soils were classified and named by some, basic elements of hydrology perceived by others. This type of knowledge accumulates as people interact with specific conditions of soil, topography and climate over a long period of time. Should they have to move to another ecosystem or experience a major change where they live (such as climate change), the people become 'environmental pioneers' who must learn all over again, modifying or replacing practices that had worked in different places in the past. Not all societies learn, and not all societies have social, political and economic structures that encourage soil conserving practices. Even a brief survey of the world's history of soil erosion and its prevention reveals examples of learners, failed learners, and politically and economically driven erosive practices.

In the late 19th century industrial societies in the northern hemisphere realized that soil loss was occurring at an increasing rate due to human activity. Once considered to be unlimited and without value, soil came to be seen as precious and in need of protection. A profession of soil conservation began to emerge. Would be protectors of soil often took one of two fundamentally different approaches to conserving soil: biological and physical. Advocates of biological control concentrated on farming systems, afforestation, and the preservation of natural vegetation, and tended to come from backgrounds in agronomy, soil chemistry and fertility, range management and forestry. Designers of structures came from the new field of agricultural engineering.

As the need for soil conservation action was recognized around the world, two fundamental questions had to be answered: would anti-erosion measures be based upon the management of vegetation or structural control, and would participation in soil conservation programs be voluntary or mandatory? The answers to these questions determined, to a large extent, popular opinion about soil conservation, and the extent to which soil conservation practices were followed after government programs ended.

In the United States, where nationally coordinated soil conservation research, legislation and programs emerged in the 20th century, participation was voluntary and great emphasis was placed on the farmer's ability to both control conservation organizations and choose the technologies to be used. Soil conservation officials designed plans for each individual farm, with the assumption that conservation plans would differ with soil, topographic and climate conditions. By the late 1940s, the initial interest in terrace construction abated. Crop management techniques emerged as more popular, more practical, more suitable to most locations, less likely to induce erosion, and much less expensive then terraces.

In southern Africa, soil erosion increased as European farms, ranches and plantations spread, reducing the amount of land for African use. While voluntary programs were offered to European settlers, most Africans were compelled to implement structures. Resentment and resistance grew. The poorly designed terraces increased erosion and took up farmland, and those who failed to comply were fined and, in some cases, imprisoned. Soil conservation came to be linked with loss of land, destruction of soil, and capricious, oppressive governments. Resisting conservation became part of independence struggles, and soil conservation officers came to be seen as enemies of the people. While these statements are generally true for east and southern Africa, they do not reflect all the world's experience of professional soil conservation. They do, however, provide a cautionary tale about the problems resulting from combining compulsion instead of education with soil protection, the serious dangers of the widespread application of an untested technology, and personal and institutional hubris that allows blame of others instead of program evaluation.

The profession of soil conservation is young. Its advocates are international environmental pioneers. They can hasten their learning curve by reviewing humanities' long history of experience with soil management, and they can learn by evaluating the soil conservation programs of the 20th century from both technical and social perspectives. Above all, soil conservationists must accept responsibility for past mistakes – however unintentional – and assume that more will be made. To protect the idea of soil conservation from contamination with social meanings of compulsion and failure, soil conservationists must find ways of genuinely involving land owners – whatever their culture or level of formal education – in the process of evaluating each landscape to determine what will be best for it and for the land user.

Suggested Further Reading

Bell, M. and J. Boardman (eds.) 1992. *Past and Present Soil Erosion: Archaeological and geographical perspectives.* Oxbow Monograph 22. Oxbow Books, Oxford.

Boardman, J. 1998. An average soil erosion rate for Europe: Myth or reality? Journal of Soil and Water Conservation 53:46–50.

Grove, A.T. 2000. The African environment, understood and misunderstood. pp. 179–206 *In* D. Rimmer and A. Kirk-Greene (eds.) The British Intellectual Engagement with Africa in the Twentieth Century. St. Martins Press, New York.

Grove, A.T. and O. Rackham. 2001. The nature of Mediterranean Europe. Yale University Press, New Haven.

Reij, C., I. Scoones, and C. Toulmin (eds.) 1996. Sustaining the soil: Indigenous soil and water conservation in Africa. Earthscan, London.

Roe, E. 1991. Development narratives, or making the best of blueprint development. World Development 19:287–300.

Spencer, J.E. and G.A. Hale. 1961. The origin, nature and distribution of agricultural terracing. Pacific Viewpoint 2:1–40.

References

Adams, W.M. and D. Hulme. undated. Community conservation research in Africa: Principles and comparative practice. Working Paper No. 4, Institute for Development Policy and Management. University of Manchester, Manchester. http://idpm.man.ac.uk/publications/archive/cc/cc_wp04.pdf.

Amborn, H. 1989. Agricultural intensification in the Burji-Konso Cluster of Southwestern Ethiopia. Azania 24:71–83.

Arnold, D. and R. Guha. 1995 (eds.) Nature, culture, imperialism: Essays on the environmental history of South Asia. 1996 edition. Oxford University Press, Delhi.

Asrat, K., K. Idris and M. Semegen. 1996. The 'flexibility' of indigenous soil and water conservation techniques: A case study of the Harerge Highlands, Ethiopia. pp. 156–162 In C. Reji, I. Scoones and C. Toulmin. Sustaining the soil: Indigenous soil and water conservation in Africa. Earthscan, London.

Beach, T. 1998. Soil catenas, tropical deforestation, and ancient and contemporary soil erosion in the Petén, Guatemala. Physical Geography 19:378–405.

Beach, T. and N.P. Dunning. 1995. Ancient Maya terracing and modern conservation in the Petén rain forest of Guatemala. Journal of Soil and Water Conservation 50:138–145.

Bell, M. and J. Boardman (eds.) Past and Present Soil Erosion: Archaeological and geographical perspectives. Oxbow Monograph 22. Oxbow Books, Oxford.

Bennett, H.H. 1929. The economics of preventing soil erosion. Agricultural Engineering 10:291–296.

Bennett, H.H. 1946. Our American land: The story of its abuse and its conservation. Misc Pub 596. USDA/SCS, Wash DC.

Bennett, H.H. and W.R. Chapline. 1928. Soil erosion a national menace. U.S.D.A. Circular No. 33. Government Printer, Washington D.C.

Bocco, G. 1991. Traditional knowledge for soil conservation in Central Mexico. Journal of Soil and Water Conservation 46:346–348.

Boardman, J. 1998. An average soil erosion rate for Europe: Myth or reality? Journal of Soil and Water Conservation 53:46–50.

Boomgaard, P. 1997. Introducing environmental histories of Indonesia. pp. 1–26 In P. Boomgaard, F. Colombijn and D. Henley (eds.) Paper landscapes: Explorations in the environmental history of Indonesia. KITLV Press, Leiden.

Browning, G.M. 1977. Developments that led to the universal soil loss equation: an historical review. pp. 3–5 In Soil erosion: prediction and control. Special Publication No. 21. Soil Conservation Society of America, Ankeny.

Drought Investigation Commission. 1923. Final Report, October 1923. U.G. 49. Government Printer, Pretoria.

Encyclopaedia Britannica. 1929. Japan. vol.12, pp. 893–954. 14th Edition. London.

Evans, R. 1995. Some methods of directly assessing water erosion of cultivated land – a comparison of measurements made on plots and in fields. Progress in Physical Geography 19:115–129.

Germond, R.C. 1967. Chronicles of Basutoland. Morija Sesuto Book Depot, Morija.

Grove, A.T. 2000. The African environment, understood and misunderstood. pp. 179–206 In D. Rimmer and A. Kirk-Greene (eds.) The British Intellectual Engagement with Africa in the Twentieth Century. St. Martins Press, New York.

Grove, R. 1989. Scottish missionaries, evangelical discourses and the origins of conservation thinking in southern Africa 1820–1900. Journal of Southern African Studies 15:163–187.

Grove, R. 1990. Colonial conservation, ecological hegemony and popular resistance: towards a global synthesis. pp. 15–50 In J.M. MacKenzie (ed.) Imperialism and the natural world. Manchester University Press, Manchester.

Grove, R.H. 1997. Ecology, climate and empire: Colonialism and global environmental history 1400–1940. White Horse Press, Cambridge.

Grove, A.T. and J.E.G. Sutton. 1989. Agricultural terracing south of the Sahara. Azania 24:113–122.

Grove, A.T. and O. Rackham. 2001. The nature of Mediterranean Europe. Yale University Press, New Haven.

Gurney, L.G.C., C. Edwards and R.O. Barnes. 1938. Notes of a visit to the Union of South Africa and Basutoland, March–April 1938. Rhodes House, Oxford: MSS, Brit. Emp. T.1(1).

Hailey, Lord. 1938. An African Survey: A study of problems arising in Africa south of the Sahara. Oxford University Press, London.

Hall, A.R. 1937. Early erosion control practices in Virginia. USDA Miscellaneous Publication 256. U.S. Government Printing Office, Washington, D.C.

Hall, A.R. 1940. The story of soil conservation in the South Carolina Piedmont, 1800–1860. USDA Misc Pub. No. 407. USDA, Washington, D.C.

Harding, T.S. 1947. Two blades of grass: A history of Scientific development in the USDA. University of Oklahoma Press, Norman.

Helms, D. 1999. Hugh Hammond Bennett. pp. 582–583 In American National Biography, vol.2. Oxford University Press, New York.

Held, R.B. and M. Clawson. 1965. Soil conservation in perspective. Johns Hopkins University Press, Baltimore.

Heusch, B. 1981. Sociological constraints in soil conservation: A case study, the Rif mountains, Morocco. pp. 419–424 In R.P.C. Morgan (ed.) 1981. Soil conservation: problems and prospects. John Wiley and Sons, London.

Hopkins, E.S. 1937. Soil conservation programs in the United States and Canada. Scientific Agriculture 17:265–269.

Huszar, Paul C. 1997. Indonesia's soil sprouts seeds of change. Forum for Applied Research and Public Policy 12:125–128.

Imperial Bureau of Soil Science. 1929. Soil erosion. Technical communication No. 5. Rothamsted Experimental Station, Harpenden.

Jacks, G.V. and R.O. Whyte. 1938. Erosion and soil conservation, Herbage Publication Series Bulletin No. 25. Imperial Bureau of Pastures and Forage Crops, Aberystwyth.

Jacks, G.V. and R.O. Whyte. 1939. The Rape of the Earth – A world survey of soil erosion. Faber, London.

Jacks, G.V. and R.O. Whyte. 1939. Vanishing Lands. Doubleday, New York.

Jackson, W.A.D. 1980. The Soviet Union. pp. 131–164 In G.A. Klee (ed.) World Systems of traditional resource management. John Wiley, New York.

Jacobsen, T. and R.M. Adams. 1958. Salt and silt in Ancient Mesopotamian agriculture. Science 128(3334): 1251–1258, Nov. 21, 1958.

Jeffries, C. 1938. The colonial empire and its civil service. Cambridge University Press, Cambridge.

Kelly, L.C. 1985. Anthropology in the Soil Conservation Service. pp. 34–45 In D. Helms and S. Flader (eds.) The history of soil and water conservation: The Agricultural History Society, Washington, D.C.

Kleine, D. 1997. Who will feed China? Journal of Soil and Water Conservation 52:398.

Lemons, J. and E. Saboski. 1994. The scientific and ethical implications of Agenda 21: Biodiversity. In N.J. Brown and P. Quibler (eds.) Ethics and Agenda 21: Moral implications of a global consensus. United Nations Environment Programme. United Nations Publications, New York.

Lipman, J.G. 1933. A quarter century progress in soil science. Journal of the American Society of Agronomy 25:9–25.

Lowdermilk, W.C. undated. Erosion control in Japan. Mimeo. Possibly reprint Oriental Engineer, March 1927. Available at Mann Library, Cornell University.

Lowdermilk, W.C. 1944. Palestine: Land of promise. Harper and Brothers, New York.

Lowdermilk, W.C. 1948. *Conquest of the land through seven thousand years.* USDA/SCS, Washington, D.C.

McConkey, O.M. 1952. Conservation in Canada. J.M. Dent & Sons (Canada), Ltd, Toronto.

McDonald, A. 1966. Early American soil conservationists. Miscellaneous Publication No. 449, Soil Conservation Service, 1941, reprinted 1966. USDA, Washington D.C.

McIlwaine, R. 1939. Report of the Commission to Enquire into the Preservation of the Natural Resources of the Colony. Government Printer, Salisbury.

McTainsh, G. and W.C. Boughton. 1993. Land degradation in Australia: An introduction. pp. 1–16 *In* G. McTainsh and W.C. Boughton (eds.) Land degradation processes in Australia. Longman Cheshire Pty, Ltd, Melbourne.

Maltby, E. 1991. Wetland management goals: wise use and conservation. Landscape and Urban Planning 20:9–18.

Meyer, L.D. and W.C. Moldenhauer. 1985. Soil erosion by water: The research experience. Agricultural History 59: 192–204.

Morgan, R.J. 1965. Governing soil conservation: Thirty years of the new decentralization. Johns Hopkins Press, Baltimore.

Morgan, R.P.C. (ed.) 1981. Soil Conservation: Problems and Prospects. John Wiley and Sons, Chichester.

Mountjoy, D.C. and S.R. Gliessman. 1988. Traditional management of a hillside agroecosystem in Tlaxcala, Mexico: An ecologically based maintenance system. The American Journal of Alternative Agriculture 3:3–10.

Mpofu, T.P.Z. 1987. History of soil conservation in Zimbabwe. *In* History of soil conservation in the SADCC region. Report No. 8. SADCC Soil and Water Conservation and Land Utilization Programme, Maseru.

Northern Rhodesia. Dept Agriculture Annual Report for the season 1929/30.

Nye, P.H. and D.J. Greenland. 1960. The soil under shifting cultivation. Technical Communication No. 51. Commonwealth Bureau of Soils, Harpenden.

Oliver, R. 1991. The African Experience. Icon Editions/HarperCollins, New York.

Oyadomari, M. 1989. The rise and fall of the nature conservation movement in Japan in relation to some cultural values. Environmental Magazine 13:22–33.

Perdue, P.C. 1987. Exhausting the earth: State and peasant in Hunan, 1500–1850. Council on East Asian Studies. Harvard University, Cambridge.

Platzky, L. and C. Walker. 1985. The surplus people: Forced removals in South Africa. Ravan Press, Johannesburg.

Powell, J.M., S. Fernández-Rivera, P. Hiernaux and M.D. Turner. 1996. Nutrient cycling in integrated rangeland/cropland systems of the Sahel. Agricultural Systems 52:143–170.

Rasmussen, W.D. 1982. History of soil conservation, institutions and incentives. pp. 3–18 *In* H.G. Halcrow, E.O. Heady and M.L. Cotner (eds.) Soil conservation: Policies, institutions and incentives. Soil Conservation Service, Ankenny.

Roe, E. 1991. Development narratives, or making the best of blueprint development. World Development 19:287–300.

Roe, E. 1995. Except-Africa: Postscript to a special section on development narratives. World Development: 23:1065–1069.

Ronan, C.A. and J. Needham. 1995. The shorter science and civilization of China: An abridgement of Joseph Needham's original text. Cambridge University Press, Cambridge.

Rooyani, F. 1982. A preliminary report on gully erosion in Lesotho. An introduction to Gully Erosion, Annex. Lesotho Agriculture College, Maseru.

Ross, J.C. 1963. Soil conservation in South Africa: A review of the problem and developments to date. Department of Agricultural and Technical Services. Government Printer, Pretoria.

Saberwal, V. 1999. Pastoral Politics: Shepherds, bureaucrats and conservation in the western Himalaya. Oxford University Press, Delhi.

Sandor, J.A., P.L. Gersper and J.W. Hawley. 1986. Soils at prehistoric agricultural terracing sites in New Mexico: I. Site placement, soil morphology and classification. Soil Science Society of America Journal 50:166–173.

Showers, K.B. *forthcoming*. Erosion awareness, international agricultural bureaucracies and state-sponsored soil conservation in Southern Africa, 1900–1930. *In* Grove, Richard and Damodaran, Vinita (eds.) Environment and Empire: The British Empire and the Natural World, Oxford University Press, Delhi. *Forthcoming*.

Showers, K.B. 1989. "Soil erosion in the Kingdom of Lesotho: origins and colonial response, 1830–1955", Journal of Southern African Studies 15:263–286.

Showers, K.B. 1994. Early experiences of soil conservation in Southern Africa: Segregated programs and rural resistance. Working Paper No. 184. African Studies Center. Boston University, Boston.

Showers, K.B. 2000. Popular participation in river conservation. pp. 459–474 *In* P.J. Boon, B.R. Davies and G.E. Petts (eds.) Global perspectives on river conservation: Science, policy and practice. John Wiley and Sons, Chichester.

Showers, K.B. 2005. Imperial Gullies: Soil erosion and conservation in the Kingdom of Lesotho. Ohio University Press, Athens.

Showers, K.B. and Malahleha, G.M. 1992. Oral evidence in historical environmental impact assessment: Soil conservation in Lesotho in the 1930s and 1940s. Journal of Southern African Studies 18:276–296.

Sobolev, S.S. 1947. Protecting the soils in the U.S.S.R. Journal of Soil and Water Conservation 2:123–132.

Spencer, J.E. and G.A. Hale. 1961. The origin, nature and distribution of agricultural terracing. Pacific Viewpoint 2:1–40.

Staunton, I. (ed.) 1990. Mothers of the Revolution. Baobab Books, Harare.

Stebelsky, I. 1974. Environmental deterioration in the Central Russian Black Earth Region: The case of soil erosion. Canadian Geographer 18:232–249.

Stockdale, F. 1937. Soil erosion in the Colonial Empire. The Empire Journal of Experimental Agriculture 5:281–297.

Stross, R.E. 1986. The stubborn earth: American agriculturalists on Chinese soil, 1898–1937. University of California Press, Berkeley.

Sutton, J.G.G. 1989. Towards a history of cultivating the fields. Azania 24:99–112.

Swaziland. Colonial Annual Report. 1956. Government Printer, Mbabane.

Temato, P. and K.R. Olson. 1997. Impacts of industrialized agriculture on land in Bafou, Cameroon. Journal of Soil and Water Conservation 52:404–405.

Tempany, H.A. 1949. The practice of soil conservation in the British Colonial Empire. Technical communication No. 45. Commonwealth Bureau of Soil Science, Harpenden.

Tlou, T. and A. Campbell. 1984. History of Botswana. Macmillan Botswana Publishing Co., Gaborone.

Trimble, S.W. 1985 Perspectives on the history of erosion control in the Eastern U.S. pp. 60–78 In D. Helms and S.L. Flader (eds.) The History of Soil and Water Conservation Engineering. Agricultural History Society, Washington, D.C.

Wilson, K.B. 1995. 'Water used to be scattered in the landscape': Local understandings of soil erosion and land use planning in southern Zimbabwe. Environment and History 1(2):281–296.

Wright, A.C.S. 1962. Comment: Some terrace systems of the western hemisphere and Pacific Islands. Pacific Viewpoint 3:97–101.

Zhu, X. and M. Ren. 2000. The Loess Plateau – Its formation, soil and water losses and control of the Yellow River. pp. 1–3 In J.M. Lafler, J. Tian, and C. Huang (eds.) Soil erosion and dry land farming. CRC Press, Boca Raton.

Zurayk, R.A. 1994. Rehabilitating the ancient terraced lands of Lebanon. Journal of Soil and Water Conservation 49:106–112.

15

Stewardship and Soil Health

E.G. Gregorich[1], G.P. Sparling[2], and L.J. Gregorich[3]

Introduction

The history of civilization is interwoven with the development of agriculture and the relationship of humans and soil. Sustaining the health of soils, which really means keeping them fit for specific uses, has proven essential to sustaining human societies. Nevertheless, humans have not always taken good care of the soil in the past, nor do they today.

The reasons for this mismanagement and abuse are many. At one time, land was plentiful and cheap, a seemingly limitless resource. When one field was spent, there was often another to turn to. People typically failed to recognize the consequences of their poor management until the soil was seriously damaged. And, as true today as in the past, economic conditions often force short-term decisions that are deleterious to the soil in the long term. But no matter how it comes about, degradation of the soil resource always carries with it significant economic and environmental costs.

The word *steward* derives from the old English word *stigweard*, meaning "guardian of the house." When we think of stewardship in relation to soils, we think not only of carefully tending the soil, but also of guarding it and protecting it from harm. Even as stewardship often connotes caring for something that does not belong to you, soil stewardship involves caring for something that has been entrusted to us by nature and whose benefits are there to be enjoyed down through the generations (*see* panel on Maori). There is the strong sense that we are preserving something for the future. In this way, a rural family thinking about stewardship and soil conservation may have in mind not only keeping the soil productive for the current generation, but also maintaining the soil's health and protecting it from degradation so that the family farm can be passed in good shape to the next generation.

---

[1] Agriculture Canada, Ottawa, Canada.

[2] Landcare Research, Hamilton, New Zealand.

[3] Gregorich Research, Ottawa, Canada.

In modern times, many agriculturalists and soil conservationists have sounded the alarm about the shortsightedness of poor soil management. Although farmers and other soil users have not always been quick to heed this alarm, today there is a widespread recognition of the relationship between good soil management and future productivity. Science has also developed a good understanding of soils and soil processes, the interactive relationships between living beings and their environment, and the interconnectedness and complex functioning of different ecosystems. This understanding has helped to promote a new, more holistic view of soil stewardship – how we should manage our soils to maintain their health in the context of ecosystems – particularly in developed countries. However, what is understood in theory is often not put into practice, and many soils throughout the world continue to be degraded. With limited land left in the world to be developed for agriculture, it is of the utmost importance that land under production be used sustainably and that degraded lands be restored to a healthy and productive state.

This chapter traces changes throughout human history in the value placed on soil and soil health and the practice of soil stewardship. It begins with a look at the importance of soils in the development of early settlements and the prosperity of ancient civilizations. It goes on to examine the perception of soil as society became more industrialized and mechanized. It then describes the development of the methods and science of soil conservation, as people grew to understand their ability to manage soils and influence soil health. And finally it comes full circle, ending where it began, with the concept that soils have a controlling influence on human endeavors, as enunciated by the modern concept of environmental sustainability.

Māori and land

Te whenua, te whenua	*The land, the land*
Te oranga o te iwi	*Is the lifeblood of the people*
No nga tipuna	*Handed down to us*
I tuku iho, I tuku iho	*By our ancestors*

New Zealand Māori oral tradition captures the importance of land for tribal and individual identity, and its role in nourishing its peoples. It also heralds the concept of intergenerational equity, with each new generation inheriting the "lifeblood" from ancestors and acting as stewards for future generations.

Early Views of Soils and Soil Health and Quality

Soils as the Cornerstone of Early Societies

Through the centuries, soils have played a central role in where and how societies have developed. A primary requirement for generating the food, fiber, fuel, and building materials needed by people, soils are the foundation of agriculture, and agriculture is the cornerstone of all early civilizations. Hunter–gatherers started domesticating plants and animals about 10,000 years ago, reducing their nomad mobility in favour of more settled societies built around agriculture (Figure 1). Soil quality controlled the success of early production, the more so because early societies had few options but to accept and use the soil in the condition in which they found it. In this chapter we use the term "soil quality" to refer to the suitability of a soil for a particular use. Matching a soil's capability to the demands placed upon it was well understood by early agricultural societies (*see* panel on Ibn al-Awan).

Human population expansion occurred independently in at least seven different regions of the world, and soil quality had a profound influence on where this development occurred. Early settlements were often sited in the broad alluvial flood plains and foothills of the Old World's biggest rivers, where fertile soils promoted crop growth and supported animal husbandry (Miller, 2002). From early settlements grew such large and sophisticated civilizations as those of ancient Egypt and India. Dependent on the annual flooding of the Nile, Indus, and Ganges rivers, agriculture in these areas developed many of the soil-sustaining practices used today in

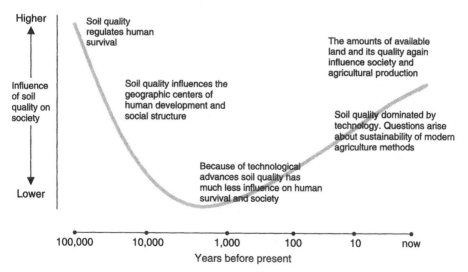

Figure 1. The influence of soil quality on human development and society has changed over time.

agriculture, including complex irrigation systems. These cultures and farmers recognized the life-giving nature of soils and the need to care for and replenish this resource regularly. Prolific agriculture allowed these cultures to diversify and grow. With only a portion of the population needed to sustain food production for the whole, others could devote their time to developing crafts and trade, building governments, and refining the arts. In fact, throughout history, burgeoning civilizations could trace their roots to a prosperous agriculture and productive soils (Hillel, 1991).

As populations grew and civilizations expanded, settlements moved into regions more removed from river settings (McNeill and Winiwarter, 2004). Among other places, societies grew up on the loess-based soils of China, the volcanic deposits of the Pacific Rim, and the glacial materials of North America. Settlements in Italy and Greece usually centered on easily cultivated soils of the coastal alluvial plains. The fertile coastal plains were also settled in Nikodemeia and spread to like regions in southern Europe. In the temperate regions, fertile and well-drained loess soils were sought out by European farmers. In East Asia the combination of fertile soil and available water in the Yangtze valley made it suitable for wetland rice, and this area became a center of population. In eastern North America, river valley locations were preferred for settlements. Development also occurred at high-altitude locations in South America, with the people selecting plants, such as potatoes, and animals, such as llamas and alpaca, that could thrive at high elevations. Early agriculturalists often had a good

Ibn al-Awan on Soil

"The first principle of agriculture is to know good soil from that which is poor. Of the good, the best is that which is like old manure, black and crumbling, not muddy, not hardening or cracking, nor given to drying out as sand does: it is a rare thing to find."

"The ancients considered that land might be known by the condition of the wild plants which grew on it. Where these are vigorous and compact, closely entwined, it is of good quality. But your judgment should not be hasty – for a soil which cracks will be good for corn even though nothing else loves it, and the pine will flourish in sand when the apple, pear and plum will fail."

From: A Moorish Calendar: The Book of Agriculture of Ibn al-Awan, Written around the late 12th Century. Version edited and translated by P. Lord, 1979.

appreciation of what factors made for a good soil and recognized that it was a scarce item (*see* panel on Ibn al-Awan).

Despite an appreciation of what constituted good soil quality, ancient societies often degraded their soils through over-exploitation. For example, it is now generally accepted that the decline of the Mayan culture was hastened by land degradation coupled with a period of extended drought that rendered the land incapable of supporting the concentration of population.

Ties to the Spiritual

The intimate relationship that humans have with the earth and its soils has, from the outset, been strongly linked with the spiritual. Early and aboriginal cultures demonstrated their spiritual attachment to a life-giving Earth through their religious beliefs and customs. Human understanding of the abundance that comes from soils is also reflected in the scriptures of the world's major religions. In the Torah, for example, the prophet Jeremiah speaks of the Israelites being brought through a wilderness of "deserts and rifts . . . droughts and darkness" into "a fertile land to eat its fruit and rich produce." (Jeremiah 2:6–7). In the Christian scripture, the parable of the sower remarks on the shallow, stony soil where the seed springs up quickly and then perishes, and the good soil that produces a crop 100 times greater than what was sown (Matthew 13:3–9). The Koran records, "we cannot bear with one food, therefore pray Lord on our behalf to bring forth for us out of what the earth grows, of its herbs and its cucumbers and its garlic and its lentils and its onions" (2.61).

Ties to Social Structure

In ancient China a system known as Tsing Tien was developed to divide up and manage the land in a productive, collaborative way based on families and cooperation. Tsing Tien means "the nine fields," which were made by dividing a large square of land into nine smaller squares. The eight outer squares of land were given to eight families for their private use, and the ninth, central square, was reserved for the government and was worked in common by the same eight families (Wrench, 1946). As few domestic animals were kept in this system, pasture did not rival the cultivation of crops. This system not only fostered good economy through cooperation, but also encouraged good stewardship of the soil and preservation of soil fertility, because families wanted to preserve their plot of land for future generations.

Also in ancient times, Greek philosopher Plato (437–327 B.C.) recognized the relationship of society to soil productivity when he suggested that Attica could support an army because the citizens didn't need to devote many men to farming. Xenophon (434–355 B.C.) believed ". . . there is nothing so good as manure" for soil and commented about a farm that had gone to ruin because "someone didn't know it was well to manure the land." Roman writer Columella (about 60 A.D.) was opposed to the general opinion of critics "that the soil, worn out by long cultivation and exhausted, is suffering from old age." He attributed soil degradation to the enforced, indifferent labour of slaves, and to bad farming practices. To those who said that frequent wars were the cause of soil deterioration, Columella responded that, although fields may be laid waste by an enemy, the fertility of the soil is not taken away or wasted; it will be renewed and will respond to good soil husbandry. He noted the importance of ploughing down plant residues and adding manure, so that the soil might "grow fat."

A feudalistic type of agriculture dominated medieval Europe. Feudal lords, along with the bishops and abbots of monasteries, held most of the land. Typically the land on a manor was divided into two fields, one field planted to grain and the other left fallow each year. Serfs, usually assigned to work a different strip of land each year, with no hope of personal gain, had little interest in carefully tending or improving the land. Pastures and meadows were held communally, with the result that they were often overgrazed. Severe erosion occurred in some fields, but farmers were more interested in getting the most off the land than in repairing damage to the fields. Communal ownership of land occurs in many parts of the world today, being common in such regions as Polynesia. This type of ownership can create problems of exploitation and non-sustainable use. Communal owners may be reluctant to invest in land improvements, as they do not own the land and they run the risk that joint owners will capture the benefits.

Farmers of the Low Countries (The Netherlands, Belgium, and areas bordering France and Germany) were the first in Europe to abandon this feudal, communal system of agriculture, leading the rest of Europe in the development of land management. Most of the land was owned by the farmers themselves or by city merchants who leased it for long periods, with the Church owning only small land holdings in this region. Farmers viewed livestock and land as capital assets that could return a profit proportional to their quality, so it paid to improve them. This system of agriculture was practical and profitable and spread throughout Europe. Modern crop rotations originated in this region in the 14th and 15th centuries. Dutch and Flemish farmers developed a system in which they grew a grain crop (such as wheat), a root crop (such as turnips), and a sod crop (such as clover) in a regular rotation. These crops were well-suited to the soils and climate of the region, and the rotation met the need for human food and livestock forage. It may not be entirely fortuitous but this system also maintained soil health, because the plant cover protected the soil from erosion, the legumes and grasses promoted good soil structure and fertility, and the animal manure applied to cultivated fields built up soil organic matter. As soils in this region tended to be acidic and legumes require neutral or alkaline soil, lime was also used extensively in this region. In Britain, where many soils also benefit from adding lime, the importance of adding manures in addition to lime was captured by the old English saying "Lime, lime, and no manure, makes the father rich, but the son poor." Again, the handing on of good-quality soil to the next generation was stressed.

Maintaining Soil Health

Industrialization and Mechanization

As a result of the industrial revolution in England during the 1700s and 1800s, labour once handled by humans and animals became increasingly mechanized, and industry grew to greater and greater proportions. It became possible for fewer people to farm larger tracts of land more intensively. This change, along with the pressures of colonialism, caused a shift in the view of natural resources, including soil.

A major consequence of industrialization was that farming gradually ceased to provide all its own inputs. Prior to industrialization farmers got seed from their own harvest of crops, manure from their own livestock, and feed for animals from grass or roots grown on their own farm. Power on the farm came from human and animal energy. Farm implements and tools were either made on the farm or bought from local craftsmen. But with the advent of industrialization, farmers bought more and more of their inputs off the farm.

In the early 1800s, Sir John Lawes laid the foundation for the development of commercial fertilizers in England when he invented a process to produce superphosphate. In the mid-1800s, the theories of Justus Liebig and Carl Sprengel in Germany became the underpinnings of agricultural chemistry, demonstrating that *"Science teaches us what elements are essential to every species of plants by an analysis of their ashes. If therefore a soil is found wanting in any of those elements, we discover at once the cause of its barrenness, and its removal may now be readily accomplished"* (Liebig, 1844).The dependence of the farmer on purchased inputs produced a fundamental shift in the way the land was managed and soil health maintained (Figure 1, approaching year 100).

Soil Stewardship and the Need for Soil Conservation

Erosion became a major problem on many farms in North America between 1620 and 1860. One reason for this erosion is that farmers who immigrated to the New World brought Old World farming practices with them, and these practices were often ill suited to the new conditions. For example, many of northwest Europe's farming methods were developed on, and suited to, relatively flat land in a climate with gentle rains. Transplanting these methods to North America had disastrous results in some areas where the land was steeply sloped and thunderstorms were common. Sudden heavy rains carried large amounts of soil down the slopes.

In general the early farmers were poorly informed about the causes of erosion, and most of them didn't notice the erosion until their land had become very degraded and barren. During this time notable advocates of soil conservation contended that ignorance was one of the main causes of erosion. Jared Eliot (1685–1763) conducted experiments and made careful observations that he incorporated into the first American book on agriculture, published in 1748. He called attention to "soil washing" and exposed some of Jethro Tull's ideas on intensive cultivation as harmful (Mcdonald, 1941). Some of Eliot's experiments showed the beneficial effects of different amendments, including manuring, adding limestone, and mixing soils of different textures. He recommended turning under buckwheat, oats, rye, and millet.

Samuel Deane (1733–1814), another experimentalist, strongly advocated erosion control, the use of different crop rotations for different soil textures, and the application of manure to fields. He recommended contour plowing on sloping land to prevent gullying and sheet erosion (Mcdonald, 1941). One of his principles was that "white" crops (exhausting crops that included oats, corn, flax, rye, and barley) should not be grown more than two years in succession. "Green" crops (soil-building crops that included legumes, root crops, and grasses) should be alternated with white crops. Deane's main idea was that the only real improvements in soil stewardship could occur with manuring. He encouraged farmers not to be ashamed of their occupation, as many of them were, and suggested that they "should toss their dung with an air of majesty."

The early pioneers of soil conservation in North America wrote detailed pamphlets describing the dangers of allowing erosion to go unchecked, as well as mechanical and agronomic methods to combat it. Going further than simply presenting solutions to the problems of erosion and soil degradation, they advocated economic and social reforms and urged the government to do something to encourage farmers to grow more soil-conserving than soil-depleting crops. This could only occur, they argued, with a reduction of interest rates and elimination of tariffs. They also lobbied for the formation of agricultural boards that could work out trade agreements whereby the farmer could get better prices for their products. These advocates of soil conservation were prescient in knowing that soil degradation would not only be ruinous for an individual but for the community or nation as a whole.

Even though erosion was identified as a degradative process and a detriment to agricultural productivity, and ways to prevent or reduce it were known, it was considered primarily a local farm problem. This view persisted well into the

1920s, because agriculture was still confined to certain tracts of land, while the greater continental areas remained undisturbed. But at the turn of the century, a soil surveyor in the United States, Hugh Hammond Bennet, became convinced that soil erosion was not only a local problem but was widespread across North America and posed a serious threat to agricultural productivity at a national scale. Starting to map soils in 1903, Bennett observed soil degradation during the course of his work over the next 20 years. He said "To visualize the full enormity of land impairment and devastation brought about by this ruthless agent is beyond the possibility of the mind. An era of land wreckage destined to weigh heavily upon the welfare of the next generation is at hand" (Bennett, 1928). Subsequent events proved his assessment correct.

During World War I, demand for farm products increased dramatically. As prices increased correspondingly, more and more marginal land, especially the vulnerable soils of the Great Plains in the United States and the Prairies in Canada, were plowed up and sown to wheat. When the war ended, an agricultural depression occurred and farmers were less concerned about their soils than they were about falling prices. Then in the late 1920s a worldwide economic depression occurred, fueled by agricultural scarcities and high prices. In the early 1930s a severe drought overtook the central part of North America, lasting for almost 10 years. During this time the infamous Dust Bowl was created.

The widespread soil degradation and devastation wrought by the Dust Bowl spurred the United States and Canadian governments into action, and both governments provided funds not only to help farmers directly, but also to set up agencies designed to promote sound soil management practices. In 1935 Canadian Parliament established the Prairie Farm Rehabilitation Administration to "develop and promote within those areas, systems of farm practice, tree culture, water supply, land utilization and land settlement that will afford greater economic security." The same year, United States passed legislation to establish the Soil Conservation Service, under the U.S. Department of Agriculture.

Urbanization and Demands on Soils

In the early 1700s nearly three-quarters of the total labour force in the Western World was made up of farmers and farm labourers. During the industrial revolution the proportion of the population employed in manufacturing, transport, and mining grew, while that employed in agriculture declined. That trend has continued, with city populations throughout the world expanding more rapidly than those in rural settings. By about 2010 more than 3 billion people will live in cities around the world. It is predicted that sometime in the next 75 to 100 years, people numbering more than our entire current global population will be living in cities.

Because of this demographic trend, the prevailing culture is largely estranged from rural landscapes, ecosystems, and values. People in rural areas who live "closer to the land" generally have a better sense of how the health and productive capacity of soils affects human lives. "One reason for this," writes Wendell Berry, "is the geographical separation that frequently exists between losses and gains. Agricultural losses occur on the farm and in farming communities, whereas the great gains of agriculture all occur in cities, just as the profits from coal are realized mainly in cities far from where the coal is mined" (Berry, 1987). If the agricultural losses are considered, they are often undervalued in terms of economics. "If in the human economy, a squash in the field is worth more than a bushel of soil, that does not mean that food is more valuable than soil; it means simply that we do not know *how* to value the soil. In its complexity and its potential longevity, the soil exceeds our comprehension; we do not know how to place a just market value on it, and we will never learn how. Its value is inestimable; we must value it, beyond whatever price we put on it, by *respecting* it" (Berry, 1987).

The importance of agriculture, and of soils in particular, is less likely to be appreciated by urban dwellers who buy their food and clothing at a remove from the farms where these products originated. The urban public may have a high interest in water quality, because they perceive that their own drinking water may be affected, but they have little interest in soil quality, because they cannot understand its connection to their own health. Yet public opinion, poorly informed as it may be, is a major driving force in developing government policy. Thus it is that much of the research on soil health today is carried out under the umbrella of protecting water quality. This is because water is more generally owned and used by the community than land, which is owned by an individual or company. Land can also be clearly delimited, whereas water flows across boundaries with no regard for legal boundaries. Nonetheless, the actions of the land owner have direct effects on surface (streams and lakes) and subsurface (groundwater and aquifers) waters that their lands adjoin or overlay. The application of agrichemicals (fertilizers and pesticides) to the soil surface carries the direct risk of those chemicals reaching receiving waters by direct transfer, mass movement, surface run-off, leaching, and subsurface flows. For this reason, it is important that nutrients and chemicals are retained on site where they will benefit the crop. Matching nutrient budgets to the soil's supplying power and to crop needs, avoiding spray drift, establishing riparian strips, and excluding stock from waterways, all contribute to those needs. Conservation tillage can help decrease soil movement and the loss of soil organic matter. An understanding of soil characteristics in relation to the demands being made upon it can help conserve the soil resource and protect the wider environment.

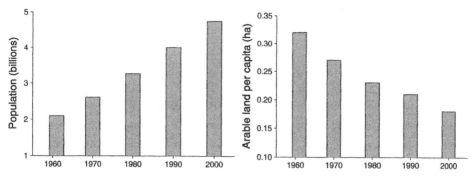

Figure 2. Trends in the population and arable land on a per capita basis in developing countries 1961–2000. (Data from FAOSTAT data, 2005)

The tension between high production and environmental risk can only be expected to rise as the Earth's population continues to grow and the demands for food and fibre escalate. This tension is particularly acute in developing countries where, in the last 30 years, population has increased by 1.8 times and the amount of arable land on a per capita basis has decreased by 1.5 times (Figure 2). The global demand for food is expected to double in the 21st century, and in most countries the best arable land is already in use. In the near East and north Africa, an estimated 97 percent of available arable land is now under production. Worldwide, 80 percent of available cropland is now cultivated. Options for expanding the land base include using marginal, less productive land or encroaching on natural ecosystems, such as grasslands or forests. Such encroachments would have to be carefully managed to avoid deforestation and cultivation of ecologically fragile lands, which would lead to soil degradation and loss of biological diversity.

Organic Farming

On philosophical grounds and because of a concern for food safety, organic farming makes up a growing share of the food growing system. Organic farmers produce crops with significant attention to husbandry and reliance on the "natural" quality of the soil. Under this system soil quality and productivity are maintained by inputs of manure and high-quality crop residues for nutrients. This type of farming is viewed as building up and maintaining soil health and marks a return to the type of agriculture practiced before the industrial revolution, particularly in "tightening the loop" of nutrient recycling. Natural inputs were the sources of nutrients before fertilizers and pesticides from the chemical industry became readily available during the 20th century.

These agrichemicals have had an enormous impact on global agriculture; they have greatly increased our ability to grow crops on impoverished soils, and to control pests and weeds.

In the 1920s Rudolf Steiner and C.A. Meir promoted "biodynamic cultivation," based on the belief that the universe is an evolving organism undergoing purposive development towards spiritual potential. Conventional soil science was thought to be a failure, because it viewed soil as an inert material rather than a sentient force. Proponents of biodynamics advocated the use of composts with activators, based on homeopathic principles, to maintain soil health.

Organic agriculture received a big boost in Britain and Europe during World War II when the chemical industry was diverted to produce munitions rather than agrochemicals. Householders in Britain were encouraged to cultivate their gardens and allotments and "dig for victory." Lady Eve Balfour, founder of the Soil Association, wrote the book *The Living Soil*, published in 1943, which promoted organic agriculture. However, when the war ceased, European government policies emphasized efficiency in agricultural production. Greater use of chemical fertilizers and machinery was advocated to avoid the possible recurrence of food shortages seen during the war years. At about this time, rural populations began to decline and agrochemical use boomed, a pattern subsequently repeated on a global scale. By the end of the 20th century farm populations in the United States dropped below 2 percent of the total population.

At the beginning of the third millennium, agricultural production has reached record highs with the use of agrochemicals and irrigation, but organic farming is growing again in popularity as a segment of the population (mainly in developed nations) becomes more aware of the problems associated with high-intensity agriculture. In terms of soil fertility, organic farming relies on organic inputs such as manures and legumes, rather than manufactured chemical fertilizers to supply plant nutrients. As a consequence, organic matter and nitrogen contents of soils in organic systems are usually higher, often significantly so, than those found in conventional systems. Mixed cropping is favoured over monoculture, with claimed benefits for production, disease control, and biodiversity.

Organic systems are geared toward maintaining soil health, but there are trade-offs. Although organic systems are more energy efficient (with higher energy output to input ratio) than conventional systems, the yields over many crops are lower. Along with rejecting the use of chemical intervention, most organic certification also now rejects the use of any genetically modified crops, animals, or microorganisms. In general, this will mean lower productivity, so for production levels to match those of conventional intensive farming, a greater amount of land will be needed. More land will be hard to obtain in regions where the food demand is high and all arable land is already being used. With around 80 percent of the world's arable land already in production, can the world support its existing population using organic farming?

Recent Trends in Stewardship and Soil Health

Ecosystem Approach

Increasingly, stewardship of soil and the preservation of soil health are considered within an ecosystem approach. Because farming takes place in the context of ecosystems and can affect not only the farm environment but also the broader environment, soil conservation and soil health are seen in the context of the environmental sustainability of agriculture. Soil health is but one environmental consideration among a host of others, including declining water quality, loss of wildlife habitat, reduced biodiversity, and emission of greenhouse gases. Most soil conservation today considers not only the health of the soil, but the environmental implications of any activity carried out to promote soil health.

Soil Quality Indicators

In the same way that indicators have been used to measure economic performance, there is a growing trend in the agriculture sector to use indicators to ensure an objective, systematic approach to assessing agricultural sustainability. Agri-environmental indicators are a subset of these sustainability indicators, and soil quality indicators specifically address soil-related aspects of the ecosystem.

Many characteristics have been proposed as soil quality indicators (Sparling, 2002), and this variety reflects the many ways in which soil quality has been defined and the many components of a soil's chemical, physical, and biological attributes that contribute to its overall character. Because it is not feasible to measure all soil quality characteristics, indicators are usually selected to address the issues of greatest concern. Three major categories of soil quality concerns include soil erosion and redistribution, chemical and biological contamination, and soil degradation and depletion.

Specific critical values for soil quality have yet to be defined and so the emphasis has usually been on assessing trends in soil quality (Gregorich, 2002). Trends away from target ranges are interpreted as a decline in soil quality; trends toward a target are interpreted as improved soil quality. It has been suggested that short-term trends away from a target range could be regarded as sustainable, provided that the trend can be reversed within an acceptable time frame, preferably within the time span of a human generation, 20–25 years.

Focus on Long-term Soil Health

Where agrochemical and technology inputs (use of fertilizers, irrigation, pesticides) are high, it is possible to produce high yields of nutritious crops on soils with low organic matter. For example, irrigation can add water as required, and nutrient deficiency can often be corrected within hours by adding chemical fertilizers. Soil physical condition can be restored over years with changes in land management and tillage practice. However, this is a case of the soil being used as an "inert medium," exactly the point that the biodynamic movement complained about. We cannot expect these energy-intensive inputs to always be available, yet the maintenance of soil organic matter is probably the most important property determining the state of soil health. Soil organic matter is the repository of nutrients in the soil and contributes to soil structure by promoting aggregation, which in turn reduces erosion and enhances the infiltration of water and gases into the soil. In contrast to the rapid "repair" that can be achieved for soil chemical and some physical characteristics through fertilizer application and tillage, the recovery of soil organic matter in depleted soils can take decades, or even hundreds of years (Figure 3).

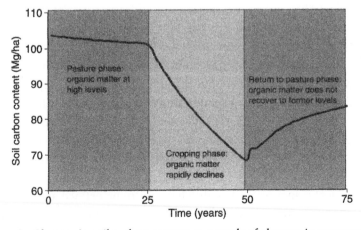

Figure 3. Changes in soil carbon content as a result of changes in management (Adapted from Sparling et al., 2001).

Soil as Part of the "Community"

The Latin word for *resource* is *surgere*, which means "to rise," calling up a picture of springs of water rising continually from the ground. Before heavy industrialization, people saw natural resources as such springs – dynamic supplies from the earth that constantly renewed themselves even when repeatedly used and consumed. This view changed as natural resources came to be seen as the inputs needed for industrial production and colonial trade. The notion grew that nature held all the raw materials needed by humans to create a more comfortable, industrious world. The raw materials were simply there for the taking, opening the door for civilization to reach its full technological potential. This industrial view of resources prevailed in the late 1800s and early 1900s. The danger this view posed, along with a remedy, can be summed up in the words of Aldo Leopold (*see* Panel on Leopold), "We abuse land because we regard it as a commodity belonging to

Aldo Leopold

Aldo Leopold was born in Burlington, Iowa. He attended Yale and attained a Master of Forestry degree. He joined the U.S. Forest Service in 1909 and subsequently worked for the Forest Service in various capacities before becoming assistant district forester in charge of operations. He recorded his observations and ideas from numerous inspection tours and proposed the Gila National Forest as a wilderness area. In 1933 he was appointed to the chair of game management in the Dept. of Agriculture Economics at the University of Wisconsin. In 1939 he became chairman of a new Dept. of Wildlife Management at the University and subsequently was appointed by the governor to the Wisconsin Conservation Commission. Leopold's best known book is A Sand County Almanac (1949), a volume of nature sketches and philosophical essays recognized as one of the enduring expressions of an ecological attitude toward people and the land. The notion of a land ethic was rooted in Leopold's perception of the environment, and that perception was deepened and clarified throughout his life. He was an internationally respected scientist and conservationist instrumental in formulating policy, promoting wilderness, and building ecological foundations for two Twentieth Century professions – forestry and wildlife ecology. He died of heart attack while helping to fight a grass fire on a neighbour's farm.

From: www.aldoleopold.org

us. When we see land as a community to which we belong, we may begin to use it with love and respect." Leopold viewed land in the broadest sense – "all the things on, over, or in the earth" and believed that conservation is the state of harmony between land and humans.

Landcare Movements: Community and Industry Support for Soil Health

A recent trend in recognizing and rectifying soil degradation has been the emergence of "landcare movements," particularly in Australia (*see* panel on Landcare Movements). Despite the growing trend towards city living in developed nations, there have emerged examples of urban-based community and industries becoming actively involved with land owners and the rural community in managing soil health issues. Australia has severe environmental problems affecting large areas of the country; examples include desertification, increasing salinity, soil acidification, pests and weeds, contaminated sites, waste disposal, loss of indigenous habitat and decreased biodiversity.

The success of the landcare movements has been to recognize the concerns of the community (often manifest through their initial awareness of poor water

Landcare Australia

The aim of Landcare Australia is to "build partnerships between communities, businesses and governments to care for Australia's environment now and for future generations." Projects are normally arranged regionally and focus on particular catchments and local concerns. Coordination is provided by Landcare Australia, which brings together local environmental organizations, farmers and land owners, local and central government, and sponsors. The sponsoring organizations participate to improve their environmental performance for triple bottom line reporting, improved marketing image, and social responsibility. Commercial organisations that have sponsored Landcare projects include telecommunications firms, power generation, rural land agents, banks, wine producers, mining, and transport. Local and federal government also support Landcare projects through catchment authorities, land regulators, and administration.

From: www.landcareaustralia.com.au

quality, salinity, and erosion) and to harness their enthusiasm to help restore the land. Land restoration is supported by community volunteers and also by funding from federal and local governments and environmental agencies. Support may take the form of resources for mapping and planning, creation of paid positions for local coordinators, publicity and education, as well as the more direct activities of recontouring land providing suitable plant species, fencing, and monitoring.

Landcare has become a preferred term for the concepts of maintaining soil health (Yallon and Arnold, 2000); it includes the important community component, all people in a community with their land.

Future Prospects

The level of resource consumption has a powerful effect on efforts to meet the needs of rising numbers of people. If the entire world were to have a standard of living equal to the average in Western Europe or North America, it would require three planet Earths to provide enough natural resources at current levels of consumption, waste, and technology (Wackernagel et al., 2002). The developed world, with only 20 percent of the world's population, consumes 80 percent of the natural resources used each year.

How do we overcome the constraints that could prevent the supply of food from keeping pace with a growing population? The view of experts on this question varies widely. Some believe that the world's capacity to produce enough food to feed a growing population is only limited by our imagination and that technical knowledge, which will develop as needed, will solve the problems of future food shortages. Other experts warn that food resource problems will emerge within the next 50 to 100 years as the population grows exponentially and as natural resources decline in quantity and quality. A third view recognizes that, although there are serious problems with food production and rising population levels, more attention should be directed to using better land-use practices and improvements in agricultural technologies that do not rely solely on fertilizers and pesticides; that is to say, that an "environmentally friendly" or a "greener" Green Revolution may overcome the constraints to food production.

No matter which scenario is favoured, the quantity and quality of productive land are key factors. Returning to Figure 1, we find ourselves at a point in human history when our development and ability to sustain ourselves as a species pivots on the health of our soils. The value we place on soil, how we tend and manage soil health, and the extent to which we practice soil stewardship will have a powerful effect on our ability to meet the needs of a growing world population.

Conclusion

Throughout history, the quality of soils has greatly influenced human development, initially in being able to supply sufficient food crops to permit stable communities and social development. As technology developed, mankind has had the capability to manipulate soils, so that inherent soil quality became less important. However, the extra technology has also given us the capability under inappropriate management to degrade soils even more rapidly. Notable pioneers have drawn our attention to the problems, and despite an increasingly urbanized population, sections of the community, particularly in some developed nations, are adopting less intensive farming methods and participating in land restoration initiatives. The amount of unused land suitable for agriculture is decreasing rapidly, and intensification in land use has resulted in land degradation, erosion, accumulation of unwanted contaminants, eutrophication of water bodies, and loss of biodiversity.

We are likely at a critical time for managing our global soil resources; in the future the principles of stewardship, soil health, and land care will become even more important. The future will place even more stresses upon the diminishing soil resource as we try to feed the expanding world population. This coincides with the end of the era of cheap energy. There is the threat of climate change disrupting agricultural systems. Globally, water quality and quantity are declining. We will need to invest more in methods to manage soils more sustainably through improved knowledge, make greater use of renewable resources, and "tread lightly upon the earth."

Literature Cited

Bennett, Hugh H. 1928. Soil erosion: A national menace. USDA Circular No. 33. U.S. Government Printing Office, Washington, D.C.

Berry, Wendell. 1987. Home economics. North Point Press, San Franciso, CA USA.

FAOSTAT data, 2005. http://faostat.fao.org. Last accessed December 28, 2005.

Gregorich, E.G. 2002. Soil quality. Pages 1058–1061 in Encyclopedia of Soil Science. R. Lal (Ed.). Marcel Dekker, Inc. New York, NY, USA.

Hillel, D.J. 1991. Out of the Earth. Macmillan, Inc. New York, NY.

Lord, P. 1979. A Moorish calendar: The book of agriculture of Ibn al-Awan (P. Lord, Ed.). The Black Swan Press, Wantage.

Liebig, J. 1844. Familiar letters on chemistry, and its relation to commerce, physiology, and agriculture. Edited by J. Gardner, M.D. Second Edition, London, UK.

McDonald, A. 1941. Early American soil conservationists. Soil Conservation Service Misc. Pub. No. 449.

McNeill, J.R. and Winiwarter, V. 2004. Breaking the sod: Humankind, history and soil. Science 1627–1629.

Miller, F.P. 2002. Soil as a heritage of planet and human society. Pages 985–989 *in* Encyclopedia of Soil Science. R. Lal (Eds). Marcel Dekker, Inc. New York, NY, USA.

Sparling G., Parfitt, R.L., Hewitt A.E., and Schipper L.A. 2001. Three approaches to define desired soil organic matter contents. J. Environ. Qual. 2003 32:760–766.

Wackernagel, M., Schulz, N.B., Deumling, D., Linares, A.C., Jenkins, M., Kapos, V., Monfreda, C., Loh, J., Myers, N., Norgaard, R., and Randers, J. 2002. Tracking the ecological overshoot of the human economy. Proc. Nat. Acad. Sci. 99:9266–9271.

Wrench, G.T. 1946. Reconstruction by way of the soil. Faber and Faber Ltd., London, UK.

Yaalon, D.H. and Arnold, R.W. 2000. Attitudes toward soils and their relevance: Then and now. Soil Science 165:5–12.

References for Further Reading

Cassman, K.G. 1999. Ecological intensification of cereal production systems: Yield potential, soil quality, and precision agriculture. Proc. Nat. Acad. Sci. 96:5952–5959.

Collins, J.P., Kinzig, A., Grimm, N.B., Fagan, W.F., Hope, D., Wu, J., and Borer, E.T. 2000. A new urban ecology. American Scientist 88:416–425.

Conford, P. 2001. The origins of the organic movement. Floris Books, Edinburgh, Scotland, UK.

Dale, T. and Carter, V.G. 1955. Topsoil and civilization. University of Oklahoma Press, Norman, OK, USA.

Diamond, J.M. 1997. Guns, germs and steel: the fate of human societies. W.W. Norton and Company, New York, NY, USA.

Gregorich, E.G. and Carter, M.R. (Eds) 1997. Soil quality for crop production and ecosystem health. Elsevier Science Publ., Amsterdam, The Netherlands.

Harris, J.M. and Kennedy, S. 1999. Carrying capacity in agriculture: Global and regional issues. Ecol. Econ. 29:443–461.

Johnson, D.G. 1999. The growth of demand will limit output growth for food over the next quarter century. Proc. Nat. Acad. Sci. 96:5915–5920.

Kendall, H.W. and Pimentel, D. 1994. Constraints on the expansion of the global food supply. Ambio Vol. 23 No. 3, May 1994.

Powlson, D.S. and Johnston, A.E. 1994. Long-term field experiments: Their importance in understanding sustainable land use. Pages 367–394 *in* Soil resilience and sustainable land use, D.J. Greenland and I. Szabolcs (Eds). CAB International, Wallingford, UK.

Richter, D.D. and Markewitz, D. 2001. Understanding soil change: Soil sustainability over millenia, centuries, and decades. Cambridge University Press, Cambridge, UK.

Schjønning, P., Elmholt, S. and Christensen, B.T. 2004. (Eds). Managing soil quality: Challenges in modern agriculture. CAB International, Wallingford, UK.

Smith, B.D. 1998. The emergence of agriculture. Scientific American Library. W.H Freeman and Co., New York, NY, USA.

Sparling, G. 2002. Soil quality indicators. Pages 1069–1072 *in* Encyclopedia of Soil Science. R. Lal (Ed.). Marcel Dekker, Inc. New York, NY, USA.

Tisdale, S.L. and Nelson, W.L. 1975. Soil fertility and fertilizers. Third Edition. Macmillan Publ. Co., Inc.; New York, NY, USA.

Vitousek, P.M., Aber, J.D., Howrath, R.W., Likens, G.E., Matson, P.A., Schindler, D.W., Schlesinger, W.H., and Tilman, D.G. 1997. Human alteration of the global nitrogen cycle: sources and consequences. Ecol. Appl. 7:737–750

Vitousek, P.M., Mooney, H.A., Lubchenco, J., and Melillo, J.M. 1997. Human domination of earth's ecosystems. Science 277:494–499.

16
Soil Nutrient Management for Plant Growth

Arnold Finck[1]

Introduction

Soils are the main basis of food production, and this will remain so even in the distant future in spite of some other possibilities. The capability of soils to produce plants resp. crops varies considerably from almost zero to very high. Under suitable climatic conditions, soils with high fertility are very productive in terms of plant growth and crop yields. Most soils of the world, however, possess only medium or low natural fertility. The present and future food security of the growing human population depends on the efforts of agricultural production which in turn depends on natural or improved high and sustained soil fertility.

Soil fertility, as a general term, is the capability of soils to produce plants resp. crop yields (Latin *fertilis* fruitful) and is based on several physical, chemical and biological soil properties of which the content of plant nutrients is especially important.

In agricultural terms, soil fertility produces crops only in combination with climatic growth factors (water, temperature etc.) and with a certain management – the resulting combined effect being called soil productivity. Colloquially, however, both terms are used interchangeably. Fertility resp. productivity varies widely; it can be improved, but can also deteriorate. This chapter summarises the problems and successes in a basic field of agricultural production, namely the supply of plant nutrients. The principles and the necessity for the present intensive soil nutrient management used in crop production can best be understood by considering the historical development of soil nutrient management.

Soils as a growth substrate for plants do not need human effort or control in order to function; in this respect they are comparable to organisms. However,

---

[1] Institute of Soil Science and Plant Nutrition, University on Kiel, D-24098 Kiel (Germany).

their efficiency for producing food can be improved by special care and corrective treatment, i.e. by adequate management procedures. Soil management, to a large extent, is soil nutrient management in a wider sense, including water and air supply.

Plant Growth without Soil Management

Soils have supported plants for millions of years without human influence. By their very nature, they provided a feeding ground (substrate) for the plant roots. All soils contain large or small amounts of all plant nutrients and enable a certain continuous nutrient supply from weathering of nutrient-containing minerals. Further, they possess a system of nutrient recycling via decomposition of organic matter and provide "automatically" additional growth conditions like adequate supply of soil water and air. This happens, because multiple transformation processes occur in soils due to physical-chemical gradients and because of the activity of soil organisms under suitable climatic conditions. Soils, and esp. highly fertile soils, possess an enormous self-regulating capacity for nutrient supply, similar to the biological cybernetic systems. During their evolution, plants have adapted themselves to different soil conditions as growth substrates.

As a result, enormous forests and grasslands developed naturally in suitable climates, which supported considerable animal populations. Some remnants of ancient soil production are fossil energy sources like coal and oil, or nutrient sources like phosphate rocks.

In broad terms of soil fertility, the soils are either rich, medium or poor substrates, depending on their natural nutrient supply, their depth and other growth conditions (as explained later on). Nevertheless, very fertile soils could have a high or low productivity, depending on the climate, ranging e.g. from luxuriant rain forest to rather arid semi-deserts. In contrast, however, even rather poor soils in humid climate could produce enormous forests if the small nutrient supply remained in close recycling. Over long periods, however, due to inevitable losses in humid climates, soils became poorer in nutrients and their fertility declined (deteriorated) slowly.

However, nature is dynamic. In the course of geological periods, some degraded soils were removed by soil erosion and fresh fertile soils developed from exposed underlying parent material or exhausted soils were buried by new materials, e.g. loamy glacial deposits. Further, many soils became periodically or even annually rejuvenated, e.g. by deposition of erosion material into river valleys washed from distant mountains, by fine soil material (loess) blown by wind across large plains, by mud deposited in tidal coastal areas or by erupted volcanic lava material. Thus, by

Box 1: Natural plant growth for millions of years

Soils always enabled natural plant growth in a suitable climate

The plant production ranged from low to high, but was often slowly declining

Many soils gained new fertility and thus productivity due to natural rejuvenation.

geological processes, even over millions of years, there was always a more or less productive soil cover, which enabled plant growth in suitable climates (Box 1).

The natural plant products of soils have been utilised by higher animals, e.g. by herds of sauriens or later by herds of mammals for their nutritive support. When humans slowly developed over many thousand years, they also utilized the natural abundance of food (symbolically called "the garden of paradise") by food gathering, e.g. of fruits, roots, nuts, leaves of nutritious plants or by hunting animals which gathered plants and converted them into animal protein. At least there was generally sufficient food for a small population.

Even at present, the prehistoric tradition of food gatherers, hunters or nomads with goats and sheep feeding on grass partly exists in semi-deserts, extensively used grasslands and forests. In any case, a considerable area of land is needed per person, e.g. about 100 hectares in forest areas or more than ten times as much in semi-deserts. The system of just utilizing natural products offered by "nature" supported a very small population for thousands of years, but was unsuitable for feeding larger numbers of people.

Empirical Soil Management

First Soil Improvement by Early Civilizations

Since the *New Stone Age*, food supply by gathering was supplemented by primitive cultivation methods, probably as early as 10.000 years ago in the Yangtze River Delta of China (Cao, 2004), about 8.000 years in the Middle East (Yaalon, 1997) or about 7.000 years ago in some areas of Southern France (Boulaine, 1992). Primitive crop cultivation probably started in small patches of gardens or fields near settlements by some top-soil loosening with wooden tools and just sowing grains as well as by watering, if required.

By clever observation resp. trial and error, early settlers seem to have learned that plant growth could be somehow improved by mixing plant and animal waste materials into the soil of their garden or field. This experience of simple soil improvement by human activities slowly developed into standard practice of manuring in the ancient cultures of China, India, Mesopotamia, Egypt. In the great river valleys, e.g. of the Nile, at least since 3000 B.C., cultivated soils have not only been improved naturally by river mud from annual flooding, but partly also by deliberate application of organic waste materials. Even the application of processed and thus more effective waste materials, like compost, has an old tradition in China and India.

The essence of this traditional knowledge of ancient soil improvement methods in the Middle East area was later included in the Greek and especially in the extensive Roman agricultural literature and thus became "standard" knowledge.

Practical Soil Nutrient Management from Antiquity Until 1800

The guideline of soil management in the Greek/Roman era and for the following 2000 years was the *Humus Theory* of the Greek philosopher ARISTOTLE, about 350 B.C. His view was based on the common observation that plants grew better with plant residues or animal manure added to the soil, the products being obviously transformed into valuable dark soil humus (Box 2). Even in ancient times, farmers were well aware that soils with a higher humus content and therefore of darker color generally proved to be more fertile and, therefore, more productive than those with only a little humus. Obviously, humus was valuable plant food.

In practical farming during antiquity, many kinds of organic waste materials were used for soil improvement, plant residues, animal and human excretions (manure) and waste products, blood ("The blood of the enemy is the best fertilizer"), compost as an improved and therefore more effective waste material. Some mineral materials, too, were found beneficial for soil fertility: ash, lime, gypsum, river mud, further mixtures of mineral and organic materials or dark topsoil from

---

Box 2: The *Humus Theory* of plant nutrition

\# Plants feed on humus particles (dark organic soil matter) taken up from the soil by their roots and built their complex body tissues out of these substances

\# Dead plants and organic waste materials are decomposed into humus particles again – an eternal cycle.

forests or waste lands. Because of urgent food requirements, there was a permanent search for "new earth" that could be applied as supplementary plant food to impoverished soils for many centuries (Finck, 1982).

In early Roman times, CATO (about 200 B.C.) stated in his famous book on agricultural practices that cropping requires good ploughing, good planting and good manuring in order to be successful. In fact, "manuring" had the broader meaning of soil nutrient management in general. As a recognition of the great importance of manuring, its legendary inventor, STERCUTIUS, was claimed to have been given immortality by the Gods. These pieces of ancient wisdom, together with Roman practical experience and always improved by minor details, were finally recorded by PLINY and more extensively by COLUMELLA, both during the mid 1st century (Tisdale et al., 1985).

Later on, the ancient Roman wisdom of soil management spread as a heritage-guideline all over Europe. It was summarized again by CRESCENTIUS in Italy (about 1300) in his popular and widely-read (Latin) treatise on agriculture. All this practical knowledge, which was also improved later and adapted to local conditions of practical farming by several regional authors in most European countries, has served the farmers for centuries (Russell, 1973).

In this respect the extensive use of marl as a local soil amendment (lime-containing loamy material found in some fields at some depth) should be particularly mentioned. When mixed into the topsoil, marl added nutrients and eliminated growth-retarding soil acidity damage as well as improved soil structure. This method was already known in old Greek and Roman times, then somehow got lost, but was rediscovered in several European countries during the 18th century and contributed to better crop yields.

Although the causal mode of action of all the manuring-rules was not understood, this practical kind of soil nutrient management of adding organic and mineral substances proved to be effective for maintaining soil fertility, and was necessary for feeding a slowly growing population.

The result of the rather simple soil management methods, for many centuries in Europe, was a low but somehow satisfactory yield level, namely for cereal grains about 3–5 times the amount sown. In absolute terms the yields ranged from 0.5 to 1 ton (500–1000 kg) per hectare; on average, they were about 0.8 tons (corresponding to grains sufficient for only one loaf of bread per 10 square meter). In favoured areas, the actual production was sometimes twice as high. However, there were also many problem soils that yielded almost nothing because growth was inhibited by natural "toxic" substances. We should also realise that yields during these centuries were not only low, but also very variable depending on the consequences of dry or wet summers causing not only poor growth and low production but also because of great losses due to diseases and pests. Promising harvests were sometimes finally reduced to a mere fraction (Finck, 1999).

During 2000 years, soil productivity remained on this rather low level. A significant progress of yield increase seemed rather impossible – in spite of the ancient desire to "grow two cereal halms where there grew only one". Unfortunately, the prospects of doubling the general yield level seemed hopeless (Box 3). The English professor MALTHUS, therefore, correctly concluded about 1800: "The population of a country will always increase much more rapidly than the food production, and hunger will be an eternal companion of humanity" (Malthus, 1803). LIEBIG later commented (1865): "With the progressive increase of people in Europe since 1800, within two generations there would have been terribly wretched conditions of food supply. Even with civilized people, hunger arouses atrocious cruelty which results in revolutions and wars." There were, indeed, repeatedly severe famines in Europe which even influenced the course of history by revolts, migration etc. A population increasing beyond the limits of the low soil productivity meant food shortage (hunger) and led to emigration and a search for new virgin land or the fight for and occupation of sparsely populated territory – the real cause of many wars in Europe.

In retrospect, it can be asked what were actually the soil management practices during this long period up to 1800? Although the scientific explanation was not known at that time, the main concept of soil management was nutrient exploitation and utilization as well as nutrient cycling and replacement of losses to the arable field by adding nutrient sources. Further, other soil parameters important for nutrient uptake, like soil structure (and thus the essential supply of water and soil air) and the soil acidity were improved, e.g. by liming materials, which also rendered harmless some growth-disturbing substances such as mobilized toxic aluminium. A remarkable progress in soil management at the end of this period

Box 3: The long period of low and stagnant soil productivity

\# During 2000 years soil fertility could be kept on a low to medium level by simple methods of soil improvement, esp. with plant nutrient sources

\# The resulting production was rather low, but sufficient for a small population

\# In years with unusual weather, poor yields led to food shortage or even famines

\# There seemed to be no way out of this production gap of no more food but more people.

(about 1770) was the introduction of legumes like clover etc. into the crop rotation which increased the yields of the following crop. The responsible mechanism of fixing nitrogen from the air, however, was discovered only hundred years later.

Throughout the centuries, due to its apparent success, the plausible *Humus Theory* always found new advocates, e.g. WALLERIUS in Sweden. One of the latest was the renowned German agronomist Albrecht THAER. Even in 1810, he emphasized that humus is the actual plant food. For practical agriculture his widely-read rules for soil management led to an extensive, orderly treatment and use of waste materials. Further, he proposed the introduction of crop sequences (rotations) according to exemplary patterns in English farming. The result of all these measures were some small definite yield increases. However, this did not prevent famines, the last one raged through Europe in 1845/46, mainly due to plant growth disturbances and plant diseases during wet summers (Finck, 1999).

Scientific Soil Nutrient Management

Progress with a New Theory (Since 1830/40)

In scientific terminology, agriculture means utilizing the ability of green plants to produce complex organic substances, contained in food, out of simple inorganic substances like carbon dioxide and water with the energy of sunlight. This could only be understood after a certain progress in natural sciences of the biochemistry of relevant gaseous exchanges, finalized by Th. de SAUSSURE in 1804. After the principles of photosynthesis were clarified, the essential preconditions came into focus, e.g., what role do salts play in this respect? Already in 1563, B. PALISSY in France had claimed that certain salts from ashes were "plant food", but his hypothesis, even specified by J. GLAUBER with emphasis on saltpeter a hundred years later, remained too vague and without useful results.

Just before the breakthrough happened, however, there was still one more practical success, namely the import of Chilean saltpeter (extracted from desert salts) and guano as fertilizers from South America to Europe. Although guano (an organic/mineral product naturally formed from excretions of fish-eating birds on the Pacific coast on South America) had been used as a nutrient source for centuries in Peru, its scientific importance was discovered only in 1804 by Alexander v. HUMBOLDT, a famous German natural scientist working in Paris (Finck, 1999). With these products the "first wave" of mineral fertilization started in Europe (first in England), because they were much more effective in nutrient supply per weight unit than the common animal manure. Their application also contributed to and was soon justified by the new theoretical incentive.

Box 4: Contributors to the mineral theory of plant nutrition (1825–1840)

\# Carl SPRENGEL, a German agricultural chemist, demonstrated the necessity of about 6 mineral nutrients for plants in exact pot experiments from 1825 onwards.

\# Jean Baptiste BOUSSINGAULT, a French agronomist in Alsace, introduced scientific field trials for plant nutritional studies and made nutrient balances from 1835 onwards.

\# John B. LAWES, an educated English farmer, tested ammonium and other salts for their differential influence on plant growth from 1837 onwards (later with J.H. GILBERT).

\# Justus LIEBIG, a famous chemist, developed his concept of plant mineral nutrition from 1837 onwards and caused the final success with his book in 1840.

The decisive progress was the development of the theory of mineral nutrition of plants, a completely new concept which was established by the work of several scientists since about 1830. In Germany, SPRENGEL analyzed the chemical composition of plants as well as soils and he first showed in pot experiments that certain mineral nutrients are essential for plant growth. In France, BOUSSINGAULT made extensive field trials showing the benefit of plant nutrients for better growth. In England, LAWES tested different mineral salts for their influence on plant growth and found phosphate rather important (Box 4).

Most studies, however, remained mainly of local importance until LIEBIG (after an informative trip to England in 1837) finalized the new concept and published it in his famous and widely-read book called *Agricultural Chemistry*, simultaneously in German, English and French, 1840 (Brock, 1997) (Box 5). Some of his students became famous, e.g. from USA E.N. HORSFORD, later at Harvard College and S.W. JOHNSON, the founder of the first Agricultural Experiment Station in USA (New Haven, Conn.) (Rossitter, 1975).

Liebig created new standards for laboratory experiments and contributed considerably to the development of organic chemistry. His training of many students and scientific publications made him a progenitor of European chemistry. "Never has somebody more clearly told the world what chemistry means", quoted from his friend F. WÖHLER. As a side-line, he dealt with plant and animal nutrition,

Box 5: The mineral theory of plant nutrition (original version)

\# Mineral substances (salts) are essential plant constituents; no plant growth without salts

\# Plants need 10 nutrients of which 7 are mineral ones and obtained from the soil

\# Some soils are deficient in some nutrients; therefore need of supplementation (fertilizers)

\# Humus substances are not plant food, but important for their nitrogen supply etc.

published his famous books and pointed out many implications of the new concept for the increase and sustainability of agricultural production, esp. the need of crop fertilization. He was a productive popularizer, as well as a scientist.

The new concept of mineral nutrition, after some initial mistakes in theory and therefore some failures in practical application, proved to be very powerful and had far-reaching effects for new approaches of soil nutrient management and led to substantial yield increases. It also gave a satisfactory explanation of the old *Humus Theory*. Humus was not really plant food as such, but, nevertheless, an important source of nitrogen supply. Why was the progress in nutrient management developed in Western Europe? In this region the challenge of food shortage existed for an ever-growing population (even in spite of some emigration). Here was a solid theoretical basis for improving this situation and here was the initiative for many soil researchers, extension service experts and leading farmers to adopt the promising new concept (Box 5).

Although the establishment of a scientific concept of plant nutrition as a basis of soil management rules was a sudden breakthrough, the further development and adoption (practical application) of the new nutrient sources to different soils in different climatic regions still remained a vast challenge. It took some generations of agricultural scientists to develop regional and local rules of adequate application of mineral fertilizers, in addition to and in combination with the traditional manures.

At the beginning of the 19th century, many European soils were indeed quite impoverished, esp. phosphate deficiency was widespread in crops and this was also limiting animal production and damaging animal as well as human health.

Therefore, the development of phosphate fertilizers (containing water soluble, i.e. easily available phosphate) was urgently needed.

Beginning of Fertilizer Production and Application of NPK-Fertilizers

The first superphosphate factory started in England in 1843, but many more soon followed all over Western Europe. The next step was the production of K-fertilizers (potash in Germany, about 1860), followed by the first N-fertilizer plant for ammonium sulphate, in 1890. But only after 1913, with the HABER-BOSCH synthesis, could ammonia (NH_3) be produced in large amounts from nitrogen in the air and became the widely used source material for N-fertilizers containing ammonium nitrate (NH_4NO_3), and later urea.

The new nutrient sources, when combined, were soon known as NPK-fertilizers, namely Nitrogen, Phosphate and Kalium (latin; engl. potassium, potash). Other fertilizers containing Mg (Magnesium), Ca (Calcium) and S (Sulfur) were required to a lesser extent. In any case, mineral fertilization proved to be an excellent means to stop the unwanted age-old exploitation of nutrients from soils.

The challenge: Although a growing number of efficient nutrient sources could be obtained by the farmers at reasonable costs, their correct application left many questions to be answered. In fact, a new system of nutrient management was required to solve the basic problems: Which fertilizers? How much? When and how applied? Further, rules of fertilizer application should not just refer to a standard soil type, but to a great number of different soils in different climates. This was an enormous task at that time and required combined efforts of agricultural science and practical know-how.

As a consequence, and promoted by farmers' unions, many agriculture research stations were founded, starting in 1843 at Rothamsted near London, England (Johnston, 1994), in 1851 at Möckern near Leipzig, Germany. Of special importance is the establishment of long-term soil management field experiments, such as Broadbalk at Rothamsted (1852), "Continuous rye plots" at Halle (1878) and the Morrow plots (1876) in Illinois, USA, for the understanding of long-term nutrient dynamics. Some plots still exist after a hundred and fifty years. Research was later extended to the European tropical colonies starting with Pasuruan Research Station for sugar cane in Java by the Netherlands in 1880.

A great additional discovery was the elucidation of the N-fixation of legumes, made by Hellriegel and Wilfahrt (1886) in Germany. They showed that symbiotic bacteria (*Rhizobia*) in the root nodules were able to convert atmospheric nitrogen (N_2) into protein which could be utilized by the plants. This changed soil management practices for the best utilization of this important natural nutrient gained from the air by clover or other legumes such as beans.

It became clear from the beginning that not just any addition of nutrients raised the yield. On the contrary, there were many disappointing fertilizer trials out of which an important rule emerged, later called the *Law of Minimum*. SPRENGEL first noticed 1828: "When a plant needs several nutrients, it will grow poorly if one of these is not available in a sufficiently large amount." LIEBIG stated more precisely 1841: "The nutrient available in the relative smallest amount limits the plant growth, even if all others are abundant." In other words: Plant growth is limited by the nutrient of lowest (minimum) supply. An impressive picture of this rule is the *Minimum Water Barrel* which cannot contain more water than the lowest stave permits and any improvement must start with this one.

In spite of early experimental work since 1840, the actual start of mineral fertilization on a statistically relevant regional scale in amounts of a few kg/ha, was from about 1880. Even in 1900, fertilizer application was still on a rather low level; in Germany a total of 15 kg, expressed as so-called "pure" nutrients $N + P_2O_5 + K_2O$ per hectare. However, compared with 1800, the grain yields were about double, i.e. 1.8 tons/ha (Table 1).

Completion of the Plant Nutrition Theory

The new theory of plant nutrition (even in 1900) was by no means complete after six decades of laboratory research and fertilizer field trials. Actually, it took almost a century until it was fully developed. One reason was that soils still supplied automatically many substances not yet known to be nutrients.

First, the original mineral theory was expanded after detecting the essential functions of seven micronutrients from about 1920 onwards. Being required only in small traces, they were also called trace elements: Iron (Fe), Manganese (Mn), Zinc (Zn), Copper (Cu), Molybdenum (Mo), Boron (B), and Chlorine (Cl). Soil nutrient management now had to care for 13 essential mineral nutrients, further some beneficial ones being additionally required by animals and humans. According to strict criteria (Arnon, 1950), "essential" means that no growth or yield is possible if the element is completely missing (Box 6). Even new functions of organic substances were discovered. Some smaller organic compounds can be taken up and may have beneficial functions, e.g. growth promoters or resistance-improving substances like antibiotics. On the other hand, some special organic substances may damage growth.

Nutrients, although usually stated as elements, are absorbed by plant roots as cations or anions. For example, K (potash) is absorbed as cation (K^+), nitrogen mainly as nitrate-anion (NO_3^-) or partly as ammonium (NH_4^+).

Box 6: Nutrient elements required for plant and animal growth and health

A. Essential nutrients for plants

6 major nutrients: N, P, K and Ca, Mg, S (required in large amounts)

7 micronutrients: 5 heavy metals (Fe, Mn, Zn, Cu, Mo) and 2 non-metals: B, Cl.

B. Beneficial nutrients for all or some plant species

Na Sodium, Co Cobalt, Si Silica, Al Aluminium, Ni Nickel (possibly others).

C. Additional essential nutrients for animals and humans

Co Cobalt, Se Selenium, Cr Chromium, I Iodine (list may not be quite complete).

Second, nutrients should not just be present in the soil, but they must be available to plant roots for uptake, i.e. either mobile in the soil solution or only loosely bound on adsorption soil surfaces. The capacity of nutrient retention (adsorption) was already discovered by Way (1850). The concept of nutrient mobility in soils proved to be an important, but difficult task (Bray, 1963). Further, nutrients are subject to transformation processes for which soil organic matter (humus) and soil life play an important role. Nutrients could be rendered either better, less or even not available at all, the latter process being called nutrient fixation in soil.

The estimation of available nutrients by soil diagnostic methods started with cumbersome pot experiments (MITSCHERLICH), followed by extraction via plant roots (NEUBAUER technique), but was later simplified by chemical extraction methods with diluted acids etc. Widely applied for P and K were the methods of TRUOG, BRAY, MORGAN, MEHLICH, OLSEN etc., later with complexing agents for heavy metals (LINDSAY). The use of isotopic methods looked promising, but is not superior in practice. However, plant analysis as an indicator of soil nutrient supply can be quite helpful as a guide for fertilization, and so are deficiency symptoms (Bergmann, 1992). In spite of all these concepts, even nowadays and with computer assistance, the determination of available nutrients in soil and the subsequent calculation of required fertilizer amounts remains only approximately possible.

Third, even if there were sufficient nutrients available, they must be accessible to the roots, more specifically to the root hairs. Therefore, soils should provide

Box 7: Requirements of plant roots in soils

\# Deep rooting volume of possibly one meter for annual crops; easily penetrated with no mechanical or chemical restrictions on root growth.

\# Sufficient available water for the supply to plants and soil life, for nutrient source transformations and for nutrient transport to the roots. Drainage of excess water.

\# Good soil aeration for the oxygen requirements of roots and for the removal of surplus carbon dioxide (largely dependent on adequate large pore space in soil).

\# Provision of sufficient available plant nutrients from soil reserves and nutrient input.

sufficient pore space, i.e. good physical structural conditions. A good pore space with large, medium and small size pores is also important for adequate water and air (oxygen) supply to the roots (Box 7). Fourth, the actual uptake of nutrients from the close surroundings of the roots hairs requires certain preconditions of the short range transport by mass flow or diffusion (Barber, 1984). Fifth, whereas water-soluble nutrients may be easily leached out of the soil with percolating rain water, the cations are loosely retained by adsorption surface forces. Sixth, the great importance of soil life, esp. bacteria, for nutrient transformation like decomposition of organic nutrient sources and for extra nitrogen gain by N-fixation gave a new insight into the importance of humus and organic manuring.

Seventh, the dominant role of soil reaction, expressed as pH-value, for nutrient availability and uptake (as well as for other soil properties) was introduced about 1920 and was of great importance. Soils should be in the optimum pH-range which depends largely on grain size distribution in the soil. Eliminating excess soil acidity by liming, e.g. with calcium carbonate, is a fundamental soil amelioration method. On the other hand, on soils with a natural or man-made pH above the optimum, some soil acidification may be of advantage for the mobilization of certain micronutrients.

Since fertilizer trials, especially large ones, were normally rather cumbersome and expensive, special mathematical growth and yield curves had already been developed at the beginning of fertilization (the first one by E. MITSCHERLICH, Germany, 1906) with the aim of predicting the yield level resulting from a certain fertilization.

Common observation shows that, with more and more fertilizer added, the yields do not increase endlessly, but increases constantly diminish until there will be no more effect. Correspondingly, the *Law of Diminishing Return* states that with an increasing supply of a nutrient, the additionally yields obtained decline according to a logarithmic or similar curve. With equal fertilizer doses given, the yield response will first be rather high, but will then be smaller and at near maximum yield will approach zero (Mitscherlich, 1909).

Although this is not a law in the physical sense and hardly precise enough for yield predictions, it is a useful rule for approximately estimating fertilization effects. Together with the *Law of Minimum* (already mentioned), it describes nutrient-induced crop yield formations. The two laws may seem to be somewhat conflicting, but they represent just different aspects of the general yield curve.

Plant nutrition management with the goal of high yields is, in fact, a continuous effort for the elimination of minimum factors and for moving the diminishing return to a higher yield level. The emphasis was not only on the supply of plant nutrients, but also on the conditions of their uptake (Box 7). It is important to realize that there are still many fields in the world with yield-limiting nutrients and with a premature diminishing return.

Development of Commercial Mineral Fertilizers

Plant nutrients are derived from a large number of sources, some available on the farm, others imported from outside (Box 8).

From the chemical viewpoint, nutrient sources are either organic substances (often with some mineral components) or inorganic (mineral) substances, i.e. mainly salts (the term mineral derived from Latin *minera* = ore). Sources of mineral nutrients occur naturally as geological deposits like K-salts from former oceans or phosphate rocks from saurian bones or Mg from dolomite mountains. All the natural sources can be used directly as nutrient input, but it is generally advisable to improve them, either by removing plant-damaging substances or by increasing the availability to plant roots, e.g. conversion of phosphate rock into water-soluble phosphate.

Because only few mineral N-substances (like saltpeter) exist in nature, most N-fertilizers are synthesized from atmospheric nitrogen (N_2) via production of ammonia by the HABER-BOSCH process (since 1913) which requires high pressure and heat. This outstanding invention of chemical technology made the plant nutrient nitrogen available on a large scale. Ammonia can be used directly or is converted to ammonium nitrate, a typical N-compound of fertilizers. Recently, ammonia has been converted more and more into urea, a common cheap N-fertilizer.

Box 8: Plant nutrient sources

Nutrients in original soil minerals.

Mineral and organic substances from outside of farm, e.g. litter, mud, imported feedstuff.

Plant materials containing nitrogen fixed by legumes, Azolla, etc.

Recycled products of farm, e.g. straw, manure, slurry, green manure, compost etc.

Recycled materials from food or fibre processing factories and industrial by-products.

Recycled waste materials from cities, e.g. refuse, sewage sludge etc.

Atmospheric nutrient input, from natural or industrial emissions.

Commercial nutrient sources, e.g. multitude of mainly mineral fertilizers.

Mineral fertilizers are use-adapted, high-grade and conveniently applicable nutrient sources with a definite composition. They are effective means for an efficient supply of nutrients to crops. Many fertilizers are water-soluble salts and are, therefore, quickly available to plants; others are only slowly available because they require mobilization in the soil. They are either single or multinutrient fertilizers containing several or even all nutrients. The majority are solid substances, some are liquid and very few gaseous, like pure ammonia (Box 9).

Box 9: Types of mineral fertilizers for major nutrients N, P, K

N-fertilizers (Nitrogen): Many different types with components of either ammonium, nitrate or urea; produced from atmospheric nitrogen via ammonia with energy input.

P-fertilizers (Phosphate): Mostly water-soluble calcium phosphates; produced from natural deposits of phosphate rock, often derived from saurian animal bones.

K-fertilizers (Kalium, potassium): Water-soluble K-salts, often K-chloride; produced from salt deposits of former oceans after removal of plant-damaging by-products.

The nutrient content of fertilizers is traditionally expressed as so-called "pure nutrients" N, P_2O_5, K_2O etc., e.g. NPK-fertilizer of 15% N + 15% P_2O_5 + 15% K_2O. The cumbersome oxide units, however, are replaced more and more by elements, e.g. 15% K_2O = 12.5% K. Fertilizer production requires not only raw materials, but also energy. It should be pointed out that the energy of 1 kg of oil is required for 1 kg of N (as nitrogen fertilizer) and that 1 kg N as fertilizer can produce 15–20 kg of grain, depending on utilization rates. Global consumption in 2000 was: N-fertilizers about 85 million tons N, the P-fertilizers about 35 million tons P_2O_5, the K-fertilizers about 25 million tons K_2O (FAOSTAT, 2000). In many countries, the effectiveness and safe use of fertilizers is regulated by a *Fertilizer Law*. Recently in developed countries, there is also a trend towards regulating some aspects of fertilizer application to decrease environmental pollution.

From NPK Fertilization to Comprehensive Soil Nutrient Management

The aim of soil management is not just care for soils, but carefully utilizing them for effective agricultural production. In fact, the general aim is much wider (Box 10). Farmers want the highest yields under the given climatic and economic conditions. In view of this pretentious goal, soil fertility management proves to be a rather complex and extensive task. This is one reason why the term fertilization should be widened to the broader concept of nutrient management.

In order to illustrate this, some shortcomings of the standard concept may be outlined. NPK fertilization in large doses by no means always produced the expected yield increases. On the contrary, there was the frequent observation that

---

Box 10: Aims and preconditions of crop production

Good and healthy crop growth with certain resistance of plants against diseases

High yields, i.e. far above the average and close to maximum obtainable yields

High quality of crop products with respect to human and animal health

Maintaining or even improving soil fertility

No avoidable negative environmental impacts

Resource-efficient, sustainable (long-term) crop production.

the usual amount of N-fertilizer gave lower yield then expected. This contributed to a common mistake, namely the tendency of farmers to use too high N-doses which led to higher N-surplus in the soil at harvest time and therefore to larger avoidable N-losses. In any case, fertilizer efficiency was often not satisfactory, e.g. for N in the first year 50–70% recovery (possible up to 80), in paddy rice often only 30–40%. For P, the actual recovery in the first year is only 10–20% or up to 25% by placing the fertilizer near the roots. However, using the concept of Apparent Recovery, the P-utilization from a practical view can be considered much higher.

Further, all over the world there are great yield gaps of the same crop between average yields of a region and those of the best farmers or maximum yields in scientific trials. In order to find the causes of differential growth, special productivity studies revealed the complexity of yield formation on high levels. A manual for the practical fertilization of many crops is published by IFA (1992).

Plant nutrition ought obviously be treated in comprehensive terms and integrated with general agronomic measures including plant protection, in other words *Integrated Plant Nutrient Management* (IPNM) was required with optimized nutrient supply, minimized pollution etc. (FAO, 1995) (Roy *et al.* 2006).

Practical Aspects of Nutrient Management

Optimizing nutrient supply: The important questions of fertilizer use are: Which type, how much, when and how to use them? The selection of a suitable fertilizer requires a sound knowledge of the main and side-effects. Fortunately, some types are, by and large, interchangeable and the cheapest form can be taken. The estimation of adequate amounts of fertilizers is a "permanent" problem, since so many and partly unknown factors are involved. For a certain yield level, one should strive for the optimum supply range of each nutrient, but avoid the deficiency or surplus supply range. Many farmers still apply "rule of thumb" based on experience, but the logical reasoning should be based on crop requirements and soil supply (Box 11).

Time of fertilizer application: The best time is often just before crop sowing or planting, in any case before the growth period of maximum requirements. There is some lag of action which is more pronounced with fertilizers that must first be mobilized in the soil. If large amounts are required, esp. with N, they should be divided into two or more portions, added at a certain interval.

Method of fertilizer application: On small fields, fertilizers are usually uniformly distributed on the soil surface and mixed into the top soil, or under special conditions placed close to the main root mass. On large fields with sub-areas of different nutrient supply, however, a differential treatment of these units is preferable. This precision fertilization requires special equipment such as global positioning

Box 11: Estimation of nutrient input requirements

How much is required by a certain crop at the achievable yield level?

Amounts derived from standard tables, often based on nutrient removal by crops.

How much is potentially supplied from available soil nutrients?

Estimation of available nutrient by diagnostic methods of soil testing or plant analysis.

Difference of A minus B is the nutrient input required, with allowance for utilization rate.

(GPS) to help define areas of fertility levels. The extra costs, however, only pay if there are substantial small-scale differences and if the differential nutrient status can easily be recorded by sensors, e.g. on plant color for additional N-application. Precision spreading is therefore less effective with P and K based on soil samples.

Nutrient management in rotations: Sometimes the efficiency of fertilization can be improved by not just treating individual crops, but the whole rotation (crop sequence over several years). For example, a surplus of (unused) available N because of an early harvest time or of lower yield than expected can be utilized by a following crop. Alternatively, green manure plants can be included in autumn which prevent a nitrate surplus from leaching and preserve it as a nutrient source for the spring crop. A good example is rotation with legumes which leave N-residues for the following crop.

Environmental Pollution Aspects and Soil Protection

During the previous periods of food shortage, the emphasis was on the production function of soils, i.e. on soil fertility. After reaching food self-sufficiency, the emphasis shifted to ecological functions of soils (nutrient filtering and retention for avoiding environmental pollution). With increasing awareness of need for sustainability, there was additional emphasis on soil protection against fertility-damaging contaminations whether by inputs or by emissions.

The most common unwanted pollution effects are losses of fertilizer nutrients to the environment. In any primary production there will be unavoidable nutrient losses. With good management practices, however, losses from soil nutrient pool (especially of N-forms) to environment (rivers, lakes, groundwater, air) can be kept below an acceptable level from the environmental point of view, e.g. below 50 kg/ha N. Leaching of phosphate is almost negligible, only about half a kg P/ha and rather independent of fertilizer amounts. P-losses by erosion may be higher, but are avoidable.

There should be no accumulation of "toxic" substances in soil which may decrease fertility and damage food quality, such as heavy metals e.g. Cadmium, organic or radioactive substances. Farmers should ensure that their soils are not slowly "poisoned" by dangerous substances imported by waste materials, esp. from cities. The introduction of tolerance levels for "toxic" elements is supposed to prevent this kind of avoidable damage.

Soil Fertility Management

The aim of soil fertility management is a highly fertile soil with ideal nutrient dynamics where possible (Box 12).

Farmers have often modified poor natural soils into highly fertile man-made soils. Typical examples are paddy rice soils in South-Asia with high fertility over thousands of years, representing a sustainable way of agricultural land use, albeit with low N-fertilizer efficiency.

Unfortunately, neither a good natural nor man-made soil fertility is a common or a stable property, but a rather fragile one and easily degraded. The result of such degradation is a reduced, defective fertility resulting in yield decreases. The causes can be loss of fertile soil components via erosion by water and wind, physical degradation, e.g. compaction of soil structure, crusting, waterlogging and/or chemical degradation: e.g. decrease of humus and soil life, loss of nutrients etc.

Box 12: Soils with ideal nutrient dynamics:

are rich in mineral and organic plant nutrient reserves.

have a strong capacity of mobilizing nutrients due to active soil life.

store nutrients in light bonds to keep them available, but protected against leaching.

are capable of immobilizing a temporary nutrient surplus without strong fixation.

are efficient in supplying all essential nutrients to plants according to their needs.

Fertility decline, in most cases, is reversible by restoration methods. Many farmers all over the world are spending enormous efforts on the improvement of their soils by amendments and nutrient sources, esp. on so-called problems soils. On very acid soils, for example, the application of rock phosphate, possibly with lime, and the use of legumes results in a greater pool of cycling nutrients and therefore in a "spiralling upwards" process towards efficient and at least medium yielding economic production. On saline soils, an excess of water-soluble salts like Na-chloride or Na–sulfate is damaging to most crops. In severe cases, the salts must be leached out of the rooting zone, whereas with slight salinity, crops can be grown by adapted nutrient management.

Primary production such as food production is a complicated process with partly unknown variable factors. Thus the main feature of primary production is a remaining and unavoidable uncertainty. Even if all production factors were quantitatively known, due to the enormous complexity of the process and due to hardly calculable growth factors like water supply by rainfall, the results of soil nutrient management on crop yields, even with the best of know-how, remain, to a certain extent, unpredictable. Therefore, the complete optimization of nutrient supply, in the scientific sense, is an unreachable goal, but its approximation is worth great efforts.

Increases of Crop Yields by Better Plant Nutrition

What is the result of these improvements of soil management during one century? Grain yields as an index are 4 times higher on the country average in Western Europe. In Germany, the rise was from 1.8 to 7.5 tons/ha (1900–2000) (Table 1).

Even more striking is the comparison between yields of good soils in 1880 at the very beginning of fertilization and of productive soils in 2000. This was a rise from about 2 to 12 tons/ha of grain, i.e. a six-fold increase. Of course, this unprecedented increase in primary production is also due to better crops, agronomic management and crop protection, but it seems that better soil nutrient management contributed to about half of it.

The decline of fertilizer amounts used from 1990 onwards, and yet rising yields, may seem puzzling. But the main "motor" of production, nitrogen, is hardly affected, especially not for cereals. For high yields, e.g. of 10 tons of grain per hectare, about 250 kg/ha N are applied.

Although mineral fertilization became common practice about 1880, the real "Era of Mineral Fertilization" only started from 1950 onwards. This was especially true in many developing countries like India and China. Whereas in Europe, the emphasis was first on phosphate, in China a better N-supply had greatest priority (1950–1965), followed by P and other nutrients. With a balanced supply of

Table 1. Mineral Fertilization and Yield Increases within the last 100 Years

(German data as an example for Western Europe of fertilizer applied on agricultural land; a) Total mineral fertilizer ($N + P_2O_5 + K_2O$); b) Nitrogen fertilizer; c) percentage N-fertilizer of total fertilizer; d) Wheat grain yield) [from Statistical Yearbooks]

Year	Mineral fertilizers (kg/ha)		% N of total fertilizer	Wheat yield tons/ha
	Total	N-fertilizer		
1880	—	—	—	0.8–1
1880	3	0.7	23	1.3
1900	15	2.2	15	1.8
1920	38	8.0	21	1.8
1940	78	20.0	25	2.4
1960	160	43.0	27	3.3
1980	288	126.0	44	5.0
1990	220	114.0	52	6.7
2000	160	109.0	68	7.5

all nutrients and continuous use of organic manures, China's grain yields increased threefold and highest yields six times, until self-sufficiency of food was reached around 1980 (Cao, 2004).

As for future prospects, even in countries with high yield levels and without new technology, there is still a considerable production gap: e.g. average grain yields of 8 tons/ha are only 2/3 of farmer's top yields (12 tons/ha) or the difference to the maximum possible yield level (about 15 t/ha?). The present production potential, however, is still far from being fully utilised. The main reason is overproduction, causing a price decline which puts brakes on nutrient inputs and production level.

Special Aspects of Soil Nutrient Management

Possibilities of Future Nutrient Supply

The nutrients consumed by plants are derived either from soil reserves, from recycled sources or from nutrient inputs. The different strategies, for use of nutrients are exploitation, utilization, replacement, or enrichment.

Exploitation of soil nutrients is the simplest and oldest method, but even today such mining of natural nutrient capital still plays an important role in many regions. In many areas and at least on a low yield level, soils have produced crops for many centuries by exploiting soil nutrients. For high-yield cropping with high demands, however, only few soils are able to supply sufficient nutrients for many years. Most soils have vast reserves of plant nutrient sources, but only a small portion of these nutrients becomes available to plants by weathering of minerals or humus decomposition during a cropping season. Whereas continuous exploitation leads to complete soil deterioration, short-term exploitation can be compensated by a following long-term recovery, esp. in warm climates with high mobilization rates. Famine was often the result of soil impoverishment due to nutrient exploitation and soil degradation which is especially damaging during climatic dry spells. Some earlier human civilizations may have perished through the far-reaching exhaustion of their soils, e.g. the *Mayas* in Central America and the *Khmers* in Cambodia.

A typical example of nutrient exploitation is the *Shifting Cultivation* in humid forest areas where the soil fertility components, accumulated over decades, are consumed during a few years of cropping (Nye and Greenland, 1960; Sanchez, 1976). In spite of frequent critical comments, this is a useful and ecologically acceptable stable form of land use, as long as the arable (exploitation) period with a few years of cropping alternates with long fallow (forest) periods for regeneration to the original fertility level. Otherwise, serious soil degradation may occur.

Utilization of soil nutrients means that, although crops annually remove a small part of the stored soil nutrients, there is no significant impoverishment of the fertility. This creates the impression of a permanence of agricultural production without input. Compare this with a well which always contains the same level of water irrespective of the amounts taken out. With nutrients this is the case if the reserves are large and the removals small. Special forms of nutrient utilization occur in modern agriculture when some nutrients are delivered from the soil in sufficient amounts during long periods and enable high yields without adding nutrients. Typical examples are the continuous supplies of some secondary or micronutrients which may last for hundreds of years without any input.

Replacement of soil nutrients means supplementing (fertilizing) nutrients that have been removed by crops or are lost from the field. On most soils with only average fertility, the replacement of major losses is essential for maintenance of the production potential. The aim is a continuous stable crop rotation with balanced nutrient supplies by recycling and supplementary input. The strategy of nutrient replacement is fully valid only in cases of a good initial soil fertility.

Enrichment with nutrients is often useful on soils with low or medium natural supply. It should concentrate on nutrients that limit production as "minimum factors". The enrichment phase is usually a transient one, followed by a replacement phase once a high yield level is reached. Many European soils, for example, are substantially enriched in N and P.

On farm scale, nutrients are partly recycled with the straw or with manures, in more or less close circulation. On a wider scale, the residues of processed crop products can be returned from factories to the field. An important consequence of increasing urbanization is an enormous plant nutrient transfer with food and other farm products from farm areas to the large market places of cities and even megacities. A diagram with nutrient pathways including recycling in shown in Fig. 1.

The degree of nutrient recycling from most cities, however, is rather low because of removal of wastes into landfills or into the sea. Example for transfer of phosphate: Humans need 1.0–1.5 g of P per day, ensured by a supply of about 1.7 g of P (phosphorus). Even a city of only 1 million people, therefore, requires 1.7 tons P per day or 620 tons P per year (1.400 t P_2O_5) for people alone. However, the P-recovery from cities varies largely, probably from 10 to 80%. The common practice of low recycling is rather wasteful and not compatible with resource-preserving nutrient management.

Further, a somewhat problematic case is the re-use of plant nutrients from cities in view of potentially "toxic" substances (mentioned above).

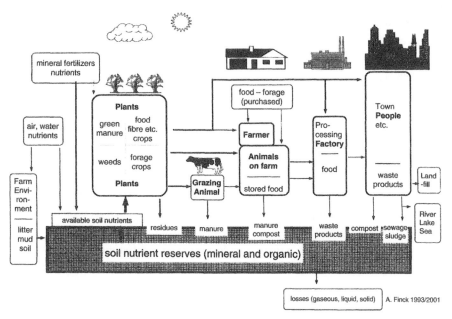

Figure 1. Cycles of plant nutrients.

Food Quality – An Additional Goal of Crop Production

In times of food shortage, the quality of the food was of minor importance, esp. since food was often inherently nutritious. In saturated markets, however, a tendency developed towards quality indices, starting with external properties, e.g. the baking index for wheat. Superior commercial quality resulted in a premium price for the farmer.

Starting from problems in animal nutrition at the beginning of the 20th century, it became more and more obvious that similar-looking grass could have quite a different feeding value with the result of different animal health. The next step was a growing awareness in human medicinal science that about half of all diseases were caused by nutritional factors which were often deficiencies. Nowadays, it is quite obvious that humans require nutritious, health sustaining food. In order to meet the requirements of humans and domestic animals, food should contain sufficient amounts of about 50 essential nutrients, i.e. valuable protein with 8–10 essential amino acids, some fatty acids, about 20 vitamins and about 20 mineral nutrients.

The nutritive value of plant products depends largely on an optimized nutrient supply to crops, therefore nutrient management must take the quality aspect into account. Typical examples of neglected quality are low vitamin contents in plants resulting from lack of certain mineral nutrients and causing human diseases due to deficiencies or even decrease of resistance against diseases due to pathogenic agents. Examples from animal nutrition are mineral deficiencies like phosphate, magnesium or copper deficiency in cows resulting in bone weakness, infertility, growth disturbance or even mental disturbances. Soil management for health promotion and protection is an often overlooked aspect. Agriculture does not only supply food against hunger, but it should be also a guardian of health for humans and domestic animals (Finck, 1982).

Crop Production with Restricted Nutrient Use ("Organic Farming")

Consumers ideas on food quality required for health have led some farmers to not make full use of the modern high production potential. The terms *Organic farming, Biofarming or Ecofarming* designate special farming systems with deliberately restricted plant nutrient input, i.e. exclusion of "chemicals" such as synthetic fertilizers, using only traditional manures, compost, etc., and with great care for nutrient cycling and natural nitrogen gains from legumes. The different terms designate a group of similar and yet different systems of nutrient supply. Whereas *Biologic-dynamic Agriculture* (the oldest, orthodox type, initiated by R. Steiner, 1929), excludes all kinds of commercial mineral chemical fertilizers, major groups like *Bioland* in Germany or European standards (EU-Law) exclude mainly water-soluble mineral fertilizers, especially N-fertilizers, but permit other major nutrient sources if they are naturally occurring products like rock phosphate, crude K-salts, lime etc. Also micronutrient fertilizers are allowed if there is an obvious deficiency. The exclusion of water-soluble N-fertilizers, esp. nitrate, but also of synthetic urea, has no scientific basis but is a purely ideological concept, specified by its adherents with personal concepts of what is an unnatural product and therefore unwanted ("*good* is what is *natural*").

Organic Farming advocates claim that it produces "superior-quality", "non-contaminated" and therefore more healthy food, further that the environment is better protected against unwanted nutrient losses. These claims have not been substantiated, because more or less all food, measured by the content of essential substances and potentially dangerous impurities is in about the same quality range. The additional claim that these types of *Bio-* resp. Ecofarming are less water polluting because of avoided chemical fertilizer input is also open to correction. Although a lower amount of N-leaching is often achieved by Ecofarming per area of land, it may not hold true per unit of crop product, because more land is required for cropping than with conventional farming. As for the economics, Organic Farming is certainly workable as long as the somewhat lower yields are compensated by higher product prices. A more complete discussion of organic farming is contained in Chapter 15.

Nutrient Resources for a Growing World Population

Since the present world population of 6,000 million people is expected to increase to about 7,500 million in 2020, the need for more efforts in soil nutrient management is obvious. The low or medium fertility of many soils must be further improved, and arable land extended to "problem soils". All this requires more nutrients. In future, for feeding more people and also their domestic animals, fertilizer use must be much higher, a total of 200 million tons of pure nutrients annually in 2030 compared with about 140 in 2000. Half of this higher amount will be needed in East and South Asia. Fortunately, soil management in combination with progress in plant breeding will secure future food security at least for a certain population growth. India and China have shown excellent examples of reaching self-sufficiency even with rising populations. As for the plant nutrient supply on a global scale, food for half of the world's population is produced via recycling of nutrients, the other half via nutrient input of mineral fertilizers with emphasis on N-fertilizers which are a powerful tool of food security.

Will there be sufficient nutrient sources available in the future? The reserves of some sources are almost unlimited (e.g., N, Ca, Mg), some are still rather abundant (K), but one source will soon be rather scarce, namely phosphate. The real problem will start from the next century onwards when the reserves will be almost exhausted. Phosphate cannot be replaced by any substitute. The finiteness of P deposits will strongly limit crop production and presumably cause serious disputes for a share long before the last deposits of phosphate rocks are exhausted. Only a stronger emphasis on P recycling can postpone this supply crisis.

Finally, what is the essence of modern soil nutrient management for the well-being of humanity? In view of growing food needs, pollution effects and dwindling reserves, humanity is well advised to strive for efficient food production methods for present and future food security. Furthermore, humanity should strive for an optimum population instead of a maximum tolerable one and undertake a better distribution of the world's wealth. As for the discrepancies between resources and demands, we should remember Mahatma GANDHI's comment: "The earth has enough (goods) for everybody, but not for everybody's greed."

Summary

1. Soils are the main basis of food production; therefore, caring for their productive capacity (fertility) is of great importance for national and global food security.
2. Although soils have produced plants without human influence, since early civilisations, farmers have been making great efforts toward an increase in crop yields.
3. Soil fertility can be improved by several management methods, but a better supply of plant nutrients is of main importance.
4. The nature of plant nutrients, discussed since antiquity, was only discovered after the rise of modern chemistry, and explained by LIEBIG and others after 1840.
5. Since the essential mineral plant nutrients are known, their supply from soils could be improved and supplemented by input of nutrient sources, i.e. fertilizers.
6. The result of rational soil nutrient management was considerable; from a low and stagnant agricultural production over many centuries the average yield levels increased many times with top grain yields even up to 10 times.
7. Whereas before, due to food shortage, periods of hunger occurred frequently in Europe, there is now food surplus causing problems of overproduction.
8. Optimisation of plant nutrition and sustaining a good soil fertility require the performance of complex nutrient management rules in the frame of agronomic know-how.
9. Not only was the yield increased by better plant nutrient management, but also the nutritional quality of food with the effect of better health of humans and domestic animals.
10. The contribution of modern soil nutrient management to global food supply is enormous and, if applied, could provide food security even for a somewhat growing world population.

References

Arnon, D.I. 1950. Criteria of essentiality of inorganic micronutrients for plants. Loytsia 3, 31–38.

Barber, S.A. 1984. Soil Nutrient Bioavailability. J. Wiley, New York.

Bergmann, W. 1992. Colour Atlas. Nutritional Disorders of Plants. Fischer, Jena.

Boulaine, J. 1992. Histoire de l'Agronomie en France. Tec Doc, Paris.

Bray, R.H. 1963. Confirmation of the nutrient mobility concept. Soil Sci. 95, 124.

Brock, W.H. 1997. Justus v. Liebig – The Chemical Gatekeeper. Cambridge Univ. Press.

Cao, Zh. 2004. Cultivation in the Stone Age; Priority of nutrients in China. Personal Communication.

FAO. 1995. Integrated Plant Nutrition Systems. Fert. Plant Nutr. Bull. 12.

FAOSTAT, FAO-Statistics 2000.

Finck, A. 1982. Fertilizers and Fertilization. Verl. Chemie, Deerfield Beach, Florida. 1992: 2nd Edit. Dünger und Düngung. VCH, Weinheim New York.

Finck, A. 1999. Research Progress in the USA for More and Better Food Production. A.v. Humboldt Foundation, New York.

Hellriegel, H. and H. Wilfahrt. 1886. Ztschr. Rübenzucker Ind. 36, 863–877.

IFA. 1992. IFA World Fertilizer Use Manual. Int. Fert. Industry Assoc., Paris.

Johnston, A.E. 1994. The Rothamsted Classical Experiments. In: CAB Interntional, Wallingford.

Liebig, J. 1840. Die Chemie in ihrere Anwendung auf Agrikultur und Physiologie. Vieweg, Braunschweig.

Liebig, J. 1865. Chemische Briefe. Winter Verlag, Leipzig.

Malthus, R.T.W. 1803. An Essay on the Principle of Population. London.

Mitscherlich, E. 1909. In: Mitscherlich, E.: Bodenkunde. Niemeyer, Halle, 1949.

Nye, P.H., and D.J. Greenland. 1960. The Soil under Shifting Cultivation. Comm. Bureau of Soils. Tech. Comm. 51.

Rossiter, M. 1975. The Emergence of Agricultural Chemistry. Yale University Press.

Roy, R.N., Finck, A., Blair, G.J., Tandon, H.L.S. 2006. Plant nutrition for food security – a guide for integrated nutrient management. FAO, Rome.

Russell, E.W. 1973. Soil Conditions and Plant Growth. Longmans, Green, England.

Sanchez, P.A. 1976. Properties and Management of Soils in the Tropics. Wiley, New York.

Steiner, R. 1929. Geisteswiss. Grundlagen zum Gedeihen der Landwirtschaft, Dornach (Switzerland).

Thaer, A. 1810. Grundsätze der rationellen Landwirtschaft. Berlin.

Tisdale, S.D., W.L. Nelson, and J.D. Beaton. 1985. Soil Fertility and Fertilizers. MacMillan, New York.

Way, J.T. 1850. J. Roy. Agri. Soc. 11, 313.

Yaalon, D.H. 1997. History of Soil Science in Context. Advances in Geoecology 29, 1–13.

17
Soils and Environmental Issues

T.M. Addiscott[1]

Introduction

Soils are fundamental to the function of the planet, but not just because they support the growth of plants. They also provide a buffer between the atmosphere and the hydrosphere and they moderate the impact of human activity on natural and agricultural ecosystems. Soils are vital too because they are living entities and provide homes for countless micro-organisms and soil animals that have key roles in processes at scales ranging from the soil crumb to the planet. This chapter examines research during the last 160 years on the combined impact of soil processes and human activity on the soil itself, on natural waters and on the atmosphere, in short, on the environment. The impacts with which we are concerned spring mainly from the use of the soil by humans, but the soil's population of soil animals, microbes and fungi also uses the soil for its own purposes and can thereby add to the problems, as do farm animals. The underlying problem is that virtually all forms of productive activity, whether by farmers, livestock or soil microbes – or industrialists – lead to some form of environmental pollution. It is a problem with a long history, a history in which perspectives on the various issues have often changed with time.

The chapter begins with the very earliest indications of water pollution from human activities. This is followed by a brief account of the emergence of 'the environment' as an issue. The beginnings of fertilizer production during the 19th century and the dramatic increase in fertilizer production during the 20th century, discussed next, were central to the nitrate issue which recently dominated agro-ecological research. Experiments made in the 19th century on losses of nitrogen from fertilizers and from organic matter still provide a substantial proportion of our current overall understanding of these processes and are described next. These are followed by recent research on soil processes leading to water and air pollution, particularly that conducted using ^{15}N. The mechanisms by which nitrate and phosphate are lost from the soil in drainage and nitrogen oxides and ammonia are emitted from the soil to

[1] Rothamsted Research, Harpenden, Herts, AL5 2JQ, UK.

the atmosphere differ in nature and importance between arable and grassland systems. These losses and their environmental consequences are discussed. The soil not only leaks environmental pollutants, it is itself the victim of pollution, notably by heavy metals, and an account of early work in this field of research is given, together with the related subject of soil remediation. The final topic covered in the chapter is a matter of great current interest, the sequestration of carbon dioxide from the atmosphere. Here again, research started towards the end of the 19th century has proved strongly relevant.

Historical Background

Early Indications of Pollution

Neolithic farmers are the earliest known polluters of water in the UK. In about 3000 BC they cleared forest around lakes in the Cumbrian Lake District (Moss, 1996), perturbing the soil and releasing nutrients that were lost to the lakes. Neolithic people elsewhere no doubt did much the same. Moss also records that the military made an early contribution to water pollution when, in 171AD, Lucius Cassius Longinius made a road out of Rome so that he could get the army into action quickly. The road went through the catchment area of a small lake which shortly afterwards produced massive growths of algae and swarms of midges as a result of the nutrient released by the soil disturbance. No doubt other incidents of this kind occurred at the time but went unrecorded. Toxic algal blooms are thought of as a contemporary problem but Bell and Codd (1996) note that deaths of animals from toxic blooms were recorded in Australia as long ago as 1853. These incidents were simply the result of soil perturbation and occurred long before the extensive use of synthetic fertilizers. There is no indication of which nutrient was involved, but as all the incidents involved fresh water, it was probably phosphate.

The Emergence of 'The Environment' as an Issue

According to the Shorter Oxford Dictionary, the word 'cology' first emerged in 1873 and the idea of an organism in, or adapting to, its environment in 1874, but it was another 80–100 years before 'the environment' came to the forefront of public awareness. An early marker of this awareness was the publication in 1962 of Rachael Carson's book 'Silent Spring'. Entries in books of quotations may also be a guide. There are no references to 'Environment' in the Penguin Book of Quotations or in my copy of the Penguin Book of Modern Quotations, which was reprinted in 1977. But we know from the Oxford Dictionary of Humourous Quotations that the environment was firmly on the political agenda before the Falklands War in 1982. Mrs Thatcher, who had then been in power for three years, said of the war that, 'It is exciting to have a real crisis on your hands, when you have spent half of your political life dealing with humdrum issues like the environment.' She was the last prime minister to refer to the environment as 'a humdrum issue'.

Fertilizers: 1843–1900

Liebig and Lawes. Early Fertilizer Research and a Famous Dispute

Baron Justus von Liebig was very distinguished as a scientist, and not only in the plant sciences (Box 1). He knew that plants needed nitrogen but, as he had established earlier that plants obtained the carbon they needed from the air rather than from the soil (as previously thought), he thought that plants derived their nitrogen from the ammonium that came down in rain. His calculations told him that no ammonium needed to be added to the soil, and this led him into his famous dispute with John Bennett Lawes (Box 1).

Dyke (1993) records the dispute as starting in 1845. Having estimated the deposition of ammonia in rainfall, Liebig wrote in 1845 that 'the supply of ammonia

Box 1. Liebig and Lawes

Baron Justus von Liebig 1803–1873

Justus von Liebig was the foremost agricultural chemist of his time, famous for the law of the minimum, the idea that plant growth is limited by the element least available in the soil. He was notable too for his realization that, although carbon dioxide makes up only 1% of the atmosphere, plants obtain their carbon from the air and not from the soil, as previously thought. Liebig was among the first scientists to achieve some kind of understanding of the nitrogen cycle. His almost correct ideas on crop mineral nutrition and his failure to realize that not enough ammonia was deposited in rain to supply the needs of crops are discussed in the main text. Liebig's contribution went far beyond agricultural chemistry. He and his friend Wöhler were among the founders of organic chemistry (Box 2.2) and provided improved methods of organic analysis. Liebig discovered chloroform and chloral and introduced the theory of radicals, the idea that a group of atoms forming part of a molecule could act as a unit and take part in chemical reactions without disintegrating, yet be unable to exist alone.

Examples include the methyl radical -CH_3 and the carboxyl radical -COOH. He was also notable for founding a chemical laboratory at Giessen in Germany which he used to train young chemists (among whom was Joseph Henry Gilbert). Liebig must be given some credit for the strength of chemistry in Germany in the latter part of the nineteenth century. Indeed, his contribution to the subject must have been part of the intellectual heritage that underpinned the development of the Haber-Bosch process for the synthesis of ammonia.

Sir John Bennet Lawes 1814–1900

Lawes is known chiefly as the founder of Rothamsted Experimental Station and for the contribution he made to agricultural science with his colleague Sir Joseph Henry Gilbert. The Broadbalk experiment brought the realization of the importance of nitrogen in the nutrition of cereals and the Drain Gauges the first recognition of the roles of mobile and immobile water in solute transport, but there was much more besides, several other field experiments and some notable work on animal nutrition. Lawes was also an entrepreneur, founding with little capital one of the first fertilizer businesses in the world and running it profitably for 30 years, before selling it and using one-third of the profits to set up the Lawes Agricultural Trust to maintain the experiments. He became in his later years a father figure in British agriculture, providing sound guidance in difficult times and looking ahead to future problems. Parish magazines of the era record him as a popular figure, and often benefactor, in Harpenden. Lawes was also a man with a social conscience, a model employer. He set aside land for allotments for workers on the estate, so that they could grow their own vegetables, and started a pig club. He also built a club-house by the allotments where those who toiled could buy a pint of beer more cheaply than elsewhere and were under no pressure to buy more than they wanted. Lawes and the club-house were celebrated by the novelist Charles Dickens in 'The poor man and his beer', an article published in *All the Year Round* magazine in April 1859.

to most of our cultivated plants is unnecessary, if only the soil contain a sufficient supply of mineral food'. By mineral food he meant phosphate, potassium, sodium and magnesium. Lawes disagreed completely and wrote in 1846, 'There cannot be a more erroneous opinion than this, or one more injurious to agriculture.' This remark was hardly designed to placate, but it was in the context of an argument with enormous consequences. Liebig claimed that the fields of England were in a state of progressive exhaustion by removal of phosphate. Lawes agreed about the progressive exhaustion but believed it was caused by ammonia carried off in the grain (Dyke, 1991). Liebig was to lose the argument because he seriously overestimated the concentration of ammonia in the atmosphere. But his ideas about minerals, based on his study of the composition of crop ash, were largely correct.

Broadbalk. The First Nitrogen Fertilizer Experiment

The Broadbalk winter wheat experiment (Box 2) was the first and most famous of the experiments made by Lawes and his lifelong collaborator Joseph Henry Gilbert. By the autumn of 1844 Lawes and Gilbert had the first results of a simplified form of the experiment, which would assume its present form in 1852. If Liebig was right, the plots given just minerals on Broadbalk would have out-yielded those without, and the 'minerals + nitrogen' plots would have done no better than the 'minerals only' plot. Table 1, based on a report by Garner and Dyke (1969), gives the results of that first experiment, which was designed to address this point, together with means from the experiments from 1856–1863 which followed the pattern started in 1852. Even the first simple experiment suggested strongly that Liebig was wrong and the later results confirmed this. Minerals alone did not give much more yield than no manure at all, while 'minerals + nitrogen' clearly out-yielded minerals alone. Lawes and Gilbert also tested Liebig's patent wheat manure but found it to have no effect (Johnston and Garner (1969).

Liebig was furious. In a confidential letter to a friend quoted by Smil (2001) Liebig wondered how 'such a set of swindlers' could produce research that is 'all humbug, most impudent humbug'. Had he known of the letter, Lawes could have

Box 2. The Broadbalk Experiment

The soil of Broadbalk field was classified as Batcombe Series by Avery and Bullock (1969). An updated soil map produced by Avery and Catt (1995) shows about 70% of the field to be 'typical Batcombe Series', a flinty clay loam to silty clay loam containing 18–27% clay, overlying mottled clay-with-flints containing about 50% clay within 80 cm of the surface. The rest of the field is either a heavier variant of the Batcombe series, Hook Series, which is a flinty silty clay loam with gravelly or chalky material in the subsoil, or Hook Series, a slightly flinty silty clay loam. The pH of the topsoil is 7.5–8.0, and it contains about 1% organic carbon in plots receiving mineral fertilizer and 2.7% in plots that have long received farmyard manure. The site has only a very slight slope but is free-draining.

Lawes and Gilbert's experiment on Broadbalk has not changed greatly since 1843, and by 1852, the present pattern had emerged and remained unchanged until 1967 (Johnston and Garner, 1969). This involved growing winter wheat continuously on plots which received either:

- No fertilizer
- Minerals (P, K, Na, Mg) only
- Minerals with four different additions of nitrogen, 43, 86, 129, or 172 lb acre^{-1} as ammonium sulphate). In metric units 48, 96, 144, 192 kg ha^{-1}. Sodium nitrate was also tested for some applications in some years.
- Farmyard manure

The main difference in the experiment today is that two extra additions of nitrogen (240 and 288 kg ha^{-1}) are now made. These reflect the much increased use of nitrogen by today's farmers, although the original additions were huge by the standards of 1843. Even by 1957, the average nitrogen application to winter wheat by farmers was only 75–90 kg N ha^{-1} (Garner, 1957).

Table 1. Yields in early experiments on Broadbalk by Lawes and Gilbert. (From Garner and Dyke, 1968). (a) The first harvest, in 1844, (b) Mean yields 1856–1863

Treatment	Yields (cwt acre-1)	
	Grain	Straw
(a)		
No manure	8.2	10.0
14 tons farmyard manure	11.4	13.2
Ash from 14 tons farmyard manure	7.9	9.9
Minerals only[1]	9.0	10.3
Minerals + 65 lb acre^{-1} ammonium sulphate[2]	11.4	
(b)		
No manure	8.8	8.3
Minerals only[1]	10.6	16.1
400 lb acre^{-1} ammonium salts only[2]	13.1	22.6
Minerals + 400 lb acre^{-1} ammonium salts[2]	21.5	37.3

[1] Minerals were P, K, Na, Mg.

[2] 65 and 400 lb acre^{-1} correspond to 73 and 447 kg ha^{-1} respectively.

retorted that one of the swindlers, Gilbert, had been trained in Liebig's own laboratory. But Lawes's attitude was entirely courteous (Dyke, 1993). He took pains to preface all criticism of Liebig with a tribute to his contributions to the subject, but the courtesy was not returned. Indeed, Dyke regarded 'some of (Liebig's) writings (as) simply spiteful', and this seems to be confirmed by Smil. Lawes did not use the 'von' when writing of Liebig.

In public Liebig maintained that his theory was supported by the Rothamsted results. The ammonium fertilizer, he said, acted as a 'facilitator' for the adsorption of the minerals by increasing their solubility. But other Rothamsted results showed that clover grown without any 'facilitation' from ammonium fertilizer removed more minerals than the grain crops did as well as taking up more nitrogen.

Lawes emerged from the controversy as a man of principal, and certainly not a swindler. He made his living from the sale of the 'Super Phosphate of Lime', but when his experiments showed that additions of nitrogen rather than phosphate caused the greatest increase in yield in wheat, he did not hesitate to publicize the fact. It is one of the more ironic aspects of the controversy that Liebig, while propagating his 'minerals' theory, had told the British prime minister Sir Robert Peel in 1843 that 'the most indispensable nourishment taken up from the soil is the phosphate of lime'. This was, of course, just what Lawes's company was selling, but Lawes would shortly be telling the world that it was nitrogen that had the greatest impact on grain yields. Liebig's comment would have been more correct if applied to a root crop, such as turnips.

Lawes's experiment on Broadbalk showed clearly the yield benefit from nitrogen fertilizer, thereby encouraging its use. The experiment was later to contribute greatly to the understanding of nitrate leaching, as described subsequently.

Nitrogen Fertilizers

Gas liquor from coal gas production must have been the first 'chemical', or non-organic, nitrogen fertilizer to be used. The Shorter Oxford Dictionary records the first use of the term 'coal gas' in 1809 and gas liquor must have been known by then and probably came into use as a fertilizer sometime between 1810 and 1820. Its use must have been well established by 1843 when it was mentioned in Lawes's first advertisement for super-phosphate of lime (Box 3). It was still in use as a fertilizer in about 1950. Ammonium salts were made by acidifying gas liquor at least by 1841 and probably earlier, and by 1900, large amounts of ammonium sulphate were produced in this way (Figure 1).

By the 1860s ammonium sulphate was also made from ammonia generated in coking ovens as well and this process had become important by 1900 (Figure 1) and continues today.

Box 3. Lawes's patent, granted 22 May 1842, and his first advertisement, in the Gardener's Chronicle of 1 July 1843

The patent, as summarized by Boalch (1978)

1. By means of sulphuric acid, decomposing bones, bone ash, bone dust, apatite, or phosphorite and other phosphoric substances, for purposes of manure.
2. Combining for manurial purposes phosphoric acid with any particular alkali, as potash or magnesia or ammonia, or any earth containing such alkalis.
3. Making a manure by combining silica (in the state of ground flint or sand) with either potash or soda, or applying as manure crystal or glass ground to a powder.

The advertisement

J.B. LAWES'S PATENT MANURES, composed of Super Phosphate of Lime, Phosphate of Ammonia, Silicate of Potass. &c., are now for sale at his factory, Deptford Creek, London, price 4s. 6d. per bushel. These substances can be had separately; the Super Phosphate of Lime alone is recommended for fixing the Ammonia of Dung-heaps, Cesspools, Gas Liquor, &c. Price 4s. 6d. per bushel.

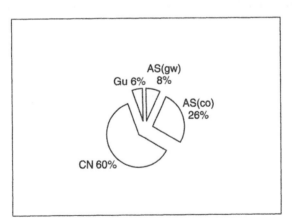

Figure 1. Relative importance of ammonium sulphate from gas-works, AS(gw), ammonium sulphate from coking ovens, AS(co), Chilean nitrate, CN, and guano, Gu, as fertilizer nitrogen sources in 1900. Based on data from Smil (2000), pp. 50–51 and p. 240.

In the 1830s and 1840s, two further sources became available, guano and Chilean nitrate. Of these, the Chilean nitrate was much the more important (Figure 1) because it produced far more fertilizer nitrate than guano and also nitrate for explosives, which guano did not provide. Also, the supply was never exhausted and some production continues to this day. The supply of guano was exploited greedily and contributed in a major way to fertilizer supplies for only about 30 years, from 1842 to 1872 (Figure 1).

Smil (2001; p. 240) calculated that during the second half of the 19th century the world's cumulative production of nitrogen fertilizer was 360 kt of N, of which 220 kt was from Chilean nitrate, 20 kt from guano, and 120 kt of ammonium sulphate from coking. Smil's table did not include gas liquor or ammonium sulphate produced from it, probably because there were no records, and such records might have been of limited use because of the variable nature of the product.

Phosphate Fertilizers

Bones, usually ground up, seem to have been the earliest form of phosphate fertilizer (apart from animal manure), but farmers soon realised that their effectiveness varied greatly according to the soil on which the crop was grown. This variability probably arose because bones were more effective on acid soils than neutral or alkaline ones. Several people, Lawes and Liebig among them, seem to have realised that acidifying the bones with sulphuric acid made the phosphate they contained more available to crops. Lawes and Liebig must have reached this conclusion at almost exactly the same time, because Liebig published the suggestion in a book in 1840, while Lawes had evidence that it was correct from at least one field trial by 1841 (Dyke, 1993). Lawes made substantial quantities in 1842 of what he first called 'Biphosphate' but within a year was calling 'Super Phosphate of Lime', a name which became 'superphosphate' and persists to the present day.

On 23 May 1842, the day Lawes made his second and third sales of superphosphate made at Rothamsted, he was granted a patent in England and Wales (Dyke, 1993) for the three manufacturing processes described in Box 3. Having received the patent, he was very anxious to upscale his production of superphosphate from the process he used at Rothamsted to a factory-scale process. So much so that having married on 28 December 1842, he cancelled his honeymoon on the continent and took his new bride on a winter boat trip along the Thames to look for a site. He found one at Deptford Creek, and it was from his new factory there that he issued his first advertisement, also shown in Box 3, in the Gardener's Chronicle of 1 July 1843.

Lawes and Liebig were not the only people to think of acidifying bones, but Lawes got his patent application in first. His patent was challenged and he had to fight to protect it, but the patent was upheld in the court, but he subsequently received a royalty on all superphosphate sold by his rivals in England and Wales. He subsequently waived his rights on the second and third items shown in Box 3. The supply of bones soon became insufficient to meet the demand for superphosphate and was supplemented by coprolites dug from shallow beds in Cambridgeshire. Coprolites were later replaced as the main source of phosphate by apatite, which Lawes had perceptively included in his patent.

Fertilizers: 1900–2000

Between 1900 and 1980 there was a remarkable increase, literally an exponential increase, in the production of nitrogen fertilizers (Figure 2). The figure also shows that this rate of increase was not sustained during the last 20 years of the century. Fertilizer use reached a plateau in the UK in the 1980s, and the same was probably broadly true in other western European countries and North America, but production in other parts of the world, notably the far-East, continued to increase. Smil's (2001) Appendix L shows that overall world production rose 238-fold, from 360 kt of fertilizer N in 1900 to 85,700 kt in 2000.

The key to the increase in the first 80 years can be found in the development of the Haber-Bosch process for the synthesis of ammonia. Smil (2001) gives a fascinating account of the development of the process and the personalities involved, together with a wealth of useful information on fertilizer production. His Appendix L shows the percentage of N fertilizer made by the Haber-Bosch process to have increased from zero in 1900 (before the process was invented) to more

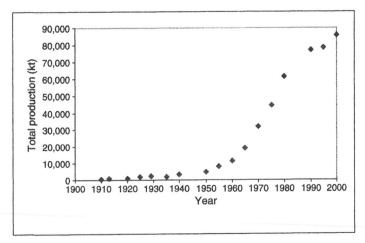

Figure 2. The exponential increase in world fertilizer production between 1900 and 1980 and the subsequent fall-off in production. Based on Appendix L of Smil (2001), p. 245.

than 99% in 2000. This latter figure not only reflects the dominance of this process in the production of nitrogen fertilizer but, viewed from a different angle, it shows how dependant the world has become on this process.

There was no explosion in the use of phosphate fertilizers in the 20th century to compare with that of nitrogen fertilizers, and the nature of phosphate fertilizers changed far less. Superphosphate continued to be manufactured and used throughout the century, but new fertilizers, based on mono- and di-ammonium phosphate came into use. The phosphate in these fertilizers was never synthesized from its elements as was ammonia. It all came from phosphate minerals.

Leaching Experiments: 1870–1980

Nitrate Leaching Experiments

Experiments at Rothamsted on losses by leaching of nitrate from the soil began in the 1870s (e.g. Lawes *et al.*, 1881, 1882; Miller, 1906). They were all concerned with the loss of nitrogen as a plant nutrient. Concern about the health and ecological implications of nitrate losses to the wider environment was not to come for nearly 100 years. Comment began to arise in about 1970 and public awareness of a 'nitrate problem' emerged in the early 1980s. The extent to which this problem is a real one is discussed below. One key event was the publication at this time by a British newspaper of an article on 'The Nitrate Time Bomb' This 'time bomb' was the downward movement of nitrate released by the ploughing-up of old permanent grassland. Two key Rothamsted leaching experiments had been under way for 100 years by then. The experiment on Broadbalk provided early information on nitrate losses from applications of fertilizer, while the Drain Gauge experiment showed how much nitrate loss could result from mineralization of soil organic matter.

The treatments on the Broadbalk experiment are outlined in Box 1. Tile drains were installed longitudinally in the centre of most of the plots in 1849 (Lawes *et al.*, 1881, 1882) and have run intermittently since then. The drains on one section of Broadbalk were replaced in the autumn of 1993. The subsoil of the field cracks sufficiently to allow free drainage and the field is not one that would necessarily require drainage to ensure trafficability. Some drainage will have occurred through the subsoil as well as through the drains, so it is not clear what proportion of each plot will have contributed to the flow through the drains. But the results are still very useful.

Goulding *et al.* (2000) recently made a very interesting comparison (Figure 3) between early nitrogen losses from the drains (1878–1883) and more recent

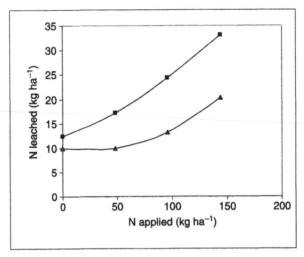

Figure 3. Comparison of losses of nitrate in drainage in 1878–1883 (squares) with that in 1990–1998 (triangles) for Broadbalk plots 5, 6, 7, and 8 (continuous wheat given 0, 48, 96, or 144 kg ha^{-1}). (From Goulding *et al.*, 2000).

ones (1990–1998). As the authors pointed out, there can be no exact comparison because of the weather differences between the two periods ('average' years rarely occur) and the difference in varieties. Goulding *et al.* were able to make some allowance for drainage differences through calculations which made allowance for the fact that the average annual drainage in 1878–1883 was 10% greater than that in the recent years. The crops made their own compensation, in that the smaller grain yields of the early years were offset by larger straw yields.

The results for 1878–1883 showed that there was a measurable loss of nitrogen by leaching even when no fertilizer was given. Subsequently, as might have been expected, the losses increased steadily with the amount of nitrogen fertilizer given. The increase was curvilinear, implying that, as more fertilizer was given, the nitrogen loss increased as a proportion of the nitrogen given as fertilizer.

The results the recent years, 1990–1998, were interesting in two respects. The overall annual losses were 43% less than in 1878–1883, and the pattern was different. The nitrogen loss from 48 kg ha^{-1} of nitrogen fertilizer was no bigger than that when no fertilizer was given at all, and the loss from 96 kg ha^{-1} was only marginally (and not significantly) greater. That from 144 kg ha^{-1} was somewhat greater but not significantly so. The changes in overall loss and the pattern of loss can be attributed to the greater grain yield, by a factor of two to three, of modern varieties of wheat and their more efficient uptake and use of nitrogen, together with better methods of crop protection.

Nitrate Leaching from the Drain Gauges

The role of the soil's microbial population in losses of nitrate became apparent when Lawes and his colleagues constructed the Drain Gauges at Rothamsted in 1870 (Lawes *et al.*, 1882). Each gauge (Figure 4) comprised a natural block of soil that was almost undisturbed during its construction. They were designed to measure the amount of rain water passing through the soil but nitrate was noticed in the drainage and from 1877 until 1915 its concentration was measured daily (Miller, 1906; Russell and Richards, 1920). These measurements, particularly those from the deepest (1.5 m) gauge, considered the most reliable, were used much later to derive information about long-term losses of nitrate from soil organic matter Addiscott (1988).

During the first seven years of these measurements (1877–1884) the soil lost an average of 45 kg ha^{-1} of nitrate-N each year, although it was not much disturbed during construction and had seven years to settle down before the measurements started. Furthermore, no crop was grown, the only form of cultivation

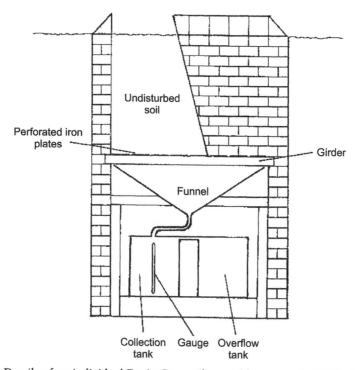

Figure 4. Details of an individual Drain Gauge (from Addiscott *et al.*, 1991). The gauges comprised natural blocks of undisturbed soil that were isolated by brick walls and undermined so that collectors for the drainage could be installed. The soil in the gauges carried no crop, received no fertilizer and was cultivated only minimally to kill weeds.

was hand-weeding and after 1868 neither fertilizer nor manure was applied. During the seven years only 3–5 kg ha^{-1} was brought in by rain each year, so practically all of the remaining 40–42 kg ha^{-1} must have come from the soil.

During the whole 38 years of the measurements about 1450 kg ha^{-1} of nitrate-nitrogen leaked from the soil in the gauge (Addiscott, 1988). This amount was almost exactly the same as the measured decrease in the soil nitrogen in the gauge, about 1410 kg ha^{-1}. Mineral nitrogen is only 1–2% of the total nitrogen in the soil, so virtually all the nitrate that leaked from the soil must have been mineralized from nitrogen in the soil organic matter. Put another way, the organic matter was the source of nearly one and a half tonnes of nitrate-nitrogen, that is, more than six tonnes of nitrate, lost per hectare during the 38 years. The leakage was a long-term process with a half-life of 41 years (Figure 5), but it leaves no doubt about the role played by mineralization in nitrate losses from the soil. Figure 5 shows the change in nitrate concentration with time.

The Drain Gauge experiment is notable historically for another reason. The results led to the first recognition of 'mobile and immobile water' in the soil. This concept has been incorporated by van Genuchten and Wierenga (1976) and Addiscott (1977) in computer models for leaching, but the principle underlying these models was enunciated almost 100 years earlier by Lawes et al. (1881). They were able to do this because they had recognised the importance of leaving the soil undisturbed when constructing the Drain Gauges.

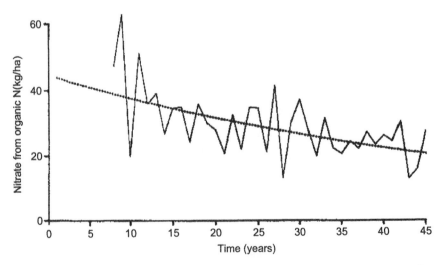

Figure 5. The leakage of nitrate derived from nitrogen in soil organic matter in the 60 inch (1.5 m) Drain Gauge at Rothamsted between 1877 and 1915. The broken line is the result of removing the effect of fluctuations in rainfall. (From Addiscott, 1988 and Addiscott et al., 1991).

Other Lysimeter Experiments

The Drain Gauges were an early form of lysimeter, but far from the earliest. Lysimetry can trace its antecedents much further back than most branches of science – all the way back to 17th century Paris, where a mathematician and meteorologist named De la Hire, employed by Louis XIV, constructed three lead-walled lysimeters in 1688. His efforts were to remain unique for more than a century until 1795–1796 when a Swiss named Maurice and John Dalton (better known as the 'father of the atomic theory') set up lysimeters independently of each other. The Drain Gauges, shown in Figure 6 as they were originally depicted by Lawes *et al.* (1882), were an advance on what had gone before in two respects, the measurement of the nitrate concentration in the drainage and the

Figure 6. The Rothamsted Drain Gauges as depicted by Lawes *et al.* (1881). Lawes *et al.* constructed three gauges, each with a surface area of 1/1000 acre (about 4m^2), in 1870 to measure the quantities of water draining through 20, 40 and 60 inches (0.5, 1.0 and 1.5 m) of soil. They also constructed a Rain Gauge with the same dimensions, which can be seen in the background.

recognition that the soil needed to be undisturbed, which had both the important theoretical implications mentioned in the previous section and some equally important practical implications.

The lysimeters before the Drain Gauges were all of the 'filled-in' type, that is, the soil was dug up and re-packed in the lysimeter. It became apparent that, however carefully this was done, the flow of water in the re-packed soil was not the same as in the undisturbed soil (Joffe, 1932). In the worst instances, rainwater was unable to penetrate the re-packed soil and remained in a pool on the surface.

One late 19th century attempt at a lysimeter with undisturbed soil deserves an honourable mention despite the fact that it never came to fruition. The Bavarian meteorologist Ebermeyer planned to undermine and collect drainage at a depth of 1.5 m from a quarter of a hectare of forest, complete with trees. His ambitious project was, sad to say, foiled by technical and financial difficulties, but his name lives on in the Ebermeyer lysimeter. This type of lysimeter involves making a trench beside the block of soil to be investigated and inserting a collecting vessel for the drainage underneath it.

The main alternative to the Ebermeyer lysimeter is the monolith lysimeter, which is constructed (Belford, 1979) by driving a metal or fibreglass pipe into the soil, partially excavating round it, cutting through the soil at the base of the pipe and lifting out the pipe and the soil it contains. Fitting a system for drainage collection to the base of the pipe completes the lysimeter. These and other recent types of lysimeter have all contributed to the assessment of nutrient losses from the soil. They were assessed critically by Addiscott et al. (1991).

One key issue in lysimetry emerged from research in the USA more than 60 years ago. The soil pores responsible for conducting drainage downwards are often reasonably continuous, so there is a 'hanging column' of water that extends to depth without a horizontal air-water interface. This 'hanging column' is broken when a lysimeter is constructed. The surface tension at the resulting air-water interface restricts drainage so that the soil above the interface has to become saturated with water before drainage can occur. This problem was identified by Richards et al. (1939) and can be overcome by applying suction to the base of the column of soil. Coleman (1946) showed that the suction controlled both the rate of drainage and the amount of water held in the soil above the air-water interface.

The influence of the air-water interface extends to nutrient concentrations in the drainage, because different pores not only drain at different suctions, they may also contain water with different nutrient concentrations. This was demonstrated in the USA by Haines et al. (1982) who compared two Ebermeyer lysimeters, both at the same depth in the soil. One had a drainage collector subjected to a 1 m hanging column of water, the other had no suction. The hanging column doubled the average flow of water from the lysimeter but decreased the nitrate concentration by a factor of three. The important thing to note is that there was a difference rather than that there was a decrease. It could have been an increase in other circumstances.

Porous Ceramic Cups

If you can estimate the amount of water moving through the soil, perhaps as the difference between rainfall and evaporation, measuring the nitrate concentration in it provides an assessment of the loss of nitrate. The porous ceramic cup affords one means of measuring this concentration. This is another form of experimentation with a century of history, mainly in the USA. The pioneer was the 'artificial root' of Briggs and McColl (1904), a porous tube through which water could be drawn under suction. The term porous ceramic cup seems to have come from Wagner (1962). The earlier users of the approach seem to have been well aware of potential problems. These included channelling of water beside the tube to the sampling unit (Wood, 1973), variability (Hansen and Harris, 1975), uncertainty about which pores were sampled (England, 1974; van der Ploeg and Beese, 1977) and distortion of flow pathways in the soil (van der Ploeg and Beese, 1977). These problems were discussed in more detail in Addiscott *et al.* (1991).

Laboratory Experiments on Leaching

The soil science literature of 40 years ago contains a number of laboratory leaching experiments made with dried and sieved soil in glass tubes (e.g. Nielsen and Biggar, 1962). Drying and sieving is a fundamental form of soil disturbance and whether these experiments bore any relation to the behaviour of solutes and water in natural field soils seems doubtful. This is particularly the case with soils that are naturally strongly structured. These experiments were useful in that they brought theory pertaining to porous media into a soils context, but whether the theory was appropriate to soils, particularly those which contain a wide range of pore sizes, is open to question. McMahon and Thomas (1974) showed major differences in leaching from dried and sieved soil in tubes and the same soil in a more natural condition.

1980–2004

Use of ^{15}N

The early experiments on losses of nitrogen fertilizer were to be augmented most effectively by experiments using the non-radioactive isotope of nitrogen ^{15}N. This research was made throughout the world and provided decisive insights into the nature of the nitrate problem (e.g. Smith *et al.*, 1984; Powlson *et al.*, 1986, 1992; Recous *et al.*, 1988). The results of Powlson *et al.* (1986, 1992), for example, were obtained with winter wheat grown at three sites: the flinty silt loam at Rothamsted, a sandy loam at Woburn in Bedfordshire and a sandy clay over a heavy clay subsoil at Saxmundham in Suffolk (Figure 7). They showed that at harvest 8–35% (average 15%) of the nitrogen applied was missing and had to be presumed lost, but 50–80% of the labelled nitrogen was in the crop and 10–25% was in the soil, almost all of which was in organic matter.

The observation that nitrogen left in the soil was nearly all in the organic matter was very important. Only about 1% of the labelled nitrogen from the fertilizer was left in the soil at harvest as nitrate. Nearly all the nitrate found was unlabelled and therefore did not come from the fertilizer. Furthermore, the amount of nitrate found in the soil bore no relation to the amount of fertilizer supplied. This seemed to refute the claim that the nitrate problem was caused by nitrogen fertilizer left unused in the soil, but we need to examine all the evidence before we can draw such a strong conclusion. It is also important to remember that these results were for winter wheat, which is recognised as parsimonious with nitrogen, at least in part because it has a well-established root system by the time the fertilizer is applied.

A substantial proportion, 8–35%, of the labelled fertilizer went missing between the spring application and the harvest. Much more tended to be missing when the application was followed by above average rainfall, and Powlson *et al.* (1992) examined the relation between the loss of ^{15}N and the rainfall during various lengths of time after application. The rainfall in the first three weeks gave the regression that accounted for the greatest percentage of the variance in ^{15}N loss

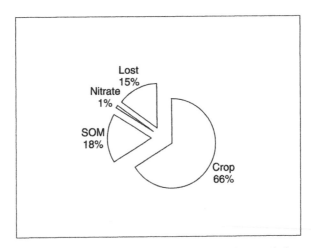

Figure 7. Fate of ^{15}N-labelled fertilizer applied to winter wheat. Likely percentage found at harvest in the crop, in soil organic matter (SOM), as nitrate in soil or lost.

(Figure 8), so they concluded that this first three weeks was the critical period. Addiscott and Powlson (1992) showed that on average one third of this loss (5% of the ^{15}N applied) occurred by leaching and the remainder by denitrification but this average conceals a wide variation between sites and seasons (Figure 9).

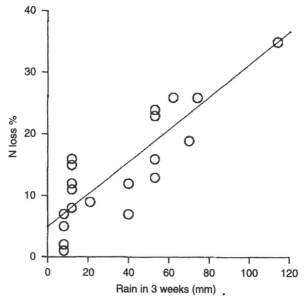

Figure 8. Relation between ^{15}N lost and rainfall in the three weeks after application in the experiments of Powlson *et al.* (1992).

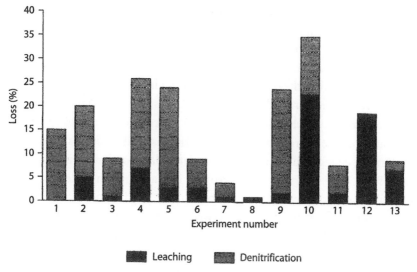

Figure 9. Partitioning of ^{15}N loss in Figures 7 and 8 between leaching and denitrification. (From Addiscott and Powlson, 1992).

Nitrate in the Environment

Nitrate was frequently blamed for algal blooms and other unwanted excess plant growth in the water environment, but we need to distinguish between its effects in freshwater and seawater, which differ considerably.

The *Cyanobacteria* that cause algal blooms in freshwater can fix nitrogen from the atmosphere. This suggests that nitrogen is unlikely to be the nutrient that limits their growth. This idea is supported by the ratios of available nitrate to available phosphate found in surface waters. Also, research reviewed by Sharpley *et al.* (1994) and Ferguson *et al.* (1996) showed that algal blooms were more sensitive in freshwater lakes to the concentration of phosphate rather than to that of nitrate. Regressions relating chlorophyll concentrations (as a proxy for the quantities of algae in the water) to the phosphate concentration were valid over five orders of magnitude of the phosphate concentration (Ferguson *et al.*, 1996). The corresponding regressions for nitrate concentration were valid over only two orders of magnitude and gave a poorer fit.

One experiment in particular was decisive in establishing that phosphate was the limiting nutrient in lakes. This was a 'whole lake' study made by an international team in Canada (Schindler, 1974) in which an oligotrophic lake with two similar basins was divided by a plastic curtain. One basin was treated with carbon and nitrogen, the other with carbon, nitrogen and phosphate. The basin that was treated with just carbon and nitrogen remained in apparently pristine condition. The other, which received phosphate as well, developed massive algal blooms.

Much of the nitrogen that gets into streams and rivers ends up in the sea, where it is a cause for concern, particularly in the estuaries and bays in which rivers meet the ocean. The large increase in the use of nitrogen fertilizer in the 20th century, together with increases in the deposition of nitrogen oxides generated by traffic and industry means that the quantity of nitrogen circulating in the global nitrogen cycle has doubled since pre-industrial times. Rivers now contribute twice as much nitrogen to marine waters as they did 100 years ago (Vitousek *et al.*, 1997). Eutrophication is therefore a problem in marine as well as in fresh-waters. It has caused startling changes in the growth of aquatic plants and algae in coastal waters (Nixon, 1995). The excessive algal growth in particular has destroyed critical aquatic habitats and lowered the concentration of dissolved oxygen, thereby killing fish and shifting the balance between marine organisms (Howarth *et al.*, 2000). This eutrophication threatens the long-term sustainability of fisheries and the use of coastal waters for recreation. Nitrate is clearly the limiting factor for these problems in coastal waters, but studies on the oceans show that no single factor can explain the complexities of all marine waters.

Nitrate and Human Health

Nitrate was widely perceived to threaten human health by causing methaemoglobinaemia in infants and stomach cancer in adults, and this perception is at the root of current legislation such as the EC directives limiting nitrate concentrations in water to 50 mg L^{-1}. But the perception and the legislation now seem to have little foundation. Medical research (e.g. Benjamin, 2000) and recent reviews of the evidence from 1950 onwards (L'hirondel and L'hirondel, 2001; Addiscott and Benjamin, 2004) have shown, not only that nitrate poses no threat to health, but that it may be beneficial.

Methaemoglobinaemia is a side effect of gastroenteritis and is not caused by nitrate but by nitric oxide, which is produced in a defensive reaction stimulated by gastroenteritis. The latter may be caused by a bacterium or a virus. The association of methaemoglobinaemia with nitrate may have arisen because early cases of the condition were associated with wells polluted with bacteria and the same pollution increased the nitrate concentration. Comly (1945) reported that, 'In every one of the instances in which cyanosis (the clinical symptom of methaemoglobinaemia) developed in infants the wells were situated near barnyards and pit privies.' In the absence of bacteria there is no link.

Several carefully conducted epidemiological studies (Forman *et al.*, 1985; Beresford, 1985; Al-Dabbagh *et al.*, 1986; van Loon, 1998) sought a link between stomach cancer and nitrate but did not find one. The incidence of this cancer has also declined during the last 30 years while nitrate concentrations in water have increased.

Nitrate does not threaten health, it preserves it. It is reduced by microbes on the tongue to nitrite, and when this nitrite is acidified it generates nitric oxide in an antibacterial defence mechanism vital to our well being. This mechanism acts with great effectiveness in the stomach against *Salmonella*, *E. coli* and other organisms that cause gastroenteritis. It also acts in our mouths against dental caries and even on our skin against fungal pathogens such as *Tinea pedis* (Athlete's foot). This mechanism is the basis of the centuries-old practice of adding nitrate or nitrite to stored meat to protect against the *Botulinum* organism, which produces the most lethal toxin known to mankind.

Issues That Emerged in the Late 20th Century

Research on all the main issues discussed so far began in the 19th century. There are, however, several issues that emerged only in the late 20th century but are an important part of the relationship between soils and the environment. These include denitrification, and especially emissions of nitrous oxide, deposition of nitrogen oxides and ammonia from the atmosphere, the emission and sequestration of carbon dioxide (in the context of concerns about global warming) and the contamination of the soil with heavy metals. One final issue also needs to be mentioned, phosphate leaching. Nitrate was known to be leached from about 1870 onwards, but the fact that phosphate was also leached did not emerge until the last decade of the 20th century.

Researchers active during the 19th century were completely unaware of denitrification, as were those working throughout much of the first 50 years of the 20th century. In 1955, the American soil scientist F.E. Allison published a noted review paper entitled 'The enigma of soil nitrogen balance sheets' (Allison, 1955). The enigma was the failure of so many experiments to balance inputs and outputs of nitrogen to and from the soil. All these experiments showed losses which could not be accounted for by leaching alone. His paper drew attention to denitrification and other microbial processes that release nitrogen gases from the soil. As with leaching, the concern was initially with the loss of a resource, soil mineral nitrogen, with time the concern transferred to the effects of the process within the environment.

Denitrification matters only as the loss of a resource if it is complete, that is, if the reduction of nitrate proceeds all the way to nitrogen gas (N_2). Incomplete denitrification produces various oxides of nitrogen, of which the most troublesome is nitrous oxide. The gas is also emitted from the soil during nitrification, i.e. the oxidation of ammonium to nitrate (Bremner, 1997). Both arable and grassland agriculture contribute to its production, as does the manufacture of the adipic acid used in the production of Nylon.

Nitrous Oxide to the Atmosphere

The release of nitrous oxide to the atmosphere is an example of a global pollution problem. Nitrate or phosphate lost to surface waters may pollute a lake, a river system or even a large area of coastal water such as the Gulf of Mexico, but nitrous oxide affects the atmosphere of the whole planet. The global nature of the problem has caused widespread concern.

Nitrous oxide contributes to two global atmospheric problems. One is the 'greenhouse effect' which is partly responsible for global warming at the earth's surface. Nitrous oxide is not the main greenhouse gas, but it is one of the most effective ones in terms of its retention of heat per kg of gas. The other atmospheric problem is the destruction of ozone in the stratosphere. It is nitric oxide rather than nitrous oxide that is responsible for the destruction, but the unreactive nitrous oxide can travel through the lower atmosphere to the stratosphere where the highly reactive nitric oxide free radical cannot exist without reacting with, and being destroyed by, other free radicals. The nitrous oxide could be said to act as a 'Trojan Horse' for the nitric oxide. In the stratosphere, the nitrous oxide is decomposed to nitric oxide by a singlet oxygen atom and the nitric oxide reacts with ozone, converting it to ordinary oxygen and thereby contributing to the destruction of the ozone layer (Granli and Bøckman, 1994).

Estimating losses of nitrous oxide from the soil to the atmosphere is a problem in both arable and grassland agriculture but for different reasons. Problems from arable tend to be spatial ones such as discontinuity (Dobbie and Smith, 2003) and spatial variation Lark et al. (2004a,b), but grassland has not only these problems but also those arising from the individual and social behaviour of cows and other farm animals (Jarvis et al., 1991; Yamulki et al., 1998) and those arising from the variety of inputs of nitrogen to grassland systems (Table 2).

Table 2. Nitrogen inputs to and outputs and losses of nitrogen from the 'model' dairy farm of Jarvis (1993)

Source	Total for farm (kg)	Kg ha^{-1}
Nitrogen inputs		
Atmosphere	1990	25
Fertilizers	19000	250
Biological fixation	760	10
Feeds	3746	52
Straw	182	
Total inputs	25588	337
Nitrogen outputs		
Milk	2940	
Protein	2160	
Total outputs	5100	67
Outputs as % of inputs	20	20
Nitrogen losses		
Leaching	4272	56
Ammonia volatilization	3483	46
Denitrification (As N_2O)	4197 (1049)	55 (14)
Total losses	11952	157
Balance not accounted for	8536	113
Balance as % of inputs	33	33

Deposition of Nitrogenous Gases on Land

The transfer of matter between the soil and the atmosphere is not a one way traffic. Up to now we have been concerned with the soil as a source of pollution either to water or to the atmosphere, but here we have the first instance of the soil as a recipient of pollution. Significant amounts of nitrogen, but not phosphate, are deposited from the atmosphere on to the soil. It comes from industry and motor traffic, each of which generates about half the nitrogen oxides, and from the farm and neighbouring farms which are mainly responsible for the ammonia.

Although this topic did not become an issue until the last two decades of the 20th century, research on it had begun a hundred years before. Rainfall at Rothamsted was first analysed for nitrate in 1877 but this was to supplement measurements of nitrate loss from the Grain Gauges so that the amount coming from the soil was known. Not much was found. Between 1877 and 1915 the total deposition of nitrogen recorded was 227 kg ha^{-1}, about 6 kg ha^{-1} per year on average, but these measurements did not include all forms of deposition.

More recent studies, such as those of Goulding (1990) at four sites in south-east England included nitrate deposited on dry matter and dry deposition of nitrogen oxides, nitric acid and ammonia. These more comprehensive estimates showed an annual deposition of 35–40 kg ha^{-1}. More recent measurements at Rothamsted (United Kingdom Review Group on Impacts of Atmospheric Nitrogen, 1994) suggested that about 37 kg ha^{-1} was deposited annually on bare soil and about 48 kg ha^{-1} on soil with a well-established crop of winter wheat. The difference is due to the larger surface area for deposition offered by the crop.

The nitrogen deposited when the wheat was grown corresponded to one quarter of the average application of nitrogen fertilizer at the time of the measurement and could be more than a quarter now. One estimate from a model (IACR, 1992) suggested that about half the nitrogen deposited was taken up by the crop and about 30% was leached. Some currently unpublished runs of another model suggested that a fairly similar proportion, 25%, was leached and that it contributed about 15% of the nitrate leached overall.

This calculation implies that nitrate deposited from the atmosphere may contribute 10–15 kg ha^{-1} to annual nitrate losses from the soil. Farmers can be encouraged to view this deposition as a resource and to make allowance for it, but they can hardly be held responsible for its consequences. Assuming that the through drainage is about 250 mm, as in many parts of eastern England, its contribution to the nitrate concentration leaving the soil is about 20 g m^{-3}, or 40% of the EC nitrate limit. This leaves farmers little room for error.

Nitrate or ammonium deposited on farmland need not be a problem, as we saw in the previous paragraph. Deposition is definitely a problem for natural vegetation.

Its effects (United Kingdom Review Group on Impacts of Atmospheric Nitrogen, 1994) can include loss of mosses and lichens and of heather, as well as damage to trees from nitrate and ammonium in fog and cloud, changes in species composition and loss of biodiversity and decreases in tolerance by plants of stresses such as frost, drought and feeding by animals and insects. The soil may undergo changes in soil solution chemistry, including acidification and loss of cations. These are problems that emerged in the last two decades of the 20th century. Two concepts, critical load and nitrogen saturation, are important for understanding the problems caused by deposition of nitrate, ammonia and nitrogen oxides from the atmosphere. A *critical load* is defined as 'an exposure below which harmful effects do not occur'. The critical load for nitrogen may relate either to its acidifying effect or to its nutrient effect. The assessment can be made by calculating the critical load for each of these and using the lower value.

Giving too much nitrogen fertilizer to a crop not only causes problems such as lodging (when the stem becomes too weak to support the weight of the crop) but also results in nitrate losses. Similarly, if more nitrogen is deposited on an ecosystem than it can use in its current state, the excess will cause some kind of change, which may well be deleterious, and nitrate will be lost from the system. Various measures of nitrogen saturation have been suggested (e.g. Grennfelt and Thornelof, 1992), but nitrate loss may be the most reliable. Perakis and Hedin (2002) provide excellent examples of ecosystems unsaturated and saturated with nitrogen. They showed that substantial amounts of dissolved organic nitrogen, but no nitrate, were lost in stream water flowing from pristine forests in the Andes mountains in Chile, whereas appreciable concentrations of nitrate were lost in streams flowing from forests in the vicinity of New York that were subject to heavy nitrogen deposition.

Deposition of phosphate from the atmosphere is not a problem, except perhaps in the vicinity of older superphosphate factories, and even there the main problem springs from the acidity rather than the phosphate content of the particles deposited.

Heavy Metals

Those elements whose total concentration in all forms in the soil is less than about 1 g kg^{-1} have long been described as trace elements (e.g. White, 1997). For much of the history of soil science it was the deficiency of some of these element that were plant nutrients that was a source of concern, but in the last two decades some of them have come to the fore as a toxicity problem and therefore as a pollution problem. They then became 'heavy metals', but this term is slightly misleading because, in terms of their atomic weights, many of them are not particularly heavy. Three categories of trace element can be identified (White, 1997).

1. *Trace elements essential to plants.* The elements Cu, Zn, Mn, B and Mo are essential for the growth of plants at normal concentrations *in the plant* but can become toxic at larger concentrations.
2. *Trace elements essential to animals.* The nutrients Cr, Se, I and Co, though not essential to plants, are essential for the normal growth and function of animals.
3. *Non-beneficial trace elements.* The elements Li, Be, As, Hg, Cd, Pb and Ni have no useful function in either plants or animals and become toxic at concentrations greater than a few mg kg^{-1}.

The ability to spot and identify trace element deficiency symptoms in plants has long been an essential skill for soil scientists who provide advice to farmers, and they are assisted by a variety of books and sets of slides showing examples of particular symptoms, some of them induced by the application of other nutrients. The topic is made more complicated by the fact that plant species differ appreciably in their susceptibility to nutrient deficiencies. Veterinary surgeons have needed to develop comparable skills for detecting deficiency symptoms in animals. But these traditional aspects of trace element science have at least partly been superceded by concerns about heavy metals.

Heavy Metals from Sewage Sludge

The toxicity of elements such as As, Cd, Hg and Pb to humans and animals has been known for some considerable time. These elements may be found in the soil because of natural, geological occurrences or from mining, sometimes in earlier centuries. Lead mining in the UK dates back to the Roman era. Not all plants are susceptible to these elements and some may accumulate concentrations that are dangerous to animals that eat them.

The discovery that made heavy metals an environmental issue did so once again because the soil itself was the recipient, or perhaps the victim, of pollution. Neither plants nor animals were involved. The discovery was entirely by chance. Rothamsted has a sub-station at Woburn in Bedfordshire, at which the Market Garden experiment was established in 1942, largely because of the Second World War. The aim (Mann, 1957) was 'to use it (the very sandy soil chosen) to judge the value of various organic manures, plus a basic dressing of phosphates and potash, with and without a further addition of soluble nitrogenous manures, in converting the area into a market garden soil.' Johnston and Wedderburn (1975) added that, 'In the early years of World War II there was considerable interest in the possibility of making this soil, which was giving a very poor yield of agricultural crops, produce large yields of market garden crops. In addition, the decision to start the experiment must have been influenced by the fact that Woburn is on the edge of the South Bedfordshire area of light land which has a tradition of growing market garden crops on soils for long manured with bulky organic manures.'

The Woburn Market Garden Experiment tested four bulky organic manures:

- Farmyard manure (FYM)
- Sewage sludge
- A compost made from vegetable waste and FYM
- A compost made from sewage sludge and straw.

The FYM was to be the standard against which the other manures were to be judged, and sewage sludge was included simply because of the interest in its use in agriculture. It was composted with straw in an effort to improve its structure. During the years from 1942 to 1949 the Chemistry Department at Rothamsted made more than one hundred experiments testing the manurial value of sewage sludge to agricultural and horticultural crops (Bunting, 1963). The Woburn experiment was not one of this series, and the test of sewage sludge lasted longer and more sludge was applied to the soil than in any other of the experiments (Johnston and Wedderburn, 1975).

The sludge used in the experiment came from a sewage works in Middlesex, and was assumed at first to be of domestic origin. But it was found later that substantial amounts of Zn had accumulated in the soils to which it was applied. This Zn could have come from the galvanized iron used is domestic water systems, or it could have implied that more of the sewage came from industry than had been realized. It was to have important consequences subsequently.

Casual conversations between scientists over coffee or in corridors have probably had more impact on progress in science than all the administrators of science put together. Perhaps only unexpected results have been more influential. My Rothamsted colleague A.E. Johnston, cited above, who held responsibility for the Woburn Market Garden experiment for many years, had noted that the plots receiving sewage sludge had a smaller proportion of the Zn in the soil in a plant-available form than the plots without sludge. He formed the hypothesis that this was because the plots with sludge contained more microbial biomass in their soil than those without it and he put forward this idea when chatting to another colleague, P.C. Brookes, an expert on microbial biomass. Brookes did a few preliminary experiments which suggested that in fact the sludge plots contained less microbial biomass than the others (P.C. Brookes, Rothamsted, 2004, personal communication).

Johnston's hypothesis had been wrong but the question he asked turned out to be immensely important. Further experiments on microbial biomass (Brookes and

McGrath, 1984) showed unequivocally that the plots with sewage sludge contained less microbial biomass than those without it and also less adenosine tri-phosphate (ATP), an independent measure of microbial biomass (Table 3). The toxic effect of the metals seemed to make the microbial biomass in the sludge-affected soils respire at a much higher rate than that in the soils given FYM (Table 3).

These results had important implications. About half a million tonnes of sewage sludge was being applied each year to soils in the UK alone and the practice was an important route for disposing of sludge. Indeed, some aspects of the results are still contested by an organization representing the water industry (P.C. Brookes, Rothamsted, 2004, personal communication). There had been some short-term investigations on the effects of metals in sludge on the soil biomass (Eiland, 1981; Zibilske and Wagner, 1982), but these gave inconclusive results. The study on the Woburn soils was the first to examine the effects of long-term applications of sewage sludge.

Perhaps the most worrying long-term effect was that the decrease in microbial biomass appeared more than 20 years after the last applications of sewage sludge. With hindsight, the continuing impact of these metals is perhaps not surprising, given that they are all cations that will be held by negatively-charged soil surfaces and will not readily be leached from the soil. They can be moved laterally when the soil is ploughed, particularly if the ploughing is always in the same direction, and McGrath and Lane (1989) showed that this lateral movement could be simulated by a two-dimensional dispersion model. When they allowed for this movement, they found that about 80% of the metals applied between 1942 and 1961 were still to be found in the topsoil.

Table 3. Effect of metals in sewage sludge on the amount, ATP concentration of and respiration by microbial biomass in soil from the Woburn plots. Adapted from Tables 1 and 3 of Brookes and McGrath (1984)

Treatment	Biomass in soil (μg g^{-1})	Concentration of ATP in soil (nmol g^{-1})	Respiration ATP in biomass by biomass (μmolg^{-1})mg CO_2 g^{-1}
Mean for plots given inorganic fertilizer or FYM[1]	409.0	2.62	6.36181
SE of mean	11.6	0.12	0.3325
Mean for plots given sludge or sludge-compost	185.0	1.34	7.18294
SE of mean	11.3	0.15	0.4136

[1] Farmyard manure.

Table 4. Effect of metals in sewage sludge on the growth of blue-green algae on Petri dishes, their nitrogen fixation and their acetylene-reduction potential. Adapted from Figure 1 and Tables 2 and 3 of Brookes *et al.* (1986)

Treatment	Development of algal colonies	Increase in soil N (mg m^{-2})	Rate of N$_2$ fixation[1] (μmol^{-2} d^{-1})	C$_2$H$_2$ peduction potential (μmol^{-2} d^{-1})
FYM	Complete cover	601	1063 ± 45.8	5136 ± 1110
Sludge	Few colonies	452	102 ± 58.8	0

[1] Assessed using ^{15}N.

This was a very important result in the context of sewage sludge application. The dispersion caused by ploughing had led to an apparent decrease in the concentration of the metals that had suggested that given time the metal contamination would be ameliorated. Correcting for the dispersion suggested that this amelioration was illusory. The 80% recovery was much higher than would otherwise have been assumed and the 20% unaccounted for had not necessarily been disposed of safely. McGrath and Lane (1989) concluded that this 20% was likely to have constituted small losses by leaching and erosion together with errors in sampling and analysis. Their conclusion is supported by the observations that only about 1% of the metals had moved below the plough layer, and only by 3 to 5 cm, and that none were to be found in samples from greater depths down to 46 cm.

Microbial biomass was only part of the story. Other experiments (Brookes *et al.*, 1986) showed the growth and nitrogen fixation by blue-green algae to be massively impaired by the heavy metals in the Market Garden soil (Table 4). The capacity of these algae to (chemically) reduce acetylene is a measure of their capacity to fix nitrogen from the atmosphere, and the rate of reduction in soils that had had sludge showed both a much longer lag before reduction began and a considerably decreased peak rate of reduction. The overall result was that the increase in soil nitrogen and was appreciably smaller in sludge-affected soils than in other soils (Table 4).

Further experiments (McGrath *et al.*, 1988) showed serious effects on nitrogen fixation by bacteria in nodules on the roots of clover. The clover itself was not directly affected by the metals from the sludge and neither was its ability to form nodules, provided nitrogen was supplied. It was the bacteria within the nodules that showed the toxic effect (Table 5). Bacteria from nodules in the healthy FYM soils were dark red or pink because of the haemoglobin they contained and also large

Table 5. Effect of metals in sewage sludge on the size, colour and number of nodules on the roots of clover plants. Adapted from Table 4 of McGrath *et al.* (1988)

Treatment	Size[1]	Nodule formation: Scores for	
		Colour[2]	Number[3]
FYM	2.5	2.9	1.0
FYM + N	2.4	2.9	1.5
Sludge	1.2	1.1	2.0
Sludge + N	1.5	1.5	3.0

[1] Scoring for size   <1 mm – 1.   1–2 mm – 2.   2–6mm – 3.
[2] Scoring for colour   White – 1.   Pink – 2.   Dark red – 3.
[3] Scoring for number   <200 per pot – 1.   200–1000 – 2.   >1000 – 3.
Each score was the mean of ten separate assessments.

and relatively few in number. Those in nodules from the soils that had received sludge were small, numerous and white and, most seriously, greatly impaired in their ability to fix nitrogen.

All the processes that were impaired in the soils that had received sewage sludge are fundamental to the function of the biosphere, but what was perhaps unexpected was that bacteria, rather than plants or animals, were affected.

With most forms of pollution, it is important to establish trends over a long period of time so that deviations from them can be identified readily. The Rothamsted long-term experiments have proved invaluable for this purpose. Lawes and Gilbert had the foresight to sample the soils and crops of the Broadbalk experiment, mentioned above, in 1865 and then at intervals subsequently. This practice was also applied to other experiments at Rothamsted, notably the Park Grass experiment, and was continued by their successors. Archived soil, grain and herbage samples from Broadbalk and Park Grass were used by Jones *et al.* (1994) to establish trends in the build-up of Cd from superphosphate applied to the experiments and in the deposition of Cd from the atmosphere on herbage.

Metals are not the only form of contamination. There is also concern about organic contaminants such as polychlorinated biphenyls (PCBs) and polyaromatic hydrocarbons (PAHs) in the atmosphere, and plant foliage is a useful indicator of ambient concentrations of these contaminants. Herbage samples from the Rothamsted Archive were used to by Jones *et al.* (1994) to show that concentrations of PCBs of lower molecular weight declined by as much as a factor of 50 in the 20 years after 1965–1969. PCBs of higher molecular weight and PAHs also declined but not by so much.

Reclamation and Monitoring of Contaminated Soils

Sewage disposal is only one of a number of processes by which soils become contaminated by metals. The influence of mining and other industrial and even natural processes was noted above, and agriculture is not blameless because superphosphate is contaminated with cadmium, which remains in the soil. The removal of these contaminants now constitute a specialist branch of Soil Science, soil clean-up, but one that is largely beyond the scope of this historical review. The techniques involved include the leaching of the soil with EDTA and the fixation of the metals by the application of bauxite residues (Lombi *et al.*, 2002). The approaches to the problem are expensive, and this has led to the interest in the use of hyper-accumulator plants.

Most plants tend to exclude heavy metals from their shoots, but there are a few species, mostly those that have evolved in naturally metal-rich environments, that not only take up substantial concentrations of these metals but also translocate them into their shoots. Hyper-accumulators for Ni, tend to be found on soils derived from Serpentine, while those for Cu often come from the copper-exporting region of the Democratic Republic of the Congo. The criterion for hyper-accumulation of a metal by a plant is the ability to absorb and retain $1000~\mu g~g^{-1}$ in the dry matter of the shoot. These hyper-accumulators are not only for cleaning up soil but as a means of extracting valuable metals from the soil. McGrath *et al.* (2002) provide a comprehensive review of the topic while McGrath and Zhao (2003) offer a useful summary.

Phosphate Leaching

Textbooks published before 1995 will probably tell you that phosphate is not leached from the soil because it is held too strongly held by soil particles. Phosphate losses in surface runoff had been studied by, for example Sharpley *et al.* (1985a,b), who were able to establish that, when rainfall caused surface runoff, it did so by interacting with a thin layer of surface soil (10–25 mm) before moving laterally. But there was an assumption that these losses, rather than those in water passing through the soil were the route of loss. Phosphate leaching is therefore a subject with a short history.

Although phosphate is strongly held by soil particles, it is not immobile. The first demonstration that there was appreciable leaching of phosphate dissolved in water draining from the soil was made on Broadbalk (Heckrath *et al.*, 1995). Their results (Figure 10) showed that leaching of phosphate was negligible from plots containing less than a certain amount of Olsen-P (phosphate extractable with the

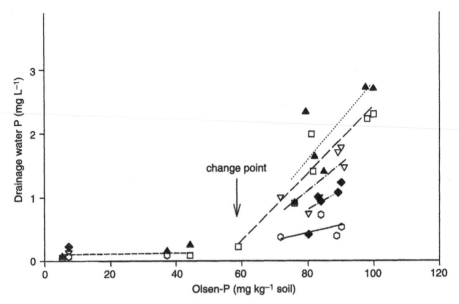

Figure 10. The relation between the phosphate concentration in drainage from the Broadbalk plots and the phosphate extractable from the soil with the reagent of Olsen *et al.* (1954), showing the 'change point'.

reagent of Olsen *et al.* (1954)) but beyond this point the concentration of phosphate in the drainage increased in proportion to the Olsen-P. The authors described this point as the 'change-point'.

Evidence that the particles to which the phosphate is bound can become detached from the soil matrix and lost in the drainage (Heckrath *et al.*, 1995; Ulèn, 1995) emerged at the same time as that for the change point. Heavy, clay-rich soils usually contain cracks and other large pores through which water can move rapidly in some circumstances, a process described as 'channelling' or 'by-pass' flow (Bouma and Dekker, 1978; Beven, 1981). The pores through which this occurs are often characterized as 'preferential pathways'. Rapid water movement through these pores can play a critical role in nitrate leaching, and they play an even more crucial part in phosphate losses from heavy soils. Large pores have a smaller surface area to volume ratio than smaller ones, lessening their capacity to sorb phosphate, and the faster flow in them shortens the time during which phosphate can become sorbed. Phosphate carried on material detached from the soil matrix, can move only through the larger pores, which further enhances their overall importance in phosphate losses.

Field drainage in heavy clay soils aims to create preferential pathways to remove water from the upper part of the soil. This improves trafficability and decreases the risk of damage to crops through water-logging, but it also increases the risk of phosphate loss and contamination of natural waters. The field drainage used on many farms comprises 'mole channels' connecting with gravel filled trenches at the bottom of which are pipe drains. The 'mole' drawn through the soil to make the mole-channels is attached to a frame by a bar or 'leg' which cuts a 'leg-slot' through the soil. This leg-slot gradually closes up but before it does so, it provides a very effective pathway from the soil surface to the mole-channel and thence to the pipe-drain and the ditch or stream. Hallard and Armstrong (1992), for example, found that a fluorescent dye applied to the surface appeared in the water in a mole channel less than an hour after rainfall started. Laubel *et al.* (1999) observed a peak in the concentration of a chloride tracer, also after less than one hour, when a tile-drained plot was irrigated, and they concluded from other data that particulate matter carrying phosphate could be transported from the topsoil into the drains within hours. There is therefore good reason for concern about phosphate losses from heavy soils with field-drains.

Addiscott *et al.* (2000) examined the nature and extent of phosphate losses at Brimstone Farm (Box 4) a site in Southern England with 0.24 ha plots of heavy,

Box 4. The Brimstone experiment

'The Brimstone experiment' is the name usually used for an important experiment at Brimstone Farm near Faringdon in Oxfordshire, UK. It is sited on a heavy clay soil of the Denchworth series (Jarvis, 1973) and has a subsoil that is impermeable to water at depth. The soil is strongly fissured, especially in late summer and early autumn, and provides pathways for preferential flow during much of the year. The 20 plots, established in 1978 (Cannell *et al.*, 1984; Harris *et al.*, 1984), are each 0.24 ha in area. They are isolated hydrologically from each other by polythene barriers inserted to a depth of 1.1 m parallel to the 2% slope and by interceptor drains at a depth of 1 m in a gravel-filled trench along their upslope margins (Figure 11). Eighteen of the plots are now drained by mole channels 0.55 m below the surface which intersect with gravel-filled trenches along the downslope margins of the plots; these have collector drains at a depth of 0.9 m. Surface runoff and downslope flow at the base of the plough layer are collected together in a deep furrow across the downslope margin of each plot to give what is described as 'cultivated layer flow'. This is usually very small compared with the flow through the drains, and is often not reported. The other two plots remain undrained.

The flow from the collector drains and the cultivated layer flow are measured by V-notch weirs and sampled on a flow-proportional basis for analysis for nitrate and phosphate and other nutrients of interest.

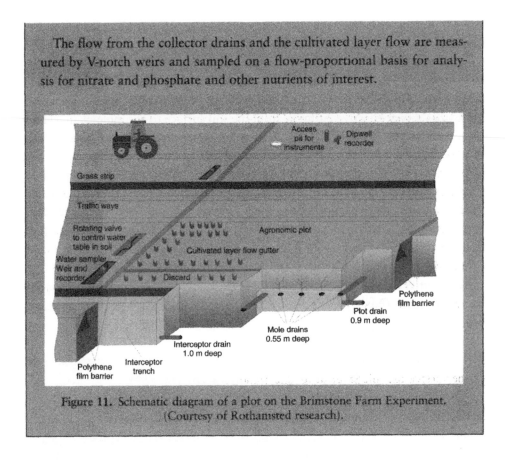

Figure 11. Schematic diagram of a plot on the Brimstone Farm Experiment. (Courtesy of Rothamsted research).

strongly-cracking clay soil with the type of interconnecting mole and pipe drains described above. Losses of total and molybdate-reactive phosphate were measured during four years in water sampled from the pipe drains. The molybdate-unreactive phosphate obtained by difference appeared from the results to be mainly phosphate carried on suspended soil material.

Losses of molybdate-reactive, molybdate unreactive- and total phosphate were all related well to the cumulative drain-flow, with the same relations covering three of the four years, designated the synoptic years. Losses were much greater in the other year, probably because phosphate was applied after, rather than before, ploughing and to very wet soil. The ceiling to annual total phosphate loss, assuming 500 mm drainage, was 0.4 kg ha^{-1} in the synoptic years, with 76% of the loss occurring as molybdate-unreactive phosphate. It was about 1 kg ha^{-1} in the high-loss year, with 88% as molybdate-unreactive phosphate, implying that this category of phosphate, probably that carried on suspended soil material, comprised 96% of the extra loss that year.

Halving the phosphate application lessened the loss of total phosphate but not conclusively, and the decrease in molybdate-reactive phosphate loss was not significant.

Restricting and thereby delaying drainage lessened losses of molybdate-unreactive phosphate, probably because suspended material carrying it was allowed to settle. Increasing the spacing between mole channels from 2 m to 4 m increased the losses of molybdate-unreactive but not molybdate-reactive phosphate (Table 6). This happened for phosphate applications of 16.5 and 33 kg P ha^{-1} and in both the synoptic years and the high-loss year, in which it was particularly marked. It was suggested to have happened because, with the wider spacing, rainwater had to travel further horizontally across the surface to meet a flow pathway connected to a mole channel and collected more phosphate-carrying soil material as it did so. This suggests an erosional process, similar to that described earlier by Sharpley et al. (1985a,b), but resulting in losses of phosphate through channels in the soil matrix rather than across its surface.

These losses involving detached soil materials were from arable soils, and it is useful to compare them with those from a heavy clay soil with field drains long managed as grassland in a rather wetter part of the UK (Haygarth et al., 1998). These authors also reported appreciable losses of phosphate, at concentrations comparable with those found at Brimstone Farm, but 50% or more of the loss they found was as dissolved molybdate-reactive phosphate. This difference is readily explicable as resulting from the cultivation of the arable soil, which makes soil material more vulnerable to detachment, and the stabilizing effect on soil structure of the surface organic layer in the grassland soil.

Table 6. The effects of mole-channel spacing and phosphate application on the flow-weighted average concentrations of molybdate-unreactive phosphate P (believed to be phosphate held on suspended material) in the synoptic years and the high-loss year. Adapted from Table 3 of Addiscott et al., 2000)

Application of phosphate P	Mole channel spacing		
(kg ha^{-1})	2 m	4 m	SED
	Concentration of molybdate-unreactive phosphate P (mg L^{-1})		
Synoptic years			
16.5	0.051	0.067	0.008
33	0.068	0.083	0.008
SED	0.008	0.008	
High-loss year			
16.5	0.12	0.25	0.03
33	0.22	0.35	0.04
SED	0.04	0.04	

The part played by of detached soil material in phosphate loss is not surprising in heavy clay soils such as those at Brimstone Farm which have cracks through which 'by-pass' flow can occur, but Hesketh *et al.* (2001) showed in lysimeters that this detached material seemed to have a role in losses of phosphate from a sandy soil with only about 10% clay. The first peak in the concentration of phosphate, like the first peaks of the pesticides triallate and chlordane, always appeared before the first peak in the concentration of bromide, which is negatively charged, very soluble in water, and most unlikely to be carried on particulate material. This strongly suggests that the phosphate was carried on colloidal or detached soil material even in this apparently structureless sandy soil with about 10% clay. There is clearly still much to be learnt about phosphate leaching.

Soils and Carbon Sequestration

At the beginning of the 21st century, the most important environmental problem is generally seen to be climate change. Soils are part of this problem as we saw in an earlier section, but they are also part of the solution. Most of us have been familiar with photosynthesis since our school-days as the mechanism by which plants trap carbon dioxide from the air and use it to form the carbohydrates which are vital building blocks for the growth of plants and algae. We and the rest of the animal kingdom depend on plants directly or indirectly for food, making photosynthesis the key to life. Like many things discussed in this chapter, photosynthesis has taken on a new significance with changing events. Awareness of climate change has made the process the key, not only to life but to the sequestration of carbon dioxide from the atmosphere, and thence to a hoped-for mitigation of climate change (Paustian *et al.*, 1997, 1998; Smith *et al.*, 1997, 2000a).

Paustian *et al.* (1997) warned that sequestration in soils alone cannot provide the answer to global warming. They estimated that such sequestration would correspond to 3 to 6% of the carbon currently emitted from fossil fuels, but pointed out that it could contribute usefully to the raft of measures proposed. Smith (2004), addressing a slightly different question, estimated that sequestration could meet at most about one-third of the current yearly increase in atmospheric carbon dioxide. Carbon dioxide is not the only 'greenhouse gas' and control measures need to keep trace gases in mind too (Smith, 1997; Smith *et al.*, 2000b).

Carbon sequestration is effected as plants grow, absorb carbon by photosynthesis and then die. Their remains return to the soil but do not remain recognisable for long because they are soon 'processed' by the soil population of microbes, earthworms and other soil animals who mineralize some of the carbon to carbon dioxide, releasing it back to the atmosphere as they do so. The more reactive carbon is processed first and, as the processing goes on, the carbon that remains includes an increasing proportion held in inert organic material. This accumulates as humus, the remains of plants, mostly long dead, that grew in the soil. The most effective form of sequestration is to immobilize carbon dioxide in this inert material. Very large amounts of carbon are already stored in this way. Soils contain more than half the carbon in all the earth's terrestrial ecosystems (Paustian, 2004). About 1500 billion tonnes of carbon are held in the inert organic forms described above, mainly in the top 25 cm of soil, and a further 900 billions are in inorganic carbonates. (Only the carbonates in the top metre of soil are counted. There are obviously vast amounts in the chalk and limestone formations which some soils overlay.)

The soil population is a mixed blessing in carbon sequestration. Its natural role is to recycle plant nutrients so that nitrogen, phosphorus and sulphur entombed in dead plant material are released as nitrate, phosphate and sulphate for use by the next generation of plants. The same is broadly true of cations such as potassium, calcium and magnesium - and of carbon in the long term. The carbon in dead material needs to be released to the atmosphere if there is not to be a gradual run-down in the supply of carbon dioxide for photosynthesis. Mineralization is an ecological imperative in normal circumstances but a challenge to those wishing to effect carbon sequestration. Its implications are that sequestration in plant remains is not the end of the story, and the process is only complete when the carbon is safely fixed in inert organic matter in the soil. Even this inert matter can be mineralized, but too slowly to be a consideration in carbon sequestration.

Optimizing the balance between sequestration and mineralization provides an example of the application of entropy theory in soil science (e.g. Addiscott, 1995). One of the main indicators of entropy production in soils is the breakdown of large, highly organized molecules with the production of small molecules such as nitrate and carbon dioxide. Entropy production is minimized when a system is allowed to settle into a steady state. It increases when the steady state is perturbed, and the system acts counter the perturbation and return to the steady state.

The system with which we are concerned is the soil-plant system. The increase in atmospheric carbon dioxide concentration that occurred during the last 100 years was a perturbation to it, albeit a very gradual perturbation, one that the system was able to accommodate and counter by taking up carbon dioxide. The steady state concentration of carbon would have increased in both soil and plant, but the steady state would have persisted. Other perturbations are not gradual and result in sharp increases in entropy production. Cultivation is a perturbation that occurs annually in many soil-plant systems, and the increase in entropy is reflected in the stimulation of mineralization and the release of carbon dioxide and nitrate. Avoiding ploughing through the use of no-till practices obviously avoids these losses.

A system that is cultivated initially and then re-cultivated in the same way every year eventually re-settles into a new steady state characterized by small, short-term perturbations. The large increase in entropy and the accompanying release of carbon dioxide occurs with the initial cultivation. In the past, this initial cultivation was often the ploughing up of old grassland, which led to substantial losses to the atmosphere of carbon as carbon dioxide and of nitrogen as nitrate in drainage from the soil (Whitmore et $al.$, 1992). If this ploughing had been preceded by forest clearance the losses would have been much greater.

The steady-state vegetation of much of Europe is thought to have been deciduous woodland and there have been two main perturbations. The first was tree clearance, which brought about a timber famine in England by the 17th century (Hoskins, 1977). Later came ploughing of the grassland which prevailed after the tree clearance because of grazing by stock.

Experiments started at two historic sites at Rothamsted about 120 years ago have proved very important in research on carbon sequestration. These sites, which had long been in arable cultivation, were left uncultivated from the 1880s to the present day and the woodland allowed to regenerate. They illustrate most effectively the return towards the steady state with the removal of the perturbation and the accompanying decline in entropy production. In 1957, Thurston (1958) wrote of the more acid of the two sites, the Geesecroft wilderness: 'The area has reverted to woodland, consisting chiefly of elm, ash and oak. The largest tree is an oak 81 inches (2.06 m) in diameter at 4 feet (1.22 m) from the ground, growing near the middle of the area that was cultivated. Of the 46 species of angiosperms present in 1957, thirty-two had been recorded previously and 14, including eight woodland species, had come in since 1913. In the same period, 55 species, all characteristic of grassland, had disappeared. All the arable weeds had already gone by 1913.' The steady state had clearly reasserted itself to a very large extent. The woodland had also reappeared on the other area, the Broadbalk wilderness, but, because of the calcareous soil, the species found were different, with hawthorn rather than oak predominant (Jenkinson, 1971).

By the time of Jenkinson's 1971 report, the acid site had accumulated 180 t ha^{-1} of trees and the calcareous site 274 t ha^{-1} of trees, including about 72 and 110 t ha^{-1} of carbon respectively. In addition, about 43 t ha^{-1} of carbon accumulated in organic matter in the soil in the calcareous area and 21 t ha^{-1} in the acid area. These two historic Rothamsted sites illustrate the very considerable potential for carbon sequestration of simply allowing the steady state vegetation to re-establish itself. Actively afforesting arable land will obviously be effective too, but the more the soil is perturbed, the longer the benefit will take to materialize.

Paustian (2004) provides a table (his Table1) showing the effectiveness of various land management practices in promoting carbon sequestration in terms of the quantity of carbon sequestered per hectare per year. Adding the re-establishment of steady-state vegetation to this list (Table 7) shows it to be among the most effective practices, at least partly because no soil perturbation is involved. The table also shows that no-till, which avoids much soil perturbation, can also be a useful option. But on some soils no till may not be so beneficial because it can increase emissions of nitrous oxide and other 'greenhouse gases' (Smith *et al.*, 2000a; Smith and Conen, 2004).

Converting arable land to permanent grassland appears from the data garnered by Paustian (2004) to be more effective than converting it to forest, but this may reflect the longer timescale needed for the latter to become effective, particularly if it involves more soil perturbation. Conversion to forest did not appear as effective as allowing the steady-state vegetation to regenerate, but this again was on a long timescale and the effect of soil acidity on the regeneration needs to be noted (Table 7). More carbon was accumulated in the regenerated vegetation than in the soil. The accumulations of carbon reported above took place over 81 years, but this is the sort of time that will be needed for mitigation.

It would be interesting to know if, in 81 years time, carbon dioxide-induced global warming will still be seen as the biggest problem that was facing mankind at the beginning of the 21st century.

Table 7. Rates of carbon sequestration in the soil for various land management practices. Adapted and extended from Table 1 of Paustian (2004)

Management practice	Sequestration rate in soil/t C ha^{-1} y^{-1}
Adopt no till on arable land	0.1–0.7
Convert arable to permanent grassland	0.5–1.5
Convert arable land to forest	0.1–0.5
Improve grazing land	0.1–1.0
Allow regeneration of steady-state vegetation – sequestration in soil[1]	
Calcareous site	0.53
Acid site	0.21
Allow regeneration of steady-state vegetation – sequestration in trees etc.[1]	
Calcareous site	1.35
Acid site	0.89

[1] Data from Rothamsted (Jenkinson, 1971).

Concluding Remarks

One recurrent theme in this chapter is the change with time in the perspective on sub-stances or processes in the soil. Denitrification is an obvious example. In the 1950s when the evidence for its occurrence was hardening, it was seen first as a nuisance, a loss of nitrogen from agricultural soils that would need to be replaced. Slightly later, an alternative perspective developed when concerns about nitrate in natural waters began to emerge. Denitrification then became a convenient means of removing nitrate from water. Its image was to change again for the worse when it became clear that the nitrous oxide produced in partial denitrification was one of the 'greenhouse gases'. Another example is that certain metals that had long been seen as entirely ben-eficial trace elements suddenly became threatening as 'heavy metals'. Nitrate itself underwent similar transitions. It was for a long time an important but rather dull plant nutrient, but it was transmogrified into something altogether more sinister by the suspicions that it caused methaemoglobinaemia in infants and stomach cancer in adults. However, very recent medical research, summarized by Addiscott and Benjamin (2004), suggests not only that these fears are unfounded but that nitrate is an integral part of our bodies' defence system against bacterial gastroenteritis, a dis-ease that kills millions in the developing world.

Research in soil science has advanced at certain key point points as a result of particular stimuli. The stimulus for the major advances in the understanding of wheat nutrition that occurred in the 1840s was the sometimes acrimonious dispute between Lawes and Liebig. Heavy metals became an issue ultimately because one Rothamsted scientist asked a question of another. The hypothesis behind it may not have been right, but the question proved to be a vital stimulus. The experiment from which the question arose (and which became a vital research resource) had been established for reasons completely unconnected with heavy metals. This example emphasizes yet again the key role of chance in science, discussed in a broader scientific context by Beveridge (1950). An earlier chance discovery can be seen in the nitrate losses from the Drain Gauges. Lawes *et al.* (1882) established the gauges to examine the balance between the amount of water entering the soil in rainfall and the amount leaving it in drainage. Regular nitrate measurements began only seven years after the establishment of the gauges when the experi-menters realised that the drainage contained appreciable concentrations of nitrate. Both examples bring out the point that a chance discovery becomes important only when it encounters an informed mind.

An interesting and sobering question that can be asked is whether the historic discoveries outlined in this chapter would be likely to have been made in today's climate of opinion. All of them were started by people who had the funding for curiosity-led research, none at the initiative of a committee. But scientists today have much less opportunity for curiosity-led research and are increasingly governed by committees and by constraints on funding. Lawes was able to make his ground-breaking experiments because he could fund them himself, but it is doubtful whether his like will appear in the future.

The reality seems to be that future soil research will be funded by the private sector, which may have a fairly limited research agenda and may understandably not always wish to publicize the results, or governments. In Europe we can add the EU. Scientists seeking funding from government or EU sources need to be conscious that in reality there is no such thing as 'government money' or 'EU money', there is only tax-payers' money.

This is not just a point of principle for scientists. Governments are aware of it too. In the UK, science suffered not only cuts in funding during the Thatcher years but also an increasingly intrusive bureaucracy to ensure that tax-payers money was well spent. Some of us were inclined to ask, 'Who's auditing the auditors?' We wondered if the bureaucracy cost more than it saved? This question was never answered.

Under the current government of the UK, the issue seems to have changed. The question, 'Is the taxpayer getting value for his or her money?' is beginning to be supplemented by a more fundamental question, 'Is the tax-payer happy with the research being done with the money?' This question is particularly live in the agricultural sector because of the recent disasters of BSE and foot-and-mouth, and the government has encouraged the formation of a 'Food Ethics Committee' intended to allow the general public to have a say in the research done. The formation of this committee raises three key questions.

- Is the committee genuinely representative of the general public or is it dominated by people from environmental pressure groups and others with an axe to grind?
- How can members of the general public, who generally have little knowledge of science, have any idea of what research needs to be done?
- There is evidence that the public has been affected by the anti-science agenda of environmental pressure groups. Is it wise to allow people so affected to comment on the future direction of science?

These questions do not seem to have been addressed.

Some commentators in the UK perceive science to be under threat. An MP for a major university city in the UK recently commented that there is a 'battle between science and anti-science'. If this battle exists, and if it exists in countries other than the UK, all scientists, including soil scientists, need to involve themselves. This is likely to be a battle in communication, and for soil science to have a future as well as a history, its proponents will have to become very good at communicating not just with each other but with the public. The soil is fundamental to our existence, and if the public is being misinformed about it, whether by pressure groups, commercial interests or even governments, soil scientists must be ready to make a stand. They must be fully prepared when they do so, with a clear and uncomplicated idea of the message they aim to get across. The soil must be managed according to the very best precepts of soil science, and that depends on us.

References

Addiscott, T.M. (1977) A simple computer model for leaching in structured soils. Journal of Soil Science 28:554–563.

Addiscott, T.M. (1988) Long-term leakage of nitrate from bare, unmanured soil. Soil Use and Management 4:91–95.

Addiscott, T.M. (1995) Entropy and sustainability. European Journal of Soil Science 46:161–168.

Addiscott, T.M. and Benjamin, N. (2004) Nitrate and human health. Soil Use and Management 20:98–104.

Addiscott, T.M. and Powlson, D.S. (1992) Partitioning losses of nitrogen fertilizer between leaching and denitrification. Journal of Agricultural Science, Cambridge 118:101–107.

Addiscott, T.M., Whitmore, A.P. and Powlson, D.S. (1991) Farming, Fertilizers and the Nitrate Problem, CABI, Wallingford.

Addiscott. T.M., Brockie, D., Catt, J.A., Christian, D.G.. Harris, G.L., Howse, K.R., Mirza, N.A. and Pepper, T.J. (2000) Phosphate losses through field drains in a heavy cultivated soil. Journal of Environmental Quality 29:522–532.

Al-Dabbagh, S., Forman, D., Bryson, D., Stratton, I. and Doll, R. (1986) Mortality of nitrate fertilizer workers. British Journal of Industrial Medicine 43:507–515.

Allison, F.E. (1955) The enigma of soil nitrogen balance sheets. Advances in Agronomy 7:213–250.

Avery, B.W. and Catt, J.A. (1995) The soil at Rothamsted. IACR-Rothamsted, Harpenden.

Belford, R.K. (1979) Collection and evaluation of large-scale monoliths for soil and crop studies. Journal of Soil Science 30:363–373.

Bell, S.G. and Codd G.A. (1996) Detection, analysis and risk assessment of cyanobacterial toxins. In Agricultural Chemicals and the Environment R.E. Hester and R.M. Harrison (eds.), Issues in Environmental Science and Technology, No. 5, Cambridge, UK: The Royal Society of Chemistry, pp. 109–122.

Benjamin, N. (2000) Nitrates in the human diet – good or bad? Annales de Zootechnologie 49:207–216.

Beresford, S.A. (1985) Is nitrate in drinking water associated with gastric cancer in the urban UK? International Journal of Epidemiology 14:57–63.

Beven, K. (1981) Micro-, meso- and macro-porosity and channelling phenomena in soils. Soil Science Society of America Journal 45:1245.

Beveridge, W.I.B. (1950) The art of scientific investigation. Heinemann: London.

Boalch, D.H. (1953) The Manor of Rothamsted. Rothamsted Experimental Station, Harpenden.

Bouma, J. and Dekker, S.W. (1978) A case study on infiltration into dry clay soil. I. Morphological observations. Geoderma 29:27–40.

Bremner, J.M. (1997) Sources of nitrous oxide in soils. Nutrient Cycling in Agroecosystems 49:7–16.

Briggs, L.J. and McColl, E.A. (1904) An artificial root for inducing capillary movement of soil moisture. Science 20:566–569.

Brookes, P.C. and McGrath, S.P. (1984) Effects of metal toxicity on the size of the soil microbial biomass. Journal of Soil Science 35:341–346.

Brookes, P.C., McGrath S.P. and Heijnen, C. (1986) Metal residues in soils previously treated with sewage-sludge and their effects on growth and nitrogen fixation by blue-green algae. Soil Biology and Biochemistry 343–353.

Bunting, A.H. (1963) Experiments on organic manures, 1942–49. Journal of Agricultural Science, Cambridge 60:121–140.

Cannell, R.Q., Goss M.J., Harris G.L, Jarvis M.G., Douglas J.T., Howse, K.R., and le Grice, S. 1984. A study of mole drainage with simplified cultivation for autumn-sown crops on a clay soil. 1. Background, experiment and site details, drainage systems, measurements of drainflow and summary of results, 1978–80. Journal of Agricultural Science, Cambridge 102:539–559.

Coleman, E.A. (1946) A laboratory study of lysimeter drainage under controlled soil moisture tension. Soil Science 62:365–382.

Comly, H.H. (1945) Cyanosis in infants caused by nitrates in well water. Journal of the American Medical Association 129:112–116.

Dobbie, K. and Smith, K.A. (2003) Nitrous oxide emission factors for agricultural soils in Great Britain: the impact of water-filled pore space and other controlling variables. Global Change Biology (in press).

Dyke, G.V. (1991) John Bennet Lawes: The Record of his Genius. Research Studies Press, Taunton.

Dyke, G.V. (1993) John Lawes of Rothamsted: Pioneer of Science, Farming and Industry. Hoos Press, Harpenden.

Eiland, F. (1981) The effects of applications of sewage sludge on microorganisms in soil. Tidsskrift for Planteavl 85:39–46.

England, C.B. (1974) Comments on 'A technique for using porous cups at any depth in the unsaturated zone' by Warren W. Wood. Water Resources Research 10:1049.

Ferguson, A.J.D., Pearson, M.J. and Reynolds, C.S. (1996) Eutrophication of natural waters and toxic algal blooms. *In* Hester, R.E. and Harrison, R.M. (eds.) Agricultural Chemicals and the Environment, Issues in Environmental Science and Technology No. 5, Royal Society of Chemistry, Cambridge.

Forman, D., Al-Dabbagh, A., and Doll, R. (1985) Nitrate, nitrite and gastric cancer in Great Britain. Nature 313:620–625.

Garner, H.V. (1957) Manures and fertilizers. MAFF Bulletin No. 36. HMSO, London.

Garner, H.V. and Dyke, G.V. (1969) The Broadbalk: yields. Report of the Rothamsted Experimental Station for 1968, Part II, pp. 26–49.

Goulding, K.W.T. (1990) Nitrogen deposition to land from the atmosphere. Soil Use and Management 6:61–63.

Goulding, K.W.T., Poulton, P.R., Webster, C.P. and Howe, M.T. (2000) Nitrogen leaching from the Broad balk Wheat Experiment, Rothamsted, UK, as influenced by fertilizer and manure inputs and the weather. Soil Use and Management 16:244–250.

Granli, T. and Bøckman, O.C. (1994) Nitrous oxide from agriculture. Norwegian Journal of Agricultural Sciences Supplement No. 12, 1–128.

Grennfelt, P. and Thornelof, E. (eds.) (1992) Critical loads for nitrogen. Nord 1992, 41, Nordic Council of Ministers, Copenhagen.

Haines, B.L., Waide, J.B. and Todd, R.l. (1982) Soil solution nutrient concentrations sampled with tension and non-tension lysimeters: Report of discrepancies. Soil Science Society of America Journal 46:658–661.

Hallard, M. and Armstrong, A.C. (1992) Observations of water movement to and within mole drainage channels. Journal of Agricultural Engineering Research 52:309–315.

Hansen, E.A. and Harris, A.R. (1975) Validity of soil-water samples collected with porous ceramic cups. Soil Science Society of America Proceedings 39:528–536.

Harris, G.L., Goss, M.J., Dowdell, R.J., Howse, K.R. and Morgan, P. (1984). A study of mole drainage with simplified cultivation for autumn-sown crops on a clay soil. 2. Soil water regimes, water balances and nutrient loss in drain water. Journal of Agricultural Science, Cambridge 102:561–581.

Haygarth, P.M., Hepworth, L. and Jarvis, S.C. (1998) Forms of phosphorus in hydrological from soil under grazed grassland. European Journal of Soil Science 49:65–72.

Heckrath, G., Brookes, P.C. Poulton, P.R. and Goulding, K.W.T. (1995) Phosphorus leaching from soils containing different phosphorus concentrations in the Broadbalk Experiment. Journal of Environmental Quality 24:904–910.

Hesketh, N., Brookes, P.C. and Addiscott, T.M. (2001) Effect of suspended soil material and pig slurry on the facilitated transport of pesticides, phosphate and bromide in sandy soil. European Journal of Soil Science 52:287–296.

Hoskins, W.G. (1977) The making of the English landscape. Hodder and Stoughton, London.

Howarth, R.W., Anderson, D.M., Church, T.M., Greening, H., Hopkinson, C.S., Huber, W.C., Marcus, N., Naiman, R.J., Segerson, K., Sharpley, A. and Wiseman (2000) Clean Coastal waters: Understanding and Reducing the Effects of Nutrient Pollution. National Academy Press, Washington DC.

IACR (1992) The nitrogen cycle. AFRC Institute of Arable Crops Reserach. Report for 1991, p. 36.

Jarvis, M.G. (1973) Soils of the Wantage and Abingdon district. Soil Survey of England and Wales, Harpenden.

Jarvis, S.C., Barraclough, D., Williams, J. and Rook, A.J. (1991) Patterns of denitrification loss from grazed grasslands: Effects of fertilizer inputs at different sites. Plant and Soil 131:77–88.

Jenkinson, D.S. (1971) The accumulation of organic matter in soil left uncultivated. Rothamsted Experimental Station Report for 1970, Part 2, pp. 113–137. Rothamsted Experimental Station, Harpenden, England.

Joffe, J.F. (1932) Lysimeter studies I Moisture percolation through the soil profile. Soil Science 34:123–143.

Johnston, A.E. and Garner, H.V. (1969) Broadbalk: Historical introduction. Report of the Rothamsted Experimental Station for 1968, Part II, pp. 12–25.

Johnston, A.E. and Wedderburn, R.W.M. (1975) The Woburn Market Garden Experiment, 1942–69. A history of the experiment, details of the treatments and the yields of the crops. Report of the Rothamsted Experimental Station for 1974, Part II, pp. 79–101.

Jones, K.C., Johnston, A.E. and McGrath, S.P. (1994) Historical monitoring of organic contaminants in soils. In R.A. Leigh and A.E. Johnston (eds.) Long-term Experiments in Agricultural and Ecological Sciences, pp. 147–163, CAB International, Wallingford.

Lark, R.M., Milne, A.E., Addiscott, T.M., Goulding, K.W.T., Webster, C.P. and O'Flaherty, S. (2004a) Analysing spatially intermittent variation of nitrous oxide emissions from soil with wavelets and the implications for sampling. European Journal of Soil Science 55:601–610.

Lark, R.M., Milne, A.E., Addiscott, T.M., Goulding, K.W.T., Webster, C.P. and O'Flaherty, S. (2004b) Scale- and location-dependent correlation of nitrous oxide emissions with soil properties: an analysis using wavelets. European Journal of Soil Science 55:611–627.

Laubel, A., Jacobsen, O.H., Kronvang, B., Grant, R. and Anderson, H.E. (1999) Subsurface drainage loss of particles and phosphorus and particles from field plot experiments and a tile-drained catchment. Journal of Environmental Quality 28:576–584.

Lawes, J.B., Gilbert, J.H. and Warington, R. (1881) On the amount and composition of the rainfall at Rothamsted. Journal of the Royal Agricultural Society of England, 2nd Series 17:241–279.

Lawes, J.B., Gilbert, J.H. and Warington, R. (1882) On the amount and composition of drainage water collected at Rothamsted. The quantity of nitrogen lost by drainage. Journal of the Royal Agricultural Society of England, 2nd Series 18, 43–71.

L'hirondel, J. and L'hirondel, J.-L. (2001). Nitrate and Man. Toxic, Harmless or Beneficial? CABI, Wallingford.

Lombi, E., Zhao, F., Wieshammer, G., Zhang, G. and McGrath, S.P. (2002) In situ fixation of metals in soils using bauxite: biological assessment. Environmental Pollution 118:445–452.

Mann, H.H. (1957) Weed herbage of slightly acid arable soils as affected by manuring. Journal of Ecology 45:149–156.

McGrath, S.P. and Lane, P.W. (1989) An explanation for the apparent losses of metals in a long-term field experiment with sewage sludge. Environmental Pollution 60:235–256.

McGrath, S.P. and Zhao, F.J. (2003) Phytoremediation of metals and metalloids from contaminated soils. Current Opinions in Biotechnology 14:277–282.

McGrath, S.P., Brookes, P.C. and Giller, K.E. (1988) Effects of potentially toxic metals in soil derived from past applications of sewage sludge on nitrogen fixation by *Trifolium Repens* L. Soil Biology and Biochemistry 20:415–424.

McGrath, S.P., Zhao, F.J. and Lombi, E. (2002) Phytoremediation of metals metalloids and radionuclides. Advances in Agronomy 75:1–56.

McMahon, M.A. and Thomas, G.W. (1974) Chloride and tritiated water flow in disturbed and undisturbed soils. Soil Science Society of America Proceedings 38:727–732.

Miller, N.H.J. (1906) The amount and composition of drainage through unmanured and uncropped land, Barnfield, Rothamsted. Journal of Agricultural Science, Cambridge 1:377–399.

Moss, B. 1996. A land awash with nutrients – the problem of eutrophication. Chemistry and Industry 3rd June 1996, 407–411.

Nielsen, D.R. and Biggar, J.W. (1962) Miscible displacement in soils. 1. Experimental information. Soil Science Society of America Proceedings 25:1–5.

Nixon, S.W. (1995) Coastal marine eutrophication: A definition, social causes and future concerns. Ophelia, International Journal of Marine Biology 41:199–219.

Olsen, S.R., Cole, C.V., Watanabe, F.S., and Dean, L.A. (1954) Estimation of available phosphorus in soils by extraction with sodium bicarbonate. USDA Circular 939, USDA, Washington, DC.

Paustian, K. (2004) Carbon Emissions and sequestration. *In* Encyclopedia of Soils in the Environment (Editor-in-Chief D. Hillel), Elsevier, Amsterdam, pp. 175–179.

Paustian, K., Andren, O., Janzen, H.H., Lal, R., Smith, P., Tian, G., Tiessen, H., van Noordwilk, M. and Woolmer, P. (1997) Agricultural soils as a sink to mitigate CO_2 emissions. Soil Use and Management 13:230–244.

Paustian, K., Cole, C.V., Sauerbeck, D., and Sampson, N. (1998) CO_2 mitigation by agriculture: an overview. Climatic Change 40:135–162.

Perakis, S.S. and Hedin, L.O. (2002) Nature 415:416–419.

Powlson, D.S., Pruden, G., Johnston, A.E. and Jenkinson, D.S. (1986) The nitrogen cycle of the Broadbalk wheat experiment: recoveries and losses of ^{15}N-labelled fertilizer applied in spring and impact of nitrogen from the atmosphere. Journal of Agricultural Science, Cambridge 107:591–609.

Powlson, D.S., Hart, P.B.S., Poulton, P.R., Johnston, A.E. and Jenkinson, D.S. (1992) Influence of soil type, crop management and weather on the recovery of 15N-labelled fertilizer applied to winter wheat in spring. Journal of Agricultural Science, Cambridge 118:83–100.

Recous, S., Fresneau, C., Faurie, G. and Mary, B. (1988) The fate of 15N-labelled urea and ammonium nitrate applied to a winter wheat crop: II Plant uptake and N efficiency. Plant and Soil 112:215–224.

Richards, L.A., Neal, O.R. and Russell, M.B. (1939) Observations on soil conditions in lysimeters II. Soil Science Society of America Proceedings 4:55–59.

Russell, E.J. and Richards, E.H. (1920) The washing out of nitrates by drainage water from uncropped and unmanured land. Journal of Agricultural Science, Cambridge 10:22–43.

Schindler, D.W. (1974) Eutrophication and recovery in experimental lakes: Implications for lake management. Science 184:897–899.

Sharpley, A.N. 1985a. Depth of surface soil-runoff interaction as affected by rainfall, soil slope and management. Soil Science Society of America Journal, 49:1010–1015.

Sharpley, A.N. 1985b. The selective erosion of plant nutrients in runoff. Soil Science Society of America Journal 49:1527–1534.

Sharpley, A.N., Chapra, S.C, Wedepohl, R., Sims, J.T., Daniel, T.C., and Reddy K.R. 1994. Managing agricultural phosphorus for protection of surface waters. Journal of Environmental Quality 23:437–451.

Smil, V. (2001) Enriching the Earth: Fritz Haber, Carl Bosch and the Transformation of World Food Production. The MIT Press, Cambridge, Massachusetts.

Smith, K.A. (1997) Soils and the greenhouse effect. Soil Use and Management 13:229.

Smith, K.A. and Conen, F. (2004) Impact of land management on fluxes of trace greenhouse gases. Soil Use and Management 20:255–263.

Smith, P. (2004) Soils as carbon sinks: the global context. Soil Use and Management 20:212–218.

Smith, K.A., Elmes, A.E., Howard, R.S. and Franklin, M.F. (1984) The uptake of soil and fertilizer nitrogen by barley growing under Scottish climatic conditions.

Smith, P., Powlson, D.S., Glendining, M.J. and Smith, J.U. (1997) Potential for carbon sequestration in European soils: preliminary estimates for five scenarios using results from long-term experiments. Global Change Biology 4:67–79.

Smith, P., Milne, R., Powlson, D.S., Smith, J.U., Falloon, P., Coleman, K. (2000a) Revised estimates of the carbon mitigation potentials of UK agricultural land. Soil Use and Management 16:293–295.

Smith, P., Goulding, K.W.T., Smith, K.A., Powlson, D.S., Smith, J.U., Falloon, P., Coleman, K. (2000b) Including trace gas fluxes in estimates of carbon mitigation potential of UK agricultural land. Soil Use and Management 16:251–259.

Thurston, J. (1958) Geescroft wilderness. Rothamsted Experimental Station Report for 1957, p. 94. Rothamsted Experimental Station, Harpenden, England

United Kingdom Review Group on Impacts of Atmospheric Nitrogen (1994) Impacts of nitrogen deposition in terrestrial ecosystems. DOE, London.

Van der Ploeg, R.R. and Beese, F. (1977) Model calculations for the extraction of soil water by ceramic cups and plates. Soil Science Society of America Journal 41:466–470.

van Genuchten, R. and Wierenga, P. (1976) Mass transfer studies in porous sorbing media. I. Analytical solutions. Soil Science Society of America Journal 40:473–480.

van Loon, A.J., Botterweck, A.A., Goldbohm, R.A., Brants, H.A. van Klaveren, J.D. and van den Brandt, P.A. (1998) Intake of nitrate and nitrite and the risk of gastric cancer: a prospective cohort study. British Journal of Cancer 78:129–135.

Vitousek, P.M., Aber, J., Howarth, R.W., Likens, G.E., Matson, P.A., Schindler, D.W., Schlesinger, W.H. and Tilman, G.D. (1997) Human alteration of the global nitrogen cycle: Causes and consequences. Issues in Ecology 1. Ecological Society of America, Washington DC.

Wagner, G.H. (1962) Use of porous ceramic cups to sample soil water within the profile. Soil Science 94:379–386.

White, R.E. (1997) Principles and practice of soil science. The soil as a natural resource, 3rd ed., Blackwell Science, Oxford, UK.

Whitmore, A.P., Bradbury, N.J. and Johnston, P.A. (1992) Potential contribution of ploughed grassland to nitrate leaching. Agriculture, Ecosystems and Environment 39:221–233.

Wood, W.W. (1973) A technique using porous cups for water sampling at any depth in the unsaturated zone. Water Resources Research 9:486–488.

Yamulki, Jarvis, S.C. and Owen, P. (1998) Nitrous oxide emissions from excreta applied in a simulated grazing pattern. Soil Biology and Biochemistry 30:491–500.

Zibilske, L.M. and Wagner, G.H. (1982) Bacterial growth and fungal genera distribution in soil amended with sewage sludge containing cadmium, chromium and copper. Soil Science 134:364–369.

18

Ancient Agricultural Terraces and Soils

Jonathan A. Sandor[1]

Introduction

Insights into soil use and soil changes through human history can be gained by studying agricultural terraces because of the storehouse of information they hold and their widespread use in time and space among many cultures. The stepped topography created by terracing is a characteristic feature of sloping lands throughout the agricultural world. Ancient and traditional agricultural terraces encompass a broad range of forms and functions, and diverse environmental settings. Archaeological and historical records indicate that terracing has been part of agriculture for millennia; it is one of the oldest forms of land modification. Terracing constitutes some of humanity's strongest and most enduring efforts to manage soil, water, and geomorphic processes in agriculture and to conserve land resources. The purpose of this chapter is to explore the history, geography, forms, and effects of traditional and ancient terrace agriculture in relation to soil.

Fundamentally, terracing involves segmentation of slopes into topographic steps to manage soil and water, and thereby provide more stable conditions for crops. Terracing comprises a diversity of methods in response to specific needs and environments. In several arid regions, ancient terracing involved relatively subtle landscape alteration in the form of small dam construction in support of runoff agriculture. Lynchets, likely created incidentally from cultivation and associated erosion processes in northwestern Europe, also represent fairly subtle terraced lands. In other arid to humid regions, terracing evolved into such remarkably engineered systems as the bench terraces of the Andes and rice terraces of eastern Asia, where entire landscapes were transformed into stepped agroecosystems.

Terracing and associated practices modify soils, so it is appropriate to explore direct and indirect soil changes resulting from terrace agriculture. Although there is substantial information on certain aspects of ancient terracing, investigations into terraced

[1] Dr. Jonathan Sandor, Agronomy Department, Iowa State University, Ames, IA, 50011-1010 USA.

soils are relatively few. The limited data on physical, chemical, and biological properties of terraced soils underscore the need for further study. Environmental consequences of ancient and traditional terracing range from cases of successful soil care to
situations in which insufficient measures and maintenance led to soil degradation.

This overview of terraced soils begins with background on the geography, history,
forms, and functions of terrace agriculture. Soil and landscape use, processes, and
changes associated with different forms of terracing are then described, followed by
examples of soil conservation and degradation in terrace agriculture. The final section presents the idea that the array of terracing strategies among past and present
agricultural societies reflects a high degree of indigenous knowledge about soil and
landscape processes.

Terracing Geography and History

Although terraced land constitutes only a small fraction of total agricultural area
(Nir, 1983), it is important in a number of regions. Agricultural terraces displaying
both a remarkable range and patterning in form and function are found on five
continents and Oceania. The geographical occurrence of agricultural terraces is
organized by continent and major region in Table 1 based on available literature.
An extensive review of the world geography and history of terracing was presented
by Spencer and Hale (1961). The question of origins and spread of terracing
through different regions is complex and beyond the scope of this chapter, but
likely early centers of origin include southeastern Asia, Papua New Guinea, southwestern Asia, and Mesoamerica or South America (Spencer and Hale, 1961).
Examples of secondary areas include Africa, Europe, and North America. Other
articles exploring the origins of terracing are Doolittle (1990) and Williams (1990).

In reviewing the literature, it is interesting that most of the earliest reasonably well-
documented ages for terracing in a number of world regions are similar at roughly
3000–4000 yr. B.P. (Table 1). While some reliable dates for terraced fields have been
obtained, most are less certain and inferred from archaeological association. A few
researchers suggest ages older than 4000 yr. B.P., especially for runoff terrace
agriculture in southwest Asia (e.g., Raikes, 1965; Bruins et al., 1986). Spencer and
Hale (1961) speculate that terracing developed 5–9,000 yr. B.P. in that region. Recent
work in Yemen indicates possible ages of 5000–6000 yr. B.P. for terrace agriculture,
with evidence from radiocarbon dating of charcoal in terraced soils and by archaeological association (Wilkinson, 1997; Wilkinson, 1998; Wilkinson and Edens, 1999).

Terracing has a long history, but the evidence for agriculture is far older. Ages
given in Table 1 comparing earliest known evidence for crops and terracing
indicate much older dates (roughly double) for crops – why? First, it is more

Table 1. Geography and approximate early ages for agriculture and terracing

Continent/Region	Subregion	Early Ages (yr B.P.)* Agriculture	Terracing	Comments on terracing ages
Eastern Asia	China	8500–11,500	3000	Uncertainty about older age
	Japan/Korea	3000–5000	2000	Rice paddy bunds
Southern to S.E. Asia and Oceania	India/Indochina	5000–7000	2300–3100	3100 yr B.P. Pakistan
	Philippines	3400–5000	1400–2000	
	Papua New Guinea	9000	?	Ancient but uncertain age
	Polynesia	1000–3600	1100	
Southwestern Asia	Near East	10,000–13,000	3000–6000	Probable 5000 and possibly 6000 yr B.P. age for terracing in Yemen. Possible 5000 yr B.P. age for gabarbands in Baluchistan. 5000–9000 yr B.P. speculated for origin of terrace agriculture
Europe	Mediterranean	8000	2500–4000	4000 yr B.P. Italy, 3700 yr B.P. Crete
	Eastern Europe	5000–7000	?	
	Western Europe	5000–7000	2000–3500	Early ages for lynchets
Africa	North Africa	6500	3000	Runoff terraces; 2450 yr B.P. Ethiopia
	SubSaharan Africa	3000–5000	500–600	Uncertainty about older age

(Continued)

Table 1. Geography and approximate early ages for agriculture and terracing—*Cont'd*

Location		Early Ages (yr B.P.)*		Comments on terracing ages
Continent/Region	Subregion	Agriculture	Terracing	
MesoAmerica		5000–10,000	2500–3000	Uncertainty about older age
South America		4000–10,000	2500–4000	Uncertainty about 3000 yr B.P. terraces agricultural funcion
North America		3000–5000	1000–3000	

*References:

Ages of early agriculture: Main references on early ages of agriculture are Cowan and Watson (1992), Harlan (1995), and Smith (1995). Other references, by region, are E. Asia (Crawford, 1992; Imamura, 1996; Science, 1998 news section 1/17/98 and 11/20/98), S. to S.E. Asia and Oceania (Golson, 1989; Kirch, 1994; Bayliss-Smith, 1996); S.W. Asia (Science news 11/20/98); Europe (Dennell, 1992); Africa (Sutton, 1985); Americas (Science news 11/20/98).

Ages of early terracing: E. Asia (Hallsworth, 1987; Barnes, 1990; Wang and Wang, 1991; Imamura, 1996); S. to S.E. Asia (Galla, 1985; Tusa, 1995 [cited in Glover and Higham, 1996]; Hudson, 1995); Oceania (van Breemen et al., 1970; Conklin, 1980; Ayres and Haun, 1985; Kirch, 1985; Sullivan et al., 1987; Bulmer, 1989; Golson, 1989; S.W. Asia (Raikes, 1965; Sherratt, 1980; Edelstein and Kisley, 1981; Evenari et al., 1982; Dennell, 1985; Hopkins, 1985; Bruins et al., 1986; Borowski, 1987; Vogel, 1987; Zurayk, 1994; Wilkinson, 1997 (and citing Ghaleb, 1990); Wilkinson, 1998; Wilkinson and Edens, 1999); Mediterranean (Courty et al., 1989; Moody and Grove, 1990; van Andel et al., 1990; Wagstaff, 1992; Zangger, 1992; James et al., 1994; Courty et al., 1998); W. and E. Europe (Bradley, 1978; Mercer, 1981; Fowler, 1983; Barker, 1985; Bell, 1992; Macphail, 1992; van Vliet-Lanoe et al., 1992); Africa (Connah, 1985; Sutton, 1978, 1985; Hallsworth, 1987; Gilbertson and Hunt, 1996; Soper, 1996); Americas (Donkin, 1979; Sandor et al., 1986; Denevan, 1988; Grieder et al., 1988; Mountjoy and Gliessman, 1988 (citing Abascal and Garcia, 1975); Toll, 1995; Zimmerer, 1995; Hard and Roney, 1998).

difficult to determine ages of ancient agricultural land (Hopkins, 1985; Sandor, 1995; Gilbertson and Hunt, 1996; Wilkinson, 1998) than crop remains, which can be directly dated and occur in several archaeological contexts. In fact, it is commonly difficult to even recognize ancient agricultural land, as in cases of subtle terrace forms with ephemeral structures and burial or erasure by later geomorphic processes and land use. On the other hand, more durable, massive terracing structures such as the monumental stone walled terraces on mountain slopes are among the most recognizable ancient agricultural fields (Fig. 1).

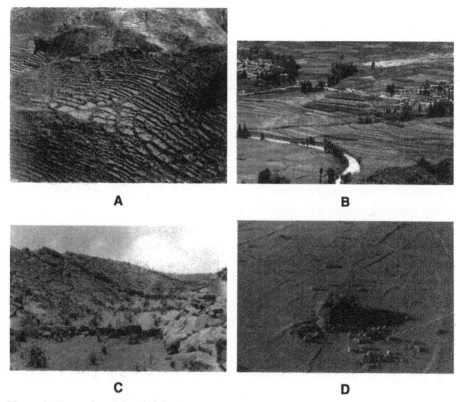

A   B

C   D

Figure 1. Examples of four kinds of ancient terrace agriculture. (A) Mountain bench terraces with irrigation canals, about 15 centuries old, southern Peruvian Andes. Photograph by Robert Shippee and George Johnson, 1931. Courtesy of the Library, American Museum of Natural History. (B) Wet-field rice terraces on lower slopes, Yunnan, China. (C) American Indian (Hopi) runoff agricultural terraces. Photograph by E.H. Morris, 1931. With permission of the University of Colorado Museum. (D) Lynchets dating to 1st to 2nd millennium B.C., southern England (Fowler, 1983), Plate IVb. Photograph AAU85, May 2, 1960. With permission of Cambridge University, Unit for Landscape Modeling. Copyright reserved Cambridge University Collection of Air photographs.

Another possibility is that conservation efforts such as terracing were in response to land degradation, so there was a lag time in development (e.g., Hopkins, 1985; van Andel et al., 1990).

Forms and Functions of Terracing

Numerous forms of terrace agriculture have been identified and classified. Definite types of terraced fields can be distinguished that correspond to certain agricultural systems and environments; however, when examined closely a myriad of variations in response to specific conditions and needs are apparent. A simplified scheme is used here to illustrate the diversity that exists in terracing (Table 2, Fig. 1). These familiar major forms of terracing and their geographic association include bench terraces in steeply sloping mountainous or hilly terrain such as in the Mediterranean region, Himalayas, and Andes; wet-field terraces of southeastern Asia, runoff terraces in arid regions, and lynchet and rideaux fields in northwestern Europe (Spencer and Hale, 1961; Donkin, 1979; Nir, 1983; Bruins et al., 1986; Centre National de la Recherche Scientifique, 1990; Wang and Wang, 1991; Treacy and Denevan, 1994; Doolittle, 2000; Denevan, 2001; Whitmore and Turner, 2001).

Functions of terracing for these major types include creation of a stable topographic base for crops, soil retention and erosion control, soil accumulation by sedimentation or construction filling, water control ranging from runoff management to irrigation, ponding, and drainage; and microclimatic modification (Spencer and Hale, 1961; Nir, 1983; Hopkins, 1985; Moody and Grove, 1990; Treacy and Denevan, 1994; Hudson, 1995; Evans and Winterhalder, 2000).

Fundamentally, terracing involves segmenting slopes into topographic steps composed of retaining walls (risers) and fields developed between the walls (treads). Basic construction elements in agricultural terraces include the terrace base, embankments or walls, and soil body (van Breemen et al., 1970). Terrace walls may be set into bedrock, sediments, or soil. In some Andean terraces, ancient walls remain securely anchored into hard soil horizons such as argillic horizons or duripans (Sandor and Eash, 1995). Embankments vary from subtle ephemeral forms to carefully crafted large walls, from single to complex series (Fig. 1). The several traditional processes for constructing the terrace fill or soil body fall into two main categories: alluvial/colluvial sedimentation and direct emplacement by hand. Construction materials for terrace walls include bedrock, stones, earth, living vegetation and other organic materials such as brush or logs etc, used alone or in combination (e.g., Spencer and Hale, 1961; van Breemen et al., 1970; Donkin, 1979; Nabhan, 1984; Patrick, 1985; Wilken, 1987; Grove and Sutton, 1989; Evans, 1992). Some terracing consists of permanent structures

Table 2. Examples of common processes, soil management practices, and soil property changes documented in different kinds of ancient and traditional terrace agriculture

Mountain and Hillslope

<u>Geomorphic processes</u>: slope decrease/leveling, some colluvial sedimentation.

<u>Soil practices</u>: relatively large walls, construction filling, fertilization, some irrigated.

<u>Morphology</u>: thickened A horizons, buried horizons, plaggen, anthropic, agric horizons; horizon reshaping, texture and rock fragment changes, structure and pore changes.

<u>Chemistry</u>: cases of organic C, N, P increase; manure residue from fertilization; CEC change, pH variable change.

<u>Biology</u>: soil fauna, enzyme activity changes.

<u>References on soils</u>: Wright, 1963; Healy et al., 1983; Pujol, 1983; Hopkins, 1985; Keeley, 1985, 1988; Sandor and Eash, 1991, 1995; Coultas et al., 1992; Dick et al., 1994; James et al., 1994; Johnson and Lewis, 1995; Wilkinson, 1997; Bull et al., 2001.

Wet-Field

<u>Geomorphic/pedogenic processes</u>: paddy soil processes, anthraquic conditions, slope leveling.

<u>Soil practices</u>: water impoundment, puddling, green manure, fertilization, soil additions/emplacement (e.g., hydraulic filling).

<u>Morphology</u>: new soil materials, constructed texture with clayey surface, soil thickening to several meters, plow pan, agric horizon.

<u>Chemistry</u>: inverted gley, ferrolysis, altered clay minerals, some organic matter increases.

<u>Biology</u>: aquatic, anaerobic organisms.

<u>References on soils</u>: van Breemen et al., 1970; Conklin, 1980; Inst. Soil Sci. Acad. Sinica, 1981, 1990; Gong, 1983; Barnes, 1990.

Runoff

<u>Geomorphic processes</u>: fluvial/colluvial episodic sedimentation, slope decrease.

<u>Soil practices</u>: relatively small walls/dams, sometimes deliberate watershed erosion.

<u>Morphology</u>: some A horizon thickening, buried horizons, increase in soil moisture and potential water retention.

<u>Chemistry</u>: examples of both increase and decrease in organic C, N, P, pH.

<u>Biology</u>: organic debris and microbial input via sedimentation.

<u>References on soils</u>: Bryan, 1929; Hack, 1942; Herold, 1965; Ron, 1966; Nabhan, 1984; Sandor et al., 1986; Sandor and Eash, 1991; Bruins, 1990; Homburg, 1992; Herold and Miller, 1995; Pawluk, 1995; Norton, 1996; Homburg and Sandor, 1997; Niemeijer, 1998; Homburg and Sandor et al., 2004; Homburg et al., 2005).

Lynchet/Rideaux

<u>Geomorphic/pedogenic processes</u>: mainly colluvial sedimentation after land clearing, especially intrafield erosion in humid climate; eolian additions, agric processes (slaking, mobilization, translocation of fine sand, silt and clay), sediment mixing.

<u>Soil practices</u>: relatively small field boundary walls, tillage, manure and other organic matter additions.

<u>Morphology</u>: some A horizon thickening, buried horizons, homogenization from plowing/mixing but also heterogeneous fabrics, agric horizons (e.g., dusty clay coatings, crust fragments), extrinsic additions (e.g., charcoal, ceramics).

<u>Chemistry</u>: examples of lower to higher organic matter and P.

<u>Biology</u>: common increase in biological activity if organic additions.

<u>References on soils</u>: Smith, 1975; Romans and Robertson, 1983; Allen, 1988; Courty et al., 1989; Bell, 1992; Macphail, 1992; Ford et al., 1994; Macphail, 1996; Simpson et al., 1998.

in fixed locations, whereas in other systems more flexibility in wall construction (e.g., permeable barriers) and placement is needed, for example in managing storm flow in runoff terrace agriculture (Gilbertson, 1986; Hudson, 1995; Sandor et al., 2002). The range of durability and permanence of terrace structures means that some survive in the archaeological record while others have long since disappeared (e.g., Treacy and Denevan, 1994; Beach and Dunning, 1995). Also, archaeological excavation of some terraced systems reveals multiple buried walls and other features indicating complex histories or incremental construction (e.g., Bruins, 1990; Wagstaff, 1992; Treacy and Denevan, 1994; Wilkinson, 1998).

Landscape Use and Modification

Landforms used for terracing are diverse at global to local scales, with specific placement often based on topographic, hydrologic, microclimatic, and soil criteria (e.g., Fedick, 1994; Sandor, 1995). Common landforms used for terracing include hillslopes and mountainsides, valley margin landforms such a colluvial footslopes, alluvial fans, and stream terraces, and drainageways on upland slopes and in larger valleys (e.g., Moorman and van Breemen, 1978; Gilbertson, 1986; Sandor, 1995). Terraces are usually associated with steeper landscape positions, but also occur on gentle slopes, especially runoff terraces, lowland wet-field terraces, and lynchets.

Multiple forms of terracing and terraced field placement in a diversity of geomorphic and ecological niches are characteristic of many indigenous agricultural systems as part of a risk reduction strategy to offset climatic and other environmental uncertainty (e.g., Ron, 1966; Turner, 1983; Hastorf and Earle, 1985; Hopkins, 1985). This agroecological diversity extends also to the numerous crops produced in terraced systems and accompanying crop and soil management practices (e.g., Mountjoy and Gliessman, 1988; Varisco, 1991; Brookfield and Padoch, 1994; Trombold and Israde-Alcantara, 2005). For runoff agriculture in arid lands, a range of local geomorphic settings are farmed in adapting to moisture and temperature constraints on crops (e.g., Hack, 1942; Woodbury, 1961; Evenari et al., 1982; Varisco, 1991; Fish, 1995; Sandor, 1995; Toll, 1995; Gilbertson and Hunt, 1996). For example, prehistoric runoff terraces in southwestern North America placed in valley margin and upland settings with small watersheds provided diversification for floodplain-based agriculture that was by itself more vulnerable to hazards of flooding or drought, saline and sodic soils, and frost from cold air drainage. These valley margin landforms used for prehistoric and historic traditional Native American runoff agriculture are often dismissed as nonarable by current standards when in fact they commonly have more favorable soils and microclimate, and are settings for fertility renewal from sediment and organic debris deposition with runoff (Sandor et al., this volume).

Specifically adapted crops and kinds of agricultural terraces are carefully matched to certain landscape positions. In high-altitude Andean terraces, barley can be grown on frost-prone, gently sloping surfaces, whereas maize is planted on steeper mountainside terraces with cold air drainage (Sandor and Eash, 1995). Agricultural terraces often conform closely to natural landforms and landscape features. In some areas they mimic and take advantage of natural stepped topography and geologic formations (e.g., Ron, 1966; Edelstein and Kislev, 1981; Hopkins, 1985). Terracing may in part have originated as attempts to approximate natural stepped features in landscapes following observed favorable effects on plant production and geomorphic processes. In some cases terracing utilizes special geomorphic features to help them function; for example, ancient Peruvian terraces constructed between parallel chute-like volcanic dikes (Sandor and Furbee, 1996) or the intriguing Moray terraces built into dolines (Donkin, 1979; Nickel, 1982; Hyslop, 1990).

Because terracing is done to manage geomorphic processes and land and water resources, landscape properties such as slope geometry, drainage patterns, and sediment transport processes correspondingly change. The stepped topography resulting from terrace wall construction and sediment filling is generally characterized by reduced field slope angle and length. In systems such as wet rice terraces, small level wetlands are created. Direct geomorphic changes spark indirect changes in other landscape and ecosystem components through feedback and complex response processes. Many ancient terraced landscapes, or portions of them, have been abandoned (Denevan et al., 1987; Denevan, 1988; Inbar and Llerena, 2000) and may be obscured or overprinted by later land use, while others remain intact. Terracing can enhance agricultural land or may create arable land where it did not previously exist (e.g., Ron, 1966; Vogel, 1987). Terraced land is commonly more fragile than adjacent agricultural lands. For example, due to their location on steeper slopes inherently more prone to instability, mountain terraces in particular can be subject to failure by mass movement (e.g., van Breemen et al., 1970; Conklin, 1980; Johnson et al., 1982; Johnson and Lewis, 1995; Sandor and Furbee, 1996).

The spatial extent of surficial change is confined in some terraced systems, while in others entire landscapes are altered, as in many mountain terraces (Sandor and

Eash, 1995) and wet-field terraces (Conklin, 1967,1980), or in lynchet areas displaying subtle but pervasive anthropogenic colluviation (Allen, 1988). In some terraced systems, watersheds above fields are modified to enhance productivity, as in the well-documented case of the ancient Negev Desert farms where surface stones were removed to increase runoff for fields downslope (Lavee et al., 1997). Deliberate watershed disturbance to promote sediment transport for filling agricultural terraces is also known (e.g., van Breemen et al., 1970; Conklin, 1980 Williams, 1990; Bocco, 1991; Simpson et al., 1998).

Although mountain and wet-field terracing have had a major impact on the landscape, many such ancient terraced systems seem to fit well within the natural topography rather than be imposed upon it. Even many of the large systems seem well-incorporated into the local environment by being shaped in accordance with existing slope contours and topography. Some terraced lands, with their mosaic of microenvironments of alternating contoured walls and small field segments, approach natural ecosystems in terms of diversity. Aesthetic and religious significance has been ascribed to terraces such as those in the Andes (Nickel, 1982; Hyslop, 1990; Treacy and Denevan, 1994). Carefully engineered, unified systems like these that have remained intact through centuries, imply a caring attitude toward the land and sustainability. Beauty in terraced landscapes has also been expressed through Eastern and Western arts and literature in forms such as Chinese landscapes (Wang and Wang, 1991), the graphics of M.C. Escher, elaborate terraced gardens (Yentsch and Krantzer, 1994), and in the writings of conservationist Aldo Leopold (1949) about runoff terraces in Mexico.

Soil Use and Alteration

The diversity of soils used for terracing matches that of terracing geography. Some cases of deliberate placement of fields on specific soils have been reported (e.g., Sandor et al., 1986; Fedick, 1994). In prehistoric runoff agricultural terraces in southwestern North America, deliberate use of soils with strongly-developed argillic horizons was inferred (Sandor et al., 1986). When combined with loamy to sandy A horizons thickened by sedimentation upslope of terrace walls, argillic horizons helped retain moisture within the crop rooting zone, a significant improvement in this semiarid region where water is a major limiting factor in agriculture. Other horizons and layers such as petrocalcic horizons, duripans, and bedrock have also been found in association with ancient and traditional runoff terraces in the region (Sandor, 1995; Homburg et al., 2004).

Geomorphic processes and agricultural management practices associated with terracing induce metapedogenic change in soil properties and development (Yaalon and Yaron, 1966). The degree of soil change varies greatly, depending on particular processes and practices, duration of use, and environmental sensitivity and response (Sandor et al., 2005). Changes resulting from terracing range from direct to indirect, deliberate to unintentional, short- to long-term, and major to minor in impact. Some terraced soils differ slightly from their natural state while others are wholly anthropogenic. Many of these terraced fields are ancient and have been abandoned, in some cases for many centuries, while others have been used almost continuously for millennia. Soil changes and causal processes and practices documented for major forms of terrace agriculture are presented in Table 2 and Fig. 2. Overlap in effects is evident, though each terrace system and environment tends to produce a characteristic set of soil changes. At the profile scale, some horizon changes are limited to the surface, while others extend deeply into subsurface horizons. Because most soil properties are interrelated, changes brought about by terracing in some properties necessarily change many others.

Methodologies for studying ancient terraced soils are important to consider because of the complexity of factors affecting data and interpretations. Studies of terraced soils are relatively few, and few of these have involved quantitative measurement of soil properties and change. Recognizing and measuring soil change from terracing, or from any ancient land use, is challenging (e.g., Smith, 1975; Hopkins, 1985; Sandor and Eash, 1991). Important information about properties, stratigraphy, and spatial variation can be gained from internal studies of terraced soils. To assess soil change; however, unterraced reference soils are needed for comparison, which usually requires a paired site sampling design. The long time elapsed since ancient terraced and unterraced soils began different formation trajectories results in a complex imprint from subsequent land history and pedogenic processes. Paired-site comparisons allow inferences about net soil change relative to the current reference soil, though not necessarily reconstruction of original soil properties. Also, it is commonly difficult to find matched uncultivated soils, especially in extensively terraced landscapes. Despite these limitations, the paired site method is one of the few available to try to actually assess long-term soil change. Controlled experiments (e.g., Macphail et al., 1990) and studies of soil change in recent terrace agriculture (e.g., Pujol, 1983) provide guidance for evaluating and understanding processes and properties of ancient agricultural soils.

Terraced soil properties reported in the literature are briefly reviewed, with emphasis on some key changes in morphology and physical, chemical, and biological properties. One of the most distinctive morphological features of

Figure 2. Examples of soils in four different kinds of agricultural terraces. (A) Wedge-shaped A horizons in bench terrace fill, southern Peruvian Andes. Buried A horizon, 31–36 cm depth, is enriched in phosphorus. (B) Lynchet (Romano-British period), with associated wall stones and colluvium, overlain by eolian sand. Scilly Isles, United Kingdom. Photograph by R.I. Macphail (1987). (C) Rice terrace soil construction by hydraulic filling. Ifugao, Philippines Photograph by H.C. Conklin, 1962 (Moorman and Van Breemen, 1978; Conklin, 1980). With permission of H.C. Conklin/International Rice Research Institute. (D) Soil thickened by sedimentation in American Indian (Zuni) runoff field. Buried horizons near base of profile (1.5 m depth). Color version (see the color plate section at the end of the book).

terraced soils is increased soil thickness resulting directly from accumulation of material upslope of terrace walls. Thickening amount within each terrace varies because of the wedge-shaped geometry common in terraced soil fills, wherein fills are thickest nearest the terraced wall and decrease upslope (Leopold and Bull, 1979; Sandor et al., 1986; Sandor and Eash, 1995). At the lower end of the range, soils thicken slightly (few centimeters to decimeters) in the case of many runoff terraces on hillslopes (e.g., Sandor et al., 1986; Gilbertson and Hunt, 1996) and lynchets involving low terrace walls and accumulation by natural sedimentation (Allen, 1988). Substantial thickening on the order of about 0.3 to several meters is common in bench terrace systems (Wright, 1963; Keeley, 1988; Sandor and Eash, 1995) and wet-field terraces (van Breemen et al., 1970; Gong, 1983). In runoff terraces located in ephemeral channels, soil thickening can also be substantial (Herold, 1965; Bruins, 1990; Gilbertson and Hunt, 1996).

The several traditional methods for filling terraces with soil or sediment by direct construction (versus natural sedimentation processes) greatly influence subsequent soil properties. In mountain/hillslope terracing, construction filling with material originating on the local slopes is common (e.g., Sandor and Eash, 1995). A few cases where material was gathered from greater distance such as valley floors have been reported (Healy et al., 1983; Hopkins, 1985), but the large volumes of soil to transport make this practice prohibitive in most circumstances. One of the most innovative filling methods is found in the Ifugao terraces, where sediment is transported in as a slurry stream from soil excavated upslope (van Breemen et al., 1970; Conklin, 1980) (Fig. 2). An analogous practice in runoff terracing is to remove vegetation and disturb soils in watersheds to encourage sedimentation in the agricultural terraces downslope (e.g., Ron, 1966; Williams, 1990). Soil thickening is often a deliberate practice to create arable land on steep slopes that otherwise lack sufficient soil, or to increase soil rooting volume for water and nutrients.

Terraced soil fill may be partly replaced periodically through erosion and deposition, thus helping to maintain fertility. In the case of runoff terraces, sedimentation is an essential component of nutrient replenishment/cycling in the agroecosystem (Nabhan, 1984; Niemeijer, 1998; Sandor et al., 2002). Soil filling can be more incidental, possibly exemplified by lynchets, in which erosion of cultivated soil accumulated at field boundaries, eventually creating a series of low terraced steps (e.g., Nir, 1983; Bell, 1992; Ford et al., 1994).

Buried horizons are characteristic of terraced soils where filling has occurred without destruction of the original soil (Keeley, 1985, 1988; Courty et al., 1989; Ford et al., 1994; Sandor and Eash, 1995; Macphail, 1996), and can be a common feature in some terraced landscapes (Table 2; Fig. 2). Where terrace filling is episodic or incremental, multiple fills (e.g., Bruins, 1990) and buried horizons may develop.

Changes in soil texture are common, especially in upper horizons directly changed by terracing activities such as filling and mixing of soil materials. In the example of hydraulic filling of Ifugao wet-field terraces, fill textures are deliberately emplaced by filling with coarser sediments first, then capping with fine sediments to achieve water ponding (van Breemen et al., 1970; Conklin, 1980). Another case of extrinsic sediments deliberately emplaced are artificial gravel layers observed at the fronts of some Andean terraces to aid drainage and reduce stress on terrace walls (Treacy and Denevan, 1994; Sandor and Eash, 1995). Other textural changes reported in terraced fills ranging from subtle to substantial may result from sediment additions in irrigation water or runoff, or from translocation (e.g., Sandor and Eash, 1995; Homburg et al., 2005). Besides textural change in terraced soil fills, translocation of clay to horizons not directly disturbed by terrace construction has also been reported, such as in agric horizons of lynchets (e.g., Courty et al., 1989; Macphail et al., 1990).

Structure and porosity changes have been observed at macro and micromorphological scales in terraced soils. In ancient American terraced soils, changes in structure shape and size, and accompanying bulk density changes, have been reported (Sandor et al., 1986; Sandor and Eash, 1991, 1995; see next section). Micromorphological changes in soil structure and pore size were documented in current agricultural terraces in Spain by Pujol (1983). Microfabric changes in lynchet soils have been described in a number of studies (Courty et al., 1989; Macphail et al., 1990; Macphail, 1996).

Changes in several chemical soil properties have been measured in terraced soils (Table 2), both in soil fills and deeper subsurface horizons. In soil fills, decreases and increases in organic carbon, nitrogen, phosphorus, some microelements, pH, carbonate, and cation exchange capacity have been documented. Processes inducing increases in elements are additions in irrigation or runoff as well as fertilization. Decreases in elements result from crop removal as well as leaching and erosion processes. In paddy soils of wet-field terraces, anaerobic biological activity and reduction/oxidation processes produce distinctive changes in soil organic and inorganic chemistry (Table 2). Some cases of subsurface chemical change, through translocation of dissolved or colloidal materials and the effects of burial by overlying terrace soil fills, have also been documented. An example is the movement of substantial phosphorus in ancient Andean terraces deep into Bt and Bk horizons (Sandor and Eash, 1995). The phosphorus was originally applied as manure and other amendments, and its deep translocation may represent an extended case of what has been observed in long-term agronomic experiments (Sandor and Eash, 1991).

Biological changes have been described in terraced soils but, with the exception of rice paddy soils (Table 2), are even less documented than physical and chemical changes. Organic chemical compounds that are biomarkers of manure fertilizer have been detected in ancient Minoan terraced soils (Bull et al., 2001). In the Peruvian Andes, Dick et al. (1994) found that soil enzyme activity was greater in terraced soils, which had greater organic matter, nitrogen, and phosphorus contents than reference uncultivated soils. Differences in faunal activity (e.g., earthworms) have been reported in comparisons of terraced and unterraced soils (Wright, 1963; Courty et al., 1989; Sandor and Eash, 1995).

The state factors of soil formation altered by terracing set further metapedogenic processes in motion (Yaalon and Yaron, 1966). Changes in soil moisture and temperature regimes from microclimatic changes accompanying terracing reverberate throughout the interconnected soil system at varying rates. Major soil transformation is exemplified in the series of miniature wetlands created in wet rice terraces, in which redoximorphic processes produce paddy soils (Table 2).

Under long-term transformation, many ancient terraced soils have acquired properties of anthropedogenic horizons recognized in soil taxonomic systems such as the World Reference Base for Soil Resources (FAO, 1998), U.S. Soil Taxonomy (Soil Survey Staff, 1999), and Chinese soil classification (Gong et al., 1998). Examples are plaggen, anthropic, anthraquic, and agric horizons that are recognized in several agricultural and archaeological contexts. Amendments, cultural debris, and other artifacts (e.g., Keeley, 1985; Sandor et al., 1986; Bulmer, 1989; Courty et al., 1989; James et al., 1994; Simpson et al., 1998) as well as high phosphorus levels (Barnes, 1990; Sandor and Eash, 1995) have been identified in ancient terraced soils. Wet-field terraced soils comprise an important part of the group of artificially hydric soils with anthraquic conditions (Table 2). In examples such as the Ifugao terraces, the original soils have been reconstituted into truly anthropogenic soils (van Breemen et al., 1970).

Cases of Soil Conservation and Degradation

Soil degradation has shadowed agriculture throughout its 10+ millennia history (e.g., Butzer, 1982; Hillel, 1991; Johnson and Lewis, 1995; Redman, 1999; Diamond, 2005). It is likely that in many cases, successful soil care was preceded by soil degradation, and that practices such as terracing evolved directly and indirectly to counteract natural resource degradation. In some instances, people were able to respond before it was too late, while in other cases, evidence of land abandonment and depopulation, coupled with stratigraphic evidence of accelerated erosion, suggest that countermeasures to degradation were not taken or failed. Some of the best examples known of long-term soil conservation involve terrace agriculture. Yet if not properly maintained, or if located in environments sensitive to disturbance, terracing can also lead to profound and chronic land degradation. This is not surprising considering that terracing can involve a relatively high degree of landscape alteration.

Documented degradation processes in terraced lands include accelerated erosion (sheet, gully, piping, landslide), compaction, and organic matter loss. Among the most sensitive environments are those that become highly erodible after loss or alteration of vegetative cover on slopes. Examples of degraded terraced land under these circumstances can be found in several geographic areas with a range of climate and geomorphology such as the Mediterranean (e.g., Naveh and Dan, 1973; van Andel et al., 1990; Gallart et al., 1994; Zurayk, 1994; Thompson and Scoging, 1995), the Andes (Inbar and Llerena, 2000), the North American Southwest (Sandor and Eash, 1991), and western Europe (Bell and Boardman, 1992). In the latter example, the lynchet form of terracing likely originated from cultivation and accelerated erosion, but in the process more stable terraced land was achieved. In the American example, clearing of relict grasslands probably established in a moister Pleistocene or early Holocene climate may have led to degradation, which has persisted for centuries after terrace agriculture ceased (Sandor et al., 1986). The degradation still observable nearly 900 years after abandonment includes accelerated sheet and gully erosion, compaction, and decreased organic carbon, nitrogen, and phosphorus. The kind and magnitude of organic matter losses and compaction are similar to that observed in modern conventional cultivation (Sandor and Eash, 1991).

Instances of successful long-term care of terraced lands are also diverse in geography, environment, and agricultural system. Two examples are the Ifugao rice terraces in the Philippines (van Breemen et al., 1970) and irrigated terraces in the Andes of southern Peru that are at least 15 centuries old (Treacy and Denevan, 1994; Sandor and Eash, 1991, 1995). Criteria for long-term soil conservation in the Peruvian example include favorable soil tilth and fertility: thickened A horizons, lower bulk density and favorable structural characteristics, and higher levels of organic carbon, nitrogen, and phosphorus relative to nearby uncultivated Mollisols that served as references. Long-term soil conservation is partly attributed to traditional management practices such as terracing, a form of conservation tillage, and fertilization with manure and other materials. Although more quantitative soil studies are needed, a number of terraced systems have been viable for millennia and are still productive under traditional management practices. This leads into the question of indigenous knowledge, namely what knowledge under-lies sustainable terraced agricultural systems in which soils retain favorable tilth and fertility after many centuries of continuous farming?

Indigenous Knowledge Expressed in Terracing

Indigenous knowledge about soil and its management is implicit throughout this dis-cussion of terracing. The placement, construction, and soil characteristics of agricul-tural terraces are the physical manifestations of what cultures have learned in working in their environments. Especially in environments with limited and vulner-able resource base, agricultural societies had to develop sustainable systems to retain long-term viability (Pawluk et al., 1992). A key element of indigenous knowledge is that it is time-tested, incorporating and linking the collective experience of people who have lived on the same land for many generations. This is true for both ancient and modern peoples, and especially applies here to traditional peoples who continue to farm and maintain agricultural terraces built by their ancestors (e.g., van Breemen et al., 1970; Conklin, 1980; Lansing and Kremer, 1995; Sandor and Furbee, 1996; Eastabrook, 1998; Sandor et al., this volume). As discussed in the previous section, both failures and successes in soil care are recorded in terraced land.

Components of indigenous knowledge particularly relevant to the history of soil care are soil management practices and traditional soil classification. Several agri-cultural terrace systems reflect a heritage of knowledge and soil stewardship, such as in the Philippine and Andean examples discussed: the way the terraces are cus-tom fitted to the landscape, the combination of soil conserving management prac-tices, the resulting productive soils, and the matching of crops and cropping strategies to soils and landscapes. In runoff agriculture with its more subtle forms of terracing, substantial knowledge of hydrology and sedimentation processes in replenishing water and nutrients is shown in the practices of past and present tra-ditional farmers in arid environments (e.g., Evenari et al., 1982; El Amami, 1983; Nabhan, 1984; Pacey and Cullis, 1986; Wilken, 1987; Bocco, 1991; Critchley et al., 1994; Pawluk, 1995; Tabor, 1995; Lavee et al., 1997; Niemeijer, 1998; Doolittle, 2000; Sandor et al., 2002; Norton et al., 2003; Homburg et al., 2005).

This knowledge and management of soil is also expressed in the culture and lan-guage of people who still farm the ancient terraced fields. Among indigenous soil classifications that have been studied, two examples of the most developed and complex involve cultures relying on terrace agriculture: the Ifugao in the Philippines (Conklin's work cited in Wilshusen and Stone, 1990) and that of farmers in the southern Peruvian Andes (Sandor and Furbee, 1996). Wilshusen and Stone (1990) point out that indigenous cultures with more intensive agricultural systems and management are likely to have corresponding soil classification systems. In the Andean example, over 50 names for soils in a four-tiered taxonomy are closely tied to an impressive terrace agriculture that incorporates several management practices conducive to maintaining soil quality. Other examples of soil classifications in traditional cultures practicing terracing are described in Williams and Ortiz-Solorio (1981), Nabhan (1984), Marten and Vityakon (1986), Pawluk (1995), Mazzacuto and Niemeijer (1997); Barrera-Bassols and Zinck, 2000; and Sandor et al., 2002.

Soil stewardship and knowledge are also expressed in traditional ceremonies involving terrace agriculture. When preparing terraces for planting each year, some Andean farmers pour liquor on the soil in a toast and offer prayers meant "to console the earth that will soon be rent by the plow" (Treacy, 1989).

The vast differences in technology and economy between modern industrial and indigenous agriculture call into question the relevance of information and lessons from ancient and traditional agriculture. Yet, most agricultural land use is similar in its basic effects on soil properties; the differences are primarily in scale, intensity, and method. The close attention to variation in soil, topography, and hydrology characteristic of many indigenous people in farming their terraced fields embodies the intentions of precision agriculture today. The record of indigenous terrace agriculture past and present is a potentially valuable resource for efforts in sustainable agriculture (Pawluk et al., 1992; Sandor et al., this volume). The importance of practices such as terracing is well-recognized in agricultural development, but the appropriate application of terracing technologies in concert with indigenous systems and farmer-based decision-making is still emerging (e.g., Mountjoy and Gliessman, 1988; Varisco, 1991; Zurayk, 1994; Brookfield and Padoch, 1994; Beach and Dunning, 1995; Hudson, 1995; Tabor, 1995; Reij et al., 1996; Mazzacuto and Niemeijer, 1997; Niemeijer, 1998; Llerena et al., 2004).

Conclusions

There is clearly a need for more pedologic studies of terrace agriculture that will add scientific knowledge about the soils and geomorphology of this venerable agricultural practice. Although there are several geographical and archaeological studies of agricultural terraces, relatively few studies consider terraced soils, and quantitative studies of the morphological, physical, biological, and chemical properties of these important anthropedogenic soils are rare. Some detailed work on biochemical soil processes in wet-field terraces and the micromorphology of lynchet soils has been done, but the world's terraced soils remain understudied. Documenting knowledge of terracing and soils among traditional cultures is at a critical point because many cultures are losing their agricultural heritage during assimilation into the modern agricultural world. Ancient agricultural terraces, their landscapes, soils, and farmers represent a library from which to learn about the history of soil use, including both successes and failures. The long time perspective on soil quality gained by studying terraced soils can help guide efforts towards sustainable agricultural systems. Ancient agricultural terraces are a link between past and present people trying to balance land use with resource conservation.

Acknowledgments

I wish to thank Dr. Dan Yaalon for the opportunity to participate in the symposium "Attitudes to Soil Care and Land Use through Human History" at the 16th World Congress of Soil Science, from which the ideas in this chapter originated. Many thanks also to colleagues, students, and generous farmers who have shared in exploring terraced soils.

References

Abascal, R., and A. Garcia. 1975. Sistemas de cultivo, riego, y control de agua en el area de Tlaxcala. *In* Arqueologia 1, 9–15, September, 1973, at Mexico, D.F. pp. 199–212.

Allen, M.J. 1988. Archaeological and environmental aspects of colluviation in southeast England. pp. 67–92. *In* W. Groenman-van Waateringe and M. Robinson (eds.) Man-made soils. British Archaeological Reports, International Series 410, Oxford.

Ayres, W.S., and A.E. Haun. 1985. Archaeological perspectives on food production in eastern Micronesia. pp. 455–473. *In* I.S. Farrington (ed.) Prehistoric intensive agriculture in the tropics. British Archaeological Reports, International Series 232, Oxford.

Barker, G. 1985. Prehistoric farming in Europe. Cambridge University Press.

Barnes, G.L. 1990. Paddy soils now and then. World Archaeology 22(1):1–17.

Barrera-Bassols, N., and J.A. Zinck. 2000. Ethnopedology in a worldwide perspective: an annotated bibliography International Institute for Aerospace Survey and Earth Sciences (ITC), Enschede, The Netherlands.

Bayliss-Smith, T. 1996. People-plant interactions in the New Guinea highlands: agricultural hearthland or horticultural backwater? pp. 499–523. *In* D.R. Harris (ed.) The origins and spread of agriculture and pastoralism in Eurasia. Smithsonian Institution Press, Washington, DC.

Beach, T., and N.P. Dunning. 1995. Ancient Maya terracing and modern terracing in the Peten rain forest of Guatemala. Journal of Soil and Water Conservation 50:138–145.

Bell, M. 1992. The prehistory of soil erosion. pp. 212–235. *In* M. Bell and J. Boardman (eds.) Past and present soil erosion: archaeological and geographical perspectives. Oxbow Monograph 22, Oxford, England.

Bell, M., and J. Boardman (eds.). 1992. Past and present soil erosion: archaeological and geographical perspectives. Oxbow Monograph 22, Oxford, England.

Bocco, G. 1991. Traditional knowledge for soil conservation in central Mexico. Journal of Soil and Water Conservation 46:346–348.

Borowski, O. 1987. Agriculture in Iron Age Israel. Eisenbrauns, Winona Lake, Indiana.

Bradley, R. 1978. Prehistoric field systems in Britain and northwest Europe: a review of some recent work. World Archaeology 9:265–280.

Brookfield, H., and C. Padoch. 1994. Appreciating agrodiversity: a look at the dynamism and diversity of indigenous farming practices. Environment 36(5):6–11, 37–45.

Bruins, H.J. 1990. Ancient agricultural terraces at Nahal Mitnan. Atiqot 10: 22–28.

Bruins, H.J., M. Evenari, and U. Nessler. 1986. Rainwater-harvesting agriculture for food production in arid zones: the challenge of the African famine. Applied Geography 6:13–32.

Bryan, K. 1929. Floodwater Farming. The Geographical Review 19:444–456.

Bulmer, S. 1989. Gardens in the south: diversity and change in prehistoric Maaori agriculture. pp. 689–705. *In* D.R. Harris and G.C. Hillman (eds.) Foraging and farming: the evolution of plant exploitation. Unwin Hyman, London.

Bull, I.D., P.P. Betancourt, and R.P. Evershed. 2001. An organic geochemical investigation of the practice of manuring at a Minoan site on Pseira Island, Crete. Geoarchaeology 16:223–242.

Butzer, K.W. 1982. Archaeology as human ecology: Method and theory for a contextual approach. Cambridge Univ. Press, Cambridge, England.

Centre National de la Recherche Scientifique. 1990. Les terrasses de cultures méditerranéennes. Méditerranee (3–4).

Conklin, H.C. 1967. Some aspects of ethnographic research in Ifugao. Transactions of the New York Academy of Sciences 130:99–121.

Conklin, H.C. 1980. Ethnographic atlas of Ifugao. Yale University Press, New Haven.

Connah, G. 1985. Agricultural intensification and sedentism in the firki of northeast Nigeria. pp. 765–795. *In* I.S. Farrington (ed.) Prehistoric intensive agriculture in the tropics. British Archaeological Reports, International Series 232, Oxford.

Coultas, C.L., M.E. Collins, and A.F. Chase. 1994. Effect of ancient Maya agriculture on terraced soils of Caracol, Belize. pp. 191–201. *In* J.E. Foss et al. (eds.)

Proceedings of the First International Conference on Pedo-archaeology. The University of Tennessee Agricultural Experiment Station, Knoxville.

Courty, M.A., H. Cachier, M. Hardy, and S. Ruellen. 1998. Soil record of exceptional wild-fires linked to climatic anomalies (inter-tropical and Mediterranean regions). Proceedings 16th World Congress of Soil Science, Montpellier, France.

Courty, M.A., P. Goldberg, and R.I. Macphail. 1989. Soils, micromorphology, and archaeology. Cambridge University Press, Cambridge, England.

Cowan, C.W., and P.J. Watson. 1992. The origins of agriculture: An international perspective. Smithsonian Institution Press, Washington, D.C.

Crawford, G.W. 1992. Prehistoric plant domestication in east Asia. pp. 7–38. In C.W. Cowan and P.J. Watson (eds.) The origins of agriculture: an international perspective. Smithsonian Institution Press, Washington, DC.

Critchley, W.R.S., C. Reij, and T.J. Willcocks. 1994. Indigenous soil and water conservation: a review of the state of knowledge and prospects for building on traditions. Land Degradation and Rehabilitation 5:293–314.

Denevan, W.M. 1988. Measurement of abandoned terracing from aerial photos: Colca Valley, Peru. Yearbook, Conference of Latin Americanist Geographers 14:20–30.

Denevan, W.M. 2001. Cultivated landscapes of Native Amazonia and the Andes. Oxford University Press, New York.

Denevan, W.M., K. Mathewson, and G. Knapp (eds.). 1987. Pre-Hispanic agricultural fields in the Andean region. British Archaeological Reports, International Series 359 (1), Oxford.

Dennell, R.W. 1985. The archaeology of check dam farming in Tauran, Iran. pp. 699–715. In I.S. Farrington (ed.) Prehistoric agriculture in the tropics. British Archaeological Reports International Series 232, Oxford.

Dennell, R.W. 1992. The origins of crop agriculture in Europe. pp. 71–100. In C.W. Cowan and P.J. Watson (eds.) The origins of agriculture: An international perspective. Smithsonian Institution Press, Washington, DC.

Diamond, J. 2005. Collapse: how societies choose to succeed or fail. Viking, Penguin Group, Ltd., New York.

Dick, R.P., J.A. Sandor, and N.S. Eash. 1994. Soil enzyme activities after 1500 years of terrace agriculture in the Colca Valley, Peru. Agriculture, Ecosystems, and Environment 50:123–131.

Donkin, R.A. 1979. Agricultural terracing in the aboriginal New World. University of Arizona Press, Tucson.

Doolittle, W.E. 1990. Terrace origins: hypotheses and research strategies. Yearbook, Conference of Latin Americanist Geographers 16:94–97.

Doolittle, W.E. 2000. Cultivated landscapes of native North America Oxford University Press, New York.

Eastabrook, G.F. 1998. Maintenance of fertility of shale soils in a traditional agricultural system in central interior Portugal. Journal of Ethnobiology 18:15–33.

Edelstein, G., and M. Kislev. 1981. Mevasseret Yerushalayim: the ancient settlement and its agricultural terraces. Biblical Archaeologist 44:53–56.

El-Amami, S. 1983. Traditional technologies and the development of African environments. Utilization of runoff waters: the 'meskats' and other techniques in Tunisia. African Environment 3:107–120.

Evans, S.T. 1992. The productivity of maguey terrace agriculture in central Mexico during the Aztec Period. pp. 92–116. In T.W. Killion (ed.) Gardens of prehistory. The University of Alabama Press, Tuscaloosa.

Evans, T.P., and B. Winterhalder. 2000. Modified solar insolation as an agronomic factor in terraced environments. Land Degradation and Development 11:273–287.

Evenari, M., L. Shanan, and N. Tadmor. 1982. The Negev: The challenge of a desert. Harvard University Press, Cambridge, MA.

FAO. 1998. World Reference Base for Soil Resources. Food and Agriculture Organization of the United Nations, Rome, Italy.

Fedick, S.L. 1994. Ancient Maya agricultural terracing in the Upper Belize River area: computer-aided modeling and the results of initial field investigations. Ancient Mesoamerica 5:107–127.

Fish, S.K. 1995. Mixed agricultural technologies in southern Arizona and their implications. pp. 101–116. In H.W. Toll (ed.) Soil, water, biology, and belief in prehistoric and traditional Southwestern agriculture. Special Publication 2. New Mexico Archaeological Council, C & M Press, Denver, CO.

Ford, S., M. Bowden, V. Gaffney, and G.C. Mees. 1994. The "Celtic" field systems on the Berkshire Downs, England. pp. 153–167 In N.F. Miller and K.L. Gleason (eds.) The archaeology of garden and field. University of Pennsylvania Press, Philadelphia.

Fowler, P.J. 1983. The farming of prehistoric Britain. Cambridge University Press, Cambridge, England.

Galla, A. 1985. Some facets of agricultural intensification in early south Asia. pp. 787–808. In I.S. Farrington (ed.) Prehistoric intensive agriculture in the tropics. British Archaeological Reports, International Series 232, Oxford.

Gallart, F., P. Llorens, and J. Latron. 1994. Studying the role of old agricultural terraces on runoff generation in a small Mediterranean mountain basin. Journal of Hydrology 159:291–303.

Ghaleb, A.O. 1990. Agricultural practice in Radman and Wadi al-Jubah. Ph.D. dissertation, University of Pennsylvania, Philadelphia.

Gilbertson, D.D. 1986. Runoff (floodwater) farming and rural water supply in arid lands. Applied Geography 6(1):5–11.

Gilbertson, D.D., and C.O. Hunt. 1996. Romano-Libyan agriculture: walls and floodwater farming. pp. 191–225. *In* G. Barker (ed.) Farming the desert: the UNESCO Libyan valleys archaeological survey. UNESCO Publishing, Paris The Dept. of Antiquities, Tripoli; Society for Libyan Studies, London.

Glover, I.C., and C.F.W. Higham, 1996. New evidence for early rice cultivation in South, Southeast, and East Asia. pp. 413–441. *In* D.R. Harris (ed.) The origins and spread of agriculture and pastoralism in Eurasia. Smithsonian Institution Press, Washington, DC.

Golson, J. 1989. The origins and development of New Guinea agriculture. pp. 678–687. *In* D.R. Harris and G.C. Hillman (eds.) Foraging and farming: the evolution of plant exploitation. Unwin Hyman, London.

Gong, Z. 1983. Pedogenesis of paddy soil and its significance in soil classification. Soil Science 135:5–10.

Gong, Z., Z. Ganlin, and L. Guobao. 1998. The Anthrosols in Chinese Soil Taxonomy, pp. 40–51 Classification, correlation, and management of anthropogenic soils. USDA-NRCS, Lincoln, Nebraska.

Grieder, T., A.B. Mendoza, C. E. Smith, Jr., and R.M. Malina. 1988. La Galgada, Peru: a preceramic culture in transition. University of Texas Press, Austin.

Grove, A.T., and J.E.G. Sutton. 1989. Agricultural terracing south of the Sahara. Azania 24:113–122.

Hack, J.T. 1942. The changing physical environment of the Hopi Indians of Arizona. Harvard University Press, Cambridge, MA.

Hallsworth, E.G. 1987. Anatomy, physiology, and psychology of erosion. Wiley, New York.

Hard, R.J. and J.R. Roney. 1998. A massive terraced village complex in Chihuahua, Mexico, 3000 years before present. Science 279:1661–1664.

Harlan, J.R. 1995. The living fields: our agricultural heritage. Cambridge University Press, New York.

Hastorf, C.A. and T.K. Earle. 1985. Intensive agriculture and the geography of political change in the upper Mantaro region of central Peru. pp. 569–595. *In* I.S. Farrington (ed.) Prehistoric intensive agriculture in the tropics. British Archaeological Reports, International Series 232, Oxford.

Healy, P.F., J.D.H Lambert, J.T. Arnason, and R. Hebda. 1983. Caracol, Belize: Evidence of ancient Maya agricultural terraces. Journal of Field Archaeology 10:397–410.

Herold, L.C. 1965. Trincheras and physical environment along the Rio Gavilan, Chihuahua, Mexico. Publications in Geography 65-1. Department of Geography, University of Denver, Denver, CO.

Herold, L.C., and R.F. Miller. 1995. Water availability for plant growth in Precolumbian terrace soils, Chihuahua, Mexico. pp. 145–153. *In* H.W. Toll (ed.) Soil, water, biology, and belief in prehistoric and traditional Southwestern agriculture. New Mexico Archaeological Council, C & M Press, Denver, CO.

Hillel, D. 1991. Out of the Earth: Civilization and the life of the soil. University of California Press, Berkeley.

Homburg, J.A. 1992. Soil fertility study. pp.145–161. *In* S.M. Whittlesley (ed.) Archaeological investigations at Lee Canyon: Kayenta Anasazi farmsteads in the Upper Basin, Coconino County, Arizona. Statistical Research, Inc., Tucson, AZ.

Homburg, J.A., and J.A. Sandor. 1997. An agronomic study of two Classic Period agricultural fields in the Horseshoe Basin. Vol. 2., pp. 127–147. *In* S.M. Whittlesey et al. (eds.) Vanishing River: Landscapes and lives of the lower Verde River, The Lower Verde Archaeological Project (Vol. 2, Agricultural, subsistence, and environmental studies, edited by J. A. Homburg and R. Ciolek-Torrello). SRI Press, Tucson, Arizona.

Homburg, J.A., J.A. Sandor, and D.R. Lightfoot. 2004. Soil investigations. pp. 62–78. *In* W.E. Doolittle and J.A. Neely (eds.) The Safford Valley grids: prehistoric cultivation in the southern Arizona desert. Anthropological Papers of the University of Arizona, Number 70. University of Arizona Press, Tucson.

Homburg, J.A., J.A. Sandor, and J.B. Norton. 2005. Anthropogenic influences on Zuni agricultural soils. Geoarchaeology 20:661–693.

Hudson, N. 1995. Soil conservation. Third ed. Iowa State University Press, Ames.

Hopkins, D.C. 1985. The highlands of Canaan. The Almond Press, Sheffield, England.

Hyslop, J. 1990. Inka settlement planning. University of Texas Press, Austin.

Imamura, K. 1996. Jomon and Yayoi: the transition to agriculture in Japanese prehistory. pp. 442–464. *In* D.R. Harris (ed.) The origins and spread of agriculture and pastoralism in Eurasia. Smithsonian Institution Press, Washington, DC.

Inbar, M., and C.A. Llerena. 2000. Erosion processes in high mountain agricultural terraces in Peru. Mountain Research and Development 20:72–79.

Institute of Soil Science, Academia Sinica. 1981. Proceedings of Symposium on Paddy Soil. Science Press, Beijing, China.

James, P.A., C.B. Mee, and G.J. Taylor. 1994. Soil erosion and the archaeological landscape of Methana, Greece. Journal of Field Archaeology 21:395–416.

Johnson, D.L., and L.A. Lewis. 1995. Land degradation: creation and destruction. Blackwell Publishers, Oxford, England.

Johnson, K., E.A. Olson, and S. Manandhar. 1982. Environmental knowledge and response to natural hazards in mountainous Nepal. Mountain Research and Development 2:175–188.

Keeley, H.C.M. 1985. Soils of prehispanic terrace systems in the Cusichaca Valley, Peru. pp. 547–568. *In* I.S. Farrington (ed.) Prehistoric agriculture in the tropics. British Archaeological Reports International Series 232, Oxford.

Keeley, H.C.M. 1988. Soils of pre-Hispanic field systems in the Rio Salado basin, northern Chile – a preliminary report. pp. 183–206. *In* W. Groeman-van Waateringe and M. Robinson (eds.) Man-made soils. British Archaeological Reports International Series 410, Oxford.

Kirch, P.V. 1985. Intensive agriculture in prehistoric Hawaii: the wet and the dry. pp. 435–454. In I.S. Farrington (ed.) Prehistoric intensive agriculture in the tropics. British Archaeological Reports, International Series 232, Oxford.

Kirch, P.V. 1994. The wet and the dry: irrigation and agricultural intensification in Polynesia. The University of Chicago Press.

Lansing, J.S., and J.N. Kremer. 1995. A socioecological analysis of Balinese water temples. pp. 258–268. *In* D.M. Warren et al. (eds.) The cultural dimension of development: indigenous knowledge systems. Intermediate Technology Publications, London.

Lavee, H., J. Poesen, and A. Yair. 1997. Evidence of high efficiency water-harvesting by ancient farmers in the Negev Desert, Israel. Journal of Arid Environments 35:341–348.

Leopold, A. 1949. A Sand County almanac and sketches here and there. Oxford University Press, London.

Leopold, L.B., and W.B. Bull. 1979. Base level, aggradation, and grade. Proceedings of the American Philosophical Society 123:168–202.

Llerena, C.A., M. Inbar, and M. Benevides (eds.) 2004. Conservacion y abandono de andenes (Conservation and abandonment of agricultural terraces). Universidad Nacional Agraria La Molina, Universidad de Haifa, Lima, Peru.

Macphail, R.I. 1987. A review of soil science in archaeology in England. pp. 332–379. *In* H.C.M. Keeley (ed.) Environmental archaeology: a regional review (Vol II). Historic Buildings and Monuments Commission Occasional Paper No 1, HBMC, London.

Macphail, R.I. 1992. Soil micromorphological evidence of ancient soil erosion. pp. 197–215. *In* M. Bell and J. Boardman (eds.) Past and present soil erosion. Oxbow Monograph 22, Oxford.

Macphail, R.I. 1996. Soil micromorphology and chemistry. pp. 198–203. *In* G. Smith, Archaeology and environment of a Bronze Age cairn and prehistoric and Romano-British field system at Chysauster, Gulval, near Penzance, Cornwall. Proceedings of the Prehistoric Society 62: 167–219.

Macphail, R.I., M.A. Courty, and A. Gebhardt, A. 1990. Soil micromorphological evidence of early agriculture in northwest Europe. World Archaeology 22(1):53–69.

Marten, G.G., and P. Vityakon. 1986. Soil management in traditional agriculture. pp. 199–225. *In* G.G. Marten (ed.) Traditional agriculture in southeast Asia. Westview Press, Boulder, CO.

Mazzucato, V., and D. Niemeijer. 1997. Beyond the development discourse: dynamic perceptions and management of soil fertility. pp. 495–504. *In* G. Renard et al. (eds.) Soil fertility management in West African land use systems. Margraf Verlag, Weikersheim, Germany.

Mercer, R. (ed.). 1981. Farming practice in British prehistory. Edinburgh University Press, Edinburgh, Scotland.

Moody, J., and A.T. Grove. 1990. Terraces and enclosure walls in the Cretan landscape. pp. 183–191. *In* S. Bottema, G. Entejes-Nieborg and W.van Zeist (eds.) Man's role in the shaping of the eastern Mediterranean landscape. A.A. Balkema, Rotterdam.

Moorman, F.R., and N. van Breemen. 1978. Rice: soil, water, land. International Rice Research Institute, Manila, Philippines.

Mountjoy, D.C., and S.R. Gliessman. 1988. Traditional management of a hillside agroecosystem in Tlaxcala, Mexico: an ecologically-based maintenance system. American Journal of Alternative Agriculture 3:3–10.

Nabhan, G.P. 1984. Soil fertility renewal and water harvesting in Sonoran Desert agriculture. Arid Lands Newsletter 20:21–28.

Naveh, Z. and J. Dan. 1973. The human degradation of Mediterrean landscapes in Israel. pp. 373–390. *In* F. di Castri and H. Mooney (eds.) Mediterranean type ecosystems. Springer-Verlag, New York.

Nickel, C. 1982. The semiotics of Andean terracing. Art Journal 42:200–204.

Niemeijer, D. 1998. Soil nutrient harvesting in indigenous *teras* water harvesting in eastern Sudan. Land Degradation and Development 9:323–330.

Nir, D. 1983. Man, a geomorphological agent. Keter Publishing House, Jerusalem.

Norton, J.B. 1996. Soil, geomorphic, and ecological factors in Zuni runoff agriculture. M.S. thesis, Iowa state University.

Norton, J.B., J.A. Sandor, and C.S. White. 2003. Hillslope soils and organic matter dynamics within a Native American agroecosystem on the Colorado Plateau. Soil Science Society of America Journal 67:225–234.

Pacey, A., and A. Cullis. 1986. Rainwater harvesting: the collection of rainfall and runoff in rural areas. Intermediate Technology Publications, London.

Patrick, L.L. 1985. Agave and Zea in highland central Mexico: the ecology and history of metepantli. pp. 539–546. *In* I.S. Farrington (eds.) Prehistoric intensive agriculture in the tropics. British Archaeological Reports International Series 232, Oxford.

Pawluk, R.R. 1995. Indigenous knowledge of soils and agriculture at Zuni Pueblo, New Mexico. M.S./ M.A. Thesis, Iowa State University.

Pawluk, R.R., J.A. Sandor, and J.A. Tabor. 1992. The role of indigenous soil knowledge in agricultural development. Journal of Soil and Water Conservation 47:298–302.

Pujol, J.L. de Olmedo. 1983. Anthropic influence on a Mediterranean brown soils. pp. 583–587. *In* P. Bullock and C.P. Murphey (eds.) Soil micromorphology. AB Academic Publishers, Berkhamsted, England.

Raikes, R.L. 1965. The ancient gabarbands of Baluchistan. East and West 15:26–35.

Redman, C.L. 1999. Human impacts on ancient environments. University of Arizona Press, Tucson.

Reij, C., I. Scoones, and C. Toulmin (eds.). 1996. Sustaining the soil: indigenous soil and water conservation in Africa. Earthscan Publications, Ltd., London.

Romans, J.C., and L. Robertson. 1983. The general effects of early agriculture on the soil. pp. 136–141. *In* G.S. Maxwell (ed.). The impact of aerial reconnaissance on archaeology. Vol. 49, CBA Research Report, London.

Ron, Z. 1966. Agricultural terraces in the Judean mountains, Israel. Israel Expl. J. 16:33–49, 111–122.

Sandor, J.A. 1995. Searching soil for clues about Southwest prehistoric agriculture. pp. 119–137. *In* H.W. Toll (ed.) Soil, water, biology, and belief in prehistoric and traditional Southwestern agriculture. Special Publication 2. New Mexico Archaeological Council, C & M Press, Denver, CO.

Sandor, J.A., and N.S. Eash. 1991. Significance of ancient agricultural soils for long-term agronomic studies and sustainable agriculture research. Agronomy Journal 83:29–37.

Sandor, J.A., and N.S. Eash. 1995. Ancient agricultural soils in the Andes of southern Peru. Soil Science Society of America Journal 59:170–179.

Sandor, J.A., and L. Furbee. 1996. Indigenous knowledge and classification of soils in the Andes of southern Peru. Soil Science Society of America Journal 60:1502–1512.

Sandor, J.A., P.L. Gersper, and J.W. Hawley. 1986. Soils at prehistoric agricultural terracing sites in New Mexico: I. Site placement, soil morphology, and classification. II. Organic matter and bulk density changes. III. Phosphorus, selected micronutrients, and pH. Soil Science Society of America Journal 50:166–180.

Sandor, J.A., J.B. Norton, R.R. Pawluk, J.A. Homburg, D.A. Muenchrath, C.S. White, S.E. Williams, C.L. Havener, and P.D. Stahl. 2002. Soil knowledge embodied in a Native American Runoff Agroecosystem. Transactions of the 17th World Congress of Soil Science, Bangkok, Thailand.

Sandor, J., C.L. Burras, and M. Thompson. 2005. Factors of soil formation: human impacts, pp. 520–532, *In* D. Hillel (ed.) Encyclopedia of soils in the environment, Vol. 1. Elsevier Ltd., Oxford, U.K.

Sherratt, A. 1980. Water, soil, and seasonality in early cereal cultivation. World Archaeology 11(3):315–330.

Simpson, I.A., R.G. Bryant, and U. Tveraabak. 1998. Relict soils and early arable land management in Lofoten, Norway. Journal of Archaeological Science 25:1185–1198.

Smith, B.D. 1995. The emergence of agriculture. Scientific American Library: W.H. Freeman, New York.

Smith, R.T. 1975. Early agriculture and soil degradation. pp. 27–37. In J.G. Evans et al. (eds.) The effect of man on the landscape: the Highland Zone. Derry and Sons, Ltd., Nottingham, England.

Soil Survey Staff. 1999. Soil Taxonomy. 2nd ed. Agriculture Handbook No. 436. USDA-Natural Resources Conservation Service, Washington, DC.

Soper, R. 1996. The Nyanga terrace complex of eastern Zimbabwe. Azania 31:1–35.

Spencer, J.E., and G.A. Hale. 1961. The origin, nature, and distribution of agricultural terracing. Pacific Viewpoint 2(1):1–40.

Sullivan, M.E., P.J. Hughes and J. Golson. 1987. Prehistoric garden terraces in the eastern highlands of Papua New Guinea. Tools and Tillage 5(4):199–213, 260.

Sutton, J.E.G. 1978. Engaruka and its waters. Azania 13:37–70.

Sutton, J.E.G. 1985. Irrigation and terracing in Africa: intensification and specialization or over-specialization? pp. 737–764. In I.S. Farrington (ed.) Prehistoric intensive agriculture in the tropics. British Archaeological Reports International Series 232, Oxford.

Tabor, J.A. 1995. Improving crop yields in the Sahel by means of water-harvesting. Journal of Arid Environments 30:83–106.

Thompson, D.A., and H. Scoging. 1995. Agricultural terrace degradation in southeast Spain. pp. 153–175. In D.F.M. McGregor and D.A. Thompson (eds.) Geomorphology and land management in a changing environment. John Wiley and Sons, New York.

Toll, H.W. (ed.). 1995 Soil, water, biology, and belief in prehistoric and traditional Southwestern agriculture. Special Publication 2. New Mexico Archaeological Council, C & M Press, Denver, Colorado.

Treacy, J.M. 1989. The fields of Coporaque: Agricultural terracing and water management in the Colca Valley, Arequipa, Peru. Ph.D. diss. Univ. of Wisconsin, Madison.

Treacy, J.M., and W.M. Denevan. 1994. The creation of cultivable land through terracing. pp. 91–110. In N.F. Miller and K.L. Gleason (eds.) The archaeology of garden and field. University of Pennsylvania Press, Philadelphia.

Trombold, C.D., and I. Israde-Alcantara. 2005. Paleoenvironment and plant cultivation on terraces at La Quemada, Zacatecas, Mexico: the pollen, phytolith and diatom evidence. Journal of Archaeological Science 32:341–353.

Turner, B.L.II. 1983. Once beneath the forest: Prehistoric terracing in the Rio Bec region of the Maya lowlands. Westview Press, Boulder, Colorado.

Tusa, S. 1990. Ancient ploughing in northern Pakistan. pp. 349–376. In M. Taddei (ed.) South Asian Archaeology 1987. Instituto per il Medio ed Extremo Oriente, Rome.

van Andel, T.H., E. Zangger, and A. Demitrack. 1990. Land use and soil erosion in prehistoric and historical Greece. Journal of Field Archaeology 17:379–396.

van Breemen, N., L.R. Oldeman, W.J. Plantinga, and W.G. Wielemaker. 1970. The Ifugao rice terraces. pp. 39–73. In N. van Breemen et al. (eds.) Aspects of rice growing in Asia and the Americas. H. Veenmen and Zonen N.V., Wageningen, The Netherlands.

van Vliet-Lanoë, B., M. Helluin, J. Pellerin, and B. Valadas. 1992. Past and present soil erosion: archaeological and geographical perspectives. pp. 100–114. In M. Bell and J. Boardman (eds.) Past and present soil erosion: archaeological and geographical perspectives. Oxbow Monograph 22, Oxford.

Varisco, D.M. 1991. The future of terrace farming in Yemen: a development dilemma. Agriculture and Human Values 8:166–172.

Vogel, H. 1987. Terrace farming in Yemen. Journal of Soil and Water Conservation 42:18–21.

Wagstaff, M. 1992. Agricultural terraces: the Vasilikos Valley, Cyprus. pp. 155–161. In M. Bell and J. Boardman (eds.) Past and present soil erosion: archaeological and geographical perspectives. Oxbow Monograph 22, Oxford.

Wang, Xing-guang, and Wang Lin. 1991. On the ancient terraced fields in China. Tools and Tillage 6:191–201.

Whitmore, T.M., and B.L. Turner II. 2001. Cultivated landscapes of middle America on the eve of conquest. Oxford University Press, New York.

Wilken, G.C. 1987. Good farmers: traditional agricultural resource management in Mexico and Central America. University of California Press, Berkeley, CA.

Wilkinson, T.J. 1997. Holocene environments of the High Plateau, Yemen. Recent geoarchaeological investigations. Geoarchaeology 12:833–864.

Wilkinson, T.J. 1998. Settlement, soil erosion, and terraced agriculture in highland Yemen: a preliminary statement. Proceedings of the seminar for Arabian Studies 28.

Wilkinson, T.J., and C. Edens. 1999. Survey and excavation in the central highlands of Yemen: results of the Dhamar. Arabian Archaeology and Epigraphy 10:1–33.

Williams, L.S. 1990. Agricultural terrace evolution in Latin America. Yearbook, Conference of Latin Americanist Geographers 16:82–93.

Williams, B.J., and C.A. Ortiz-Solorio. 1981. Middle American folk soil taxonomy. Annals of the Association of American Geographers 71:335–358.

Wilshusen, R.H. and G.D. Stone. 1990. An ethnoarchaeological perspective in soils. World Archaeology 22(1):104–114.

Woodbury, R.B. 1961. Prehistoric agriculture at Point of Pines. Memoirs of the Society for American Archaeology 17. American Antiquity 26 (3, part 2).

Wright, A.C.S. 1963. The soil process and the evolution of agriculture in northern Chile. Pacific Viewpoint 4:65–74.

Yaalon, D.H. and B. Yaron. 1966. Framework for man-made soil changes – an outline of metapedogenesis. Soil Science 102:272–277.

Yentsch, A.E., and J.M. Kratzer. 1994. Techniques for excavating and analyzing buried eighteenth-century garden landscapes. pp. 168–201. *In* N.F. Miller and K.L. Gleason (eds.) The archaeology of garden and field. University of Pennsylvania Press, Philadelphia.

Zangger, E. 1992. Neolithic to present soil erosion in Greece. pp. 133–147. *In* M. Bell and J. Boardman (eds.) Past and present soil erosion: archaeological and geographical perspectives. Oxbow Monograph 22, Oxford.

Zimmerer, K.S. 1995. The origins of Andean irrigation. Nature 378:481–483.

Zurayk, R.A. 1994. Rehabilitating the ancient terraced lands of Lebanon. Journal of Soil and Water Conservation 49:106–112.

Colour Plate Section

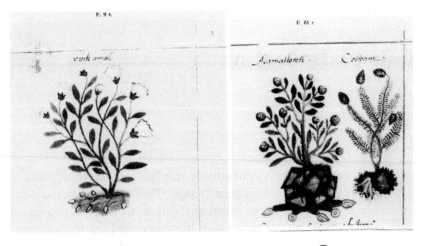

A   **B**

Color Plate 1. Illustrations of medicinal plants in the Libellus de medicinalibus indorum herbis (1552) show that soil and plant domains were closely related in Aztec ecology. Not only are key features of plant foliage depicted but also the roots and their surrounding environment. The drawings blend indigenous ideographic art with evolving European naturalism. For example, flowing water characterizing the habitat of certain plants is rendered by a semi-naturalistic representation of the Aztec water glyph in A (LM, f. 9r) in contrast to the classic form of the water symbol in B (LM, f. 61r). The vividly-colored, blocky material within which the Acamallotetl roots are wedged is abstract but in a distinctively Europeanized painting style. Acamallotetl labels both a plant and a stone in the herbal and is of uncertain etymology.

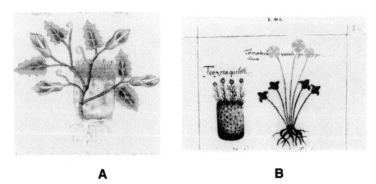

A   **B**

Color Plate 2. Illustrations of plants growing atop framed soil columns seem to depict soil horizons. Rather than modeled from actual soil profiles, these drawings quite likely represent plants growing in seed beds or planters in an artificially-produced plant medium. In A, at the base of the soil column is a depiction of small pebbles, pictographically alluding to the accompanying medicinal remedy ingredient, "river gravels of diverse colors" (LM, f. 18v). In B, the plant medium appears to be of granular structure or sandy/gravelly texture (LM, f. 30r). Other than the rock and possibly sand signs, the LM artist did not use the corpus of native graphemes for Aztec soil classes.

B　　　　　　　　　　　　　　　**D**

Color Plate 3. Arid and Semiarid Lands: (B) Zuni runoff maize field on footslope, with forested watershed runoff source in background. (D) Zuni agricultural soil thickened by runoff deposition (note buried horizon) in same runoff field shown in Fig. 8c. Field is approximately 1000 years old or more, based on archaeological evidence.

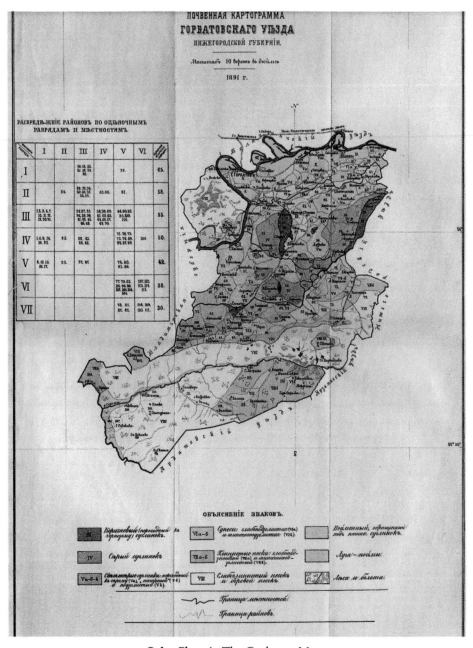

Color Plate 4. The Gorbatov Map.

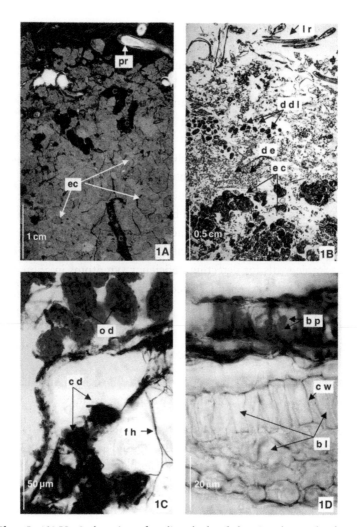

Color Plate 5. (A) Vertical section of undisturbed soil showing the result of earthworm activity with formation of channels (c) and production of casts (e c) that constitute the major part of the humus form Mull layer of a meadow, in a deep loamy soil with plant residue (p r) at the surface. Photo corresponds to a upper part of A-horizon, (photography U. Babel). (B) Soil thin section showing superposition of droppings of diptera larvae (d d l), of enchytraeids (d e) and earthworm casts (e c) in a humus (name Tangel in german) with leaf (needle) residue (l r) at the surface. The lower part is the A horizon of a spruce forest soil. (photography U. Babel). (C) Soil thin section of the OF layer in the upper part of a humus form Tangel showing on the top oribate droppings (o d) lying inside a spruce needle residue. From center to bottom and right, dark cell residue from droppings of collembola (c d) and dark fungal hyphae (f h). (photography U. Babel). (D) Soil thin section observed in transmitted light. In the upper part one beech leave residue with brown pigments (b p) and in the lower part another leaf bleached by white-rot fungi which have degraded the brown pigments. Cell wall structures are still preserved (c w). (humus form Moder) (photography U. Babel).

A   D

Color Plate 6. Examples of soils in four different kinds of agricultural terraces.
(A) Wedge-shaped A horizons in bench terrace fill, southern Peruvian Andes. Buried
A horizon, 31–36 cm depth, is enriched in phosphorus. D) Soil thickened by
sedimentation in Native American runoff field, southwest U.S. Buried horizons near
base of profile (1.5 m depth).

Author Index

Subject Index

Printed and bound by CPI Group (UK) Ltd, Croydon, CR0 4YY

03/10/2024

01040419-0014